NEW FRONTIERS IN NANOCHEMISTRY

Concepts, Theories, and Trends

Volume 2
Topological Nanochemistry

Edited by

Mihai V. Putz, PhD, Dr.-Habil., MBA

Professor, Faculty of Chemistry, Biology, Geography,
Laboratory of Computational and Structural Physical Chemistry for
Nanosciences and QSAR, West University of Timișoara, Romania

PI-1, Laboratory of Renewable Energies - Photovoltaics,
National Research and Development Institute for Electrochemistry
and Condensed Matter (INCEM), Timișoara, Romania

APPLE
ACADEMIC
PRESS

Apple Academic Press Inc.
4164 Lakeshore Road
Burlington ON L7L 1A4
Canada

Apple Academic Press Inc.
1265 Goldenrod Circle NE
Palm Bay, Florida 32905
USA

© 2020 by Apple Academic Press, Inc.

First issued in paperback 2021

Exclusive worldwide distribution by CRC Press, a member of Taylor & Francis Group

No claim to original U.S. Government works

New Frontiers in Nanochemistry: Concepts, Theories, and Trends, Volume 2: Topological Nanochemistry
ISBN 13: 978-1-77463-175-1 (pbk)
ISBN 13: 978-1-77188-778-6 (hbk)

New Frontiers in Nanochemistry, 3-volume set
International Standard Book Number-13: 978-1-77188-780-9 (Hardcover)
International Standard Book Number-13: 978-0-36707-782-2 (eBook)

Library and Archives Canada Cataloguing in Publication

Title: New frontiers in nanochemistry : concepts, theories, and trends / edited by Mihai V. Putz, PhD, Dr.-Habil., MBA.

Names: Putz, Mihai V., editor.

Description: Includes bibliographical references and indexes. | Contents: Volume 1. Structural nanochemistry -- Volume 2. Topological nanochemistry -- Volume 3. Sustainable nanochemistry.

Identifiers: Canadiana (print) 20190106808 | Canadiana (ebook) 20190106883 | ISBN 9781771887779 (v. 1 ; hardcover) | ISBN 9781771887786 (v. 2 ; hardcover) | ISBN 9781771887793 (v. 3 ; hardcover) | ISBN 9780429022937 (v. 1 ; ebook) | ISBN 9780429022944 (v. 2 ; ebook) | ISBN 9780429022951 (v. 3; ebook)

Subjects: LCSH: Nanostructured materials. | LCSH: Nanochemistry.

Classification: LCC TA418.9.N35 N49 2019 | DDC 541/.2—dc23

Library of Congress Cataloging-in-Publication Data

CIP data on file with US Library of Congress

Apple Academic Press also publishes its books in a variety of electronic formats. Some content that appears in print may not be available in electronic format. For information about Apple Academic Press products, visit our website at **www.appleacademicpress.com** and the CRC Press website at **www.crcpress.com**

About the Editor

Mihai V. Putz, PhD, Dr.-Habil., MBA

Professor, Laboratory of Computational & Structural Physical Chemistry for Nanosciences & QSAR, West University of Timişoara; Laboratory of Renewable Energies-Photovoltaics, National Research & Development Institute for Electrochemistry & Condensed Matter (INCEM), Romania

Mihai V. Putz, PhD, MBA, Dr.-Habil, is a laureate in physics (1997), with a post-graduation degree in spectroscopy (1999), and a PhD degree in chemistry (2002). He bears with many postdoctoral stages: in chemistry (2002–2003) and in physics (2004, 2010, 2011) at the University of Calabria, Italy, and Free University of Berlin, Germany, respectively. He is currently a Full Professor of theoretical and computational physical chemistry at his alma mater, West University of Timisoara, Romania.

He has made valuable contributions in computational, quantum, and physical chemistry through seminal works that appeared in many international journals. He is an Editor-in-Chief of the *International Journal of Chemical Modeling* and *New Frontiers in Chemistry*. In addition, he has authored and edited many books. He is a member of many professional societies and has received several national and international awards from the Romanian National Authority of Scientific Research (2008), the German Academic Exchange Service DAAD (2000, 2004, 2011), and the Center of International Cooperation of Free University Berlin (2010). He is the leader of the Laboratory of Computational and Structural Physical Chemistry for Nanosciences and QSAR at the Biology-Chemistry Department of West University of Timisoara, Romania, where he conducts research into the fundamental and applicative fields of quantum physical chemistry and QSAR. He is also a PI-1 at the Laboratory of Renewable Energies – Photovoltaics, National Research and Development Institute for Electrochemistry and Condensed Matter (INCEM), Timişoara, Romania. Among his numerous awards, in 2010 Mihai V. Putz was declared, through a national competition, the Best Researcher of Romania, while in 2013 he was recognized among the first Dr.-Habil. in Chemistry in Romania. In 2013, he was appointed

Scientific Director of newly founded Laboratory of Structural and Computational Physical Chemistry for Nanosciences and QSAR at his alma mater university. In 2014, he was recognized by the Romanian Ministry of Research as Principal Investigator of the first degree at the National Institute for Electrochemistry and Condensed Matter (INCEMC), Timisoara, and was also granted full membership in the International Academy of Mathematical Chemistry.

Recently, Mihai V. Putz expanded his interest to strategic management in general and to nanosciences and nanotechnology strategic management in particular. In this context, between 2015–2017, he attended and finished as the promotion leader of the MBA in Strategic Management of Organizations – The Development of the Business Space specialization program at the West University of Timişoara, the Faculty of Economics and Business Administration. While from 2016, he was engaged in the doctoral school of the same faculty, advancing new models of strategic management in the new economy based on frontier scientific inclusive ecological knowledge.

Contents

Contributors

Yaser Alizadeh
Department of Mathematics, Hakim Sabzevari University, Sabzevar, Iran,
E-mail: y.alizadeh@hsu.ac.ir

Giovanni Attolini
IMEM-CNR Institute, Parco Area delle Scienze 37A 43124 PARMA, Italy,
E-mail: giovanni.attolini@imem.cnr.it

Mahdieh Azari
Department of Mathematics, Kazerun Branch, Islamic Azad University, P.O. Box: 73135-168,
Kazerun, Iran

Radu Bănică
National Institute of Research and Development for Electrochemistry and Condensed Matter,
Timisoara, 300569, Romania, Tel. +40752197759, E-mail: radu.banica@yahoo.com

Sorana D. Bolboacă
Iuliu Hațieganu University of Medicine and Pharmacy Cluj-Napoca, Romania,
E-mail: sbolboaca@gmail.com

Matteo Bosi
IMEM-CNR Institute, Parco Area delle Scienze 37A 43124 PARMA, Italy,
E-mail: matteo.bosi@imem.cnr.it

Bogdan Bumbăcilă
Laboratory of Computational and Structural Physical Chemistry for Nanosciences and QSAR,
Biology-Chemistry Department, Faculty of Chemistry, Biology, Geography at West University of
Timişoara, Pestalozzi Street No.16, Timişoara, RO-300115, Romania

Franco Cataldo
Actinium Chemical Research Institute, Via Casilina 1626A, 00133 Rome, Italy; Università degli Studi
della Tuscia, Via S.Camillo de Lellis, Viterbo, Italy, Tel.: +39 06-94368230, Fax: +39 06-94368230,
E-mails: franco.cataldo@fastwebnet.it, cataldo.franco@fastwebnet.it

N. I. Chekunaev
N.N. Semenov's Institute of Chemical Physics, Russian Academy of Sciences, Kosygin Street 4,
Moscow 119991, Russian Federation

Edgar G. DuCasse
Mathematics Department, Pace University, New York, NY 10038, USA, E-mail: egducasse@pace.edu

Harish Dureja
Faculty of Pharmaceutical Sciences, Maharshi Dayanand University, Rohtak, 124001, India

Selçuk Gümüş
Yuzuncu Yıl University, Faculty of Sciences, Department of Chemistry, Van, Turkey

Asma Hamzeh
Department of Mathematics, Faculty of Mathematical Sciences, Tarbiat Modares University,
P.O. Box: 14115-137, Tehran, Iran

Mirela I. Iorga
National Institute for Research and Development in Electrochemistry and Condensed Matter,
Department of Applied Electrochemistry, 300569 Timisoara, Romania

Ali Iranmanesh
Department of Pure Mathematics, Faculty of Mathematical Sciences, Tarbiat Modares University,
P. O. Box: 14115-137, Tehran, Iran, Tel: +982182883447; Fax: +982182883493;
E-mail: iranmanesh@modares.ac.ir

Lorentz Jäntschi
Technical University of Cluj-Napoca, Romania, Tel.: +40-264-401775,
E-mail: lorentz.jantschi@gmail.com

A. M. Kaplan
N.N. Semenov's Institute of Chemical Physics, Russian Academy of Sciences, Kosygin Street 4,
Moscow 119991, Russian Federation, E-mail: amkaplan@mail.ru

Paola Lagonegro
IMEM-CNR Institute, Parco Area delle Scienze 37A 43124 PARMA, Italy;
Università degli studi di Parma, via Università, 12 - 43121 Parma, Italy

Petrica Linul
National Institute of Research and Development for Electrochemistry and Condensed Matter,
Timisoara, 300569, Romania; Politechnica University Timisoara, 300006, Romania

A. K. Madan
Faculty of Pharmaceutical Sciences, Pt. B. D. Sharma University of Health Sciences, Rohtak–124001,
India, Tel.: +91-9896346211, E-mail: madan_ak@yahoo.com

Ioan-Cezar Marcu
Laboratory of Chemical Technology & Catalysis, Department of Organic Chemistry,
Biochemistry & Catalysis, Faculty of Chemistry, University of Bucharest, 4-12,
Blv. Regina Elisabeta, 030018 Bucharest, Romania, Tel.: +40213051464, Fax: +40213159249,
E-mail: ioancezar.marcu@chimie.unibuc.ro; ioancezar_marcu@yahoo.com

Rakesh Kumar Marwaha
Faculty of Pharmaceutical Sciences, Maharshi Dayanand University, Rohtak, 124001, India

Marius C. Mirica
National Institute for Research and Development in Electrochemistry and Condensed Matter,
Department of Applied Electrochemistry, 300569 Timisoara, Romania

Cristina Moşoarcă
National Institute of Research and Development for Electrochemistry and Condensed Matter,
Timisoara, 300569, Romania, Tel. +40752197759, E-mail: mosoarca.c@gmail.com

Marco Negri
IMEM-CNR Institute, Parco Area delle Scienze 37A 43124 PARMA, Italy; Department of
Mathematical, Physical and Computer Sciences, Parma University, Viale delle Scienze 7/A,
I-43124 Parma, Italy; Axcelis Technologies, 108 Cherry Hill Drive, Beverly, MA, USA,
E-mail: negri.m1@gmail.com

Ottorino Ori
Actinium Chemical Research, 00133 Rome, Italy; Laboratory of Computational and Structural Physical
Chemistry for Nanosciences and QSAR, West University of Timişoara, 300115 Timioara, Romania;
Laboratory of Renewable Energies-Photovoltaics, National Institute for R&D in Electrochemistry and
Condensed Matter INCEMC-Timisoara, 300569 Timisoara, Romania,
E-mail: Ottorino.Ori@gmail.com

Octavian D. Pavel
University of Bucharest, Faculty of Chemistry, Department of Organic Chemistry,
Biochemistry and Catalysis, 4-12 Regina Elisabeta Av., 030018, Bucharest, Romania

Ana-Maria Putz
Institute of Chemistry Timisoara of the Romanian Academy, 24 Mihai Viteazul Bld.,
Timisoara 300223, Romania

Mihai V. Putz
Laboratory of Computational and Structural Physical Chemistry for Nanosciences and QSAR,
Biology-Chemistry Department, Faculty of Chemistry, Biology, Geography at West University of
Timişoara, Pestalozzi Street No. 16, 44, Timişoara, RO-300115, Romania; Laboratory of Renewable
Energies-Photovoltaics, R&D National Institute for Electrochemistry and Condensed Matter,
Dr. A. Paunescu Podeanu Str. No. 144, RO-300569 Timişoara, Romania,
Tel.: +40-256-592-638; Fax: +40-256-592-620; E-mail: mv_putz@yahoo.com; mihai.putz@e-uvt.ro

Louis V. Quintas
Mathematics Department, Pace University, New York, NY 10038, USA, E-mail: lvquintas@gmail.com

James Sawyer
Six Dimension Design, Buffalo, N.Y., USA

Monika Singh
Anand College of Pharmacy, Keetham, Agra–282007, India

Marina A. Tudoran
Laboratory of Structural and Computational Physical Chemistry for Nanosciences and QSAR,
Biology-Chemistry Department, West University of Timisoara, Pestalozzi Street No. 44, Timisoara,
RO-300115, Romania; Laboratory of Renewable Energies-Photovoltaics, R&D National Institute
for Electrochemistry and Condensed Matter, Dr. A. Paunescu Podeanu Str. No. 144, Timisoara,
RO-300569, Romania

Lemi Türker
Department of Chemistry, Middle East Technical University, 06531 Ankara, Turkey

Adriana Urdă
Laboratory of Chemical Technology & Catalysis, Department of Organic Chemistry,
Biochemistry & Catalysis, Faculty of Chemistry, University of Bucharest, 4-12,
Blv. Regina Elisabeta, 030018 Bucharest, Romania

Marla Wagner
Six Dimension Design, Buffalo, N.Y., USA

Abbreviations

3C-SiC	cubic silicon carbide
3D	three dimensions
3DT	three dimensions triangular plane
6D	six dimensions
ABC	atom bond connectivity
Abr	abruptness
AcenFac	acentric factor
AIB	atom-in-bonding
AIM	atoms-in-molecule
ALE	atomic layer epitaxy
AM1	Austin model 1
AN	acrylonitrile in the crystalline state
AZI	augmented Zagreb index
BBU	basic building blocks in zeolite structure
BFS-graphs	breadth-first searching graphs
BMI	brain-machine interfaces
BN	boron-nitride
BMA	butyl methacrylate
BP	boiling point
CAC	classic active center
CFD	compact finite difference
ChSR	chain solid-state reaction
CKC	Ciesielski-Krygowski-Cyrański
Cl-VPE	chloride vapor phase epitaxy
CNT	carbon nanotube
DAE	diffusion activation energy
DC	direct current
df	degree of freedom
DFT	density functional theory
DHVAP	standard enthalpy of vaporization
DMS	diamante mechano-synthesis
DOX	dominating oxide network
DSC	differential scanning calorimetry
DSL	dominating silicate network

EAc	immovable electron acceptor
EDS	eccentric distance sum
ELF	electron localization function
EMA	Edmond matching algorithm
FET	field emission transistor
FGF	fries structure generating function
FZs	friable zones
GA	geometric-arithmetic
GAP	groups, algorithms, and programming
GF	geometric frustration
GR	general relativity
GRC	growth rate constant
HFORM	enthalpy of formation
HIV	human immunodeficiency virus
HL	HOMO-LUMO
HOMO	highest occupied molecular orbital
HTSGr	high-temperature process of polymers shear grinding
HVAP	enthalpy of vaporization
H-VPE	hydride vapor phase epitaxy
IBM	international business machines
IETr	initial-electron trap
IR	infrared spectroscopy
LAC	living active center
LDH	layered double hydroxides
LPG	liquefied (or liquid) petroleum gas
LRP	living radical polymerization
LUMO	lowest unoccupied molecular orbital
MAS	magical angle spinning
MATLAB	matrix laboratory
MCI	multicenter index of aromaticity
MD	mobile structural defect
MD	mobile defect
MEDV-13	molecular electronegativity distance vector
MEMS	micro electro mechanical systems
MENTr	mobile electron nanotraps
MI	Miller index
MN	magic numbers
MND	mobile nanodefects
MOL	molecular
MOSFET	metal oxide semiconductor field-effect transistors

MOVPE	metal-organic vapor phase epitaxy
MTG	methanol-to-gasoline process
MTH	methanol-to-hydrocarbons processes
MTO	methanol-to-olefins process
MV	mobile vacancies
MW	molecular weight
NEMS	nanoelectromechanical systems
NIST	National Institute of Standards and Technology
NMR	nuclear magnetic resonance
NPCs	nanoparticles of the complex C
NTKP	non-trivial kinetic particularity
NW	nanowire
PAHs	polycyclic aromatic hydrocarbons
PBHs	polybenzenoid hydrocarbons
PC	pulse current
PCDDs	polychlorinated Dibenzo-P-dioxins
PP	post-polymerization
PRC	pulse reverse current
QCT	quantum chemical topology
QD	quantum dot
QDDSSC	quantum double dots bondotic sensitized solar cell
QSAR	quantitative structure-activity relationship
QSPR	quantitative structure-property relationship
QTMS	quantum topological molecular similarity
RE	resonance energy
REY	zeolite Y partially exchanged with rare-earth ions
RI	retention indices
RQR	ratio of quadratic mean of residuals
RRR	reduced reciprocal Randić index
RTOX	regular triangulane oxide network
SAIB	Specific Adjacency in Bonding
SBU	secondary building blocks in zeolite structure
SCI	six-center index
SDS	six-dimensional string theory
SLR	spectral-like resolution
SLS	solid-liquid-solid growth mechanism
SP	solid phase polymerization
SR	special relativity
SS	structure sensitivity
SSA	specific surface area of a catalyst ($m^2\,g^{-1}$)

STW	Stone–Thrower–Wales defect
STY	space-time yield (mol L^{-1} h^{-1})
SW	Stone-Wales defect
TDCS	tetrahedral dual coordinate system
TEM	transmission electron microscopy
TG/DTA	thermogravimetry/differential thermal analysis
TI	topological index
TM-EC	Timisoara eccentricity index
TOF	turnover frequency (time-1)
TON	turnover number (dimensionless)
TPD	temperature programmed desorption
TPD	thermo-programmed desorption
TPR	thermo-programmed reduction
UChSR	unbranched chain reaction in solids
UV-VIS	ultraviolet-visible spectroscopy
Vac	vacancy
VE	vector equilibrium
VEDCS	vector equilibrium dual coordinate system
VLS	vapor-liquid-solid growth mechanism
VS	vapor-solid growth mechanism
VSS	vapor-solid-solid growth mechanism
W	Wiener index
WW	Wiener weights
XPS	x-ray photoelectron spectroscopy
XRD	x-ray diffraction
XRF	x-ray fluorescence
ZSM-5	Zeolite Socony-Mobil-5

Symbols

$\{b_{ik}\}$	Wiener weights
$(FL)_i$	fractional loss
$(LiCl)g$	glassy matrix of 9.5M aqueous LiCl solution
A	classic active center
A	frequency factor in the Arrhenius equation
b	adsorption coefficient (dependent on temperature) in the Langmuir model of adsorption
B	number of graph edges
$BV(ELF_\pi)$	bifurcation values of the electron localization function
C	complex(A...D)
D	structural defect
D	system dimensionality
D_{METr}	the diffusion coefficient of METr
d_W	Wiener dimensionality
E_a	activation energy of the chemical reaction (kJ mol^{-1})
e_{tr}	electron, trapped by (IETr)
f_W	compression factor of the Wiener index
G	direct space
G^*	dual space
K	Koordinate systeme Galilean
k	rate constant
K'	Koordinate systeme geodesic
$logK_{OW}$	structural-experimental octano-water partition coefficients
LogP	lipophilicity
M	graph diameter
M	metallic ions
n	number of graph vertices
p	partial pressure of the adsorbate
P	phenomenology
R	gas constant
S	entropy
$SW_{q|r}$	arbitrary SW operator
T	reaction temperature (K)
T_{PCr}	immobile precritical cracks

T_{SCr}	mobile supercritical cracks
WC	Wiener index for periodic closed lattices
w_i	contribution to Wiener index from atom i
η	size of the SW defect
Θ	surface coverage, defined as the fraction of the surface covered with adsorbed species
ρ	topological efficiency
ρ^E	extreme topological efficiency
T_a	stopped supercritical cracks

General Preface to Volumes 1–3

The nanosciences, just born on the dawn of the 21st century, widely require dictionaries, encyclopedias, handbooks to fulfill its consecration, and development in academia, research, and industry.

Therefore, this present editorial endeavor springs from the continuous demand on the international market of scientific publications for having a condensed yet explicative dictionary of the basic and advanced up-to-date concepts in nanochemistry. It is viewed as a combination (complementing and overlapping) on various notions, concepts, tools, and algorithms from physical-, quantum-, theoretical-, mathematical-, and even biological-chemistry. The definitions given in the integrated volumes are accompanied by essential examples, applications, open issues, and even historical notes on the persons/subjects, with relevant literature and scholarly contributions.

The current mission is to prepare a premiere referential work for graduate students, PhDs and post-docs, researchers in academia, and industry dealing with nanosciences and nanotechnology. From the book format perspective, the volumes are imagined as practical and attractive as possible with about 130 essential terms as coming from about 60 active scholars from all parts of the globe, explaining each entry with a minimum of five pages (viewed as scientific letters/essays or short course about it), containing definitions, short characterizations, uses, usefulness, limitations, etc., and references – while spanning more than 1,600 pages in the present edition. This effort resulted in this unique and up-to-date *New Frontiers in Nanochemistry: Concepts, Principles, and Trends*, a must for any respected university and the individual updated library! It will also support the future more than necessary decade-editions with new additions in both entries and contributors worldwide!

On the other side, the broad expertise of the editor – in the non-limitative fields of quantum physical-chemical theory, nanosciences and quantum roots of chemistry, computational and theoretical chemistry, quantum modeling of chemical bonding, atomic periodicity and scales of chemical indices, molecular reactivity by chemical indices, electronegativity theory of atoms in molecule, density functionals of electronegativity, conceptual density functional theory, graphene and topological defects of graphenic ribbons, quantitative structure-activity/property relationships (QSAR/QSPR), effector-receptor complex interaction and quantum/logistic enzyme kinetics, and many other related scientific branches – assures that both the educated and generally interested reader in science and technology will be benefited from the book in many ways. They may be listed as:

- an introduction to general nanochemistry;
- an introduction to nanosciences;
- a resource for fast clarification of the basic and modern concepts in multidisciplinary chemistry;
- inspiration for a new application and transdisciplinary connection;
- an advocate for unity in natural phenomena at nano-scales, merging between mathematical, physics and biology towards nanochemistry;
- a reference for both academia (for lessons, essays, exams) as well in R&D industrial sectors in planning and projecting new materials, with aimed properties in both structure and reactivity;
- a compact yet explicated collection of updated scientific knowledge for general, university, and personal libraries; and
- a self-contained book for personal, academic, and technological instruction and research.

Accordingly, the specific aims of the present book are:

- to be a concise and updated source of information on the nanochemistry fields;
- to comprehensively cover nanosystems: from quantum theory to atoms, to molecules, to complexes and chemical materials;
- to be a necessary resource for every-day use by students, academics, and researchers;
- to present informative and innovative contents alike, presented in a systematical and alphabetically manner;
- to present not only definitions but consistent explicative essays on nanochemistry in a short essay or paper/chapter format; and
- to be written by leading and active experts and contributors in nanochemistry worldwide.

All in all, the book is aimed to give supporting material for relevant multi- and trans-disciplinary courses and disciplines specific to nanosciences and nanotechnology in all the major universities in Europe, the Americas, and Asia-Pacific, among which nanochemistry is the core for the privileged position in between physics (elementary properties of quantum particles) and biology (manifested properties of bodies by the environment/forces/ substances influences). They may be non-exclusively listed as:

- *Nanomaterials: Chemistry and Physics*
- *Introduction to Nanosciences*
- *Bottom-Up Technology: Nanochemistry*
- *Nanoengineering: Chemistry and Physics*
- *Sustainable Nanosystems*
- *Ecotoxicology*
- *Environmental Chemistry*
- *Quantum Chemistry*
- *Structural Chemistry*
- *Physical Chemistry–Chemical Physics*

With the belief that, at least partly, this present editorial endeavor succeeds in the above mission and purposes. The editor of this publishing event heartily thank all the contributors for their dedication, inspiration, insight, and generosity, along with the truly friendly and constructive Publisher, Apple Academic Press (jointly with CRC Press at Taylor & Francis), and especially to the President, Mr. Ashish Kumar, and the rest of the AAP team.

Special Preface to Volume 2: Topological Nanochemistry

The nanochemistry research *is addressed to fundamental physicochemical phenomena with bio-eco-pharmaco and even technological impact and even exotics*, meaning that those phenomena are "pushed to the limit" of their manifesting (and knowledge), so that can be introduced new phenomena adaptive and functional controlled (energy- and information-harvesting feedback type response), which should stay as the basis for exploiting and recycling technology to integrate in the polynomial human knowledge-production/social utility-resource economy-interaction design with the medium-nature equilibrium.

With this nanochemistry mission of explaining innovative knowledge by fundamental scientific research, specific missions about designing and controlling the nanotechnology by structural and reactive topology are added. For instance, the *quantum modeling of chemical reactivity* is targeting the unitary understanding based on quantum mechanics, with the eventual involvement of the bosonic-bondonic phenomenology, in explaining the reactivity mechanisms at the molecular level, respectively, in atoms-in-molecules aggregation and nanocomposites. This way, the consecrate principle of electronegativity and chemical hardness, which stays at the base of explaining the chemical's potential equalization of subsystems (quantum atomic pool) in molecules (through frontiers delimited by the annihilation of the electronic density gradient in molecules, same as in orbital hierarchy in atoms), but also respectively of the chemical reactions with the hard bases and acids paradigm, may be reformed and generalized from the combined perspective of electrophilicity (about the activation energy) and chemical power (about the maximum number of electrons interchanged in a chemical interaction, intra- or inter-molecular) concepts; see for instance the present volume entry on "Specific Adjacency in Bonding." Such types of studies create a physical-mathematic universal model in order to treat the "chemical atom," meaning the atom engaged in the chemical bond and prepared for reactivity–ulterior interaction; this approach permits the quantum control of new projected molecules with specific properties of reactivity and specific responses (on specific atoms or other molecules with specific "recognized" molecular zone). This way, the "memory" effects are combined in this direction with modeling the quantum information and the quantum cryptography, with Bohmian effects of distance interaction (about the electrons delocalization in polyenes and polymers, as example); even further, this way a sort of "teleportation of the chemical information and the chemical bond in general" can be achieved, a phenomenon that is comprehensible from the perspective of up-named bondon-quantum particles of the chemical bond; the impact in economy and in quantum information transport. Hence, the storage energy, in different nano- and mesoscopic processes is immediate, but with applicability at nanosystems that are still in study (fullerenes, endo-fullerenes, ionic liquids, composite systems inorganic-organic type, etc.), to which the present volume aims to contribute.

In the same spirit, the other included voices of this second volume of the multi-volume package unfolds *the topological* (so applicative) themes relevant to it, accordingly with the specific nanotopological research (see the first paragraph). The entries all came from eminent international scientists and scholars and span from A-to-Z nanochemistry topological and experimental entries, such as: connectivity, catalysis, topopolynomials, topostructures, topological indices from Wiener to Zagreb, Randic, centric and eccentricity ones, nanocompounds including fullerenes, metal nanoparticles, nanospheres, nanocones, nanoporous carbons, zeolites, electrodeposition, topological reactivity, vapor phase epitaxy, and many more.

This second volume of the three-volume *New Frontiers in Nanochemistry: Concepts, Theories, and Trends* contains about 46 entries, with contributors coming from three continents (Europe, America, and Asia), from seven countries (USA, India, Turkey, Iran, Russia, Italy, and Romania), in multiple first-rate explicative dictionary voices. Let's hope they will be heard worldwide and will positively (in an ecolo-progressist manner) influence the 21st-century macro-destiny of the Earth from the nano-topology and reactivity perspective!

Fugit irreparabile tempus!

Heartily Yours,
—**Mihai V. PUTZ**
(Timisoara, Romania)

CHAPTER 1

Atom-Bond Connectivity Index

MAHDIEH AZARI[1] and ALI IRANMANESH[2]

[1]*Department of Mathematics, Kazerun Branch, Islamic Azad University, P.O. Box: 73135-168, Kazerun, Iran*

[2]*Department of Pure Mathematics, Faculty of Mathematical Sciences, Tarbiat Modares University, P.O. Box: 14115-137, Tehran, Iran, Tel: +982182883447; Fax: +982182883493; E-mail: iranmanesh@modares.ac.ir*

1.1 DEFINITION

The atom-bond connectivity index of a simple graph G is defined as

$$ABC = ABC(G) = \sum_{uv \in E(G)} \sqrt{\frac{d_u + d_v - 2}{d_u d_v}}, \tag{1}$$

where the summation goes over all edges of G, d_u, and d_v are the degrees of the terminal vertices u and v of the edge uv, and $E(G)$ is the edge set of G.

1.2 HISTORICAL ORIGIN(S)

Mathematical chemistry is the area of research engaged in novel applications of mathematics to chemistry; it concerns itself principally with the mathematical modeling of chemical phenomena (Helm, 1897; Gutman & Polansky, 1986). The major areas of research in mathematical chemistry include chemical graph theory (Trinajstić, 1992), which deals with topologies such as the mathematical study of isomerism and the development of topological descriptors; and chemical aspects of group theory (Cotton, 1990), which finds applications in stereochemistry and quantum chemistry.

Chemical graphs, particularly molecular graphs, are models of molecules in which atoms are represented by vertices and chemical bonds by edges of a graph. Physico-chemical or biological properties of molecules can be predicted by using the information encoded in the molecular graphs, eventually translated in the adjacency or connectivity matrix associated to these graphs. This paradigm is achieved by considering various graph theoretical invariants of molecular graphs (also known as topological indices or structural descriptors/measures) and evaluating how strongly are they correlated with various molecular properties. In this way, chemical graph theory plays an important role in the mathematical foundation of QSAR and QSPR research (Diudea, 2001; Todeschini & Consonni, 2000).

A topological index is any function calculated based on a molecular graph (i.e., a connected graph with maximum vertex degree at most 4 and whose graphical representation may resemble a structural formula of a molecule), irrespective of the labeling of its vertices. Hundreds of topological indices have been introduced and studied, starting with the seminal work by Wiener (1947a,b), in which he used the sum of all shortest-path distances of a (molecular) graph for modeling physical properties of alkanes. Topological indices can be divided into various categories such as graph entropy (Bonchev, 1983; Dehmer, 2008; Dehmer et al., 2009; Dehmer & Mowshowitz, 2011) representing information-theoretic indices, eigenvalue-based measures (Randić et al., 2001; Estrada, 2002; Dehmer et al., 2012b), distance-based measures (Balaban, 1982; Khalifeh et al., 2009; De Corato et al., 2012; Putz et al., 2013; Alizadeh et al., 2013; 2014; Xu et al., 2014; Azari & Iranmanesh, 2013a, 2015a; Azari et al., 2016), symmetry-based descriptors (Todeschini & Consonni, 2000), and degree-based invariants (Nikolić et al., 2003; Vukičević & Gašperov, 2010; Furtula et al., 2013; Gutman, 2013; Gutman & Tošović, 2013; Zhong & Xu, 2014; Azari & Iranmanesh, 2011, 2013b, 2015, 2015b; Azari, 2014; Falahati-Nezhad et al., 2015).

Topological indices based on end-vertex-degrees of edges have been used over 40 years. Among them, several indices are recognized to be useful tools in QSPR/QSAR studies. Some of the oldest and the most thoroughly investigated are the first and second Zagreb indices (Gutman & Trinajstić, 1972; Gutman et al., 1975), Randić connectivity index (Randić, 1975), and harmonic index (Fajtlowicz, 1987).

Probably, the best-known such descriptor is the Randić connectivity index, which is suitable for measuring the extent of branching of the carbon-atom skeleton of saturated hydrocarbons (Randić, 1975). However, many physicochemical properties are dependent on factors rather different than branching.

The atom-bond connectivity index defined in Eq. (1) was put forward in 1998 by Estrada et al. (1998) in an attempt to provide an improved version of the Randić connectivity index. Estrada et al. (1998) showed that the *ABC* index is well correlated with the heats of formation of alkanes and that it thus can serve for predicting their thermodynamic properties. This work attracted little attention and has for a long time been ignored by theoretical chemists. Ten years later, Estrada (2008) published another paper on the *ABC* index in which he elaborated a novel quantum theory-like justification for this topological index, showing that it provides a model for taking into account 1,2-, 1,3-, and 1,4-interactions in the carbon–atom skeleton of saturated hydrocarbons, and that it can be used for rationalizing steric effects in such compounds. Contrary to Estrada et al. (1998), this work did attract the attention of mathematical chemists and a long series of mathematical investigations on the *ABC* index appeared in the literature.

1.3   NANO-SCIENTIFIC DEVELOPMENT(S)

The *ABC* index has attracted a lot of interest in the last several years both in mathematical and chemical research communities, and numerous results and structural properties of the *ABC* index were established. The graph theoretical concepts and notations not defined in this essay can be found in the standard books of graph theory such as West (1996).

It can be easily recognized that *ABC* is just a representation of a class of topological indices of the form

$$ABC_{general} = ABC_{general}(G) = \sum_{uv \in E(G)} \sqrt{\frac{Q_u + Q_v - 2}{Q_u Q_v}}, \qquad (2)$$

where Q_u is some quantity that in a unique manner can be associated with the vertex u of the graph G and the summation goes over all edges uv of G. Six members of *ABC* group topological indices have been put forward up to now. The atom-bond connectivity index *ABC* is the first member of this group which is obtained from Eq. (2) by setting Q_u to be the degree d_u of the vertex u in the graph G. So, in the literature, *ABC* is also called the first atom-bond connectivity index and denoted by ABC_1. The second member of this class, denoted by ABC_2, was considered by Graovac and Ghorbani (2010) by setting Q_u to be the number n_u of vertices of G lying closer to the vertex u than to the vertex v for the edge uv of the graph G. The third member of this class, denoted by ABC_3, was introduced in Farahani (2012a) by setting Q_u to be the number m_u of edges of G lying closer to the vertex

u than to the vertex v for the edge uv of the graph G. The fourth member of this class, denoted by ABC_4, was proposed by Ghorbani and Hosseinzadeh (2010) by setting Q_u to be the quantity S_u defined as the sum of degrees of all neighbors of the vertex u in the graph G. The fifth member of this class, denoted by ABC_5, was introduced in Farahani (2013a) by setting Q_u to be the eccentricity ε_u of the vertex u in the graph G which is the largest distance between u and any other vertex of G. Another member of this class, which we denote it by ABC_6, was proposed by Yang and Deng (2012) by setting Q_u to be the quantity D_u defined as the sum of distances between the vertex u and all other vertices in the graph G.

The edge version of the atom-bond connectivity index $ABC_e(G)$ of a given graph G was introduced by Farahani (2013b) to be the ABC index of the line graph $L(G)$ of G, where $L(G)$ is the graph whose vertices correspond to the edges of G with two vertices being adjacent if and only if the corresponding edges in G have a vertex in common.

A descriptor is called degenerate if it possesses the same value for more than one graph. Dehmer et al. (2012a) evaluated the uniqueness, discrimination power or degeneracy of some degree, distance, and eigenvalue-based descriptors for investigating graphs holistically. In fact, the uniqueness of structural descriptors has been investigated in mathematical chemistry and related disciplines for discriminating the structure of isomeric structures and other chemical networks. A highly discriminating graph measure is desirable for analyzing graphs; hence, measuring the degree of its degeneracy is important for understanding its properties, limits, and quality. For the degree-based descriptors such as ABC, it is not surprising that these measures have only little discrimination power, as many graphs can be realized by identical degree sequences. This effect is even stronger if the cardinality of the underlying graph set increases. However, among the degree-based descriptors considered in Dehmer et al. (2012a), the ABC index had the highest discrimination power. In order to tackle the question of what kind of degeneracy the indices possess, Dehmer et al. (2012a) plotted their characteristic value distributions. For a graph class, they used the class of exhaustively generated non-isomorphic, connected and unweighted graphs with nine vertices. In the case of ABC, it was found that by using this index there exist many degenerate graphs possessing quite similar index values where the hull of the distributions forms a Gaussian curve.

The bounds of a topological index are important information of a molecular graph in the sense that they establish the approximate range of the index in terms of molecular structural parameters. Computing upper and lower bounds on the ABC index has been the subject of many papers.

Furtula et al. (2009) studied the mathematical properties of the *ABC* index of trees. They found chemical trees (i.e., trees with maximum degree at most 4 used to model carbon skeletons of acyclic hydrocarbons) with maximal and minimal values of the *ABC* index. In addition, they proved that, among all trees, the star tree has the maximal *ABC* value. Finding the minimum *ABC* value of trees, in the case of general trees, remained an open problem.

Xing et al. (2010) presented the best upper bound for the *ABC* index of trees with a perfect matching and characterized the unique extremal tree, which is a chemical tree. They also gave upper bounds for the *ABC* index of trees with fixed number of vertices and maximum degree, and of chemical trees with fixed numbers of vertices and pendent vertices, and characterized the extremal trees. The problem of finding the minimum *ABC* index and characterizing the trees with the minimum *ABC* index remained to be resolved in the future.

Das (2010) presented some lower and upper bounds on the *ABC* index of graphs and trees using parameters such as the number of vertices and edges, number of pendant vertices, maximum vertex degree, minimum nonpendant vertex degree, and the second modified Zagreb index [see Nikolić et al., (2003) for the definition of the modified Zagreb indices]. He also characterized graphs for which these bounds are best possible.

Das and Trinajstić (2010) compared the *ABC* index and the geometric-arithmetic index *GA* [see Vukičević & Furtula (2009) for the definition of *GA*] for chemical trees and molecular graphs. Besides chemical trees and molecular graphs, general graphs were treated. It was also shown in Das & Trinajstić (2010) that the *ABC* index is smaller than the *GA* index for the difference between the maximum and minimum degree, and less than or equal to three. Comparison between these two indices, in the case of general trees and general graphs, remained an open problem.

Das et al. (2011) established several properties of the *ABC* index. In particular, they showed that if a new edge is inserted into *G*, then *ABC* necessarily increases. By means of this result, they characterized the graphs extremal with respect to *ABC*.

Horoldagva and Gutman (2011) obtained an upper bound on the *ABC* index of graphs with *n* vertices and *m* edges in terms of the number of vertices, number of edges, and second Zagreb index. They applied these results to compute more upper bounds on the *ABC* index in terms of other graph parameters such as radius and minimum vertex degree.

Zhou and Xing (2011) gave upper bounds for the *ABC* index using the number of vertices, number of edges, Randić connectivity index, and first Zagreb index. They also determined the unique tree with the maximum *ABC* index among trees with given numbers of vertices and pendant vertices, and the n-vertex trees with the maximum, and the second, the third, and the fourth maximum ABC indices, for $n \geq 6$.

Fath-Tabar et al. (2011) gave some inequalities for the ABC index of a series of graph operations. They also proved that the obtained bounds are tight. As an application, the ABC indices of C_4-nanotubes and C_4-nanotori were computed.

Chen and Guo (2011) characterized the catacondensed hexagonal systems with the maximum and minimum ABC indices. They also proved that the ABC index of a graph decreases when any edge is deleted. This result implies that Chen & Guo (2011), among all n-vertex graphs, the complete graph K_n has the maximal ABC value and among all n-vertex connected graphs, the graph with the minimal ABC index is a tree.

Gan et al. (2011) presented some sharp lower and upper bounds on the ABC index. In addition, they gave a characterization of the maximum and minimum ABC index and the corresponding extremal graphs among all unicyclic graphs and unicyclic chemical graphs.

Chen and Liu (2011) obtained a sharp upper bound on the ABC index of bicyclic graphs.

Xing et al. (2011) gave an upper bound for the ABC index of connected graphs with a fixed number of vertices, number of edges, and maximum degree, and characterized the extremal graphs. From this, they obtained an upper bound and extremal graphs for the ABC index of molecular graphs with a fixed number of vertices and edges. The n-vertex unicyclic graphs with the maximum, the second, the third and the fourth maximum ABC indices, and the n-vertex bicyclic graphs with the maximum and the second maximum ABC indices, for $n \geq 5$, were also determined in Xing et al. (2011).

Chen et al. (2012a) obtained some sharp lower bounds on the ABC index of chemical unicyclic graphs. Chen and Guo (2012) obtained sharp upper and lower bounds on the ABC index of chemical bicyclic graphs. Das et al. (Das et al., 2012) established some upper bounds and Nordhaus–Gaddum type results for ABC.

Chen et al. (2012b) characterized the unique graph with the maximum ABC index among all n-vertex graphs with vertex connectivity k. Furthermore, they determined the maximum ABC index of a connected graph with n vertices and matching number β and characterized the unique extremal graph.

Xing and Zhou (2012) characterized the extremal trees with fixed degree sequence that maximize and minimize the ABC index, respectively. They also provided the algorithms to construct such trees.

Gan et al. (2012) characterized the trees which have the maximum or minimum *ABC* index with given degree sequences.

Lin et al. (2013a) introduced the breadth-first searching graphs (BFS-graphs for short), and proved that for any degree sequence π, there exists a BFS-graph with minimal *ABC* index in the class of connected graphs with degree sequence π. This result is applicable to obtain the minimal value or lower bounds on the *ABC* index of connected graphs.

Chen et al. (2013b) presented lower and upper bounds on the *ABC* index of line and total graphs, and characterized graphs for which these bounds are tight.

Li and Zhou (2013) studied the *ABC* index of unicyclic graphs with perfect matchings.

Gutman et al. (2014b) studied the relation between the *ABC* and harmonic indices.

Zhong and Xu (2014) obtained several inequalities between the *ABC* index and some vertex-degree-based topological indices such as the Randić connectivity index, harmonic index, and sum connectivity index (see Zhou & Trinajstić (2009) for the definition of the sum connectivity index).

Palacios (2014) gave a new upper bound for the *ABC* index that uses some ideas pertinent to the Kirchhoff and other resistive descriptors. Using this general bound, it is possible to obtain many other particular bounds and asymptotic maximal results for the *ABC* index with elementary proofs (Palacios, 2014).

Cruz et al. (2014) found the extremal values of the *ABC* index over the set of Kragujevac trees with the central vertex of fixed degree.

A connected non-acyclic graph G is called a quasi-tree graph if there exists a vertex u of G such that $G-u$ is a tree. Dehghan-Zadeh & Ashrafi (2014) computed the first, the second, and the third maximum *ABC* index in the class of all quasi-tree graphs.

Dehghan-Zadeh et al. (2014) computed the first and the second maximum values of the *ABC* index in the class of all n-vertex tetracyclic graphs.

A connected graph G is called a cactus if any two of its cycles have at most one common vertex. Dong & Wu (2014) gave sharp bounds for the *ABC* index of cacti among the set of cacti with n vertices and r cycles, and the set of cacti with n vertices and p pendant vertices. They also characterized the corresponding extremal cacti.

Li (2015) determined the unique cactus with the maximum *ABC* index among all cacti with n vertices, and the unique cactus with the maximum *ABC* index among all cacti with n vertices and r cycles, and among all cacti with n vertices and p pendent vertices, respectively, where $0 \le r \le \left\lfloor \dfrac{n-1}{2} \right\rfloor$ and $0 \le p \le n-1$.

Ashrafi et al. (2015) obtained the first and the second maximum values of the *ABC* index among all *n*-vertex cacti.

Hemmasi & Iranmanesh (2015) gave some lower and upper bounds for the *ABC* index of all connected *n*-vertex graphs in terms of the Randić connectivity index, Zagreb indices, and second modified Zagreb index. As an application, the *ABC* index of a family of molecular graphs was computed in Hemmasi & Iranmanesh (2015).

Du (2015) studied the relationship between the *ABC* index and radius of connected graphs. In particular, he determined upper and lower bounds on the difference between the *ABC* index and radius of connected graphs.

Ashrafi et al. (2016) presented the first, the second, and the third maximum values of the *ABC* index in the class of all *n*-vertex tricyclic graphs.

Raza et al. (2016) proved that the *ABC* index is smaller than the *GA* index for the line graphs of molecular graphs, for general graphs in which the difference between the maximum degree Δ and the minimum degree δ ($\delta > 1$) is less than or equal to $(2\delta-1)^2$, and for some families of trees; giving partial solution to an open problem proposed in Das & Trinajstić (2010). The complete solution of the said problem was left as a task for the future.

Azari & Falahati-Nezhad (2015, 2017) presented exact formulae for the *ABC* index of two composite graphs called splice and link of graphs and applied the results to compute the *ABC* index of a class of dendrimer graphs.

When examining a topological index, one of the important questions that need to be answered is that for which graphs this index assumes minimal and maximal values, and what are these extremal values. Usually, one first tries to resolve this problem for trees with a fixed number of vertices. In the case of the *ABC* index, finding the tree for which this index is maximal was relatively easy (Furtula et al., 2009): this is the star. Eventually, the trees with the second, the third, and the fourth maximum *ABC* indices were also determined (Zhou & Xing, 2011). On the other hand, the characterization of trees with the smallest *ABC* index, in spite of numerous attempts, is still an open problem. What only is known is that the connected graph with minimal *ABC* index must be a tree, and this minimal-*ABC* tree needs not to be unique (Chen & Guo, 2011; Das et al., 2011). Some structural features of such trees have been determined up to now and several conjectures on the structure of the minimal-*ABC* trees were disproved by counterexamples. For more information see Gutman & Furtula (2012); Furtula et al. (2012); Vassilev & Huntington (2012); Gutman et al. (2012a); Ahmadi et al. (2013a,b, 2014); Dimitrov (2013, 2014); Lin et al. (2013a,b, 2014, 2015); Hosseini et al. (2014); Liu & Chen (2014); Goubko et al. (2015); Magnant et al. (2015), and specially the recent survey of Gutman et al. (2013).

In recent years, there has been considerable interest in the general problem of determining topological indices of nanostructures. There exist a large number of papers on computing *ABC* index of chemical graphs and nanostructures.

Khaksar et al. (2010) computed the *ABC* index of two special kinds of carbon nanocones. The *ABC* index of *V*-phenylenic nanotube and nanotorus were given in Ghorbani et al. (2011). The *ABC* index of $TUC_4C_8(S)$ nanotube and $TUC_4C_8(R)$ nanotorus were computed in Asadpour et al. (2011). Farahani (2012b) computed the *ABC* index of zig-zag and armchair polyhex nanotubes, $HAC_5C_7[p,q]$ nanotube (Farahani, 2013c), $HAC_5C_6C_7[p,q]$ nanotube (Farahani, 2013d), and circumcoronene series of benzenoid (Farahani, 2013e). Nikmehr et al., (Nikmehr et al., 2014; 2015) computed the *ABC* index of $VAC_5C_6C_7[p,q]$ nanotube, *H*-phenylenic nanotube, *H*-naphthylenic nanotube, and *H*-anthracenic nanotube. Soleimani et al. (2015) gave exact formulae for computing the *ABC* index of linear [*n*]-tetracene, *V*-tetracenic nanotube, *H*-tetracenic nanotube, and tetracenic nanotorus. Kürkçü & Aslan (2015) computed the *ABC* index of generalized carbon nanocones and proposed an algorithm for the evaluation of the *ABC* index of chemical graphs. The *ABC* indices of some classes of nanostar dendrimers were determined in Ahmadi & Sadeghimehr (2010); Ghorbani & Songhori (2010); Husin et al. (2013, 2015); Imran et al. (2014a). Asadpour (2012) obtained the *ABC* index of an infinite family of nanostructures derived from a class of composite graphs called bridge graphs.

The *ABC* index of polyomino chains of *k* cycles and triangular benzenoid graphs were computed in Ghorbani & Ghazi (2010). The *ABC* index of the zig-zag chain polyomino graphs was given in Chen et al. (2013a). A sharp upper bound on the *ABC* index of catacondensed polyomino graphs with *n* squares and the corresponding extremal graphs were also given in Chen et al. (2013a).

Yang et al. (2011) obtained the *ABC* index of benzenoid systems and phenylenes and established a simple relation between the *ABC* index of a phenylene and the corresponding hexagonal squeeze. Ke (2012) computed the *ABC* index of fluoranthene congeners and characterized the extremal catacondensed benzenoid systems with the maximal and minimal *ABC* index. Cruz et al. (2013) studied the maximal and minimal values of the *ABC* index over the class of all hexagonal systems with a fixed number of hexagons. Farahani, 2015) obtained the *ABC* index for a type of benzenoid system called jagged-rectangle.

Rajan et al. (2012) computed the *ABC* index of silicate, honeycomb, and hexagonal networks. The *ABC* indices of oxide and chain silicate networks were given in Hayat & Imran (2014). Imran et al. (2014b) extended the results obtained by Hayat & Imran (2014) to interconnection networks and derived analytical closed results of the *ABC* index for butterfly and Benes networks. The *ABC* index of dominating oxide network (DOX), dominating silicate network (DSL), and regular triangulene oxide network (RTOX) were computed in Baig et al. (2015). They studied a carbon nanotube network that is motivated by the molecular structure of a regular hexagonal lattice and computed the *ABC* index of this class of networks.

Farahani (2013f) computed the *ABC* index of polycyclic aromatic hydro-carbons (PAHs). The *ABC* index of some members of isomers of decane, namely *n*-decane, 3,4,4-trimethyl heptanes, and 2,4-dimethyl-4-ethyl hexane were obtained in Raut (2014). Imran & Hayat (2014) obtained the *ABC* index of the Aztec diamond. The *ABC* index of graphene was given in Sridhara et al. (2015). The *ABC* indices of some special molecular graphs were given in Gao (2015).

1.4 NANOCHEMICAL APPLICATION(S)

The *ABC* index has been applied up to now to study the stability of linear and branched alkanes as well as the strain energy of cycloalkanes (Estrada et al., 1998; Estrada, 2008). In all mathematical studies of the *ABC* index, its applicability for modeling thermochemical data is being emphasized. A critical re-examination performed in Gutman et al. (2012b) confirmed that *ABC* is well reproducing the heats of formation of alkanes (ΔH_f). Moreover, the simple empirical formula $\Delta H^{\circ}_f = -(65.98 + 20.37ABC)$ reproduces the heats of formation with an accuracy comparable to that of high-level *ab initio* and DFT (MP2, B3LYP) quantum chemical calculations. By this, *ABC* happens to be the only topological index for which a theoretical, quantum-theory-based, foundation, and justification has been found.

A comparative test was performed by Gutman & Tošović (2013) of how well a class of 20 vertex-degree-based topological indices, including *the* ABC index, are correlated with two simple physicochemical parameters of octane isomers. These parameters were chosen to be the standard heats of formation (representative for thermo-chemical properties) and the normal boiling points (representative for intermolecular, van-der-Waals-type, interactions). It was found by Gutman & Tošović (2013) that the correlation ability of many of these indices is either rather weak or nil. The only vertex-degree-based topological index that had correlation coefficients over 0.9 and could successfully pass the test was the augmented Zagreb index *AZI* [see Furtula et al., (2010) for the definition of *AZI*]. Gutman & Tošović (2013) also observed that, among the examined vertex-degree-based molecular structure-descriptor, the second-best index appeared to be the *ABC* index. This observation showed that, the *AZI* and *ABC* indices should be preferred to other examined vertex-degree-based indices in designing quantitative structure-property relations.

1.5 MULTI-/TRANS-DISCIPLINARY CONNECTION(S)

As explained before, most of the results on the *ABC* index have been achieved by using mathematical; particularly graph/group theoretical, methods. Some mathematical software such as GAP, MATLAB, MATHEMATICA, etc. has also been applied in solving computational problems on the *ABC* index. As far as we know, the applications of the *ABC* index in other fields of science have not yet been reported in the literature.

1.6 OPEN ISSUES

In order to reduce the arbitrariness in the production of novel topological indices, several criteria are put forward that such a molecular structure descriptor should be required to satisfy. One of these criteria is always that a topological index "should change gradually with a gradual change in (molecular) structure." This property may be called the smoothness of the topological index in question. In order to quantify this concept, two measures thereof have been introduced (Furtula et al., 2013), called "structure sensitivity" (*SS*) and "abruptness" (*Abr*). These are defined as follows (Furtula et al., 2013):

Let *G* be a molecular graph and let *TI* be the topological index considered. The set $\Gamma(G)$ consists of all connected graphs obtained from *G* by replacing one of its edges by another edge. Intuitively, the elements of $\Gamma(G)$ are those graphs that are structurally most similar to *G*. Then

$$SS(TI,G) = \frac{1}{|\Gamma(G)|} \sum_{\gamma \in \Gamma(G)} \left| \frac{TI(G) - TI(\gamma)}{TI(G)} \right|, \qquad (3)$$

$$Abr(TI,G) = \max_{\gamma \in \Gamma(G)} \left| \frac{TI(G) - TI(\gamma)}{TI(G)} \right|. \qquad (4)$$

According to Eq. (3), *SS* is the average relative sensitivity of *TI* to small changes in the structure of the graph *G*. According to Eq. (4), *Abr* shows how much a small structural change may cause a jump-wise (large) change in the considered topological index. From a practical point of view, the best is if the structure sensitivity is as large as possible, but the abruptness is as small as possible.

Furtula et al. (2013) examined the smoothness of 12 vertex-degree-based topological indices, including *the* ABC index, recently. The test was done for the set of all trees with *n* = 6,7,...,13 vertices. Computational details and the other calculated values for *SS* and *Abr* can be found in Furtula et al. (2013). The results showed that, the *ABC* index has the least structure sensitivity, which disagrees with the claims (Estrada et al., 1998; Estrada, 2008; Gutman et al., 2012b) that *ABC* is a very good measure of branching dependent thermodynamic properties of alkanes. The degree-based topological indices with the greatest structure sensitivity were found by Furtula et al. (2013) to be the augmented Zagreb index and the second Zagreb index. Thus, at least in the case of trees, the degree-based topological indices with best structure sensitivity are the augmented and the second Zagreb indices, and these appear to be superior to the other examined indices studied such as *ABC*. As a sort of unpleasant surprise, the very same indices were found to have the greatest abruptness.

Another comparative test performed by Gutman et al. (2014a) on the set of all isomeric octanes showed that for the standard enthalpy of formation and normal boiling points, the predictive ability of *ABC* index is slightly lower than the predictive ability of two other examined degree-based indices such as *AZI* and the recently-introduced reduced reciprocal Randić index *RRR* [see Manso et al. (2012) for the definition of *RRR*].

KEYWORDS

- **chemical graph theory**
- **Estrada**
- **heat of formation of alkanes**
- **molecular graph**
- **topological index**

REFERENCES AND FURTHER READING

Ahmadi, M. B., & Sadeghimehr, M., (2010). Atom-bond connectivity index of an infinite class $NS_1[n]$ of dendrimer nanostars. *Optoelectron. Adv. Mat.-Rapid Commun., 4*(7), 1040–1042.

Ahmadi, M. B., Dimitrov, D., Gutman, I., & Hosseini, S. A., (2014). Disproving a conjecture on trees with minimal atom-bond connectivity index. *MATCH Commun. Math. Comput. Chem.*, *72*, 685–698.

Ahmadi, M. B., Hosseini, S. A., & Salehi, N. P., (2013a). On trees with minimal atom bond connectivity index. *MATCH Commun. Math. Comput. Chem.*, *69*, 559–563.

Ahmadi, M. B., Hosseini, S. A., & Zarrinderakht, M., (2013b). On large trees with minimal atom-bond connectivity index. *MATCH Commun. Math. Comput. Chem.*, *69*, 565–569.

Alizadeh, Y., Azari, M., & Došlić, T., (2013). Computing the eccentricity-related invariants of single-defect carbon nanocones. *J. Comput. Theor. Nanosci.*, *10*(6), 1297–1300.

Alizadeh, Y., Iranmanesh, A., Došlić, T., & Azari, M., (2014). The edge Wiener index of suspensions, bottlenecks, and thorny graphs. *Glas. Mat. Ser. III.*, *49*(69), 1–12.

Asadpour, J., (2012). Computing some topological indices of nanostructures of bridge graph. *Dig. J. Nanomater. Bios.*, *7*(1), 19–22.

Asadpour, J., Mojarad, R., & Safikhani, L., (2011). Computing some topological indices of nanostructures. *Dig. J. Nanomater. Bios.*, *6*(3), 937–941.

Ashrafi, A. R., Dehghan-Zadeh, T., & Habibi, N., (2015). External atom bond connectivity index of cactus graphs. *Commun. Korean Math. Soc.*, *30*(3), 283–295.

Ashrafi, A. R., Dehghan-Zadeh, T., Habibi, N., & John, P. E., (2016). Maximum values of atom-bond connectivity index in the class of tricyclic graphs. *J. Appl. Math. Comput.*, *50*(1/2), 511–527.

Azari, M., (2014). Sharp lower bounds on the Narumi-Katayama index of graph operations. *Appl. Math. Comput.*, *239C*, 409–421.

Azari, M., & Falahati-Nezhad, F., (2015). On vertex-degree-based invariants of link of graphs with application in dendrimers. *J. Comput. Theor. Nanosci.*, *12*(12), 2611–5616.

Azari, M., & Falahati-Nezhad, F., (2017). Splice graphs and their vertex-degree-based invariants. *Iranian J. Math. Chem.*, *8*(1), 61–70.

Azari, M., & Iranmanesh, A., (2011). Generalized Zagreb index of graphs. *Studia Univ. Babes Bolyai Chem.*, *56*(3), 59–70.

Azari, M., & Iranmanesh, A., (2013a). Computing the eccentric-distance sum for graph operations. *Discrete Appl. Math.*, *161*(18), 2827–2840.

Azari, M., & Iranmanesh, A., (2013b). Chemical graphs constructed from rooted product and their Zagreb indices. *MATCH Commun. Math. Comput. Chem.*, *70*, 901–919.

Azari, M., & Iranmanesh, A., (2014). Harary index of some nanostructures. *MATCH Commun. Math. Comput. Chem.*, *71*, 373–382.

Azari, M., & Iranmanesh, A., (2015). Some inequalities for the multiplicative sum Zagreb index of graph operations. *J. Math. Inequal.*, *9*(3), 727–738.

Azari, M., Iranmanesh, A., & Došlić, T., (2016). Vertex-weighted Wiener polynomials of subdivision-related graphs. *Opuscula Math.*, *36*(1), 5–23.

Bača, M., Horváthová, J., Mokrišová, M., Semaničová-Feňovčíková, A., & Suhányiova, A., (2015). On topological indices of carbon nanotube network. *Can. J. Chem.*, *93*, 1–4.

Baig, A. Q., Imran, M., & Ali, H., (2015). On topological indices of polyoxide, polysilicate, DOX, and DSL networks. *Can. J. Chem.*, *93*(7), 730–739.

Balaban, A. T., (1982). Highly discriminating distance-based topological index. *Chem. Phys. Lett.*, *89*, 399–404.

Bonchev, D., (1983). *Information Theoretic Indices for Characterization of Chemical Structures*. Research Studies Press: Chichester, England.

Chen, J., & Guo, X., (2011). Extreme atom-bond connectivity index of graphs. *MATCH Commun. Math. Comput. Chem.*, *65*, 713–722.

Chen, J., & Guo, X., (2012). The atom-bond connectivity index of chemical bicyclic graphs. *Appl. Math. J. Chinese Univ., 27*(2), 243–252.

Chen, J., & Liu, J., (2011). On atom-bond connectivity index of bicyclic graphs. *J. Guangxi Teachers Education Univ., 28*, 8–12.

Chen, J., Liu, J., & Guo, X., (2012a). The atom-bond connectivity index of chemical unicyclic graphs. *Journal of Zhejiang University (Science Edition), 39*(4), 377–380.

Chen, J., Liu, J., & Guo, X., (2012b). Some upper bounds for the atom-bond connectivity index of graphs. *Appl. Math. Lett., 25*, 1077–1081.

Chen, J., Liu, J., & Li, Q., (2013a). The atom-bond connectivity index of catacondensed polyomino graphs. *Discrete Dyn. Nat. Soc.*, 598517.

Chen, Z., Meng, J., & Tian, Y., (2013b). On atom-bond connectivity index of line and total graphs. *Operations Research Transactions, 17*(3), 1–10.

Cotton, F. A., (1990). *Chemical Applications of Group Theory* (3ʳᵈ edn.). John Wiley & Sons: Canada.

Cruz, R., Giraldo, H., & Rada, J., (2013). Extremal values of vertex–degree topological indices over hexagonal systems. *MATCH Commun. Math. Comput. Chem., 70*, 501–512.

Cruz, R., Gutman, I., & Rada, J., (2014). Topological indices of Kragujevac trees. *Proyecciones, 33*(4), 471–482.

Das, K. C., (2010). Atom-bond connectivity index of graphs. *Discrete Appl. Math., 158*, 1181–1188.

Das, K. C., & Trinajstić, N., (2010). Comparison between first geometric-arithmetic index and atom-bond connectivity index. *Chem. Phys. Lett., 497*, 149–151.

Das, K. C., Gutman, I., & Furtula, B., (2011). On atom-bond connectivity index. *Chem. Phys. Lett., 511*, 452–454.

Das, K. C., Gutman, I., & Furtula, B., (2012). On atom-bond connectivity index. *Filomat., 26*(4), 733–738.

De Corato, M., Benedek, G., Ori, O., & Putz, M. V., (2012). Topological study of Schwarzitic junctions in 1D lattices. *Int. J. Chem. Model, 4*(2/3), 105–113.

Dehghan-Zadeh, T., & Ashrafi, A. R., (2014). Atom-bond connectivity index of quasi-tree graphs. *Rend. Circ. Mat. Palermo, 63*, 347–354.

Dehghan-Zadeh, T., Ashrafi, A. R., & Habibi, N., (2014). Maximum values of atom-bond connectivity index in the class of tetracyclic graphs. *J. Appl. Math. Comput., 46*(1), 285–303.

Dehmer, M., (2008). Information processing in complex networks: Graph entropy and information functional. *Appl. Math. Comput., 201*, 82–94.

Dehmer, M., & Mowshowitz, A., (2011). A history of graph entropy measures. *Inf. Sci., 1*, 57–78.

Dehmer, M., Grabner, M., & Furtula, B., (2012a). Structural discrimination of networks by using distance, degree and eigenvalue-based measures. *PLoS One, 7*(7), e38564.

Dehmer, M., Sivakumar, L., & Varmuza, K., (2012b). Uniquely discriminating molecular structures using novel eigenvalue-based descriptors. *MATCH Commun. Math. Comput. Chem., 67*, 147–172.

Dehmer, M., Varmuza, K., Borgert, S., & Emmert-Streib, F., (2009). On entropy-based molecular descriptors: Statistical analysis of real and synthetic chemical structures. *J. Chem. Inf. Model., 49*, 1655–1663.

Dimitrov, D., (2013). Efficient computation of trees with minimal atom-bond connectivity index. *Appl. Math. Comput., 224*, 663–670.

Dimitrov, D., (2014). On structural properties of trees with minimal atom-bond connectivity index. *Discrete Appl. Math., 172*, 28–44.

Diudea, M. V., (2001). *QSPR/QSAR Studies by Molecular Descriptors*. Nova: New York, USA.

Dong, H., & Wu, X., (2014). On the atom-bond connectivity index of cacti. *Filomat., 28*(8), 1711–1717.

Du, Z., (2015). On the atom-bond connectivity index and radius of connected graphs. *J. Inequal. Appl.,* 188.

Estrada, E., (2002). Characterization of the folding degree of proteins. *Bioinformatics, 18*, 697–704.

Estrada, E., (2008). Atom-bond connectivity and the energetic of branched alkanes. *Chem. Phys. Lett., 463*, 422–425.

Estrada, E., Torres, L., Rodríguez, L., & Gutman, I., (1998). An atom-bond connectivity index: Modeling the enthalpy of formation of alkanes. *Indian J. Chem., 37A*, 849–855.

Fajtlowicz, S., (1987). On conjectures on graffiti-II. *Congr. Numer., 60*, 187–197.

Falahati-Nezhad, F., Iranmanesh, A., Tehranian, A., & Azari, M., (2015). Upper bounds on the second multiplicative Zagreb coindex. *Util. Math., 96*, 79–88.

Farahani, M. R., (2012a). New version of atom-bond connectivity index of $TurC_4C_8(S)$ nanotube. *Int. J. Chem. Model., 4*(4), 527–532.

Farahani, M. R., (2012b). Some connectivity indices and Zagreb index of polyhex nanotubes. *Acta Chim. Slov., 59*, 779–783.

Farahani, M. R., (2013a). Eccentricity version of atom-bond connectivity index of benzenoid family $ABC_5(H_k)$. *World Appl. Sci. J., 21*(9), 1260–1265.

Farahani, M. R., (2013b). The edge-version of atom-bond connectivity index of connected graphs. *Acta Univ. Apulensis, 36*, 277–284.

Farahani, M. R., (2013c). Atom-bond connectivity and geometric-arithmetic indices of HAC_5C_7[p,q] nanotube. *Int. J. Chem. Model, 5*(1), 127–132.

Farahani, M. R., (2013d). On the geometric-arithmetic and atom-bond connectivity index of $HAC_5C_6C_7$[p,q] nanotube. *Chemical Physics Research Journal, 6*(1), 21–26.

Farahani, M. R., (2013e). Computing randic, geometric-arithmetic and atom-bond connectivity indices of circumcoronene series of benzenoid. *Int. J. Chem. Model, 5*(4), 485–493.

Farahani, M. R., (2013f). Some connectivity indices of polycyclic aromatic hydrocarbons (PAHs). *Adv. Mat. Corrosion, 1*, 65–69.

Farahani, M. R., (2015). On atom bond connectivity and geometric-arithmetic indices of a benzenoid system. *International Journal of Engineering and Technology Research, 3*(2), 1–4.

Fath-Tabar, G. H., Vaez-Zadeh, B., Ashrafi, A. R., & Graovac, A., (2011). Some inequalities for the atom-bond connectivity index of graph operations. *Discrete Appl. Math., 159*, 1323–1330.

Furtula, B., Graovac, A., & Vukičević, D., (2009). Atom-bond connectivity index of trees. *Discrete Appl. Math., 157*, 2828–2835.

Furtula, B., Graovac, A., & Vukičević, D., (2010). Augmented Zagreb index. *J. Math. Chem., 48*(2), 370–380.

Furtula, B., Gutman, I., & Dehmer, M., (2013). On structure-sensitivity of degree-based topological indices. *Appl. Math. Comput., 219*, 8973–8978.

Furtula, B., Gutman, I., Ivanović, M., & Vukičević, D., (2012). Computer search for trees with minimal ABC index. *Appl. Math. Comput., 219*, 767–772.

Gan, L., Hou, H., & Liu, B., (2011). Some results on atom-bond connectivity index of graphs. *MATCH Commun. Math. Comput. Chem., 66*, 669–680.

Gan, L., Liu, B., & You, Z., (2012). The ABC index of trees with given degree sequence. *MATCH Commun. Math. Comput. Chem., 68*, 137–145.

Gao, W., (2015). Edge average Wiener index and atom-bond connectivity index of molecule graphs with special structure. *Global Journal of Mathematics, 2*(1), 84–98.

Ghorbani, M., & Ghazi, M., (2010). Computing some topological indices of triangular benzenoid. *Dig. J. Nanomater. Bios., 5*(4), 1107–1111.

Ghorbani, M., & Hosseinzadeh, M. A., (2010). Computing ABC_4 index of nanostar dendrimers. *Optoelectron. Adv. Mater.-Rapid Commun.*, *4*(9), 1419–1422.

Ghorbani, M., & Songhori, M., (2010). Some topological indices of nanostar dendrimers. *Iranian J. Math. Chem.*, *1*(2), 57–65.

Ghorbani, M., Mesgarani, H., & Shakeraneh, H., (2011). Computing *GA* index and *ABC* index of V-phenylenic nanotube. *Optoelectron. Adv. Mater.-Rapid Commun.*, *5*(3), 324–326.

Goubko, M., Magnant, C., Salehi, N. P., & Gutman, I., (2015). ABC index of trees with fixed number of leaves. *MATCH Commun. Math. Comput. Chem.*, *74*, 697–702.

Graovac, A., & Ghorbani, M., (2010). A new version of atom-bond connectivity index. *Acta Chim. Slov.*, *57*, 609–612.

Gutman, I., (2013). Degree-based topological indices. *Croat. Chem. Acta.*, *86*, 351–361.

Gutman, I., & Furtula, B., (2012). Trees with smallest atom-bond connectivity index. *MATCH Commun. Math. Comput. Chem.*, *68*, 131–136.

Gutman, I., & Polansky, O. E., (1986). *Mathematical Concepts in Organic Chemistry.* Springer: Berlin, Germany.

Gutman, I., & Tošović, J., (2013). Testing the quality of molecular structure descriptors. Vertex-degree-based topological indices. *J. Serb. Chem. Soc.*, *78*, 805–810.

Gutman, I., & Trinajstić, N., (1972). Graph theory and molecular orbitals. Total π–electron energy of alternant hydrocarbons. *Chem. Phys. Lett.*, *17*, 535–538.

Gutman, I., Furtula, B., & Elphick, C., (2014a). Three new/old vertex-degree-based topological indices. *MATCH Commun. Math. Comput. Chem.*, *72*, 617–632.

Gutman, I., Furtula, B., & Ivanović, M., (2012a). Notes on trees with minimal atom-bond connectivity index. *MATCH Commun. Math. Comput. Chem.*, *67*, 467–482.

Gutman, I., Furtula, B., Ahmadi, M. B., Hosseini, S. A., Salehi, N. P., & Zarrinderakht, M., (2013). The ABC index conundrum. *Filomat.*, *27*, 1075–1083.

Gutman, I., Ruščić, B., Trinajstić, N., & Wilcox, C. F., (1975). Graph theory and molecular orbitals. XII. Acyclic polyenes. J. Chem. Phys., 62, 3399–3405.

Gutman, I., Tošović, J., Radenković, S., & Marković, S., (2012b). On atom-bond connectivity index and its chemical applicability. *Indian J. Chem.*, *51A*, 690–694.

Gutman, I., Zhong, L., & Xu, K., (2014b). Relating the *ABC* and harmonic indices. *J. Serb. Chem. Soc.*, *79*(5), 557–563.

Hayat, S., & Imran, M., (2014). Computation of topological indices of certain networks. *Appl. Math. Comput.*, *240*, 213–228.

Helm, G. H., (1897). *The Principles of Mathematical Chemistry: The Energetics of Chemical Phenomena* (pp. 228). 1st ed.; John Wiley & Sons: New York, USA.

Hemmasi, M., & Iranmanesh, M., (2015). Some inequalities for the atom-bond connectivity index of graph. *J. Comput. Theor. Nanosci.*, *12*(9), 2172–2179.

Horoldagva, B., & Gutman, I., (2011). On some vertex-degree-based graph invariants. *MATCH Commun. Math. Comput. Chem.*, *65*, 723–730.

Hosseini, S. A., Ahmadi, M. B., & Gutman, I., (2014). Kragujevac trees with minimal atom-bond connectivity index. *MATCH Commun. Math. Comput. Chem.*, *71*, 5–20.

Husin, N. M., Hasni, R., & Arif, N. E., (2013). Atom-bond connectivity and geometric arithmetic indices of dendrimer nanostars. *Austral. J. Basic Appl. Sci.*, *7*(9), 10–14.

Husin, N. M., Hasni, R., & Arif, N. E., (2015). Atom-bond connectivity and geometric arithmetic indices of certain dendrimer nanostars. *J. Comput. Theor. Nanosci.*, *12*(2), 204–207.

Imran, M., & Hayat, S., (2014). On computation of topological indices of Aztec diamonds. *Sci. Int. (Lahore)*, *26*(4), 1407–1412.

Imran, M., Hayat, S., & Mailk, M. Y. H., (2014b). On topological indices of certain interconnection networks. *Appl. Math. Comput.*, *244*, 936–951.

Imran, M., Hayat, S., & Shafiq, M. K., (2014a). On topological indices of nanostar dendrimers and polyomino chains. *Optoelectron. Adv. Mater.-Rapid Commun.*, *8*(9/10), 948–954.

Iranmanesh, A., & Azari, M., (2015a). Edge-Wiener descriptors in chemical graph theory: A survey. *Curr. Org. Chem.*, *19*(3), 219–239.

Iranmanesh, A., & Azari, M., (2015b). The first and second Zagreb indices of several interesting classes of chemical graphs and nanostructures. In: Putz, M. V., & Ori, O., (eds.), *Exotic Properties of Carbon Nanomatter* (Vol. 8, pp. 153–183). Springer: Dordrecht, Netherlands (Chapter 7).

Ke, X., (2012). Atom-bond connectivity index of benzenoid systems and fluoranthene congeners. *Polycycl. Aromat. Comp.*, *32*(1), 27–35.

Khaksar, A., Ghorbani, M., & Maimani, H. R., (2010). On atom bond connectivity and GA indices of nanocones. *Optoelectron. Adv. Mater.-Rapid Commun.*, *4*(11), 1868–1870.

Khalifeh, M. H., Yousefi-Azari, H., Ashrafi, A. R., & Wagner, S. G., (2009). Some new results on distance-based graph invariants. *European J. Combin.*, *30*(5), 1149–1163.

Kürkçü, Ö. K., & Aslan, E., (2015). Atom-bond connectivity index of carbon nanocones and an algorithm. *Applied Mathematics and Physics*, *3*(1), 6–9.

Li, J., (2015). On the ABC index of cacti. *International Journal of Graph Theory and its Applications*, *1*(1), 57–66.

Li, J., & Zhou, B., (2013). Atom-bond connectivity index of unicyclic graphs with perfect matchings. *Ars. Combin.*, *109*, 321–326.

Lin, W., Chen, J., Chen, Q., Gao, T., Lin, X., & Cai, B., (2014). Fast computer search for trees with minimal ABC index based on tree degree sequences. *MATCH Commun. Math. Comput. Chem.*, *72*, 699–708.

Lin, W., Gao, T., Chen, Q., & Lin, X., (2013a). On the minimal ABC index of connected graphs with given degree sequence. *MATCH Commun. Math. Comput. Chem.*, *69*, 571–578.

Lin, W., Lin, X., Gao, T., & Wu, X., (2013b). Proving a conjecture of Gutman concerning trees with minimal ABC index. *MATCH Commun. Math. Comput. Chem.*, *69*, 549–557.

Lin, W., Ma, C., Chen, Q., Chen, J., Gao, T., & Cai, B., (2015). Parallel search trees with minimal ABC index with MPI + OpenMP. *MATCH Commun. Math. Comput. Chem.*, *73*, 337–343.

Liu, J., & Chen, J., (2014). Further properties of trees with minimal atom-bond connectivity index. *Abstr. Appl. Anal.*, 609208.

Magnant, C., Salehi, N. P., & Gutman, I., (2015). Which tree has the smallest ABC index among trees with k leaves? *Discrete Appl. Math.*, *194*, 143–146.

Manso, F. C. G., Júnior, H. S., Bruns, R. E., Rubira, A. F., & Muniz, E. C., (2012). Development of a new topological index for the prediction of normal boiling point temperatures of hydrocarbons: The *Fi* index. *J. Mol. Liquids*, *165*, 125–132.

Nikmehr, M. J., Soleimani, N., & Agha, T., (2015). Computing some topological indices of carbon nanotubes. *Proceedings of IAM*, *4*(1), 20–25.

Nikmehr, M. J., Soleimani, N., & Veylaki, M., (2014). Topological indices based end-vertex degrees of edges on nanotubes. *Proceedings of IAM*, *3*(1), 89–97.

Nikolić, S., Kovačević, G., Miličević, A., & Trinajstić, N., (2003). The Zagreb indices 30 years after. *Croat. Chem. Acta*, *76*, 113–124.

Palacios, J. L., (2014). A resistive upper bound for the ABC index. *MATCH Commun. Math. Comput. Chem.*, *72*, 709–713.

Putz, M. V., Ori, O., Cataldo, F., & Putz, A. M., (2013). Parabolic reactivity "coloring" molecular topology: Application to carcinogenic PAHs. *Curr. Org. Chem.*, *17*(23), 2816–2830.

Rajan, B., William, A., Grigorious, C., & Stephan, S., (2012). On certain topological indices of silicate, honeycomb and hexagonal networks. *J. Comp. & Math. Sci.*, *3*(5), 530–535.

Randić, M., (1975). On characterization of molecular branching. *J. Am. Chem. Soc.*, *97*, 609–6615.

Randić, M., Vračko, M., & Novič, M., (2001). Eigenvalues as molecular descriptors. In: Diudea, M. V., (ed.), *QSPR/QSAR Studies by Molecular Descriptors* (pp. 93–120). Nova: New York, USA.

Raut, N. K., (2014). Degree based topological indices of isomers of organic compounds. *International Journal of Scientific and Research Publications*, *4*(8), p. 4.

Raza, Z., Ali, A., & Bhatti, A. A., (2016). More on comparison between first geometric-arithmetic index and atom-bond connectivity index. *Miskolc. Math. Notes.*, *17*(1), 561–570.

Soleimani, N., Nikmehr, M. J., & Agha, T. H., (2015). Computation of the different topological indices of nanostructures. *J. Natn. Sci. Foundation Sri Lanka, 43*(2), 127–133.

Sridhara, G., Rajesh, K. M. R., & Indumathi, R. S., (2015). Computation of topological indices of graphene. *J. Nanomater.,* 969348.

Todeschini, R., & Consonni, V., (2000). *Handbook of Molecular Descriptors* (Vol. 11, pp. 668). Wiley-VCH: Weinheim, Germany.

Trinajstić, N., (1992). *Chemical Graph Theory* (pp. 352). 2nd ed.; CRC Press: Boca Raton, FL.

Vassilev, T. S., & Huntington, L. J., (2012). On the minimum ABC index of chemical trees. *Appl. Math.*, *2*(1), 8–16.

Vukičević, D., & Furtula, B., (2009). Topological index based on the ratios of geometrical and arithmetical means of end-vertex degrees of edges. *J. Math. Chem.*, *46*, 1369–1376.

Vukičević, D., & Gašperov, M., (2010). Bond additive modeling 1. Adriatic indices. *Croat. Chem. Acta, 83*(3), 243–260.

West, D. B., (1996). *Introduction to Graph Theory* (pp. 512). 1st ed.; Prentice Hall: Upper Saddle River, USA.

Wiener, H., (1947a). Structural determination of paraffin boiling points. *J. Am. Chem. Soc., 69*, 17–20.

Wiener, H., (1947b). Correlation of heats of isomerization and differences in heats of vaporization of isomers among the paraffin hydrocarbons. *J. Am. Chem. Soc., 69*(11), 2636–2638.

Xing, R., & Zhou, B., (2012). External trees with fixed degree sequence for atom-bond connectivity index. *Filomat., 26*(4), 683–688.

Xing, R., Zhou, B., & Dong, F., (2011). On atom-bond connectivity index of connected graphs. *Discrete Appl. Math.*, *159*, 1617–1630.

Xing, R., Zhou, B., & Du, Z., (2010). Further results on atom-bond connectivity index of trees. *Discrete Appl. Math.*, *158*, 1536–1545.

Xu, K., Liu, M., Das, K. C., Gutman, I., & Furtula, B., (2014). A survey on graphs extremal with respect to distance-based topological indices. *MATCH Commun. Math. Comput. Chem.*, *71*, 461–508.

Yang, J., & Deng, H., (2012). A novel atom-bond connectivity index based on the distance of a graph. *J. Nat. Sci. Hunan Norm. Univ.*, *35*(3), 6–9.

Yang, J., Xia, F., & Cheng, H., (2011). The atom-bond connectivity index of benzenoid systems and phenylenes. *Int. Math. Forum, 6*(41), 2001–2005.

Zhong, L., & Xu, K., (2014). Inequalities between vertex-degree-based topological indices. *MATCH Commun. Math. Comput. Chem.*, *71*, 627–642.

Zhou, B., & Trinajstić, N., (2009). On a novel connectivity index. *J. Math. Chem.*, *46*, 1252–1270.

Zhou, B., & Xing, R., (2011). On atom-bond connectivity index. *Z. Naturforsch., 66a*, 61–66.

CHAPTER 2

Bondonic Topo-Reactivity

MIHAI V. PUTZ[1,2]

[1]*Laboratory of Structural and Computational Physical Chemistry for Nanosciences and QSAR, Biology-Chemistry Department, West University of Timisoara, Pestalozzi Street No. 44, Timisoara, RO-300115, Romania*

[2]*Laboratory of Renewable Energies-Photovoltaics, R&D National Institute for Electrochemistry and Condensed Matter, Dr. A. PaunescuPodeanu Str. No. 144, Timisoara, RO-300569, Romania*

2.1 DEFINITION

The bondonic topo-reactivity is referring to a molecular characterization method based on combining reactivity indices and adjacency matrix to which the value of bondonic (or its inverse information, the antibondonic value) contribution is added in the calculation formula.

2.2 HISTORICAL ORIGIN(S)

The matter quantification proposed by De Broglie states that any wave function is associated with a quantum particle, but over the past years, the chemical bonding was modeled exclusively as being governed by Molecular Orbitals. In this context, the quantum particle of chemical bonding, also called bondon, should also exist being responsible of the chemical interaction (Putz & Tudoran, 2014b). The bondonic particle is present on the wave functions, on the covalent bonds, in dispersive–weak interactions (as in the ADN), on the mechanism of action between ligand and receptor, and even on the ionic interactions.

In physics, on the proximal vicinity of two electrons is created a photon, which also presents the antiparticle behavior, the anti-photon, i.e.,

an excited electronic state is created through photon absorption and annihilated through an equal photon emission, these two entities being equal
(Figure 2.1). When this process occurs in a molecular system, the photon
becomes a bondon. The difference between chemistry and physics is that,
in chemistry, the bonding and antibonding are not annihilated with each
other, meaning that, from the same bondon, there are two different states
which are created. Through the bondonic interpretation, the molecule is
not an electronic distribution on levels, but it represents an interaction
between electrons and an interaction between levels. In other words, the
bonding-antibonding mechanism determines a continuous movement of
the molecule (Putz & Tudoran, 2014a,b).

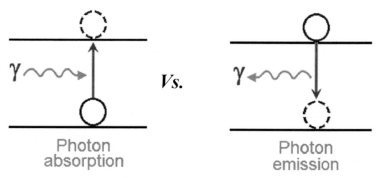

FIGURE 2.1   Schematic representation of the photon emission and absorption. Redrawn
and adapted after Putz and Tudoran (2014b).

On the other hand, using graph theory models in chemistry applications
offers an alternative way in obtaining qualitative predictions for molecular
structure and reactivity without being necessary complicate methods (Tudoran
& Putz, 2015), and consists in defining a graph G (i.e., the graphical structure
of a molecule; see Figure 2.2) consisting in a set of vertices (representing the
atoms) and a set of edges (representing the bonds).

Using the topological/chemical graph chemistry along with QSAR
modeling, one can determine the optimal correlation between the activities
of a certain series of N-compounds and their structural properties. Starting
from these premises, the chemical orthogonal space of chemical bonding
(Putz, 2012c) is constructed in terms of the principle of electronegativity
and chemical hardness as primary reactivity indices, being completed by
the chemical power and electrophilicity as their related quantities (Putz &
Dudaş, 2013, Putz, 2016a).

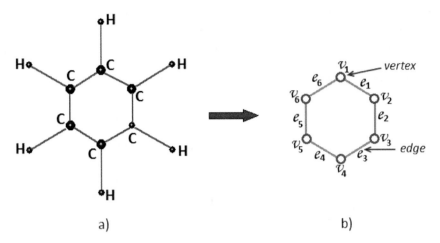

a)   b)

FIGURE 2.2   Benzen: (a) molecular representation from Hyperchem[1]; (b) graph representation.

2.3   NANO-SCIENTIFIC DEVELOPMENT(S)

Starting from the work of Putz on bondon particle (Putz, 2010), the idea of introducing the bondonic contribution in the topo-reactivity studies arise from the necessity of improving the SAIB method (Tudoran & Putz, 2015), in which three type of simple and double carbon-carbon bonds (carbon bond with two hydrogen atoms in vicinity, carbon bond with one hydrogen atom in vicinity and carbon bond with carbon atoms in vicinity). The new bondonic topo-reactivity method made a step further and introduces the bondonic contribution (Table 2.1) for each type of bonding in the general formula of electronegativity and chemical hardness calculations as a way of distinguishing between simple and double bond.

TABLE 2.1   Values of Bondonic (mass) Correction for Carbon Bonds and Carbon-Hydrogen Bond (Putz, 2010, 2012a, 2012b, 2015b; Putz & Tudoran, 2014a)

	C–C	C = C	C–H
β	0.45624	0.33286	0.7455

The chemical bonding is described by atoms-in-molecules paradigm considering the reactivity indices for bonding formation from the orthogonal

---

[1]Program Package, *HyperChem* 7.01; Hypercube, Inc.: Gainesville, FL, USA, 2002.

approach of reactivity. This way, the electronegativity value with bondonic contribution is calculated with the specific-bond-in-adjacency (SBA) by dividing the total number of bonds with the summation of the type of bond corrected by bondonic ($\mathcal{B}$) contribution, with formulas also considering the specific type (Putz & Tudoran, 2014a):

$$\chi_{Specific-Bond-in-Adjacency} = \frac{n_{Bonds-in-Adjacency}}{\sum\limits_{Type-of-Bond} \frac{1}{\mathcal{B}} \frac{n_{Type-of-Bond}}{\chi_{Type-of-Bond}}} \tag{1}$$

The same formula can be further applied to calculate the *bondonic* chemical hardness:

$$\eta_{Specific-Bond-in-Adjacency} = \frac{n_{Bonds-in-Adjacency}}{\sum\limits_{Type-of-Bond} \frac{1}{\mathcal{B}} \frac{n_{Type-of-Bond}}{\eta_{Type-of-Bond}}} \tag{2}$$

Applying the consecrated rule for calculated *bondonic chemical power* one will obtain:

$$\pi_{SBA} = \frac{\chi_{SBA}}{2\eta_{SBA}} \tag{3}$$

and also *bondonic electrophilicity*:

$$\omega_{SBA} = \pi_{SBA} \times \chi_{SBA} \tag{4}$$

The anti-bondonic behavior was also considered in the topo-reactivity study on polycyclic aromatic hydrocarbons by considering the inverse of bondonic information:

$$\tilde{\mathcal{B}} \rightarrow \mathcal{B}^{-1} \tag{5}$$

In the same way, the reactivity indices are calculated considering the anti-bondonic contribution with respecting the SBA rule, by dividing the total number of bonds with the summation of the type of bond corrected, this time, by the anti-bondonic ($\tilde{\mathcal{B}}$) information (Putz & Tudoran, 2014b). In this context, the *anti-bondonic electronegativity* can be calculated by applying the formula:

$$\tilde{\chi}_{Specific-Bond-in-Adjacency} = \frac{n_{Bonds-in-Adjacency}}{\sum\limits_{Type-of-Bond} \frac{1}{\tilde{\mathcal{B}}} \frac{n_{Type-of-Bond}}{\chi_{Type-of-Bond}}} \tag{6}$$

The *anti-bondonic chemical hardness* can be calculated using the same rule with the formula:

$$\tilde{\bar{\eta}}_{Specific-Bond-in-Adjacency} = \frac{n_{Bonds-in-Adjacency}}{\displaystyle\sum_{Type-of-Bond} \frac{1}{\tilde{\bar{B}}} \frac{n_{Type-of-Bond}}{n_{Type-of-Bond}}} \tag{7}$$

The *anti-bondonic electronegativity* and *chemical hardness* are further used to calculate *anti-bondonic chemical power* (8) and *electrophilicity* (9):

$$\tilde{\bar{\pi}}_{SBA} = \frac{\tilde{\bar{\chi}}_{SBA}}{2\tilde{\bar{\eta}}_{SBA}} \tag{8}$$

$$\tilde{\bar{\omega}}_{SBA} = \tilde{\bar{\pi}}_{SBA} \times \tilde{\bar{\chi}}_{SBA} \tag{9}$$

One can say that, in the bondonic/anti-bondonic topo-reactivity approach, in order to calculate the reactivity indices one needs to consider not only the type of bond and the electronic contribution of each atom, but also the bondonic/anti-bondonic information which is different for simple and double C-C bonds and for C-H bond.

2.4 NANOCHEMICAL APPLICATION(S)

The effect of bondonic contribution may be directly applied in the reactivity study on a series of 16 polycyclic aromatic hydrocarbons (PAHs). For each molecule, the adjacency matrix is generated, and the values "1" are replaced with the bondonic-reactivity values calculated for each specific type of bonds-in-adjacency (Tudoran & Putz, 2015). In the next step, for each PAH was calculated the specific eigenvalue spectra from which were extracted the global bondonic-topo-reactivity indices averaging over a max-min differ-ence of the positive entries. The correlation between the lipophilicity (LogP) values collected from previous work (Putz et al., 2013b) and bondonic-topo-reactivity indices were obtained next using the Pearson correlation factor. All possible paths which connect the correlation were considered, and their Euclidian lengths were computed. From the obtained results, the most prob-able way of chemical reactivity causes was selected, being associated with the shortest distance of two-points of correlated lipophilicity activities.

The bondonic effect on chemical reactivity dogma (χηπω/[xipo]) can be seen from the fact that the succession of indices is changed (Figure 2.3). In this case, the first electron is moved from the $HOMO_L$ to $LUMO_L$ due to the chemical power, than the remaining electron is transferred to the LUMO$_{R0}$, after that the middy levels of HOMO-LUMO (Highest Occupied Molecular Orbital-Lowest Unoccupied Molecular Orbital) gaps for both ligand and receptor are equalized, the process being closed by chemical hardness influence. In other words, one can say that the bondonic approach, where the chemical power is the first "force in action," can be associated with the electronic interactions of the HOMO level (Putz & Tudoran, 2014a).

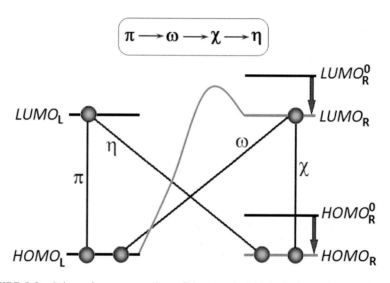

FIGURE 2.3 Schematic representation of the chemical-biological reaction-activity when the bondonic contribution is used in calculating the reactivity indices. Redrawn and adapted after Putz and Tudoran (2014a).

The anti-bondonic contribution was also studied by Putz and Tudoran (2014b), the anti-bondonic-topo-reactivity values being determined for each type of specific bonds-in-adjacency in the same algorithm as in bondonic topo-reactivity method. For the same PAHs molecules, the specific eigen-value spectra were computed, and after that, the global antibondonic-topo-reactivity indices were determined by the average of max-min difference from the positive entries. In the next step, the obtained anti-bondonic topo-reactivity indices were correlated with the log P values collected from previous work (Putz et al., 2013b) in terms of Pearson correlation factors and the Euclidian lengths of all possible paths connecting the previous correlations were computed. As the final step, the most probable way of chemical reactivity is selected with respecting the principle associated with the shortest distance of two-points of the correlated lipophilicity activities (Putz & Tudoran, 2014b).

The anti-bondonic influence also has the effect of changing the natural order of forces from the chemical reactivity dogma (Figure 2.4). In this case, the electrophilicity action determines the movement of the first electron from $HOMO_L$ to $LUMO_{R0}$, while the remaining electron is propelled by the chemical power (the second movement) to the $LUMO_L$. The energetic stabilization of HOMO-LUMO is assured by the electronegativity, and the chemical hardness assures reaching the maximum stability, meaning that the HOMO-LUMO energetic gap is maxim so that will no longer occur any electronic transitions. In other words, one can say that in the antibondonic approach, the electronic interactions are associated with the LUMO level (Putz & Tudoran, 2014b).

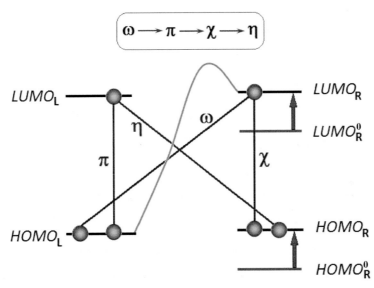

FIGURE 2.4 Schematic representation of the chemical-biological reaction-activity when the anti-bondonic contribution is considered in reactivity indices calculations. Redrawn and adapted after Putz and Tudoran (2014b).

2.5 MULTI-/TRANS-DISCIPLINARY CONNECTION(S)

The Wiener index $W(N)$ based on the distance matrix has an important role in influencing the chemical properties of molecules consisting of similar units, such as hexagonal rings, and can be calculated by adding the minimum topological distance d_{ij} for a graph $G(N)$ as in formula (Putz et al., 2013a, 2016c; Putz & Ori, 2015):

$$W(N) = \frac{1}{2}\sum_{ij} d_{ij}, \ d_{ij} = 0, \ i,j = 1,2,...,N-1,N \qquad (10)$$

In other words, one can say that the Wiener index can be used to measure the topological compactness average for a molecular graph. Applied on defective graphenic planes with nanocones and on C_{66} fullerene, by superimposing the minimum principle on the $W(N)$ the stable system can be selected among them (Putz et al., 2016c; Vukicevic et al., 2011; Cataldo et al., 2010, 2011).

Considering a chemical compound with its associated graph G, the topological distance can also be analyzed by counting the graphs chemical bondings. In this context, given a vertex v_i of the graph G, the largest distance between v_i and any other vertex is called the eccentricity ε_i, having the property that $M = \max\{\varepsilon_i\}$, with M as the graph diameter. In their work, Sharma and his co-workers (1997) define, for a graph G, the eccentric connectivity index $\xi(N)$ as in formula:

$$\xi(N) = \frac{1}{2}\sum_i b_{i1}\varepsilon_i \qquad (11)$$

with b_{il} bonds number of atom v_i, with the property:

$$b_{ik} = 0 \ \Leftrightarrow \ \varepsilon_i < k \leq M \qquad (12)$$

In their work, Putz and his co-workers (2016c) use the eccentric connectivity index in order to measure the bosonic chemical field range of action which is spread around a specific chemical structure and is determined by its bondons movement, here seen as delocalized quanta of the chemical field (Putz & Ori, 2012, 2014). Using the spontaneous symmetry breaking at numerical and analytical levels, they manage to reveal the π-bondonic and σ-bondonic behaviors along with the massive Higgs-like bondons which depends on the chemical hardness and quasi-massless Goldenstone like bondons which depends on the topologically extreme sphericity (Putz et al., 2016c).

2.6   OPEN ISSUES

The bondot as a chemical bond is a new particle proposed by Putz (2016a) obtained by extracting the physical characteristic from the electronic propagation in bondons (as pairs) on a heterojunction, with application in producing new materials with increased electrochemical efficiency. In other words, the bondot can be seen as bondon variant from the electrochemical systems whit optical lattice (Putz et al., 2016b), characterized by the bondon to electron mass (Eq. 13), bondon to electron charge (Eq. 14), and bondon to electron velocity (Eq. 15) as following (Putz et al., 2015, 2016a; Putz, 2010, 2012a-b, 2016c):

$$\varsigma_m = \frac{m_B}{m_0} = \frac{87.8603}{\left(E_{bond}\left[kcal/mol\right]\right)\left(X_{bond}\left[\dot{A}\right]\right)^2} \tag{13}$$

$$\varsigma_e = \frac{e_B}{e} \sim \frac{1}{32\pi} \cdot \frac{\left(E_{bond}\left[kcal/mol\right]\right)\left(X_{bond}\left[\dot{A}\right]\right)}{\sqrt{3.27817 \times 10^3}} \tag{14}$$

$$\varsigma_v = \frac{v_B}{c} = \frac{100}{\sqrt{1 + \dfrac{3.27817 \times 10^6}{\left(E_{bond}\left[kcal/mol\right]\right)^2 \left(X_{bond}\left[\dot{A}\right]\right)^2}}}\left[\%\right] \tag{15}$$

The spin information in quantum dots can be used in an innovative way, with further applications in the photovoltaic area, using the photo-electro-chemical circuit in controlling an electronic pair with anti-parallel (coupled) spin, in order to produce the double quantum dots bondotic sensitized solar cell (QDDSSC) (Putz et al., 2016b).

ACKNOWLEDGMENT

This contribution is part of the research project "Fullerene-Based Double-Graphene Photovoltaics" (FG2PV), PED/123/2017, within the Romanian National program "Exploratory Research Projects" by UEFISCDI Agency of Romania.

KEYWORDS

- adjacency matrix
- anti-bondon
- bondon
- boson
- fermion

REFERENCES AND FURTHER READING

Cataldo, F., Ori, O., & Gravoac, A., (2011). Topological efficiency of C_{66} fullerene. *International Journal of Chemical Modeling, 3*, 45–63.

Cataldo, F., Ori, O., & Iglesias-Groth, S., (2010). Topological lattice description of graphene. *Molecular Simulation, 36*, 341–353.

Putz, M. V., (2010). The bondons: The quantum particles of the chemical bond. *International Journal of Molecular Sciences, 11*(11), 4227–4256.

Putz, M. V., (2012a). *Quantum Theory: Density. Condensation and Bonding.* Apple Academics, Toronto, Canada.

Putz, M. V., (2012b). Density functional theory of Bose-Einstein condensation: Road to chemical bonding quantum condensate. *Structure and Bonding, 149*, 1–50.

Putz, M. V., (2012c). Chemical orthogonal spaces. In: *Mathematical Chemistry Monographs* (Vol. 14). University of Kragujevac, Serbia.

Putz, M. V., (2015). *Quantum Nanochemistry: A Fully Integrated Approach* (Vol. IV). Quantum solids and orderability. Apple Academic Press & CRC Press.

Putz, M. V., (2016a). *Quantum Nanochemistry: A Fully Integrated Approach* (Vol. V). Quantum structure-activity relationships (Qu-SAR). Apple Academic Press & CRC Press, Toronto-New Jersey, Canada-USA.

Putz, M. V., (2016b). *Quantum Nanochemistry: A Fully Integrated Approach* (Vol. III). Quantum molecules and reactivity. Apple Academic Press & CRC Press, Toronto–New Jersey, Canada–USA.

Putz, M. V., & Dudaş, N. A., (2013). Variational principles for mechanistic quantitative structure–activity relationship (QSAR) studies: Application on uracil derivatives' anti-HIV action. *Structural Chemistry, 24*, 1873–1893.

Putz, M. V., & Ori, O., (2012). Bondonic characterization of extended nanosystems: Application to graphene's nanoribbons. *Chemical Physics Letters, 548*, 95–100.

Putz, M. V., & Ori, O., (2014). Bondonic effects in group-IV honeycomb nanoribbons with stone-wales topological defects. *Molecules, 19*, 4157–4188.

Putz, M. V., & Ori, O., (2015). Predicting bondons by Goldstone mechanism with chemical topological indices. *International Journal of Quantum Chemistry, 115*(3), 137–143.

Putz, M. V., & Tudoran, M. A., (2014a). Bondonic effects on topo-reactivity of PAHs. *International Journal of Chemical Modeling, 6*(2/3), 311–346.

Putz, M. V., & Tudoran, M. A., (2014b). Anti-bondonic effects on topo-reactivity of PAHs. *International Journal of Chemical Modeling, 6*(4), 475–505.

Putz, M. V., Duda-Seiman, C., Duda-Seiman, M., & Bolcu, C., (2015). Bondonic chemistry: Consecrating silanes as metallic precursors for silicenes materials. In: Putz, M. V., & Ori, O., (eds.), *Exotic Properties of Carbon Nanomatter, Advances in Physics and Chemistry.* Dordrecht: Springer; Series "Carbon Materials: Chemistry and Physics", Vol. 8, Ed. Cataldo, F.; Milani, P.; pp. 323–345.

Putz, M. V., Ori, O., Cataldo, F., & Putz, A. M., (2013a). Parabolic reactivity "Coloring" molecular topology: Application to carcinogenic PAHs. *Current Organic Chemistry, 17*(23), 2816–2830.

Putz, M. V., Ori, O., Diudea, M. V., Szefler, B., & Pop, R., (2016c). Bondonic chemistry: Spontaneous symmetry breaking of the topo-reactivity on graphene. In: Diudea, M. V., & Ashrafi, A. R., (eds.), *Distance, Symmetry, and Topology in Carbon Nanomaterials, Carbon Materials: Chemistry and Physics* (Vol. 9, pp. 345–389). Springer, Chapter 20.

Putz, M. V., Tudoran, M. A., & Mirica, M. C., (2016a). Quantum dots searching for bondots: Towards sustainable sensitized solar cells. In: Putz, M. V., & Mirica, M. C., (eds.), *Sustainable Nanosystems Development, Properties, and Applications* (pp. 261–327). IGI Global, Chapter 9.

Putz, M. V., Tudoran, M. A., & Mirica, M. C., (2016b). Bondonic electrochemistry: Basic concepts and sustainable prospects. In: Putz, M. V., & Mirica, M. C., (eds.), *Sustainable Nanosystems Development, Properties, and Applications* (pp. 328–411). IGI Global, Chapter 10.

Putz, M. V., Tudoran, M. A., & Putz, A. M., (2013b). Structure properties and chemical-bio/ecological of PAH interactions: From synthesis to cosmic spectral lines, nanochemistry, and lipophilicity-driven reactivity. *Current Organic Chemistry, 17*, 2845–2871.

Sharma, V., Goswami, R., & Madan, A. K., (1997). Eccentric connectivity index: A novel highly discriminating topological descriptor for structure-property and structure-activity studies. *Journal of Chemical Information and Modeling, 37*, 273–282.

Tudoran, M. A., & Putz, M. V., (2015). Molecular graph theory: From adjacency information to colored topology by chemical reactivity. *Current Organic Chemistry, 19*, 359–386.

Vukicevic, D., Cataldo, F., & Gravoac, A., (2011). Topological efficiency of C_{66} fullerene. *Chemical Physics Letters, 501*, 442–445.

CHAPTER 3

Carbon Onions

LEMI TÜRKER[1] and SELÇUK GÜMÜŞ[2]

[1]*Department of Chemistry, Middle East Technical University, 06531 Ankara, Turkey*

[2]*Yuzuncu Yıl University, Faculty of Sciences, Department of Chemistry, Van, Turkey*

3.1 DEFINITION

The idea of onion-like carbon nanostructures has not been long to emerge, after the discovery of Buckminsterfullerene, C_{60}, which has 60 carbon atoms forming a truncated-icosahedral structure with 12 pentagonal rings and 20 hexagonal rings. These types of structures are considered as new allotropes of carbon. Carbon-onions are endohedrally fullerene doped systems; the smallest ball is located in the most inner hole, and the cover gets bigger and bigger. In the literature of multi-layer fullerenes, C_{60}-C_{240}-C_{540}-C_{960}-C_{1500}-C_{2160}-C_{2940}-C_{3840}-C_{4860} onion has been studied, recently.

3.2 HISTORICAL ORIGIN(S)

Diamond and graphite are the traditional forms of crystalline carbon familiar to people dealing with science. Diamond has four-coordinate sp^3 carbon atoms forming an extended three-dimensional network whose motif is the chair conformation of cyclohexane, a puckered six-membered ring molecule. Graphite, on the other hand, has three-coordinate sp^2 hybridized carbons forming planar sheets whose orientation is the flat six-membered ring (Rao et al., 1995).

The new carbon allotropes, the fullerenes, are closed-cage carbon molecules with three-coordinate carbon atoms tiling spherical or nearly-spherical surfaces. Carbon-based fullerenes were discovered in 1985 (Kroto et al., 1985). They observed the presence of even-numbered clusters of carbon atoms in the molecular range of C_{30}-C_{100} while performing mass-spectroscopy analysis of carbon vapor.

The best known of these molecules is Buckminsterfullerene, C_{60}, which has 60 carbon atoms forming a truncated-icosahedral structure with 12 pentagonal rings and 20 hexagonal rings (Figure 3.1). Other spherical carbon molecules, such as C_{180}, C_{240}, and C_{320}, were also obtained over the next decades (Prinzbach et al., 2000; Piskoti et al., 1998; Kroto, 1987). The structures are essentially that of a soccer ball. The coordination at every carbon atom is not planar but rather slightly pyramidalized at every carbon atom. In other words, some sp^3 character is present in the essentially sp^2 carbons of fullerenes. While regular hexagons can tile a plane, pentagons can tile a sphere. The simplest example of pentagons tiling a sphere is a pentagonal dodecahedron with twelve pentagons. The structure of C_{60} can be visualized as being obtained by spacing apart the pentagons of the pentagonal dodecahedron with hexagons. The key feature of the fullerenes is the presence of five-membered rings which provide the curvature necessary for forming a closed-cage molecule. Such structural motifs are not new to chemistry – particularly in the chemistry of elemental boron, C_{60}-like orientations are ubiquitous (Rao et al., 1995).

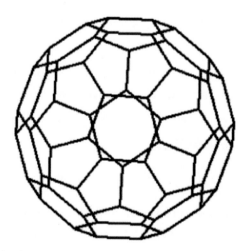

FIGURE 3.1 C_{60} structure.

Unlike the previously mentioned fullerenes that have a cage structure, the carbon multi-shell fullerenes have cage-inside-cage concentric structures that are known as double-shell or triple-shell fullerenes. These double-shell and triple-shell structures were predicted by Kroto and Mckay in 1988 (Kroto and Mckay, 1988). However, the first observation of several cage-inside-cage molecules was reported by Mordkovich et al. in 1999 and 2000 (Mordkovich et al., 1999; Mordkovich, 2000). In those studies, the double-shell C_{60}-C_{240}, C_{240}-C_{560}, and triple-shell C_{60}-C_{240}-C_{560} were found in the products of 3000°C high-temperature treatments of laser pyrolysis carbon blacks. Theoretically, carbon onions as big as C_{60}-C_{240}-C_{540}-C_{960}-C_{1500}-C_{2160}-C_{2940}-C_{3840}-C_{4860} were considered (Ghavanloo and Fazelzadeh, 2015).

3.3 NANO-SCIENTIFIC DEVELOPMENT(S)

The electron irradiation experiments by Ugarte (1992) encouraged scientists to explore alternative routes for the formation of carbon anions and the simplest of which was the heat treatment of preformed carbon nanostructures. Today, various techniques are known to produce nearly sphere-shaped and concentric carbon multi-shell fullerenes, such as high temperature annealing of diamond nanoparticles (Tomita et al., 2002), high pressure transformation of single-crystal graphite (Blank et al., 2007) or radio frequency plasma synthesis from coal (Fu et al., 2007). It was shown that the transformation of nanostructures in carbon soot into carbon anions could occur through heating under vacuum at 2250°C for 1 h (de Heer and Ugarte 1993) and that diamond nanoparticle annealed at 1500°C in 1 h (Kuznetsov et al., 1994; Butenko et al., 2000; Tomita, et.al., 2002) in vacuum to yield carbon anions. Cabioc'h et al. (1995, 1997, 1998) developed a production method of carbon anions based on carbon implantation into a metal matrix (Ag, Cu). Sano et al. (2001, 2002) reported the production of carbon anions using an arc discharge between two graphite electrodes submerged in water or liquid nitrogen. Du et al. (2007) by using radio frequency and microwave plasmas produced carbon anions with a minimum of structural defects. Xu et al. (2006, 2008) managed the synthesis of metal particles covered by graphitic layers using arc discharge between graphite electrodes in water solutions of metal salts. The size and the number of layers of carbon onion fullerenes produced with these methods appear to be limited. The most likely reason for this size restriction is the mechanical instability that is introduced by the formation of an additional layer (Zwanger et al., 1996; Todt et al., 2011). These nanosized fullerene structures exhibit superior physical, mechanical, chemical, electronic, and electrical properties. Because of these properties, fullerenes have attracted significant research interest due to their potential applications in electronic and optical nanodevices as well as in nanoelectro-mechanical systems. The mechanical strength of a nanodevice component is a critical parameter that determines the stability of a nanodevice. Due to the very small size of fullerenes, it is very complicated to study their mechanical properties via experimental techniques such as transmission electron microscope or atomic force microscopy. Therefore, computer simulations can be used for further investigations of the mechanical properties of carbon fullerenes. The classical molecular dynamics (MD) and quantum mechanics (QM) methods are convenient for theoretically analyzing the mechanical properties of fullerenes. QM calculations can be used to obtain more accurate mechanical properties than MD simulations.

Ahangaria et al. (2013) studied the total and average energies of fullerenes and onions at three different temperatures. The average energy of the atoms is obtained by dividing the total energy by the total number of atoms. For example, the total energy of fullerenes with a diameter of 12.187 Å at 300 K is−310.97 a.u., and the total number of atoms is 180; thus, the average energy of atoms in the fullerene is−1.73 a.u./atom (−310.97 a.u./180). The results (a) indicate that the total energy decreases as the diameter of the fullerenes increases. (b) The average energy of the atoms rapidly decreases and then slowly decreases as the diameter increases. Therefore, an individual fullerene molecule with a larger diameter is more stable than an individual fullerene molecule with a smaller diameter. The studies showed that the fullerenes become energetically more stable with increasing size. The average energy of the C_{60}-C_{180} molecule is lower than that for the C_{60}-C_{240} molecule (Figure 3.2). Therefore, the stability of the double-shell fullerene (C_{60}-C180 and C_{60}-C_{180}) decreases as the distance between the two shells in the complex increases. According to the SCC-DFTB-MD calculations, the energy of the optimized fullerene structure is sensitive to temperature and decreases as the temperature increases.

Next, the bulk modulus of fullerenes and onions were calculated as a function of the diameter as well as the number of shells at various temperatures, to evaluate their hardness. Similar calculation procedures have been performed for the selected systems

The influence of the number of walls on the bulk modulus of the selected fullerenes is evaluated. From the obtained results, it is concluded that the bulk modulus of double-shell fullerenes is larger than both the internal and external nanocages. For example, at 300 K, the bulk modulus follows the trend where C_{60}-C_{240} (1085.03 GPa) > 240 (880.63 GPa) > 60 (720.46 GPa). Shen (2007) investigated the mechanical properties of C_{60}, C_{180}, and C_{60}-C_{180} using classical MD simulations and found that the mechanical stability of C_{60} fullerene is much higher than those of the C_{180} and C_{60}-C_{180} fullerenes. In addition, the effect of the distance between the fullerene shells on the bulk modulus of the double-shell fullerenes has been investigated. For this purpose, two different shell distances (d = 2.548 and 3.487 Å) were selected for the double-shell fullerene. The results show that the bulk modulus of a double-shell fullerene decreased from 1202.39 to 1085.03 GPa with an increase in the shell distance between the C_{60}-C_{180} and C_{60}-C_{240} fullerenes, respectively. The previous computational papers have studied the effect of an increased layer distance on Young's modulus of double-walled carbon nanotubes and observed the same trend in the onion results (Ganji et al., 2012).

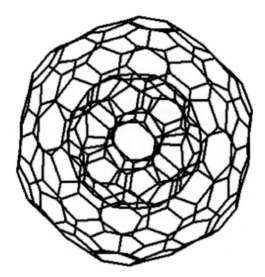

FIGURE 3.2 C_{60}-C_{180} onion structure.

These results for the double-layer CNTs and fullerenes can be attributed to the interlayer bonding that arises from the π–π interaction between the aromatic systems. Indeed, graphite-like materials, such as CNTs and fullerenes, consist of cyclic aromatic carbon rings. The C atoms in these aromatics are arranged such that electrons fill the in-plane sp^2 hybrid orbitals, resulting in a resonant bond shared by all of the C atoms. In addition, there is a set of π-orbitals perpendicular to the surface and delocalized on the molecular plane. Therefore, a π–π interaction is possible between the aromatic systems (such as graphite, multi-wall CNTs, and onions), which is long ranged and includes van der Waals forces. This π–π interaction leads to stacking, which is apparent in the interlayer bonding in onions with desirable interlayer distances resulting in a higher modulus compared to the single layer and double layer fullerenes/CNTs with large interlayer distances.

During the synthesis of the double-shell C_{60}-C_{240}, double-shell C_{240}-C_{560} and triple-shell C_{80}-C_{240}-C_{560} were found in the products of 3000°C high-temperature treatment of laser pyrolysis carbon blacks. The content of multi-shell fullerenes was less than 0.01%, while most of the material was dominated by hollow graphitic particles of ~20 nm size and ordinary, single-shell fullerenes.

Though the multi-shell fullerenes possibly possess interesting and attractive properties, no measurement could have been done until these fullerenes were isolated in milligram amounts, at least. So the first priority of the research in this field was to find an efficient synthetic route to the production of multi-shell fullerenes.

It is hard to improve the method of high-temperature treatment of laser pyrolysis carbon black. Also, it seems reasonable to suggest that actual growth of additional shells around C_{60} cores took place not during 3000°C treatment but during cooling down after the heat treatment. If this suggestion is correct, a successful synthetic route should comprise at least two steps: (1) formation of carbon vapor which contains many C_{60} molecules; and (2) cooling down at the appropriate substrate. Another attractive feature of this method is that it allows flexible control of different key parameters of the process.

Mordkovich et al. (1999, 2000) performed a work to find an efficient synthetic route to multishell fullerenes by use of laser vaporization method. The multi-shell fullerene fraction has been easily separated by vacuum sublimation though it requires at least 1200°C in a vacuum of 10^{-7} Torr. So high sublimation temperature is in sharp contrast with that of C_{60}, which can be easily sublimated at 400°C. The sublimated film has a yellow color and is comprised predominantly of multishell fullerenes (Mordkovich et al., 2001).

The literature also contains theoretical work on the structure and electronic properties of carbon-onions. The bucky-onion-like systems C_{20}-C_{60} has been considered for theoretical analysis. It is found that there is through space interaction between C_{20} and C_{60} shells. The enlargement of C_{60} structure in C_{20}-C_{60}, as compared to isolated C_{60} structure, the narrowing of the inner frontier molecular orbital energy gap, the formation of new electrostatic potential energy pattern, the existence of dipole moment in the composite system while its components were nil, are all the indicators of strong inner-shell interaction operative through space in this onion (Turker, 2001).

In addition to C_{20}-C_{60}, AM1 type (RHF) semi-empirical calculations are performed on endohedrally C_{60} doped C_{180} (C_{60}-C_{180}) structure to examine the electronic and structural properties (Turker and Gumus, 2004). They found that the structure is found to be stable in terms of total and binding energies but endothermic. The molecule has a dipole moment directing from the outer sphere to the center of the molecule. The result indicates that two fullerenes have not been located concentrically. The calculations (within the limitations of the level of calculations) have revealed that only the outer sphere, and in a very limited extent, contribute to the HOMO and LUMO.

3.4 NANOCHEMICAL APPLICATION(S)

Owing to the three different orbital hybridizations carbon can adopt, the existence of various carbon nano-allotropes differing also in dimensionality has been already affirmed with other structures predicted and expected to emerge in the future. Despite numerous unique features and applications of 2D graphene, 1D carbon nanotubes, or 0D fullerenes, nano-diamonds, and carbon quantum dots, which have been already extensively explored, any of the existing carbon allotropes do not offer competitive magnetic properties. For challenging applications, carbon nano-allotropes are functionalized with magnetic species, especially of iron oxide nature, due to their interesting magnetic properties (super para-magnetism and strong magnetic response under external magnetic fields), easy availability, biocompatibility, and low cost. In addition, a combination of iron oxides (magnetite, maghemite, hematite) and carbon nanostructures brings enhanced electrochemical performance and (photo) catalytic capability due to synergetic and coop-erative effects. Various architectures of carbon/iron oxide nanocomposites, their synthetic procedures, physicochemical properties, and applications are observed. A special attention is devoted to hybrids of carbon nanotubes and rare forms (mesoporous carbon, nanofoam) with magnetic iron oxide carriers for advanced environmental technologies. Among various discussed medical applications, magnetic composites of zero-dimensional fullerenes and carbon dots are emphasized as promising candidates for complex ther-anostics and dual magneto-fluorescence imaging.

KEYWORDS

- **carbon onions**
- **endohedral doping**
- **multi-layer fullerenes**

REFERENCES AND FURTHER READING

Ahangaria, M. G., Fereidoon, A., Ganji, M. D., & Sharifi, N., (2013). Density functional theory based molecular dynamics simulation study on the bulk modulus of multi-shell fullerenes. *Physica B., 423*, 1–5.

Blank, V. D., Denisov, V. N., Kirichenko, A. N., Kulnitskiy, B. A., Martushov, S. Y., Mavrin, B. N., & Perezhogin, I. A., (2007). High-pressure transformation of single-crystal graphite to form molecular carbon onions. *Nanotechnology, 18*, 345601–345604.

Butenko, Y. V., Kuznetsov, V. L., Chuvilin, A. L., Kolomiichuk, V. N., Stankus, S. V., Khairulin, R. A., & Segall, B., (2000). Kinetics of the graphitization of dispersed diamonds at "low" temperatures, *J. App. Phys., 88*, 4380–4388.

Cabioch, T., Girard, J. C., Jaouen, M., Denanot, M. F., & Hug, G., (1997). Carbon onions thin film formation and characterization. *Europhys. Lett., 38*, 471–475.

Cabioch, T., Jaouen, M., Denanot, M. F., & Bechet, P., (1998). Influence of the implantation parameters on the microstructure of carbon onions produced by carbon ion implantation. *Appl. Phys. Lett., 73*, 3096–3098.

Cabioch, T., Kharbach, A., Roy, A. L., Rivière, J. P., & Titov, V. M., (1998). Fourier transform infra-red characterization of carbon onions produced by carbon-ion implantation. *Chemical Physics Letters, 285*, 216–220.

Cabioch, T., Riviere, J. P., & Delafond, J., (1995). A new technique for fullerene onion formation. *J. of Mat. Sci., 30*, 4787–4792.

De Heer, W. A., & Ugarte, D., (1993). Carbon onions produced by heat treatment of carbon soot and their relation to the 217.5 nm interstellar absorption feature. *Chemical Physics Letters, 207*, 480–486.

Du, A. B., Liu, X. G., Fu, D. J. X. G., Han, P. D., & Xu, B. S., (2007). Onion-like fullerenes synthesis from coal. *Fuel, 86*, 294–298.

Fu, D., Liu, X., Lin, X., Li, T., Jia, H., & Xu, B., (2007). Synthesis of encapsulating and hollow onion-like fullerenes from coal. *J. Mater. Sci., 42*, 3805–3809.

Ganji, M. D., Ahangari, M. G., Fereidoon, A., & Jahanshahi, M., (2012). Investigation of the mechanical properties of multi-walled carbon nanotubes using density functional theory calculations. *J. Comput. Theor. Nanosci., 9*, 980–985.

Ghavanloo, E., & Fazelzadeh, S. A., (2015). Continuum modeling of breathing-like modes of spherical carbon onions. *Physics Letters A., 379*, 1600–1606.

Kroto, H. W., & Mckay, K., (1988). The formation of quasi-icosahedral spiral shell carbon particles. *Nature, 331*, 328–331.

Kroto, H. W., (1987). The stability of the fullerenes C_n, with $n = 24, 28, 32, 36, 50, 60$, and 70. *Nature, 329*, 529–531.

Kroto, H. W., Heath, J. R., ÓBrien, S. C., Curl, R. F., & Smalley, R. E., (1985). C_{60}: Buckminsterfullerene. *Nature, 318*, 162–163.

Kuznetsov, V. L., Chuvilin, A. L., Butenko, Y. V., Mal'kov, I. Y., & Titov, V. M., (1994). Onion-like carbon from the ultra-disperse diamond. *Chemical Physics Letters, 222*, 343–348.

Kuznetsov, V. L., Chuvilin, A. L., Moroz, E. M., Kolomiichuk, V. N., Shaikhutdinov, S. K., Butenko, Y. V., & Malkov, I. Y., (1994). Effect of explosion conditions on the structure of detonation soots: Ultradispersed diamond and onion carbon. *Carbon, 32*, 873–882.

Mordkovich, V. Z., (2000). The observation of large concentric shell fullerenes and fullerene-like nanoparticles in laser pyrolysis carbon blacks. *Chem. Mater., 12*, 2813–2818.

Mordkovich, V. Z., Shiratori, Y., Hiraoka, H., Umnov, A. G., & Takeuchi, Y., (2001). A path to larger yields of multishell fullerenes. *Carbon, 39*, 1929–1941.

Mordkovich, V. Z., Umnov, A. G., Inoshita, T., & Endo, M., (1999). The observation of multiwall fullerenes in thermally treated laser pyrolysis carbon blacks. *Carbon, 37*, 1855–1858.

Piskoti, C., Yarger, J., & Zettl, A., (1998). C_{36}, a new carbon solid. *Nature, 393*, 771–774.

Prinzbach, H., Weiler, A., Landenberger, P., Wahl, F., Wörth, J., Scott, L. T., Gelmont, M., Olevano, D., & Issendorff, B. V., (2000). Gas-phase production and photoelectron spectroscopy of the smallest fullerene, C_{20}. *Nature, 407*, 60–63.

Rao, C. N. R., Seshadri, R., Govindaraj, A., & Sen, R., (1995). Fullerenes, nanotubes, onions and related carbon structures. *Materials Science and Engineering, 95*, 209–262.

Sano, N., Wang, H., Alexandrou, I. I., Chhowalla, M., Teo, K. B. K., Amaratunga, G. A. J., & Iimura, K., (2002). Properties of carbon onions produced by an arc discharge in water. *J. Appl. Phys., 92*, 2783–2788.

Sano, N., Wang, H., Chhowalla, M., Alexandrou, I. I., & Amaratunga, G. A. J., (2001). Nanotechnology: Synthesis of carbon 'onions' in water, *Nature, 414*, 506–507.

Shen, H., (2007). Mechanical properties and electronic structures of compressed C_{60}, C_{180} and C_{60}-C_{180} fullerene molecules. *J. Mater. Sci., 42*, 7337–7342.

Todt, M., Rammerstorfer, F. G., Fischer, F. D., Mayrhofer, P. H., Holec, D., & Hartmann, M. A., (2011). Continuum modeling of van der Waals interactions between carbon onion layers. *Carbon, 49*, 1620–1627.

Tomita, S., Burian, A., Dore, J. C., Le Bolloch, D., Fujii, M., & Hayashi, S., (2002). Diamond nanoparticles to carbon onions transformation: X-ray diffraction studies. *Carbon, 40*, 1469–1474.

Turker, L., & Gumus, S., (2004). An AM1 study on C_{60}-C_{180} system. *J. Mol. Struct. (THEOCHEM), 674*, 15–18.

Turker, L., (2001). A bucky onion from C_{20} and C_{60}: an AM1 Treatment. *J. Mol. Struct. (THEOCHEM), 545*, 207–214.

Ugarte, D., (1992). Curling and closure of graphitic networks under electron-beam irradiation. *Nature, 359*, 707–709.

Xu, B., (2008). Prospects and research progress in nano-onion-like fullerenes. *New Carbon Mat., 23*, 289–301.

Xu, B., Guo, J., Wang, X., Liu, X., & Ichinose, H., (2006). Synthesis of carbon nanocapsules containing Fe, Ni or Co by arc discharge in aqueous solution. *Carbon, 44*, 2631–2634.

Zwanger, M. S., Banhart, F., & Seeger, A., (1996). Formation and decay of spherical concentric-shell carbon clusters. *J. Cryst. Growth, 163*, 445–454.

CHAPTER 4

Catalysis

ADRIANA URDĂ and IOAN-CEZAR MARCU

Laboratory of Chemical Technology & Catalysis, Department of Organic Chemistry, Biochemistry & Catalysis, Faculty of Chemistry, University of Bucharest, 4-12, Blv. Regina Elisabeta, 030018 Bucharest, Romania, Tel: +40213051464, Fax: +40213159249, E-mail: ioancezar.marcu@chimie.unibuc.ro; ioancezar_marcu@yahoo.com

4.1 DEFINITION

Catalysis represents the acceleration of thermodynamically possible chemical reactions by activating the reactants in the presence of compounds called catalysts. The acceleration comes from the lowered activation energy, which occurs when the reactant molecule interacts with the catalyst. Although the catalyst interacts with the reactant(s), it is not consumed in the reaction; therefore, it needs to be present only in very small amounts for the reaction to take place. Catalysis is one of the processes that support life through enzymatic reactions and, as a key concept in chemistry, it is involved in the synthesis of a large number of compounds used in everyday life.

4.2 HISTORICAL ORIGIN(S)

Catalysis was already used in antiquity, even without knowing it: the alcoholic fermentation and the conversion of ethanol to acetic acid are chemical reactions catalyzed by enzymes (Bornscheuer & Buchholz, 2005; Hagen, 2006). In medieval times, the use of the philosopher's stone to transform base metals into "gold" was a catalytic concept (Lohse, 1945).

In 1834, Mitscherlich introduced the term "contact" in his report discussing the conversion of alcohol into ether and water in the presence of dilute sulfuric acid (Davis, 2008). The concept of "catalysis" was coined by the Swedish chemist Berzelius in 1835 (Lohse, 1945; Somorjai, 1994), and in

1894, Ostwald defined it as "the acceleration of a slow chemical process by the presence of a foreign material" (Ertl, 2009), while the first definition of "catalyst" was given by Armstrong in 1885 (Davis, 2008). Later, in 1901, he defined a catalyst as "a material that changes the rate of a chemical reaction without appearing in the final product" (Ertl, 2009). Sabatier proposed in 1923 the temporary formation of unstable chemical compounds between catalysts and reactants, believing them to be intermediate steps in the reaction (Védrine, 2014). Kinetic studies in adsorption/desorption phenomena, crucial in heterogeneous catalyzed processes, were initiated by Langmuir and developed by Hinshelwood (Davis, 2008).

In the 1920s, two of the main theories in heterogeneous catalysis were formulated: the first one by Taylor, who proposed that catalytic reactions take place on "active sites" (which he called "aristocratic sites") on the surface of solid catalysts (Schlögl, 2015), and the second one by Balandin, who proposed the "multiplet theory" for the adsorption of a reactant molecule simultaneously on several atoms on the catalyst's surface (Védrine, 2014). The "transition state" theory was developed by Polanyi in 1935, assuming the presence of free homopolar valencies that protrude from the surface of a catalyst, where the reaction takes place (Davis, 2008).

Several Nobel prizes for chemistry recognized pioneer work in different fields of catalysis: W. Ostwald in 1909, P. Sabatier in 1912, F. Haber in 1918, C. Bosch and F. Bergius in 1931, I. Langmuir in 1932, C. N. Hinshelwood and N. N. Semenov in 1956, W. S. Knowles, R. Noyori and K. B. Sharpless in 2001, G. Ertl in 2007, R. F. Heck, E. Negishi, and A. Suzuki in 2010.

The history of industrial chemistry and catalysis are linked together: the lead chamber process for the synthesis of sulfuric acid using NO_x catalysts developed by Désormes and Clement in 1806, chlorine production by HCl oxidation with $CuCl_2$ catalyst dispersed on a porous and heat-resistant support by Deacon in 1868, the contact process for sulfuric acid synthesis with Pt or V_2O_5 catalysts by Winkler in 1875 and Knietsch in 1890, the synthesis of nitric acid by ammonia oxidation with catalytic Pt nets by Ostwald in 1906, ammonia synthesis from nitrogen and hydrogen on iron catalysts by Mittasch, Haber and Bosch in 1908, the methanol synthesis on ZnO/Cr_2O_3 catalysts by Mittasch in 1923 and the Fischer-Tropsch synthesis of hydrocarbons on iron, cobalt or nickel catalysts (1925) are just some of the early examples (Lohse, 1945; Somorjai, 1994; Hagen, 2006; Ertl, 2009; Lloyd, 2011). Many more catalytic processes were developed in the second part of the 20th century and are still in use now, and this development continues today.

Most chemical industrial processes (over 90% of the newly developed ones) and almost all biological reactions require catalysts (Hagen, 2006; Dumesic et al., 2008). Many compounds used in the production of plastics, pharmaceuticals, synthetic fibers, but also most of the crude oil processing, petrochemistry, and environmental protection methods use heterogeneous catalysts (Hagen, 2006).

4.3   NANO-SCIENTIFIC DEVELOPMENT(S)

Catalysis is the nanoscience and technology of influencing the rates of chemical reactions that are thermodynamically possible, by activating the reactants in the presence of compounds called catalysts (Schlögl, 2015). The catalysts do not modify the thermodynamics of a reaction, just its kinetics, by favoring a reaction path with smaller activation energy. In the Arrhenius equation:

$$k = A \exp\left(-\frac{E_a}{RT}\right) \tag{1}$$

where k is the rate constant, A is the frequency factor, E_a is the activation energy, R is the gas constant, and T is the reaction temperature (in K), when the activation energy is smaller, the rate constant (therefore the reaction rate) increases. The decrease of E_a occurs through the formation of transient weak bonds between the reactants and the catalyst, changing the electron distribution in the reactant molecules. This lowers the energy of the transition state, by partly compensating for the energy required to break the old bonds until the formation of the new bonds takes place (Roduner, 2014). After the formation and desorption of products, the catalyst is ready for a new reaction cycle.

The concept of "autocatalysis" refers to the case of catalytic reaction with a slow start during the initial induction period, followed by a rapid acceleration due to the catalytic action exhibited by one of the reaction products. One example is the reduction of metal oxides with hydrogen, provided that the metal catalyzes the reduction (Davis, 2008).

Catalytic processes that take place with the reactants and catalysts in a single phase (either gas or liquid) are called homogeneous catalytic processes. Examples of homogeneous catalysts are organometallic complexes and some metallic carbonyls (e.g., Co, Fe, and Rh). When the reactants and products are in a separate phase (gas or liquid) than the catalyst (usually a solid, with the reaction taking place on its surface), the reaction is called heterogeneously catalyzed. Common heterogeneous catalysts are metals, oxides, sulfides, and ion exchangers (see Chapter 5: Catalytic Material).

As Ostwald's definition states, the main function of a catalyst is to accelerate a chemical reaction, while being regenerated to its original form at the end of each reaction cycle (Dumesic et al., 2008). The most important properties for a catalyst are its activity, selectivity, and stability. These properties depend on many factors, such as catalyst preparation, chemical composition and, for heterogeneous catalysts, activation, bulk and/or surface crystalline structures, electronic effects, interaction between active phase and support, etc. (Védrine, 2014).

The catalytic activity is a measure of how fast the reaction takes place in the presence of the catalyst and can be measured as the reaction rate r, measured in the kinetic regime:

$$r = \frac{converted\ amount\ of\ reactant}{catalyst\ volume\ (or\ mass) \times time} \quad (\text{in } mol\ L^{-1}\ h^{-1} \text{ or } mol\ kg^{-1}\ h^{-1}) \quad (2)$$

If the catalyst mass is used instead of volume, then the reaction rate is called a specific rate. The reaction rate can also be normalized to the surface area of the active phase, e.g., the metal surface area for the supported metal catalysts (Deutschmann et al., 2009).

In certain conditions, the turnover number (TON) or turnover frequency (TOF) can be used. TON defines the number of reaction cycles taking place at an active site up to the total decay of activity, while TOF counts the number of catalytic cycles at the active site per unit time. Therefore, TON = TOF (time^{-1}) × catalyst lifetime (time). While for homogeneous catalytic reactions, the number of active sites is easily determined, for heterogeneous catalysts this is difficult, and values are obtained from chemisorption experiments. This leads to ambiguities; therefore, extended definitions for TON and TOF were proposed (Kozuch & Martin, 2012), taking into consideration all the reaction conditions in order to properly judge the catalytic efficiency. For relevant industrial applications, TOF values fall between 10^{-2} and 10^{2} s^{-1}, while TON should be 10^{6}–10^{7} (Thomas & Thomas, 1997; Hagen, 2006).

In practice, for comparative measurements, other activity measures can be used, such as conversion under constant reaction conditions, the temperature required for a certain conversion value, space-time yield (STY), etc. (Hagen, 2006). The conversion represents the ratio between the amount of converted reactant and the amount that was introduced in reaction, while the STY is defined as (Deutschmann et al., 2009):

$$STY = \frac{amount\ of\ desired\ product}{catalyst\ volume \times time} \quad (\text{in } mol\ L^{-1}\ h^{-1}) \quad (3)$$

When a comparison between catalysts uses the temperature required for a certain conversion, the catalyst that gives the desired conversion value at the lowest temperature is the best one (Hagen, 2006).

Another important property of catalysts is to influence the *selectivity* of chemical reactions: if different products can be obtained, the catalyst should be able to direct the reaction towards the desired products. Thus, selectivity measures the fraction of converted reactant that was transformed into the desired product(s) (Hagen, 2006). In industrial processes, high selectivity is more important than a good activity and has implications for the separation and purification of products (Prieto & Schüth, 2015; Hagen, 2006). Another way for measuring catalytic performance is the yield, calculated as the product between selectivity and conversion. It represents the fraction of the reactant introduced in the process that was transformed into the desired product(s).

An ideal catalyst remains unchanged during the chemical reaction it catalyzes, but in practice, it suffers changes that lower its activity (catalyst deactivation) (Hagen, 2006). *Stability* is an important issue, especially in industrial catalysis, where the chemical, thermal, and mechanical stability determines the catalyst lifetime in the reactor and, consequently, the economics of the process. The deactivation of catalysts can occur by poisoning, coking, losing the active component, thermal degradation, etc. (Dumesic et al., 2008). Some deactivation processes can be reversed by regeneration (e.g., coking), but ultimately the catalysts have to be replaced with fresh ones.

Catalysts can be classified according to many criteria: state of aggregation (heterogeneous or solid catalysts and homogeneous ones), structure (bulk and supported catalysts), chemical composition (metal catalysts, oxides, sulfides, etc.), and many other (see Chapter 5: Catalytic Material).

In homogeneous catalysis, a substance is considered to be a catalyst if the reaction rate depends on its concentration, but its concentration does not change in the course of the reaction (and it is not in such a large excess that the modification in concentration is not observable) (Roduner, 2014). In heterogeneous catalysis, the substance is considered to be a catalyst if, in its absence, the reaction does not occur, or it occurs much slower. A special case of heterogeneous catalysis, electrocatalysis involves oxidation or reduction by electrons transfer across the interface between the electrode (catalyst) and electrolyte, while photocatalysis (homogeneous or heterogeneous) uses light absorbed either by the catalyst or by the reactant during the reaction (Deutschmann et al., 2009). Electrocatalysis is used, for example, in all types of fuel cells for delivering stationary, mobile or portable power, and has rapidly developed in recent decades (Krischer and Savinova, 2008). In photocatalysis, materials based on transition metal oxides and semiconductors (e.g., TiO_2) are used, aiming at an efficient utilization of visible and/or solar light energy in different catalytic processes (Takeuchi et al., 2008).

Catalysis is a discipline that deals with the fundamental principles of catalytic reactions, and also with the preparation, properties, and applications of catalytic materials (Deutschmann et al., 2009).

4.4   NANOCHEMICAL APPLICATION(S)

Catalysis has a key importance in the chemical and petrochemical industries, in the production of fuels, foodstuff, pharmaceuticals and many other products used in everyday life, but also in environmental processes that are becoming increasingly important (Thomas & Thomas, 1997).

Three main domains of catalytic applications exist, namely homogeneous, enzymatic and heterogeneous catalysis, with the latter representing the majority of known applied processes. Although there are significant differences between the catalysts used in each domain, Deutschmann et al., (2009) cited a statement by David Parker in a meeting at the University of St. Andrews in 1998: "... at the molecular level, there is little to distinguish between homogeneous and heterogeneous catalysis, but there are clear distinctions at the industrial level."

4.4.1   HOMOGENEOUS CATALYSTS

Homogeneous catalysts are well-defined chemical compounds or coordination complexes, molecularly dispersed in the reaction environment with the reactants (Hagen, 2006). Examples of such compounds are mineral acids and organometallic complexes (compounds with metal-carbon bonds), such as rhodium carbonyl complexes.

Homogeneous catalysts have a high degree of dispersion; therefore, each individual species is potentially a active catalytic site that can be approached from any direction. Hence, they exhibit a high activity per unit mass of metal (Hagen, 2006). The reactions are controlled mainly by kinetics and not by transport, and their mechanisms can be investigated readily by spectroscopic methods.

The thermal stability of the organometallic complexes limits the industrially feasible processes to temperatures below 200°C. An important disadvantage of these processes is the separation of catalyst from the reaction products, which can only be done by distillation, extraction, ion exchange, or other time and energy consuming methods. In recent years separation was improved by using complexes that are soluble in both organic and aqueous phases, removable by transfer into water at the end of the process (Hagen, 2006). Attempts for immobilizing homogeneous catalysts on solid supports have not been successful enough to be used industrially due to active phase leaching (partial transfer into the solvent) and low stability. Besides that, the immobilization process occurs through strong chemical bonds, which affect the chemistry and catalytic properties of the organometallic complex (Védrine, 2014).

The catalytic activity of the organometallic compounds is explained by the reactivity of the organic ligands, which are activated upon bond formation with the d orbitals of the transition metal center (Hagen, 2006). These transition metals can exist in various oxidation states and, therefore, they can exhibit different coordination numbers. The ligands can be either ionic (e.g., HO^-, H^-, CH_3CO^-, alkyl$^-$, aryl$^-$) or neutral (e.g., CO, alkene, H_2O, amine) species. The most important reactions involve ligands located in the coordination sphere of the same metal center: reactants coordinate loosely to the active center, the reaction takes place by rearranging bonds between the ligands, then reaction products are released from the coordination sphere of the metal center, with the formation of a vacant coordination site for the catalytic cycle to take place again. The organometallic complexes are required to have labile bonds with the ligands in order for the reactions to occur with an activation energy as low as possible (Hagen, 2006).

Numerous reaction types can take place, such as exchange of ligands, complexation of a substrate at the transition metal center, acid-base reactions, redox reactions (such as oxidative additions and reductive eliminations), insertions (such as CO or alkenes insertion into metal-alkyl or metal-hydride bonds), or reactions at coordinated ligands (van Leeuwen, 2004; Hagen, 2006).

In a simplified general mechanism, the catalyst precursor dissociates a ligand to form the active catalyst, M^n, capable to coordinate the substrate S (see Scheme 1). The reactant is then coordinated at the metal center (oxidative addition), followed by a rearrangement (e.g., insertion of the coordinated reactant molecule into a metal-hydride bond), and finally, the reductive elimination of the reaction product takes place, re-forming the starting active catalyst (Hagen, 2006).

Homogeneous catalysis is used industrially in processes such as the selective oxidation of ethane to acetaldehyde (the Wacker process) with Pd^{II}/Cu^{II} chloride solutions (Brégeault & Launay, 2002; Hagen, 2006), the Monsanto methanol carbonylation to acetic acid by rhodium or iridium complexes (Dingerdissen et al., 2008), the synthesis of aliphatic aldehydes by the addition of CO and H_2 to alkenes in the presence of Co carbonyls, known as the Oxo synthesis (Deutschmann et al., 2009).

$$M^n \xrightarrow{\;+S\;} M^n - S \xrightarrow[\substack{\text{Oxidative} \\ \text{addition}}]{\;+R\;} R - M^{n+2} - S \xrightarrow[\substack{\text{Reductive} \\ \text{elimination}}]{} M^n + R - S$$

SCHEME 1   Simplified representation of a homogeneous catalytic process as an oxidative addition followed by a reductive elimination (S = substrate; R = reactant).

4.4.2   ENZYMATIC CATALYSIS

Enzymes are protein molecules of colloidal size, usually composed of more than 100 amino acids, able to catalyze biological reactions with remarkable activities and selectivities (Hagen, 2006). They work either in a dissolved state in cells, or chemically bound to cell membranes or on surfaces (Deutschmann et al., 2009). They can be classified somewhere between homogeneous and heterogeneous catalysts.

The main advantages of enzymes are their very high activity and selectivity (including enantioselectivity) for the conversion of reactant molecules (substrates), and the fact that they work in mild conditions: room temperature, in aqueous solution at physiologically pH values (Hagen, 2006). Consequently, they are destroyed by extreme reaction conditions, such as high temperatures. The ratio between the catalyzed rate of reaction and the uncatalyzed rate is called the catalytic power of an enzyme, and can reach $10^{15}–10^{16}$. Enzymes have high specificity, meaning they are very selective both on the substrates with which they interact and the reactions they catalyze, while no wasteful by-products are produced. Sometimes, enantioselectivities of more than 99% can be achieved.

The active center in an enzyme is, in most cases, a metal ion with only restricted access of selected substrates with a certain orientation relative to the catalytic site (Roduner, 2014). Substrates bind with relatively weak bonds at a specific site on the enzyme, forming stoichiometric complexes that release the product easily after reaction (Hagen, 2006). The specificity of an enzyme depends on the complementarity between the structures of substrate and active site (called the "key and lock" or "induced-fit" model). In many oxidation/reduction enzymatic reactions (especially in biological reactions) a second substrate, called coenzyme or cofactor, is necessary to activate the enzyme.

After systematic research in the 19th century that marked the fundamentals of enzymatic processes (Bornscheuer & Buchholz, 2005), immobilized enzymes were produced in the 1950s for their economic advantages related to the ease of separation and reuse. Their first industrial application was in the production of amino acids, and the two most important processes are the hydrolysis of penicillin and the isomerization of glucose. All these processes replaced previous chemical synthesis methods that were more complicated and produced environmental problems.

The most important areas for the use of enzymatic catalysis are food manufacturing (starch processing, bread, cheese, beer, fruit juice manufacturing, etc.), fine chemicals (such as amino acids, vitamins), and detergent formulations (Bornscheuer & Buchholz, 2005). Enzymes are applied in the production of basic chemicals such as bioethanol and acrylamide (for polyacrylamide fibers), but also pharmaceuticals (like antibiotics), agrochemicals, optically active compounds, and many more. The production scale ranges from more than ten million to hundreds of tons per year, and it is estimated that more than 5% of the fine chemicals are produced by processes partly or entirely enzymatic. The success of industrial enzymatic processes is related to their reduced production costs, as less reaction steps are required in comparison with traditional chemical processes, and improved environmentally friendly and sustainable technologies are used.

4.4.3   HETEROGENEOUS CATALYSIS

Heterogeneous catalysis takes place between several phases: usually, the catalyst is a solid (see Chapter 5: Catalytic Material), while the reactants are gases or liquids (Hagen, 2006). Examples of such catalysts are the three-way catalysts for automobiles, zeolites in petrochemical conversions, etc.

Due to the presence of a phase boundary, heterogeneous catalysts exhibit active sites only at their surface and, hence, they have a lower activity per unit mass of active phase than homogeneous catalysts, and need more severe reaction conditions. Mass transfer is often present due to the presence of pores in the solid, and the mechanistic investigations are difficult compared to homogeneous reactions, due to the nature of reaction intermediates on the catalyst's surface.

The main advantage of the heterogeneous processes compared to the homogeneous ones is the simple separation of the catalyst from the reaction products in fixed-bed reactors, or by a filtration/centrifugation method when suspensions are involved.

The identification of active sites in heterogeneous catalysis was and still is a challenging task, due to the fact that not all sites on the surface have the same activity (Roduner, 2014; Schlögl, 2015). The first to discuss "active sites" was Taylor, who attributed special activity to the surface atoms (on terraces, steps, kinks, etc.), due to lower coordination to other catalyst atoms. This concept emphasized the heterogeneity of the solid surface, and the fact that not all sites are equally active in catalytic reactions (Davis, 2008; Schlögl, 2015).

Shortly after Taylor's theory, Balandin proposed the multiplet theory, explaining the bond scission and formation by the geometrical concordance between the spacing of the catalyst surface atoms and the adsorbate molecule atoms (Davis, 2008). He postulated that the active sites are not randomly placed on the surface, but in small groups (multiple sites) reflecting the geometry of the catalyst's crystalline framework. The catalytic reaction will take place only if the distance between the atoms on the catalyst surface will match the distance between the atoms in the reactant molecule, in order for the molecule to be activated. The multiplet concept was later extended by Kobozev, who introduced the hypothesis of "active ensembles," in order to explain the high activity of highly dispersed supported metal catalysts.

Wolkenstein introduced the electron theory of catalysis on semiconductors (Wolkenstein, 1960), based on previous work by Pisarzhevsky and Roginsky. This approach tried to elucidate the relation between the catalytic and electronic properties of the semiconductor catalysts, in order to understand the elementary electronic mechanisms of catalytic reactions.

There are many methods and techniques that are used for the characterization of solid catalysts. Some of them, such as XRD, SEM/TEM, Raman, IR, UV-Vis, MAS-NMR, EXAFS/XANES, etc., give information on the whole material, while others such as XPS, AES, AFM, LEED, etc. characterize only the top surface layers of the solid, where the catalytic action takes place (Védrine, 2014). For active sites characterization, transient techniques such as temporal analysis of products (TAP) or steady state isotopic transient kinetic analysis (SSITKA) give valuable information. Theoretical methods of modeling are used to simulate solid materials and to determine the most probable mechanisms and reaction intermediates. *In situ* and *operando* experiments are used to characterize real active sites in working conditions and under real catalytic reaction conditions (Schlögl, 2015).

The heterogeneous catalytic reactions take place through a sequence of elementary steps that include (Somorjai, 1994; Dumesic et al., 2008):

- diffusion of the reactant(s) through the boundary layer surrounding the catalyst particle;
- intraparticle diffusion of the reactant(s) into the catalyst pores to the active sites;
- adsorption of reactant(s) on the surface, possibly including diffusion on the surface;
- chemical rearrangement (bond breaking, bond forming, molecular rearrangement) of the reaction intermediates in adsorbed state;
- desorption of products;
- intraparticle diffusion of the products through the catalyst pores;
- diffusion of the products through the boundary layer surrounding the catalyst particle.

Each of these steps can be the rate controlling one, therefore different regimes of rate control can exist (Dumesic et al., 2008): film diffusion control (diffusion of reactants and/or products across the boundary layer as rate determining step), pore diffusion control (diffusion of reactants and/or products in the catalyst pores), and intrinsic reaction kinetics control (adsorption on/desorption from the surface and the reaction on the surface). Additionally, besides mass transfer effects, heat transfer effects can occur for highly exothermic on endothermic reactions.

Adsorption on an ideal surface is usually described with the Langmuir adsorption isotherm, linking the amount of adsorbed reactant (expressed as the surface coverage Θ) and its partial pressure P at a specific temperature (Freund, 2008):

$$\Theta = \frac{b \cdot p}{1 + b \cdot p} \tag{4}$$

where b is the Langmuir constant, temperature dependent.

Two types of adsorption exist: chemisorption and physisorption (Davis, 2008), the latter being transformed into chemisorption by activated adsorption. Older definitions of the two adsorption forms differentiated them via the liberated energy, placing a rather artificial limiting energy value between chemisorption and physisorption regimes at 40 kJ mol^{-1} (Freund, 2008). The modern accepted definition of chemisorption is independent of thermochemical data, relying on the concept of a direct intermixing of the substrate and adsorbant charge densities, meaning a short-range chemical bond.

The adsorption of reactant molecules can be either associative (as molecule) or dissociative, after bond breaking upon interaction with the surface. Interaction (attractive or repulsive) between adsorbed species is also possible, resulting in either cooperative or competitive adsorption on the active sites (Freund, 2008). Due to repulsive interactions, the adsorption energy decreases at medium and high surface coverage on the surface.

If the reactant adsorbs too strongly on the catalyst, then the active site is blocked because the product will not be desorbed from the surface, leading to catalyst poisoning and low activity; if the adsorption energy is too low, then the activity of the catalyst for that reaction is low because the reactant is not activated enough (Deutschmann et al., 2009; Roduner, 2014). This principle, known as "the Sabatier principle," is usually represented with volcano curves of activity versus energy of adsorption (Védrine, 2014).

Diffusion of the adsorbate molecules on the surface means that their mobility on the surface is high enough for them to be considered in a state of a two-dimensional gas (Davis, 2008). The surface diffusion plays an important role, enabling the system to achieve an equilibrium structure (Freund, 2008). The process is believed to occur through hops between adjacent sites, from an occupied to an empty site.

After surface reaction and desorption of the products, the catalyst should be able to carry the conversion of a new reactant molecule, and this repetition (called catalytic cycle) should ideally take place numerous times (Roduner, 2014).

The role of the catalyst to accelerate the reaction rate can be depicted by the potential energy curves versus the reaction coordinate (see Figure 4.1). A comparison between the gas phase reaction and the surface reaction shows that the catalyst alters the reaction pathway: instead of a single homogeneous (gas phase) step that requires high activation energy, the surface reaction has several steps, each with a lower value for the activation energy.

The catalysts can catalyze both the forward and the backward reaction, and that means the same transition state is formed; therefore, the activation energy for the reverse reaction will be reduced by the same amount as for the forward reaction (Thomas & Thomas, 1997; Roduner, 2014).

Heterogeneous catalysis is a surface process, therefore the higher the active surface area of the catalyst, the greater the number of reactant molecules that are converted per unit time. Numerous methods are used to obtain catalysts with high specific surface areas (SSA, defined as the total surface area per unit of mass; usually measured in $m^2 g^{-1}$; see Chapter 5: Catalytic Material). Some of these catalysts are single or multicomponent bulk catalysts, others contain a support phase, usually with large SSA, onto which another phase is deposited with high dispersion (the latter being defined as the ratio between the number of atoms exposed to reactants, and the total number of atoms in the supported phase). Supports such as alumina, silica, titania, carbon, mixed oxides, etc., with SSA between 100 and 400 m² g⁻¹, are commonly used in catalysis. In many reactions, a synergy effect was observed between separate phases of the catalyst, or between the support and the active phase (Védrine, 2014). If both the support and the deposited phase are catalytically active (but in different stages of the desired reaction sequence), the catalyst is called bifunctional (or multifunctional, if more than one active phase is present).

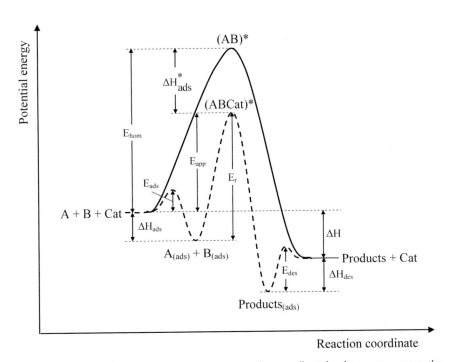

FIGURE 4.1 Potential energy curves versus the reaction coordinate in a homogeneous reaction between reactants A and B (solid curve) compared to the surface reaction in the presence of a solid catalyst Cat (dotted curve) showing lower activation energy steps for the latter ((AB)* = activated state for the homogeneous reaction; (ABCat)* = activated state for the surface reaction; E_{hom} = activation energy for the homogeneous reaction; E_{app} and E_r = apparent and real activation energy, respectively, for the surface reaction; E_{ads} and E_{des} = activation energies for the reactants adsorption and products desorption, respectively; ΔH = heat of reaction; ΔH_{ads} and ΔH_{des} = heats of adsorption and desorption, respectively; ΔH^*_{ads} = heat of adsorption of the activated complex).

In heterogeneous catalysis, the porous solids or fine powders with large specific surface areas offer different types of sites for catalytic reactions (heterogeneity of the surface due to crystalline anisotropy), provided that reactants have access to the centers located inside the pores (Somorjai, 1994).

A large part of the observed catalytic activity may be caused by a small fraction of the sites ("active sites") present on the solid surface (Dumesic et al., 2008). Besides the fact that small particles have a larger fraction of their atoms on the surface and exposed to the reaction environment, the surface atoms of a small metal particle have significantly different electronic properties from those on the bulk or surface of large particles (Roduner, 2014). Catalysts that contain nanoparticles, which have numerous atoms with low coordination (partly unsaturated due to positioning at edges, corners, or as adatoms on surfaces) possess higher activity. This is caused by a better bonding with reactants, and also because the nanoparticles have a different electronic structure, with electron affinity and ionization potential between those of bulk and individual atoms, and these properties control bond formation with the reactant (Roduner, 2014).

The classical view of solid catalysts considers the active sites as being static; however, the surface may change under reaction conditions (Védrine, 2014), and even oscillations occur on the surface during catalytic reactions (Ertl, 2008; Védrine, 2014). Because of the contact with reactants, the surface layers of the catalyst often exhibit different chemical composition and crystallization state than the bulk of the solid: defects of the lattice or new active sites may be generated, and sometimes the outermost layers may even be considered amorphous material deposited on a crystalline phase (Davis, 2008; Védrine, 2014). All these observations led to a redefinition of the catalyst as a "functional material that continually creates active sites with its reactants under reaction conditions" (Schlögl, 2015).

Heterogeneous catalytic reactions can be divided into "structure sensitive" and "structure insensitive," based on the influences on the reaction rate of the particle size of the catalyst or the crystallographic face exposed to reactants (Thomas & Thomas, 1997; Védrine, 2014). For example, in C–C or N–N bonds breaking the rate depends on which face of the crystal is exposed, while the formation of H–H, C–H or O–H bonds is structure insensitive.

Several reaction mechanisms are proposed for heterogeneous catalytic reactions (Ertl, 2008; Védrine, 2014). For the case when two reactants are involved, the Langmuir–Hinshelwood mechanism considers both of them adsorbed, fully equilibrated with the surface and reacting after activation on the surface. Most of the catalytic reactions occur through this mechanism. In the Eley–Rideal type, one of the reactants is adsorbed and activated on the solid surface, while the other reacts from gas phase. In selective oxidation reactions over oxide-based catalysts, the Mars-van Krevelen redox mechanism implies the oxidation of the reactant molecule with lattice oxygen coming from the catalyst, followed by the re-oxidation of the reduced catalyst surface by co-fed oxygen; if the first step is too facile, the reaction becomes non-selective; if it is not facile enough, the catalytic activity is too low (Thomas & Thomas, 1997).

The activity, selectivity, and stability of solid catalysts are influenced by modifiers (promoters and poisons), deliberately introduced or accidentally present (e.g., from impurities in the reactant mixture) on the catalyst surface (Koel & Kim, 2008). One compound can act as a modifier for a reaction, while being a poison in another case. A promoter is a modifier that increases the catalytic performance, while a poison has detrimental effects. Their action occurs at even trace amounts, because they act on the very small percentage of the surface atoms that are active sites. Some promoters, called "structural" or "textural" (Koel & Kim, 2008), e.g., alumina in the ammonia synthesis iron catalyst, enhance the catalytic activity by stabilizing the surface area against sintering, affecting pore size and distribution, dispersion of active components on the support, etc. Others, called "electronic" promoters (like potassium oxide in ammonia synthesis iron catalyst), influence the strength of the chemisorptive bond between the reactants, intermediates or products and the active surface (Brosda et al., 2006). Promoters are usually added to the catalyst during the preparation step (chemical promotion, also called *ex-situ* promotion), or *in situ*, via electrical potential or current application if the active phase is in contact with an ionic or mixed ionic-electronic conducting support (electrochemical promotion).

Catalyst deactivation is a complex phenomenon, caused by either a decrease in the number of active sites, degradation in the quality of these sites, or the decrease of accessibility in the pores, and can be reversible or irreversible in reaction conditions (Moulijn et al., 2008). The main causes for deactivation are poisoning, formation of deposits, thermal degradation (sintering, evaporation), mechanical damage, and corrosion/ leaching by the reaction mixture. Deactivation processes occur in very different time scales, from less than a second in the fluid catalytic cracking process to several years in ammonia synthesis.

Poisons have a deactivating effect even in very small amounts, by strong chemisorption on the active sites (Davis, 2008). In the case of temporary poisoning, the adsorption of the poison is weak and reversible, and removing it from the reactants flow restores the original catalytic activity; for permanent poisons, a strong and irreversible adsorption occurs (Moulijn et al., 2008). For example, sulfur often acts as a poison, because it forms strong bonds with the active phase (e.g., metals) of the catalyst, blocking the sites and changing the electronic structure in their vicinity (Koel & Kim, 2008).

Another example is lead, which poisons the Pt-Pd catalyst of the automobile catalytic converter, forcing the removal of antiknocking agent tetraethyl-lead from gasoline.

Deactivation can also take place by formation of deposits, covering the active sites on the surface (e.g., coking–depositing carbon from organic reactions at high temperatures). Thermal degradation is often encountered, caused by high-temperature sintering, chemical transformations, evaporation, etc. (Moulijn et al., 2008). In sintering, crystallite growth of either the active phase or the support leads to a loss of the active surface of the catalyst, through surface diffusion or coalescence at high temperatures. Chemical reactions can also take place between support and active phase components, e.g., formation of inactive spinel compounds in the case of alumina-supported catalysts.

Mechanical deactivation refers to crushing of the catalyst pellets and attrition resulting in the formation of fine particles carried away with the product flow, both processes leading to loss of catalytic activity (Moulijn et al., 2008). Leaching (mainly for liquid-phase processes) implies the disso-lution of the active component in the reaction mixture, causing not only deactivation, but also contamination of the product.

Sometimes deactivating impurities are used in small amounts, to improve the selectivity of a catalyst (Somorjai, 1994; Moulijn et al., 2008) or its stability (Deutschmann et al., 2009). Regeneration of deactivated catalysts is often possible, depending on the cause of deactivation; e.g., coke can be burned off the surface in air or oxygen, reactant flow can be purified to reverse some poisoning effects, etc. However, high temperatures during regeneration can cause sintering, with loss in SSA; therefore, these proce-dures must be applied with caution. At the end of its lifetime, the catalyst is recycled or disposed (if more economical) (Moulijn et al., 2008).

Here are some examples of large scale heterogeneous catalytic processes:

Sulfuric acid production is one of the oldest chemical processes (first patent filed in 1831). The process consists of several steps, the catalytic one being SO_2 oxidation to SO_3. It first used homogeneous catalysts (nitrogen oxides, in the lead chamber process), then platinum heterogeneous catalysts. As Pt was easily poisoned, it was replaced in the 1930s by alkali-promoted vanadium oxide, a catalyst that is liquid under reaction conditions on the silica support (Dingerdissen et al., 2008, Näumann & Schulz, 2008; Prieto & Schüth, 2015). Sulfuric acid is used in all fields of chemistry, mainly the fertil-izer industry, synthesis of dyes, pigment, plastics, in petroleum processing, extraction of non-ferrous metals, etc. (Näumann & Schulz, 2008).

Ammonia synthesis represents the most important catalytic process ever developed, two Nobel prizes being awarded for Fritz Haber in 1918 and Carl

Bosch in 1931 (together with Friederich Bergius). Nitrogen from air and hydrogen from natural gas are used in this key process in nitrogen chemistry; all other nitrogen-containing chemicals are produced from ammonia (Schlögl, 2008). After extensive experiments, an iron-based catalyst, doubly promoted with potassium- (electronic promoter), calcium- and aluminum oxides (structural promoters) was discovered and is still used today with minor modifications. Ruthenium-based catalysts are also used, with higher activity and in milder conditions (Schlögl, 2008; Prieto & Schüth, 2015). Ammonia is oxidized to nitric acid (to be further used for fertilizers) on catalytic gauzes made of platinum-rhodium and platinum-palladium-rhodium alloys (Dingerdissen et al., 2008).

ZnO/Cr_2O_3 catalysts were used at high pressure and temperature for the production of methanol from synthesis gas (a mixture of carbon monoxide and hydrogen) from, 1920s; being replaced after 40 years by $Cu/ZnO/Al_2O_3$ catalysts, which are still in use today (Dingerdissen et al., 2008; Hansen & Nielsen, 2008; Prieto & Schüth, 2015).

Fischer-Tropsch conversion of synthesis gas to synthetic fuels or oxygenated compounds occurs over modified iron or cobalt catalysts, which replaced earlier cobalt, nickel, or manganese oxides (Dingerdissen et al., 2008). Depending on process conditions, high-quality diesel fuel, gasoline, linear alkenes, alkanes, and oxygenated compounds such as alcohols are obtained (Dry, 2008).

Catalytic cracking and fluid catalytic cracking (FCC) represent the core of the petroleum refining industry since 1930s (Cheng et al., 2008; Prieto & Schüth, 2015). The first solid catalysts were acid-leached clays, soon replaced with synthetic amorphous silica-alumina and then Y zeolites (see Zeolite). Many other refining processes use catalysts: hydrocracking, hydrotreating, reforming, isomerization, and alkylation (Dingerdissen et al., 2008).

Base chemicals like butadiene, isoprene, styrene, and vinyl chloride monomers, alcohols, ethers, ketones and aldehydes, ethylene oxide, are just a few of the compounds produced using heterogeneous catalysis (Dingerdissen et al., 2008, Schlögl, 2015).

Propylene oxide, obtained by propylene epoxidation with hydrogen peroxide, is used in numerous syntheses for a wide range of chemicals and polymers, and was produced since a century ago by the non-catalytic dehydrochlorination of chlorohydrins. In the 1970s, a catalytic process was introduced, using oxygen as oxidant and organic hydroperoxides as intermediate oxygen couriers and mesoporous titanosilicates as catalysts (Brégeault & Launay, 2002; Prieto & Schüth, 2015). More recently, direct epoxidation with H_2O_2 became commercially available, with water as the only by-product, using a titanosilicate with zeolitic structure, called TS-1 (titanium silicalite-1).

Environmental catalysis represents roughly one-third of the worldwide market for solid catalysts (Dingerdissen et al., 2008). The main applications used today on very large scale are automotive exhaust emissions treatment (Lox, 2008), removal of harmful pollutants from gaseous effluents in chemical industries and power stations (Gabrielsson & Pedersen, 2008), and wastewater treatment (Prüße et al., 2008).

4.5   MULTI-/TRANS-DISCIPLINARY CONNECTION(S)

The field of catalysis (homogeneous, enzymatic, or heterogeneous) is highly interdisciplinary in nature, requiring the cooperation between chemists, physicists, surface scientists, reaction engineers, etc. The chemists are called to prepare the catalysts and to study them in the catalytic reaction for which they are designed. Physicists and surface scientists develop methods and perform careful characterization of these catalysts, in order to obtain meaningful relationships between their properties and catalytic behavior. For industrial applications, reaction engineers are needed to design and optimize the reaction equipment. This team effort explains the large amount of work and long time necessary for the development of catalytic processes.

4.6   OPEN ISSUES

As a key concept in chemistry, catalysis is likely to remain of central importance in the future, for the safer and cheaper manufacture of many desirable products.

In enzymatic catalysis, much effort is devoted for the improvement of stability in extreme reaction conditions (temperature, pH, organic solvents), both for existing and new applications of biocatalysts. The high stability is necessary for repeated or continuous use over extended time scales.

In heterogeneous catalysis, continuous research is required to solve the increasingly complex environmental and energy issues facing our industrialized society. The green catalysis and chemistry concept implies more environmentally benign processes, with less undesired by-products, using lower-cost, benign feedstocks and less energy, aiming for closer to 100% yields. For example, using oxygen as an oxidant instead of hydroperoxides or H_2O_2 in oxidation processes is still a topic of research, or the development of solid catalysts for enantioselective conversions to replace homogeneous catalysts for the production of fine chemicals and pharmaceuticals.

The energy demand is increasing worldwide, and catalysis is very much involved in research for practical applications regarding new fuels using non-conventional feedstocks, such as hydrogen from photocatalytic water splitting, biofuel from renewable sources and waste, etc., and also for the improvement of older processes, with new, high-performance catalysts for the more efficient utilization of resources. Catalyst recovery, regeneration and recycle should be performed on a routine basis, since many of the catalysts used industrially are based on expensive (and sometimes toxic) components.

KEYWORDS

- **enzymatic catalysis**
- **heterogeneous catalysis**
- **homogeneous catalysis**
- **Langmuir**
- **Ostwald**

REFERENCES AND FURTHER READING

Bornscheuer, U. T., & Buchholz, K., (2005). Highlights in biocatalysis–Historical Landmarks and Current Trends. *Engineering in Life Sciences, 5*, 309–323.

Brégeault, J. M., & Launay, F., (2002). Catalyse homogène d'oxydation. *L'actualité Chimique, 56*, 45–53.

Brosda, S., Vayenas, C. G., & Wei, J., (2006). Rules of chemical promotion. *Applied Catalysis B: Environmental, 68*, 109–124.

Cheng, W. C., Habib, Jr., E. T., Rajagopalan, K., Roberie, T. G., Wormsbecher, R. F., & Ziebarth, M. S., (2008). Fluid catalytic cracking. In: Ertl, G., Knözinger, H., Schüth, F., & Weitkamp, J., (eds.), *Handbook of Heterogeneous Catalysis* (2nd edn., Vol. 6, pp. 2741–2778). Wiley-VCH Verlag GmbH & Co. KGaA: Weinheim, Germany.

Davis, B. H., (2008). Development of the science of catalysis. In: Ertl, G., Knözinger, H., Schüth, F., & Weitkamp, J., (eds.), *Handbook of Heterogeneous Catalysis* (2nd edn., Vol. 1, pp. 16–37). Wiley-VCH Verlag GmbH & Co. KGaA: Weinheim, Germany.

Deutschmann, O., Knözinger, H., Kochloefl, K., & Turek, T., (2009). Heterogeneous catalysis and solid catalysts. In: *Ullmann's Encyclopedia of Industrial Chemistry* (pp. 1–110). doi:

10.1002/14356007.a05_313.pub2, Wiley-VCH Verlag GmbH & Co. KGaA: Weinheim, Germany.

Dingerdissen, U., Martin, A., Herein, D., & Wernicke, H. J., (2008). The development of industrial heterogeneous catalysis. In: Ertl, G., Knözinger, H., Schüth, F., & Weitkamp, J., (eds.), *Handbook of Heterogeneous Catalysis* (2nd edn., Vol. 1, pp. 37–56). Wiley-VCH Verlag GmbH & Co. KGaA: Weinheim, Germany.

Dry, M. E., (2008). The Fischer-Tropsch (FT) synthesis processes. In: Ertl, G., Knözinger, H., Schüth, F., & Weitkamp, J., (eds.), *Handbook of Heterogeneous Catalysis* (2nd edn., Vol. 6, pp. 2965–2994). Wiley-VCH Verlag GmbH & Co. KGaA: Weinheim, Germany.

Dumesic, J. A., Huber, G. W., & Boudart, M., (2008). Principles of heterogeneous catalysis. In: Ertl, G., Knözinger, H., Schüth, F., & Weitkamp, J., (eds.), *Handbook of Heterogeneous Catalysis* (2nd edn., Vol. 1, pp. 1–15). Wiley-VCH Verlag GmbH & Co. KGaA: Weinheim, Germany.

Ertl, G., (2008). Dynamics of surface reactions. In: Ertl, G., Knözinger, H., Schüth, F., & Weitkamp, J., (eds.), *Handbook of Heterogeneous Catalysis* (2nd edn., Vol. 3, pp. 1462–1479). Wiley-VCH Verlag GmbH & Co. KGaA: Weinheim, Germany.

Ertl, G., (2009). Wilhelm Ostwald: Founder of physical chemistry and Nobel laureate 1909. *Angewandte Chemie International Edition, 48*, 6600–6606.

Freund, H. J., (2008). Principles of chemisorption. In: Ertl, G., Knözinger, H., Schüth, F., & Weitkamp, J., (eds.), *Handbook of Heterogeneous Catalysis* (2nd edn., Vol. 3, pp. 1375–1415). Wiley-VCH Verlag GmbH & Co. KGaA: Weinheim, Germany.

Gabrielsson, P., & Pedersen, H. G., (2008). Flue gases from stationary sources. In: Ertl, G., Knözinger, H., Schüth, F., & Weitkamp, J., (eds.), *Handbook of Heterogeneous Catalysis* (2nd edn., Vol. 5, pp. 2345–2385). Wiley-VCH Verlag GmbH & Co. KGaA: Weinheim, Germany.

Hagen, J., (2006). *Industrial Catalysis–A Practical Approach* (2nd edn., pp. 1–221). Wiley-VCH Verlag GmbH & Co. KGaA, Weinheim, Germany.

Hansen, J. B., & Nielsen, P. E. H., (2008). Methanol synthesis. In: Ertl, G., Knözinger, H., Schüth, F., & Weitkamp, J., (eds.), *Handbook of Heterogeneous Catalysis*, (2nd edn., Vol. 6, pp. 2920–2949). Wiley-VCH Verlag GmbH & Co. KGaA: Weinheim, Germany.

Koel, B. E., & Kim, J., (2008). Promoters and poisons. In: Ertl, G., Knözinger, H., Schüth, F., & Weitkamp, J., (eds.), *Handbook of Heterogeneous Catalysis* (2nd edn., Vol. 3, pp. 1593–1624). Wiley-VCH Verlag GmbH & Co. KGaA: Weinheim, Germany.

Kozuch, S., & Martin, J. M. L., (2012). "Turning over" definitions in catalytic cycles. *ACS Catalysis, 2*, 2787–2794.

Krischer, K., & Savinova, E. R., (2008). Fundamentals of electrocatalysis. In: Ertl, G., Knözinger, H., Schüth, F., & Weitkamp, J., (eds.), *Handbook of Heterogeneous Catalysis* (2nd edn., Vol. 4, pp. 1873–1905). Wiley-VCH Verlag GmbH & Co. KGaA: Weinheim, Germany.

Lloyd, L., (2011). *Handbook of Industrial Catalysts, Fundamental and Applied Catalysis* (pp. 23–71). Springer US.

Lohse, W. H., (1945). *Catalytic Chemistry* (pp. 4–8). Chemical Publishing Co., Brooklyn, USA.

Lox, E. S. J., (2008). Automotive exhaust treatment. In: Ertl, G., Knözinger, H., Schüth, F., & Weitkamp, J., (eds.), *Handbook of Heterogeneous Catalysis* (2nd edn., Vol. 5, pp. 2274–2345). Wiley-VCH Verlag GmbH & Co. KGaA: Weinheim, Germany.

Moulijn, J. A., Van Diepen, A. E., & Kapteijn, F., (2008). Deactivation and regeneration. In: Ertl, G., Knözinger, H., Schüth, F., & Weitkamp, J., (eds.), *Handbook of Heterogeneous Catalysis* (2nd edn., Vol. 4, pp. 1829–1846). Wiley-VCH Verlag GmbH & Co. KGaA: Weinheim, Germany.

Näumann, F., & Schulz, M., (2008). Oxidation of sulfur dioxide. In: Ertl, G., Knözinger, H., Schüth, F., & Weitkamp, J., (eds.), *Handbook of Heterogeneous Catalysis* (2nd edn., Vol. 5, pp. 2623–2635.). Wiley-VCH Verlag GmbH & Co. KGaA: Weinheim, Germany.

Prieto, G., & Schüth, F., (2015). The Yin and Yang in the development of catalytic processes: Catalysis research and reaction engineering. *Angewandte Chemie International Edition, 54*, 3222–3239.

Prüße, U., Thielecke, N., & Vorlop, K. D., (2008). Catalysis in water remediation. In: Ertl, G., Knözinger, H., Schüth, F., & Weitkamp, J., (eds.), *Handbook of Heterogeneous Catalysis* (2nd edn., Vol. 5, pp. 2477–2500). Wiley-VCH Verlag GmbH & Co. KGaA: Weinheim, Germany.

Roduner, E., (2014). Understanding catalysis. *Chemical Society Reviews, 43*, 8226–8239.

Schlögl, R., (2008). Ammonia synthesis. In: Ertl, G., Knözinger, H., Schüth, F., & Weitkamp, J., (eds.), *Handbook of Heterogeneous Catalysis* (2nd edn., Vol. 5, pp. 2501–2575). Wiley-VCH Verlag GmbH & Co. KGaA: Weinheim, Germany.

Schlögl, R., (2015). Heterogeneous catalysis. *Angewandte Chemie International Edition, 54*, 3465–3520.

Somorjai, G., (1994). Catalysis by surfaces. In: John Wiley & Sons, (eds.), *Introduction to Surface Chemistry and Catalysis* (pp. 442–595). Inc., New York, USA.

Takeuchi, M., Kitano, M., Matsuoka, M., & Anpo, M., (2008). Photocatalysis: Development of highly functional titanium oxide photocatalysts. In: Ertl, G., Knözinger, H., Schüth, F., & Weitkamp, J., (eds.), *Handbook of Heterogeneous Catalysis* (2nd edn., Vol. 4, pp. 1958–1968). Wiley-VCH Verlag GmbH & Co. KGaA: Weinheim, Germany.

Thomas, J. M., & Thomas, W. J., (1997). *Principles and Practice of Heterogeneous Catalysis* (pp. 1–64). VCH Verlagsgesellschaft mbH, Weinheim, Germany.

Van Leeuwen, P. W. N. M., (2004). *Homogeneous catalysis–Understanding the Art* (pp. 1–73). Kluwer Academic Publishers, Dordrecht, the Netherlands.

Védrine, J. C., (2014). Revisiting active sites in heterogeneous catalysis: Their structure and their dynamic behavior. *Applied Catalysis A: General, 474*, 40–50.

Wolkenstein, T., (1960). The electron theory of catalysis on semiconductors. *Advances in Catalysis, 12*, 189–264.

CHAPTER 5

Catalytic Material

ADRIANA URDĂ and IOAN-CEZAR MARCU

Laboratory of Chemical Technology & Catalysis, Department of Organic Chemistry, Biochemistry & Catalysis, Faculty of Chemistry, University of Bucharest, 4-12, Blv. Regina Elisabeta, 030018 Bucharest, Romania, Tel: +40213051464, Fax: +40213159249, E-mail: ioancezar.marcu@chimie.unibuc.ro, ioancezar_marcu@yahoo.com

5.1 DEFINITION

Materials can be defined as solids with specific properties and therefore intended for certain applications. CMs are thus solids whose specific properties are their activity and their selectivity for a certain chemical reaction. They are also called solid catalysts or heterogeneous catalysts. The catalytic activity represents their ability to increase the rate of a thermodynamically possible chemical reaction without being consumed by the reaction itself. Indeed, in the presence of a solid catalyst the chemical reaction, which requires a high activation energy in the absence of the catalyst, takes place via a new mechanism involving several elementary steps, each of them requiring a lower activation energy. According to the Arrhenius equation:

$$k = A \exp\left(-\frac{E_a}{RT}\right) \tag{1}$$

where k is the rate constant, A is a constant called frequency factor, E_a is the activation energy, R is the gas constant, and T is the reaction temperature in K, a decrease of the activation energy will lead to an increase of the rate constant and, consequently, of the reaction rate. The catalyst selectivity represents its ability to choose a certain reaction among several reactions that could occur in the considered system of reactants. This catalyst function allows the desired reaction products to be obtained with minimal amounts of undesirable byproducts. Responsible for these catalytic properties are the active sites (or active centers) at the surface of the CM. They can be associated with atomic vacancies, single atoms or ensembles of atoms. Usually, CMs are complex high-surface area solids, amorphous or crystalline (in most of the cases polycrystalline), generally containing more than one component, with surface compositions and structures different from those of the bulk and susceptible to change during the catalytic process. The catalytically active materials include metals and metallic alloys, variable-valence metal oxides, acid-base oxides, sulfides, etc., and can be unsupported (bulk) or supported on porous carriers. For practical applications, in addition to its high activity and good selectivity, a solid catalyst must satisfy the requirement of stability, defined as its capacity to keep its catalytic performances for a long time under the reaction conditions. Notably, at present, 90% of all chemical processes use heterogeneous catalysts.

5.2   HISTORICAL ORIGIN(S)

The development of CMs is strongly related to the development of the science of catalysis and, particularly, to the development of industrial heterogeneous catalysis. These two interrelated histories are clearly and concisely presented by Davis (2008) and Dingerdissen et al., (2008), respectively, while the first industrial catalysts together with the corresponding catalytic processes are described by Lloyd (2011). Berzelius introduced the concept of catalysis in 1835, calling it "catalytic force," and fifty years later Armstrong called the substance responsible for the catalytic action "catalyst" and pointed out the importance of studying not only the catalytic reaction but also the catalyst (Davis, 2008). Notably, even before the Berzelius' definition of catalysis, in 1831, Phillips patented the oxidation of SO_2 to SO_3 over a Pt-based catalyst (Dingerdissen et al., 2008). In 1868, one of the first industrial catalysts was patented to be especially designed rather than discovered empirically, the Deacon catalyst, consisting in $CuCl_2$ dispersed on a porous and heat-resistant support (Lloyd, 2011). It was used in the catalytic oxidation of hydrochloric acid with air to produce chlorine. An important step forward in understanding the catalytic phenomena has been done at the end of the 19th century with Ostwald that performed systematic studies in catalysis and for which he was awarded the Nobel Prize for chemistry in 1909. He showed that a catalyst can speed up only a thermodynamically possible reaction and cannot shift the equilibrium (Davis, 2008). Also, Ostwald designed the first industrial process of oxidation of ammonia to nitric acid over platinum gauze catalysts (Dingerdissen et al., 2008). A key breakthrough in developing CMs was marked in 1910 by Mittasch and coworkers from BASF that obtained after systematic and extensive investigations the iron-based industrial catalyst used for the ammonia synthesis, which remains basically the same today (Davis, 2008). This is a complex CM containing both electronic and textural promoters in addition to the active phase. Notably, after these studies, Mittasch explained the effects of different catalyst modifiers and clearly defined the concepts of promoter, co-activator, support, and poison (Davis, 2008). Great advancements in both fundamental and applied heterogeneous catalysis are due to the work of Sabatier on the hydrogenation of different organic compounds in the presence of finely divided metal catalysts for which he was awarded the Nobel Prize for chemistry in 1912 (Davis, 2008). One of the main findings of Sabatier was one of the principles of heterogeneous catalysis which state that the intermediate compound formed between the reactant(s) and the catalyst surface must be stable enough to be formed and labile enough to decompose into the reaction product(s). He also demonstrated the selectivity of catalytic action and introduced the use of supports showing the enhanced activity of the supported catalysts. Several years later Langmuir's studies on chemisorption highlighted its role in heterogeneous catalysis and showed that the catalytic act takes place on the catalyst surface in a single layer and involves the same chemical forces that held molecules and solids together (Davis, 2008). He was awarded the Nobel Prize for chemistry in 1932 "for his discoveries and investigations in surface chemistry." In 1925–26 Taylor introduced the concept of the active site and evidenced the heterogeneity of the surface of a solid catalyst (Davis, 2008). With this starts the study of the CM, as suggested earlier by Armstrong, in order to establish correlations between its physicochemical characteristics and its catalytic properties. In the first half of the 20th century, a high number of CMs was developed for a large variety of chemical processes, some of them remaining basically the same today. The later advancements in the solid-state chemistry, physical chemistry, and theoretical physics, as well as the development of physical techniques for the characterization of solids, allowed the deep understanding of the heterogeneous catalytic phenomena and the improvement on scientific bases of the existing CMs as well as the creation of new highly active and selective CMs.

5.3 NANO-SCIENTIFIC DEVELOPMENT(S)

Most of CMs are complex systems, amorphous or crystalline, consisting of several components and phases. Traditionally they are classified according to the chemical nature of the active component, in metals, metal oxides, and metal sulfides. Division of metal oxide-based catalysts according to their acid-base and redox properties in acid-base and redox (variable-valence) metal oxides is also used. Yet, all of them can be unsupported (bulk) or supported on porous carriers. Although widely used, this traditional classification of heterogeneous catalysts does not include polymers (ion exchangers), carbides, nitrides and phosphides, metal-organic frameworks and heterogenized molecular catalysts including metal complexes, organometallic compounds, small organic molecules (organocatalysts) or enzymes immobilized on porous solid supports.

The metals active in catalysis are the transition metals having partially filled d orbitals that allow the chemisorption of reactants on their surfaces (Hagen, 2006). Among these, particularly good catalysts are the group VIIIb transition metals, as they are fully satisfying the Sabatier's principle which states that for a good solid catalyst the reactant or surface complex chemisorption should have an intermediate strength (Hagen, 2006).

However, metals like gold, which are usually bad catalysts, were shown to be good catalysts for certain reactions when used as nanoparticles (Haruta, 2004). Compared to bulk metals, metallic nanoparticles provide a larger fraction of low-coordinated surface atoms associated with the chemisorption and catalytic sites. Also, nanoparticles have a quite different electronic structure with ionization potential and electron affinity, properties that control bond formation, between those of individual atoms and bulk metal (Roduner, 2006).

Metals and metallic alloys can be used in unsupported form or, mainly, supported on porous carriers. There is only a small number of bulk metallic catalysts, but they are industrially important. They are prepared by the fusion of the precursors (metals or even oxides). A typical example is the Pt-Rh gauze used for the oxidation of ammonia to nitric oxide. Another relevant example is the multiple promoted iron catalyst for the ammonia synthesis. In this case, the fusion of magnetite with a few percent K_2O, Al_2O_3, and CaO are followed by the reduction in a H_2/N_2 flow (Thomas & Thomas, 1996). The crystalline α-Fe phase with the (111) face being prominent is the active component of the resulting CM (Thomas & Thomas, 1996). Notably, this CM shows a remarkable longevity likely due to its quasi-liquid nature allowing a constant regeneration of the surface atoms with continuous removal of traces of surface poisons during catalysis (Thomas & Thomas, 1996). It is also worth mentioning the skeletal or Raney-type metallic catalysts which are mainly based on Ni, Cu, and Co with or without promoters, being specifically applied for liquid-phase hydrogenation reactions (Smith &

Wainwright, 2008). They are prepared by the selective removal of aluminum from an active metal–aluminum alloy by alkali leaching. In this way, small particles of skeletal metallic catalysts composed of agglomerated nanometric crystallites are obtained. They are highly active but have a relatively low lifetime (Smith & Wainwright, 2008).

In most of the metal-based CMs, the metallic active component is dispersed on porous materials in order to obtain a high active surface area per unit volume of catalyst. Thus, supported metal catalysts consist of metal nanoparticles dispersed on the surface of a support, which in most of the cases is a high surface area oxide material. The most frequently used method for the preparation of supported catalysts is impregnation, including ion-exchange (Marceau et al., 2008). Silicas, including ordered mesoporous silicas, aluminas, silica-aluminas, zeolites and zeotypes, magnesia, titania, different forms of carbon, etc., are widely used supports. They not only provide a large surface area but also can stabilize the active surface area by preventing the sintering of metallic particles. Moreover, the oxide support can have several effects on the characteristics of metallic nanoparticles, finally affecting their catalytic properties (Ruppert & Weckhuysen, 2008). This metal-support interaction strongly depends on both the nature of the supported metal and the oxide support. For a given metal, the nature of the support is crucial. For example, in the case of metals supported on reducible oxides like titania, reduced TiO_x particles migrate from the support, partially covering the metal particles and blocking the active sites, with consequences on their catalytic performances (Ruppert & Weckhuysen, 2008). This effect, called strong metal-support interaction, was not observed for non-reducible oxides supports, such as silica, alumina, zirconia, etc. Also, there are supports containing acidic or basic sites having their own catalytic activity for a certain acid- or base-catalyzed reaction step of the overall process (see below). In this case, the CM is named bifunctional catalyst. It is to be noted that in some cases, such as alumina-supported Ag catalyst used for the selective oxidation of ethylene into ethylene oxide, this effect is undesirable as the desired product can be isomerized to acetaldehyde on the protonic acid sites present on the support, the latter being then rapidly oxidized on Ag into CO_2 and H_2O with negative consequences on the reaction selectivity. In this case, for suppressing the total oxidation reaction and thus improving the epoxide selectivity, the acidity of alumina is neutralized with alkali metal ions (van Santen & Neurock, 2006).

In principle, increasing the dispersion of the metallic particles on the support surface, the particle size decreases, leading to an increased active surface area and, thus, to a higher number of active sites (Ruppert &

Weckhuysen, 2008). Moreover, by decreasing the size of a nanoparticle, the surface atoms are placed on edges and corners rather than terraces, being more coordinatively unsaturated and thus more reactive, resulting in an increased catalytic activity (van Santen & Neurock, 2006). In other cases, the active sites can be located at the metal particle–support interface, which increases by decreasing the particle size and leads to an increased catalytic activity (van Santen & Neurock, 2006). This is called a positive particle size effect (Semagina & Kiwi-Minsker, 2009). However, the large nanoparticles can be more active than the small ones and thus decreasing the particle size can lead to a decreased catalytic activity. This is called a negative particle size effect. In this case, the catalytic reaction takes place on active sites composed of specific ensembles of surface atoms, which only appear for particle sizes above a certain limit. Finally, in some cases, the catalytic activity passes through a maximum for an optimum particle size (Semagina & Kiwi-Minsker, 2009). Notably, for particle sizes in the nanometer size scale, a shift from the band structure of the metal to a molecular-like structure with discrete energy differences between molecular orbitals can take place. This corresponds to a metal to insulator transition being known as a quantum size effect (van Santen & Neurock, 2006). This effect allowed to explain the unique reactivity of 3.5 nm Au particles supported on TiO_2 (van Santen & Neurock, 2006).

The metal oxide-based catalysts form the largest group of heterogeneous catalysts, being used for their both acid-base and redox properties.

The variable-valence metal oxides can easily form mixed-valence and non-stoichiometric compounds, which together with the variability of the metal oxidation state are responsible for their redox catalytic properties. They are semiconducting materials. The origin of their non-stoichiometry is a direct result of the existence of point defects, the surface point defects being associated with the active catalytic sites (Misono, 2013). They are used in catalytic oxidation and dehydrogenation, mainly as supported catalysts and mixed metal oxides. Notably, the CMs industrially used for selective oxidation are based mostly on vanadium and molybdenum. Vanadia-based catalysts are remarkable redox CMs. Bulk V_2O_5, yet promoted, is used for the oxidation of naphthalene into naphthoquinone (Haber, 2008). V_2O_5 is mainly used as an active component dispersed on appropriate supports, such as titania, alumina, or silica. Vanadia-based catalysts were used in the first industrial organic gas-phase oxidation processes, including the oxidations of benzene to maleic anhydride and naphthalene to phthalic anhydride, and are still used today, i.e., for the oxidation of *o*-xylene to phthalic anhydride. The current catalyst used in the latter reaction is V_2O_5 supported on titania (anatase) promoted with alkali ions, Sb, and P (Busca, 2014). Structurally, it consists of a vanadium oxide monolayer enveloping the grains of the support (Haber, 2008) and belongs to the group of monolayer oxide catalysts. It is worth mentioning the silica-supported V_2O_5-K_2SO_4 catalyst that replaced the more easily poisoned Pt in the contact process. In this case, under operating conditions, the active composition is present as a melt in the pores of the silica support (Näumann & Schulz, 2008). It is assumed that two different V^{4+} species are present in the CM, one being an active soluble species which participates in the redox reaction and the other, an inactive, insoluble species (Näumann & Schulz, 2008). Vanadium remains a key component of the mixed oxide catalysts used for catalytic oxidation. One of these is the so-called VPO catalyst used for the selective oxidation of *n*-butane to maleic anhydride (Schunk, 2008). It mainly consists of orthorhombic vanadyl pyrophosphate $(VO)_2P_2O_7$, but a slight excess of P with respect to the stoichiometry is required for high catalytic performances (Busca, 2014). Another mixed oxide containing V as a key activating element is the complex MoVNbTeO system with outstanding performances in the oxidation of propane to acrylic acid as well as the ammoxidation of propane to acrylonitrile. It has the general empirical formula $MoV_{0.25}Nb_{0.20}Te_{0.17}O_x$ and consists of at least two phases, M1 and M2, which are capable of symbiotic interaction resulting in enhanced acrylic acid yields (Grasselli & Burrington, 2008). Molybdenum-containing industrially relevant redox CMs are bismuth molybdate-based systems employed for the selective oxidation of propane and isobutene to acrolein and methacrolein, respectively, and for the ammoxidation of propane to acrylonitrile (Busca, 2014). The chemical formula of bismuth molybdates is $Bi_2O_3 \cdot nMoO_3$ where n = 3, 2 or 1, corresponding to the α, β and γ phases, respectively, each of them with different structures and, consequently, different catalytic performances for a given reaction. Actually, the industrial catalysts are multicomponent molybdates also containing Fe, Co, Ni, and other elements. Notably, different metallic cations play different roles during catalysis. For example, it has been shown that Bi^{3+}, Sb^{3+} and/or Te^{4+} are involved in the activation of propane, Mo^{6+} and/or Sb^{5+} play the role of oxygen and nitrogen insertion sites, and the presence of Fe^{2+}/Fe^{3+} or Ce^{4+}/Ce^{3+} redox couples favor the diffusion of lattice oxygen (Busca, 2014). Polyoxometallates containing vanadium, molybdenum or tungsten and different other elements acting as promoters represent another class of CMs used in selective oxidation. For example, silica-supported $Cu_aV_b(Sn,Sb)_cW_dMo_{12}O_x$ catalyst is currently used for the oxidation of acrolein into acrylic acid (Grasselli & Burrington, 2008).

The oxide-based CMs are also used for the complete oxidation of volatile organic compounds (VOCs). They mainly contain manganese, copper, and/or chromium, the manganese-based ones, such as MnO_x/Al_2O_3, being commercially used for catalytic combustion of oxygenated VOCs (Busca, 2014). Transition metal-containing complex oxides, such as perovskites and spinels, as well as ceria-based materials were shown to be active catalysts for total oxidation, some of them having outstanding activities in soot combustion (Busca, 2014). It is worth mentioning the copper chromite spinel $(CuCr_2O_4)$, which is a highly active catalyst for CO oxidation, exhibiting comparable activity to that of noble metals (Prasad & Singh, 2012).

The redox CMs are operating according to a heterogeneous redox mechanism proposed by Mars and van Krevelen (1954) for the oxidation of aromatic hydrocarbons over V_2O_5-based catalysts and then confirmed for most of variable-valence metal oxide-based CMs. According to this mechanism, the substrate molecule (S) is oxidized by the catalyst (Cat-O), and then the reduced catalyst (Cat) is reoxidized by gas-phase dioxygen:

$$S + Cat\text{-}O \rightarrow S\text{-}O + Cat$$
$$Cat + \tfrac{1}{2}O_2 \rightarrow Cat\text{-}O$$

In fact, the Mars-van Krevelen mechanism involves the consumption by the substrate of surface lattice oxygen from the catalytic site able to activate it, and the oxygen vacancy formed being replenished either directly by gas-phase oxygen or by diffusion of bulk lattice oxygen to the surface. In the latter case, gas-phase oxygen is not incorporated into the oxide at the site reduced by the substrate, but at a different one associated to an oxygen vacancy formed by diffusion of lattice oxygen (Haber, 2008). However, not all the surface transformations take place via the Mars-van Krevelen redox mechanism. The oxide surface may be populated by different chemisorbed oxygen species, which may attack all the reaction intermediates leading to total oxidation and, therefore, to a decrease of selectivity in partial oxidation (Haber, 2008). Several key factors, such as the lattice oxygen mobility, the metal-oxygen bond strength, the host structure, the redox properties, the multifunctionality of the active sites, the active site isolation and the phase cooperation, determine the catalytic behavior of the metal oxides and, consequently, have to be considered in the design of oxidation catalysts (Grasselli, 2002).

Some transition metal oxides stable under reducing atmospheres, such as Cr_2O_3 and Fe_2O_3, are used as active components of the dehydrogenation catalysts (Misono, 2013). For example, Fe_2O_3 promoted with K_2O, which increases both the activity and the durability, is industrially used in the dehydrogenation of ethylbenzene in the presence of steam (Rase, 2000).

Also, alumina-supported Cr_2O_3 is used for the dehydrogenation of butanes into butenes (Rase, 2000). In this case, alumina is not only a catalyst support, but also forms mixed oxide phases with the active component (Hagen, 2006a). It is worth noting that, as in the case of supported metallic catalysts, the oxide active phase is in close interaction with the support, the nature of the latter having a significant effect on the catalytic properties of supported oxide catalysts (Bergwerff & Weckhuysen, 2008).

Amorphous silicas, aluminas, silica-aluminas, zeolites, magnesia as well as layered double hydroxides and the mixed oxides obtained by their thermal decomposition are representative *acid-base catalysts and catalyst supports.*

Because of the low Brønsted acidity of their surface hydroxyl groups, the amorphous silicas are not used as catalysts, but they are widely used as supports for various active components including metals, oxides, and molecular species. The latter are immobilized on the pore walls of silica by direct interaction with surface silanol groups (grafting) or via a "linker" first attached to the surface (anchoring) (Averseng et al., 2008). As a function of the method of preparation used, different amorphous silicas, including silica gel, precipitated silica and pyrogenic silica, with different physico-chemical characteristics can be obtained (Setzer et al., 2002). For example, the pyrogenic silicas are nonporous while silica gels and precipitated silicas are highly porous, but with different pore size distributions: very broad for the first and narrow for the latter. Yet, xerogels, obtained by a simple evaporation of the solvent, have much lower mean pore sizes than aerogels, obtained by supercritical solvent release (Setzer et al., 2002). It is worth noting that ordered mesoporous silicas can be obtained via a sol-gel process in the presence of different micellar aggregates as structure-directing agents. These materials contain ordered mesopores, with amorphous silica walls, having different topologies and tunable narrow size distributions (Kleitz, 2008). They are also characterized by very high surface areas and large pore volumes. Due to their large pore sizes, the ordered mesoporous silicas are excellent supports for catalysts used in processes involving large molecules (Taguchi & Schüth, 2004).

Among the different aluminas known, the so-called transition aluminas are used as catalysts and, mainly, as catalyst supports (Euzen et al., 2002). The transition χ-, κ-, γ-, δ-, θ- and η-Al_2O_3 are high surface area, poorly crystalline transitional oxide phases in the process of thermal transformation of aluminum trihydroxides and oxyhydroxides into the crystalline, nonporous, thermodynamically stable α-Al_2O_3 (corundum). The latter can also be used as a catalyst support in applications where low surface areas are desired, as in partial oxidation reactions (Schüth et al., 2008). The surface of aluminum oxides contains coordinatively unsaturated oxygen and aluminum atoms as well as hydroxyl groups being, respectively, associated to Lewis basic, Lewis acid and Brønsted acid sites (Euzen et al., 2002). In some cases, such as Pt/γ-Al$_2$O$_3$ reforming catalysts, alumina plays not only the role of support but also acts as an active component for the isomerization and cyclization reactions due to its surface acidity enhanced by chloride. The other active phase, Pt, introduced by impregnation of alumina support, provides the active sites for the dehydrogenation of alkanes to alkenes and hydrogenation of alkenes back to alkanes. This is a well-known example of a bifunctional catalyst (Euzen et al., 2002). On the other hand, the addition of basic compounds to alumina results in solid base catalysts, such as MgO-Al$_2$O$_3$, Na/Al$_2$O$_3$, Na/NaOH/Al$_2$O$_3$, KF/Al$_2$O$_3$, NaNH$_2$/Al$_2$O$_3$ (Misono, 2013), active for a series of organic reactions (Hattori, 2015). It is to be noted that organized mesoporous aluminas with tailored surface areas, pore volumes and sizes have been successfully synthesized. They were mainly used as catalysts supports for different active components, the CMs thus obtained being more active than those prepared with conventional aluminas supports (Márquez-Alvarez et al., 2008).

The coprecipitation of silica and alumina leads to amorphous silica-alumina mixed oxides having an acidic character. They replaced the first solid acid catalysts used in the catalytic cracking, i.e., the acid-treated clay-type materials, for being then substituted for more acidic and, therefore, more active faujasite zeolites (Corma & Martinez, 2002). However, they are still used as acid supports in some hydrocracking processes (Corma & Martinez, 2002). They possess mainly low to medium strength Brønsted acidity associated with the presence of tetrahedrally coordinated aluminum, which creates a negative charge that can be compensated by a proton (Corma & Martinez, 2002). Thus, their acidity depends on the alumina content. Notably, a high surface area mesoporous silica-alumina (MSA) with a narrow pore-size distribution has been successfully prepared and used as support in order to obtain Pt/MSA bifunctional catalysts. They gave good results in the hydroisomerization-hydrocracking of long-chain *n*-paraffins (Corma & Martinez, 2002).

Zeolites (see ZEOLITE) are microporous crystalline aluminosilicates possessing both Brønsted and Lewis acid sites. They have a three-dimensional structure formed by corner-sharing SiO$_4^{4-}$ and AlO$_4^{5-}$ tetrahedra. The excess negative charge of AlO$_4^{5-}$ tetrahedra is compensated by an extra framework, exchangeable cations. When these cations are protons, OH groups bridging Si and Al species are formed being associated with the Brønsted acid sites (Karge, 2008). The Lewis acid sites mainly correspond to coordinatively unsaturated Al species, i.e., threefold-coordinated, obtained by the dehydroxylation of the hydrogen form of a zeolite (Karge, 2008). Due to their strong acidity, zeolites are important acid catalysts for several oil refining and petrochemical processes including catalytic cracking, catalytic reforming (where they play both the role of support for the hydrogenating-dehydrogenating Pt component and the role of acid component in the bifunctional Pt/zeolite catalysts), methanol-to-gasoline process, etc. Zeolites also contain low-strength Lewis basic sites associated with the lattice oxygen species. Strong Lewis basic sites can be created in zeolites by ion-exchange with alkali cations, such as Na^+, K^+, Cs^+, or by additional loading with oxide particles, such as Cs_2O (Hattori, 2015). Notably, due to their well-defined pore sizes, zeolites are shape-selective catalysts: the selectivity of a reaction is determined by the pore size or pore architecture of the catalyst (Traa & Weitkamp, 2002). There are three types of shape selectivity: reactant, product, and restricted transition state selectivity. The first two selectivity effects are mass transfer shape selectivities as they are due to the hindered diffusion of reactant and product molecules, respectively. The last selectivity effect consists in completely suppressing of a chemical reaction involving bulky transition states and intermediates due to the limited space around the intracrystalline active sites (Traa & Weitkamp, 2002; Csicsery, 1986).

Aluminum and silicon, together with phosphorus, can form other zeolite-like materials such as aluminophosphates (AlPOs) and silicoaluminophosphates (SAPOs), named zeotypes (Corma & Martinez, 1995). Also, crystalline silicas analogs of silica-rich zeolites like ZSM-5 or ZSM-11, called silicalite-1 and -2, respectively, are known (Flanigen et al., 1978; Bibby et al., 1979). AlPOs and silicalites have no acidic properties, but due to partial substitution of P by Si in AlPOs, acidity appears in SAPOs. Due to its optimal structure for isomerization and mild acidity, a member of the latter family is an acidic carrier of the bifunctional Pt or Pd/SAPO-11 catalysts industrially used in the iso-dewaxing process (van Veen et al., 2008). On the other hand, transition metals like Ti, V, Cr, Co, etc. can be incorporated in all these structures, including typical zeolites (Clerici & Domine, 2013; Selvam & Sakthivel, 2013; Kholdeeva, 2013), leading to highly active and selective liquid-phase oxidation catalysts such as titanium-silicalite-1 (TS-1) (Smith & Notheisz, 1999).

Magnesia (MgO) is a typical basic solid, the surface hydroxyl groups as well as O^{2-} species being the basic sites. The thermal dehydroxylation of the surface leads to coordinatively unsaturated O^{2-} species which are very strong basic sites. The lower the coordination number, the higher the basic strength. The coordination number of O^{2-} ions on corners, edges and flat surfaces is three, four, and five, respectively. It is to be noted that on the magnesia surface there are also coordinatively unsaturated Mg^{2+} ions acting as weak Lewis acid sites. The basic and Lewis acid sites mostly work synergistically, being thus more relevant to talk about acid-base pairs, i.e., Mg^{2+}-O^{2-}. Apart from magnesia, strong solid bases are also SrO, BaO, and lanthanide oxides such as La_2O_3 and Sm_2O_3 (Hattori, 2015). For all solid base catalysts, the catalytic activity passes through a maximum as a function of pretreatment temperature; firstly, the basicity increases due to the removal of water and carbon dioxide from the surface, then it decreases because of the loss of basic sites due to the surface rearrangement of atoms (Hattori, 2015). These oxides together with other base materials are used as base catalysts for several organic reactions such as alkylation, isomerization, condensation, esterification, etc. (Ono & Baba, 1997; Tichit et al., 2006; Hattori, 2015). They are also used in bifunctional catalysts for processes involving base-catalyzed steps such as metathesis (Lummus olefin conversion technology), ethanol conversion into butadiene, etc. (Hattori, 2015). It is to be noted that mesoporous MgO and MgAl mixed oxides were prepared using methods similar to that used for obtaining mesoporous silicas (Sun et al., 2015).

The layered double hydroxides (LDHs) or hydrotalcite-like materials are anionic clay minerals with general formula $M^{2+}_{1-x}M^{3+}_x(OH)_2A^{n-}_{x/n}\cdot mH_2O$ with M^{2+} = Mg^{2+}, Cu^{2+}, Ni^{2+}, Co^{2+}, Zn^{2+}, Fe^{2+} and Mn^{2+}, M^{3+} = Al^{3+}, Ga^{3+}, Fe^{3+}, Cr^{3+}, Co^{3+}, etc., A^{n-} = any anionic species and $0.2 \leq x \leq 0.4$ (see Layered Double Hydroxide). In the naturally occurring parent hydrotalcite material M^{2+} = Mg^{2+}, M^{3+} = Al^{3+}, A^{n-} = CO_3^{2-}, $x = 0.25$, and $m = 4$. LDHs can contain more than two different cations homogeneously distributed and intimately mixed together. Depending on their composition, which is highly flexible, they can possess both acid-base and redox properties. They can be used in catalysis as such or, mainly, in the form of mixed oxides obtained by their controlled thermal decomposition. The latter is highly homogeneous, with large surface areas and small crystallite sizes, possess high basicity and memory effect (Cavani et al., 1991). The transition metal-containing LDHs-derived mixed oxides also show redox properties and, by their reduction, small and thermally stable metal crystallites dispersed in a mixed oxide matrix can be formed (Cavani et al., 1991). They are used as solid base catalysts in several organic reactions (Sels et al., 2001; Tichit & Coq, 2003; Fan et al., 2014) and as redox catalysts in selective or total oxidation reactions (Marcu et al., 2017). They are also used as a catalyst supports and bifunctional catalysts (Zhang et al., 2008; Fan et al., 2014).

It is to be noted that the oxides of elements with valence five or higher, such as vanadia, niobia, molybdena, and tungsta present strong to very strong Brønsted acidity (Busca, 2006) which make them good candidates as solid acid catalysts (Mitran et al., 2013, 2015). Also, very strong acidity (superacidity) was reported for sulfated zirconia and titania, as well as for heteropoly compounds like $H_4SiW_{12}O_{40}$ and $Cs_{2.5}H_{0.5}PW_{12}O_{40}$ (Misono, 2013). For example, due to its superacidity sulfated zirconia was shown to be active for *n*-butane isomerization at temperatures as low as room temperature (Jentoft, 2008).

The *sulfide-based catalysts* constitute a small class of heterogeneous catalysts but of high industrial importance as they are used in the hydrotreating processes in the petroleum refining industry. They are alumina-supported metal oxides of Co-Mo, Ni-Mo and Ni-W transformed into sulfides *in situ* in a stream of H_2/H_2S, thiophene, or a feed containing sulfur compounds reacting relatively easily (Rase, 2000a). They provide hydrogenation and hydrogenolysis activity for the reactions taking place during the hydrotreating process including hydrodesulfurization, hydrodenitrogenation, hydrodeoxygenation, hydrodemetallization, hydrodearomatization, and olefin hydrogenation (Rase, 2000a; Prins, 2008). For example, the Co-Mo and Ni-Mo active catalysts contain Co_9S_8, Ni_3S_2, and MoS_2 phases supported on alumina. MoS_2 crystallites represent the active phase promoted by Co and Ni ions, the latter contributing to the activity itself as sulfides (Rase, 2000a). The catalytic sites are associated to surface Mo cations located at edges and corners of MoS_2 crystallites neighboring surface sulfur vacancies (Prins, 2008).

In the discussion above, we only considered the "workhorses" of heterogeneous catalysis, i.e., metals, metal oxides, and metal sulfides (Alexander & Hargreaves, 2010). However, it is to be noted that ion-exchange resins, such as sulfonated crosslinked polystyrene, are industrially used in acid-catalyzed processes and are finding increasing application as replacements for soluble mineral acid catalysts (Gates, 2008). Also, other materials such as carbides, nitrides, and phosphides (Oyama, 2008; Alexander & Hargreaves, 2010), metal-organic frameworks (Chughtai et al., 2015), etc., are intensively studied as alternative heterogeneous catalysts.

5.4 NANOCHEMICAL APPLICATION(S)

The solid catalysts are traditionally used in petrochemical and chemical industries, most of the large-scale industrial processes being carried out only in their presence. The most important industrial applications of heterogeneous catalysts are summarized in Table 5.1 (Rase, 2000b).

TABLE 5.1   The Most Important Industrial Applications of CMs

Process	Catalytic material
Hydrogenations	Unsupported (Raney) or supported base metals (Ni, Cu, Co) and supported precious metals (Pt, Pd, Rh, Ru)
Catalytic reforming of naphthalene	γ-alumina-supported Pt, Pt-Re, Pt-Sn, Pt-Ir
Steam reforming of methane	Ni-supported on ceramic materials
Methanation of carbon oxides	Ni on refractory carriers
Ammonia synthesis	Multiply promoted Fe
Ammonia oxidation to NO in nitric acid technology	Pt-Rh wire gauze
Ethylene oxidation into ethylene oxide	α-alumina-supported Ag
Methanol oxidation into formaldehyde	Ag or Fe_2O_3-MoO_3
High and low-temperature shift conversion	Fe_2O_3-Cr_2O_3 and Cu-ZnO, respectively
Sulfur dioxide oxidation to SO_3 in sulfuric acid technology	Silica-supported V_2O_5-K_2SO_4
Propylene ammoxidation into acrylonitrile	Multicomponent molybdates
Catalytic cracking of heavy petroleum fractions	Zeolites
Alkylations	Solid acids: zeolites, BF_3/Al_2O_3, $H_3PO_4/$Kieselguhr
Hydrotreating of different petroleum fractions	γ-alumina-supported Co-Mo-S, Ni-Mo-S, Ni-W-S

Nowadays, CMs are also used in new processes from different industrial areas such as environmental protection, power generation, fuel cell technology, fine chemicals for cosmetics and pharmaceuticals, etc.

5.5   MULTI-/TRANS-DISCIPLINARY CONNECTION(S)

Usually, CMs are constituted by nanometer-sized particles with different compositions, structures, and shapes. These characteristics strongly influence their catalytic properties and are determined by the method of catalyst preparation, generally involving a series of physical and chemical steps. In order to obtain information about the physicochemical characteristics of the prepared CM, it is characterized by using a series of physical techniques, such as XRF,

XRD, XPS, nitrogen adsorption at –196°C, TPD, TPR, chemisorption studies, electron microscopy, and different spectroscopies, etc. Catalytic tests on the laboratory scale allow obtaining information about their catalytic properties for a certain reaction. For describing the catalytic site and understanding the mechanism of their functioning during catalysis, CMs are characterized by using advanced *in situ* and *operando* techniques. Understanding the heterogeneous catalytic phenomenon at the molecular and nanometric level allows improving the CM in order to be industrially employed. At the industrial level, the CM in the form of extrudates, spheres, pellets or other shapes possessing required mechanical properties is used in a chemical reactor operated in the optimum reaction conditions in order to obtain the desired performances. Thus, from their design and preparation to their industrial application, several scientific disciplines are involved, such as materials chemistry, physical chemistry, physics, spectroscopy, computational chemistry, materials science, and chemical engineering, pointing out the complexity of CMs.

5.6 OPEN ISSUES

High-precision control of nanoscale characteristics and structure by using synthesis and material-characterization methods with nanometer accuracy represents the way of choice for improving the existing CMs and to develop new highly active and selective ones (Zečević et al., 2015).

KEYWORDS

- catalyst support
- catalytic activity
- catalytic material
- heterogeneous catalyst

REFERENCES AND FURTHER READING

Alexander, A. M., & Hargreaves, J. S. J., (2010). Alternative catalytic materials: Carbides, nitrides, phosphides, and amorphous boron alloys. *Chemical Society Reviews, 39,* 4388–4401.

Averseng, F., Vennat, M., & Che, M., (2008). Grafting and anchoring of transition metal complexes to inorganic oxides. In: Ertl, G., Knözinger, H., Schüth, F., & Weitkamp, J., (eds.), *Handbook of Heterogeneous Catalysis* (2nd edn., Vol. 1, pp. 522–539). Wiley-VCH: Weinheim, Germany.

Bergwerff, J. A., & Weckhuysen, B. M., (2008). Oxide-support interactions. In: Ertl, G., Knözinger, H., Schüth, F., & Weitkamp, J., (eds.), *Handbook of Heterogeneous Catalysis* (2nd edn., Vol. 2, pp. 1188–1197) Wiley-VCH: Weinheim, Germany.

Bibby, D. M., Milestone, N. B., & Aldridge, L. P., (1979). Silicalite–2, a silica analogue of the aluminosilicate zeolite ZSM–11. *Nature, 280,* 664–665.

Busca, G., (2006). The surface acidity and basicity of solid oxides and zeolites. In: Fierro, J. L. G., (ed.), *Metal Oxides: Chemistry and Applications* (pp. 247–318). CRC Press: Boca Raton, Florida, USA.

Busca, G., (2014). *Heterogeneous Catalytic Materials: Solid State Chemistry, Surface Chemistry and Catalytic Behavior* (pp. 375–419)*.* Elsevier: Amsterdam, Netherlands.

Cavani, F., Trifirò, F., & Vaccari, A., (1991). Hydrotalcite-type anionic clays: Preparation, properties, and applications. *Catalysis Today, 11,* 173–301.

Chughtai, A. H., Ahmad, N., Younus, H. A., Laypkov, A., & Verpoort, F., (2015). Metal-organic frameworks: Versatile heterogeneous catalysts for efficient catalytic organic transformations. *Chemical Society Reviews, 44,* 6804–6849.

Clerici, M. G., & Domine, M. E., (2013). Oxidation reactions catalyzed by transition-metal-substituted zeolites. In: Clerici, M. G., & Kholdeeva, O. A., (eds.), *Liquid Phase Oxidation via Heterogeneous Catalysis: Organic Synthesis and Industrial Applications* (pp. 21–94). Wiley: Hoboken, New Jersey, USA.

Corma, A., & Martinez, A., (1995). Zeolites and zeotypes as catalysts. *Advanced Materials, 7,* 137–144.

Corma, A., & Martinez, A., (2002). Catalysis on porous solids. In: Schüth, F., Sing, K. S. W., & Weitkamp, J., (eds.), *Handbook of Porous Solids* (Vol. 5, pp. 2825–2922). Wiley-VCH: Weinheim, Germany.

Csicsery, S. M., (1986). Catalysis by shape selective zeolites–science and technology. *Pure and Applied Chemistry, 58,* 841–856.

Davis, B. H., (2008). Development of the science of catalysis. In: Ertl, G., Knözinger, H., Schüth, F., & Weitkamp, J., (eds.), *Handbook of Heterogeneous Catalysis* (2nd edn., Vol. 1, pp. 16–37). Wiley-VCH: Weinheim, Germany.

Dingerdissen, U., Martin, A., Herein, D., & Wernicke, H. J., (2008). The development of industrial heterogeneous catalysis. In: Ertl, G., Knözinger, H., Schüth, F., & Weitkamp, J., (eds.), *Handbook of Heterogeneous Catalysis* (2nd edn., Vol. 1, pp. 37–56). Wiley-VCH: Weinheim, Germany.

Euzen, P., Raybaud, P., Krokidis, X., Toulhoat, H., Le Loarer, J. L., Jolivet, J. P., & Froidefond, C., (2002). Alumina. In: Schüth, F., Sing, K. S. W., & Weitkamp, J., (eds.), *Handbook of Porous Solids* (Vol. 3, pp. 1591–1677). Wiley-VCH: Weinheim, Germany.

Fan, G., Li, F., Evans, D. G., & Duan, X., (2014). Catalytic applications of layered double hydroxides: Recent advances and perspectives. *Chemical Society Reviews, 43,* 7040–7066.

Flanigen, E. M., Bennett, J. M., Grose, R. W., Cohen, J. P., Patton, R. L., Kirchner, R. M., & Smith, J. V., (1978). Silicalite, a new hydrophobic crystalline silica molecular sieve. *Nature, 271,* 512–516.

Gates, B. C., (2008). Catalysis by ion-exchange resins. In: Ertl, G., Knözinger, H., Schüth, F., & Weitkamp, J., (eds.), *Handbook of Heterogeneous Catalysis* (2nd edn., Vol. 1, pp. 278–285). Wiley-VCH: Weinheim, Germany.

Grasselli, R. K., & Burrington, J. D., (2008). Oxidation of low-molecular-weight hydrocarbons. In: Ertl, G., Knözinger, H., Schüth, F., & Weitkamp, J., (eds.), *Handbook of Heterogeneous Catalysis* (2nd ed., Vol. 7, pp. 3479–3489). Wiley-VCH: Weinheim, Germany.

Grasselli, R. K., (2002). Fundamental principles of selective heterogeneous oxidation catalysis. *Topics in Catalysis, 21*, 79–88.

Haber, J., (2008). Fundamentals of hydrocarbon oxidation. In: Ertl, G., Knözinger, H., Schüth, F., & Weitkamp, J., (eds.), *Handbook of Heterogeneous Catalysis* (2nd edn., Vol. 7, pp. 3359–3384). Wiley-VCH: Weinheim, Germany.

Hagen, J., (2006). *Industrial Catalysis: A Practical Approach* (2nd edn., pp. 116–131). Wiley-VCH: Weinheim, Germany.

Hagen, J., (2006a). *Industrial Catalysis: A Practical Approach* (2nd edn., pp. 143–179). Wiley-VCH: Weinheim, Germany.

Haruta, M., (2004). Gold as a novel catalyst in the 21st century: Preparation, working mechanism, and applications. *Gold Bulletin, 37*, 27–36.

Hattori, H., (2015). Solid base catalysts: Fundamentals and their applications in organic reactions. *Applied Catalysis A: General, 504*, 103–109.

Jentoft, F. C., (2008). Oxo-anion modified oxides. In: Ertl, G., Knözinger, H., Schüth, F., & Weitkamp, J., (eds.), *Handbook of Heterogeneous Catalysis* (2nd edn., Vol. 1, pp. 262–278). Wiley-VCH: Weinheim, Germany.

Karge, H. G., (2008). Concepts and analysis of acidity and basicity. In: Ertl, G., Knözinger, H., Schüth, F., & Weitkamp, J., (eds.), *Handbook of Heterogeneous Catalysis* (2nd edn., Vol. 2, pp. 1096–1122). Wiley-VCH: Weinheim, Germany.

Kholdeeva, O. A., (2013). Selective oxidations catalyzed by mesoporous metal silicates. In: Clerici, M. G., & Kholdeeva, O. A., (eds.), *Liquid Phase Oxidation via Heterogeneous Catalysis: Organic Synthesis and Industrial Applications* (pp. 127–220). Wiley: Hoboken, New Jersey, USA.

Kleitz, F., (2008). Ordered mesoporous materials. In: Ertl, G., Knözinger, H., Schüth, F., & Weitkamp, J., (eds.), *Handbook of Heterogeneous Catalysis* (2nd edn., Vol. 1, pp. 178–219). Wiley-VCH: Weinheim, Germany.

Lloyd, L., (2011). *Handbook of Industrial Catalysts* (pp. 23–71). Fundamental and applied catalysis. Springer US.

Marceau, E., Carrier, X., Che, M., Clause, O., & Marcilly, C., (2008). Ion exchange and impregnation. In: Ertl, G., Knözinger, H., Schüth, F., & Weitkamp, J., (eds.), *Handbook of Heterogeneous Catalysis* (2nd edn., Vol. 1, pp. 467–484). Wiley-VCH: Weinheim, Germany.

Marcu, I. C., Urdă, A., Popescu, I., & Hulea, V., (2017). Layered double hydroxides-based materials as oxidation catalysts. In: Putz, M. V., & Mirica, M. C., (eds.), *Sustainable Nanosystems Development, Properties and Applications* (pp. 59–121). IGI Global: Hershey, PA, USA.

Márquez-Alvarez, C., Žilková, N., Pérez-Pariente, J., & Čejka, J., (2008). Synthesis, characterization and catalytic applications of organized mesoporous aluminas. *Catalysis Reviews, 50*, 222–286.

Mars, P., & Van Krevelen, D. W., (1954). Oxidations carried out by means of vanadium oxide catalysts. *Chemical Engineering Science Special Suppl., 3*, 41–59.

Misono, M., (2013). Chemistry and catalysis of mixed oxides. *Studies in Surface Science and Catalysis, 176*, 25–65.

Mitran, G., Pavel, O. D., & Marcu, I. C., (2013). Molybdena–vanadia supported on alumina: Effective catalysts for the esterification reaction of acetic acid with *n*-butanol. *Journal of Molecular Catalysis A: Chemical, 370*, 104–110.

Mitran, G., Yuzhakova, T., Popescu, I., & Marcu, I. C., (2015). Study of the esterification reaction of acetic acid with *n*-butanol over supported WO_3 catalysts. *Journal of Molecular Catalysis A: Chemical, 396*, 275–281.

Näumann, F., & Schulz, M., (2008). Oxidation of sulfur dioxide. In: Ertl, G., Knözinger, H., Schüth, F., & Weitkamp, J., (eds.), *Handbook of Heterogeneous Catalysis* (2nd edn., Vol. 5, pp. 2623–2635). Wiley-VCH: Weinheim, Germany.

Ono, Y., & Baba, T., (1997). Selective reactions over solid base catalysts. *Catalysis Today, 38*, 321–337.

Oyama, S. T., (2008). Transition metal carbides, nitrides, and phosphides. In: Ertl, G., Knözinger, H., Schüth, F., & Weitkamp, J., (eds.), *Handbook of Heterogeneous Catalysis* (2nd edn., Vol. 1, pp. 342–356). Wiley-VCH: Weinheim, Germany.

Prasad, R., & Singh, P., (2012). A review on CO oxidation over copper chromite catalyst, *Catalysis Reviews: Science and Engineering, 54*, 224–279.

Prins, R., (2008). Hydrotreating. In: Ertl, G., Knözinger, H., Schüth, F., & Weitkamp, J., (eds.), *Handbook of Heterogeneous Catalysis* (2nd edn., Vol.6, pp. 2695–2718). Wiley-VCH: Weinheim, Germany.

Rase, H. F., (2000). *Handbook of Commercial Catalysts: Heterogeneous Catalysts* (pp. 67–89). CRC Press: Boca Raton, Florida, USA.

Rase, H. F., (2000a). *Handbook of Commercial Catalysts: Heterogeneous Catalysts* (pp. 315–343). CRC Press: Boca Raton, Florida, USA.

Rase, H. F., (2000b). *Handbook of Commercial Catalysts: Heterogeneous Catalysts*. CRC Press: Boca Raton, Florida, USA.

Roduner, E., (2006). Size matters: Why nanomaterials are different. *Chemical Society Reviews, 35*, 583–592.

Ruppert, A. M., & Weckhuysen, B. M., (2008). Metal-support interactions. In: Ertl, G., Knözinger, H., Schüth, F., & Weitkamp, J., (eds.), *Handbook of Heterogeneous Catalysis* (2nd edn., Vol. 2, pp. 1178–1188.). Wiley-VCH: Weinheim, Germany.

Schunk, S. A., (2008). Oxyfunctionalization of alkanes. In: Ertl, G., Knözinger, H., Schüth, F., & Weitkamp, J., (eds.), *Handbook of Heterogeneous Catalysis* (2nd edn., Vol. 7, pp. 3400–3425.). Wiley-VCH: Weinheim, Germany.

Schüth, F., Hesse, M., & Unger, K. K., (2008). Precipitation and coprecipitation. In: Ertl, G., Knözinger, H., Schüth, F., & Weitkamp, J., (eds.), *Handbook of Heterogeneous Catalysis* (2nd edn., Vol. 1, pp. 100–119). Wiley-VCH: Weinheim, Germany.

Sels, B. F., De Vos, D. E., & Jacobs, P. A., (2001). Hydrotalcite-like anionic clays in catalytic organic reactions. *Catalysis Reviews, 43*, 443–488.

Selvam, P., & Sakthivel, A., (2013). Selective catalytic oxidation over ordered nanoporous Metallo-aluminophosphates. In: Clerici, M. G., & Kholdeeva, O. A., (eds.), *Liquid Phase Oxidation via Heterogeneous Catalysis: Organic Synthesis and Industrial Applications* (pp. 95–126). Wiley: Hoboken, New Jersey, USA.

Semagina, N., & Kiwi-Minsker, L., (2009). Recent advances in the liquid-phase synthesis of metal nanostructures with controlled shape and size for catalysis. *Catalysis Reviews, 51*, 147–217.

Setzer, C., Essche, C., & Pryor, N., (2002). Silica. In: Schüth, F., Sing, K. S. W., & Weitkamp, J., (eds.), *Handbook of Porous Solids* (Vol. 3, pp. 1543–1591). Wiley-VCH: Weinheim, Germany.

Smith, A. J., & Wainwright, M. S., (2008). Skeletal metal catalysts. In: Ertl, G., Knözinger, H., Schüth, F., & Weitkamp, J., (eds.), *Handbook of Heterogeneous Catalysis* (2nd edn., Vol. 1, pp. 92–100). Wiley-VCH: Weinheim, Germany.

Smith, G. V., & Notheisz, F., (1999). *Heterogeneous Catalysis in Organic Chemistry* (pp. 229–245). Academic Press: San Diego, USA.

Sun, L. B., Liu, X. Q., & Zhou, H. C., (2015). Design and fabrication of mesoporous heterogeneous basic catalysts. *Chemical Society Reviews, 44*, 5092–5147.

Taguchi, A., & Schüth, F., (2004). Ordered mesoporous materials in catalysis. *Microporous and Mesoporous Materials, 77*, 1–45.

Thomas, J. M., & Thomas, W. J., (1996). *Principles and Practice of Heterogeneous Catalysis* (pp. 548–559). VCH: Weinheim, Germany.

Tichit, D., & Coq, B., (2003). Catalysis by hydrotalcites and related materials. *CATTECH, 7*, 206–217.

Tichit, D., Iborra, S., Corma, A., & Brunel, D., (2006). Base-type catalysis. In: Derouane, E. G., (ed.), *Catalysts for Fine Chemical Synthesis: Microporous and Mesoporous Solid Catalysts* (Vol. 4, pp. 171–205). John Wiley & Sons: Chichester, UK.

Traa, Y., & Weitkamp, J., (2002). Characterization of the pore width of zeolites and related materials by means of molecular probes. In: Schüth, F., Sing, K. S. W., & Weitkamp, J., (eds.), *Handbook of Porous Solids* (Vol. 2, pp. 1015–1057). Wiley-VCH: Weinheim, Germany.

Van Santen, R. A., & Neurock, M., (2006). *Molecular Heterogeneous Catalysis* (pp. 47–61). Wiley-VCH: Weinheim, Germany.

Van Veen, J. A. R., Minderhoud, J. K., Huve, L. G., & Stork, W. H. J., (2008). Hydrocracking and catalytic dewaxing. In: Ertl, G., Knözinger, H., Schüth, F., & Weitkamp, J., (eds.), *Handbook of Heterogeneous Catalysis* (2nd edn., Vol. 6, pp. 2778–2808). Wiley-VCH: Weinheim, Germany.

Zečević, J., Vanbutsele, G., De Jong, K. P., & Martens, J. A., (2015). Nanoscale intimacy in bifunctional catalysts for selective conversion of hydrocarbons. *Nature, 528*, 245–248.

Zhang, F., Xiang, X., Li, F., & Duan, X., (2008). Layered double hydroxides as catalytic materials: Recent development. *Catalysis Surveys from Asia, 12*, 253–265.

CHAPTER 6

Centric Index: Topological Shape

BOGDAN BUMBĂCILĂ[1] and MIHAI V. PUTZ[1,2]

[1]*Laboratory of Computational and Structural Physical Chemistry for Nanosciences and QSAR, Biology-Chemistry Department, Faculty of Chemistry, Biology, Geography at West University of Timişoara, Pestalozzi Street No. 16, Timişoara, RO-300115, Romania*

[2]*Laboratory of Renewable Energies-Photovoltaics, R&D National Institute for Electrochemistry and Condensed Matter, Dr. A. Paunescu Podeanu Str. No. 144, RO-300569 Timişoara, Romania, Tel.: +40-256-592-638; Fax: +40-256-592-620; E-mail: mv_putz@yahoo.com or mihai.putz@e-uvt.ro*

6.1 DEFINITION

The centric index is a distance index, expressed as a value (number) that is reflecting the closeness of the vertices in the hypermolecule to the center of the hypermolecule's network of vertices. It assesses the degree of compactness of the hypermolecule, showing a structure is more eccentric than others (Todeschini, 2009).

The most known centric index is a Balaban centric index (BAC). To calculate this specific centric index, the vertices with a vertex degree of a unity are stepwise removed until no more vertices can be removed.

$$BAC = \sum_{g=1}^{R} n_g^2 \tag{1}$$

where n_g is the number of atoms with a vertex degree of a unity removed at a step g and R is the number of removal steps (U.S. Environmental Protection Agency, 2008).

6.2   HISTORICAL ORIGINS

The value of BAC and its calculus were introduced in 1979 by Alexandru Balaban, and it was discussed in the following years in papers by himself or together with his working team (Balaban, 1979, 1983).

6.3   NANOCHEMICAL IMPLICATIONS

For the evaluation of the structural similarity or diversity of the molecules and for building QSAR models, it is necessary to obtain some molecular descriptors associated with the molecular structure in research. There are many numerical molecular descriptors available in chemistry like physical-chemical parameters, topological indices, and 3D descriptors. The topological indices, however, can offer distance values for assessing branching, molecular size, shape, cyclicity, centricity, symmetry, etc. (Basak, 1999; Hu et al., 2003).

One compound's molecule (or a hypermolecule drew from analog molecules) can be viewed as a molecular graph. Its atoms can represent vertices and its edges, covalent bonds between atoms within it. For simplifying the studies but also because most of the QSAR studies are made on compounds with an organic structure, the following assumptions are made: vertices are in fact carbon atoms (the molecular graph is a carbon skeleton), an edge represents a bond between atoms, therefore, a pair of shared electrons, the distance between two vertices is expressed as a number of edges in the shortest path, two vertices with 1 edge are called adjacent (Balaban, 1983).

Describing, analyzing, or modeling the stereochemistry of a molecule involves computational skills and methods, so it's a quite challenging domain. Because most of the studies (today all of them) are conducted using computer software but also for a better correlation between a structure and its characteristics/properties, each molecular graph is transformed in a number (index) or a sequence of numbers (an adjacency/distance matrix, a polynomial) (Balaban, 1980).

Topological descriptors/indices are numbers or sequences of numbers that are associated with chemical constitutions in order to correlate a structure with properties of that structure (Balaban, 1983).

One of these topological indices is the centric index. Based on Jordan Camille's curve theorem in topology, there is a statement consisting in the idea that an acyclic graph ("tree") has an unique center (a central vertex) or a bicenter (two adjacent vertices joined by an edge). By "looping" or "pruning" step-by-step all of the degree one vertices (end-points, "leafs of the tree") in the graph, together with their incident edges, finally, the center/bicenter is all that is left. n_g or δ_3 is the number of vertices depleted/ pruned at each step of the pruning process, noted as S (Balaban, 1980; Merris, 2001).

In Figure 6.1, an example is given: at step 1 (S_1), five vertices can be pruned, at step 2 (S_2), three vertices can be pruned, and a step 3 (S_3) is reached, where one vertex is left.

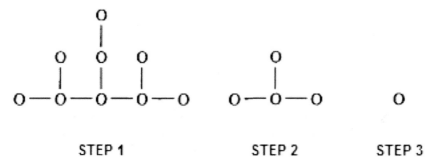

| STEP 1 | STEP 2 | STEP 3 |

FIGURE 6.1 Pruning a graph (acyclic C skeleton) (Merris, 2001).

As the BAC is calculated using the formula:

$$B = \sum_i \delta_i^2 \tag{2}$$

with the value: **B** $= 5^2 + 3^2 + 1^2 = \mathbf{35}$

A normalized centric index was afterward proposed:

$$C = 1/2 \left(B - 2n + U \right) \tag{3}$$

where δ_i is the number of vertices pruned at step i and U is the Kronecker delta depending on the parity of the numbers of vertices n (Hu et al., 2003).

The Kronecker delta,

$$U = \left[1 - (-1)^n \right] \tag{4}$$

derives from the fact that a linear chain of carbons has a pruning sequence that involves a $(2n-U)$ deletions of vertex pairs (Devillers, 1999). So U is 0 if n is even and U is 1 if n is odd (Todeschini, 2009).

A binormalized centric index was again proposed,

$$C' = \left(B - 2n + U \right) / \left[(n-2)^2 - 2 + U \right] \tag{5}$$

This centric index can be applied to a star-shaped graph with a center vertex adjacent to $(n-1)$ vertices of degree 1 (Devillers, 1999).

As it was stated before, these centric indices were developed for acyclic chemical structures (graphs). Therefore, in 1980, Bulgarian Professor Danail Bonchev et al., defined the polycentric graphs. The polycenter of the graph is formed by a nucleus of vertices. The vertices can be adjacent on non-adjacent (Bonchev, 1980).

Four criteria are described below, respecting these precise levels of importance, for every vertex based on the distance matrix of the graph (unlike the pruning process, where for tree-like graphs it leads to the center of the graph step-by-step, the minimal distance process leads to the graph central nucleus). The central nucleus is represented either by a vertex or by a pair of adjacent vertices.

1. Select the vertices with the maximal d_{ij} value. d_{ij} is the number of edges between two vertices (i and j) following the shortest path, as in the following example (Figure 6.2).

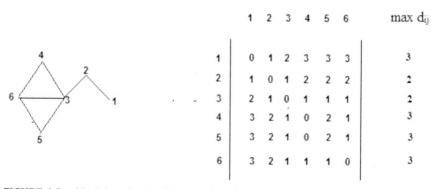

FIGURE 6.2   Obtaining d_{ij} value for a graph molecule from the distance matrix (Bonchev, 1980).

In Figure 6.2, vertices 2 and 3 could be the center of the graph.

2. Then, select the vertices with the smallest distance S_i. A central vertex/adjacent vertices should be less distant to all vertices in the graph than a non-central vertex (the sum of distances from central vertex to the other vertices should be minimum) as in Figure 6.3.

In Figure 6.3, vertices 2 and 5 could be the center of the graph.

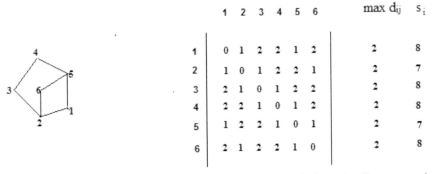

	1	2	3	4	5	6	max d_{ij}	s_i
1	0	1	2	2	1	2	2	8
2	1	0	1	2	2	1	2	7
3	2	1	0	1	2	2	2	8
4	2	2	1	0	1	2	2	8
5	1	2	2	1	0	1	2	7
6	2	1	2	2	1	0	2	8

FIGURE 6.3 Obtaining d_{ij} and s_i values for a graph molecule from the distance matrix (Bonchev, 1980).

3. Then, select the vertices with the smallest number of the largest d_{ij} values (the vertices for which the minimum number of times the largest distance occurs). The number of the largest d_{ij} values for a vertex i is defined as the distance code of i vertex.

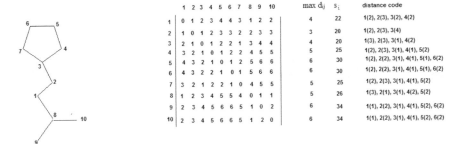

	1	2	3	4	5	6	7	8	9	10	max d_{ij}	s_i	distance code
1	0	1	2	3	4	4	3	1	2	2	4	22	1(2), 2(3), 3(2), 4(2)
2	1	0	1	2	3	3	2	2	3	3	3	20	1(2), 2(3), 3(4)
3	2	1	0	1	2	2	1	3	4	4	4	20	1(3), 2(3), 3(1), 4(2)
4	3	2	1	0	1	2	2	4	5	5	5	25	1(2), 2(3), 3(1), 4(1), 5(2)
5	4	3	2	1	0	1	2	5	6	6	6	30	1(2), 2(2), 3(1), 4(1), 5(1), 6(2)
6	4	3	2	2	1	0	1	5	6	6	6	30	1(2), 2(2), 3(1), 4(1), 5(1), 6(2)
7	3	2	1	2	2	1	0	4	5	5	5	25	1(2), 2(3), 3(1), 4(1), 5(2)
8	1	2	3	4	5	5	4	0	1	1	5	26	1(3), 2(1), 3(1), 4(2), 5(2)
9	2	3	4	5	6	6	5	1	0	2	6	34	1(1), 2(2), 3(1), 4(1), 5(2), 6(2)
10	2	3	4	5	6	6	5	1	2	0	6	34	1(1), 2(2), 3(1), 4(1), 5(2), 6(2)

FIGURE 6.4 Obtaining the distance code for a graph molecule from the distance matrix.

In Figure 6.3, the "shortest" distance code is for vertex 2. So the central vertex is vertex no. 2.

Supposedly we have a graph with two shortest distance codes for two vertices, like in the following example: vertex 1: 1(2), 2(4), 3(1) and vertex 2: 1(3), 2(2), 3(2). The largest distance (max $d_{ij} = 3$) occurs 1 time for vertex 1 and 2 times for vertex 2. In this case, vertex 1 will be the center of the graph.

4. According to the criteria 1–3, we obtain two or more same "shortest" distance codes. The graph obtained by conserving only these vertices (with same distance codes) will undergo these three operations again

till the same graph is obtained, which is called the pseudo-center or the polycenter of the graph (e.g., see, Figure 6.5). The pseudo-center is represented by vertices 1 and 2.

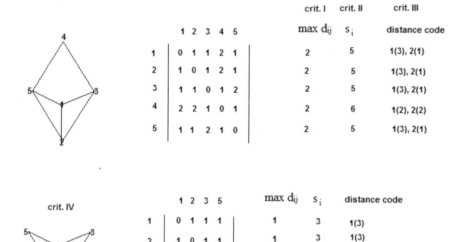

FIGURE 6.5 Sequentially applied criteria 1–3 for determining the center of a graph molecule from its distance matrix (Bonchev, 1980).

6.4 TRANS-DISCIPLINARY CONNECTIONS

Correlations were made for alkanes having 4 to 8 carbon atoms in the molecule between their octane numbers and 14 topological descriptors. The Balaban Centric Index, especially the binormalized one, gave the best correlations with the octane numbers (Horvath, 1992).

Topological correlations between the molecular structure of the polycyclic aromatic hydrocarbons (catafusenes) and their carcinogenic activity were made by Balaban, A.T., and Harary, F. before the centric index was defined (Balaban et al., 1968). Their structure can be resembled with a "dualist graph," where the vertices are the centers of the aromatic hexagon cycles, and the edges are representing two condensed hexagons (Voiculetz, 1991). Any catafusene can be coded according to Bonchev and Balaban, using the determined topological center of the dualist graph (Bonchev, 1993). Later, it was found that substitution by methyl or hydroxyl groups in the position with maximal distance sums is correlated with the highest carcinogenic activity. Ring fusion (the degree of the polycondensation–each addition of a benzene ring) can also be directly correlated with the increase of the carcinogenic effects. The value of the *vertex distance complexity* (the measure of the information content of a vertex is based on the distances to all other vertices in the graph) and the *graph distance complexity* is also directly correlated with the carcinogenic activity of the polycondensed aromatic hydrocarbon (Raychaudhury, 1984; Ray et al., 1982, 1983).

In 2009, Cheng, Z. and his co-workers showed that 5HT receptor ligands have a greater selectivity on the $5HT_{1A}$ receptor subtype if their centricity is more symmetric. So, selective $5HT_{1A}$ molecules can be best described using BAC (Cheng, 2009).

6.5 OPEN ISSUES

In time, more and more variants of centric indices or derivate descriptors were introduced by scientists. We will name some of them in the following:

– Lopping centric information index $(\bar{I}_B)$ – the mean information content obtained from the pruning process of a molecular graph

   o $$\bar{I}_B = -\sum_{k=1}^{R} \frac{n_k}{A} \cdot \log_2 \frac{n_k}{A} \tag{6}$$

   n_k is the number of vertices removed at step k, R is the total number of steps in order to remove all graph vertices (Balaban et al., 1980; Todeschini, 2009).

– Information content based on center (IBC) – defined for acyclic graphs, it represents the total information content based on the "shells" of vertices around the center of the graph:

   o $$IBC = 2W \cdot \log_2 2W - \sum_{k} q_k \cdot \log_2 q_k \tag{7}$$

   W is the Wiener index (the sum of all distances in the graph), q_k is the vertex distance degree (the sum of all distances from a vertex to the others), located at a topological distance k from the central vertex (Balaban et al., 1994).

– Average information content based on center (AIBC)

   o defined only for acyclic graphs and substituents, it is the average of the information content based on center, IBC:

- $$AIBC = \frac{IBC}{W} \tag{8}$$

  ○ IBC is divided by W, the Wiener index (Todeschini, 2009).

– Bonchev centric indices are the centric indices derived from the vertex, the distance matrix, and the edge distance matrix based on the concept of graph center and calculated as mean of the information content (Balaban et al., 1980; Bonchev et al., 1980). For vertex-centric indices, the number of equivalent vertices in each equivalence class is calculated applying the center distance-based criteria 1–4 to the graph vertices. So, the subsequent application of these criteria increases the discrimination of graph vertices. Analogously, for edge-centric indices, the number of equivalent edges in each equivalence class is calculated applying the center distance-based criteria to the graph edges. Once the polycenter of the graph has been found, four other centric information indices, called generalized centric information indices, are calculated on the vertex (edge) and the atoms of polycenter. An increasing discrimination of the graph vertices (edges) is obtained by subsequently applying the criteria steps 2 to 4 (Todeschini, 2009).

  Other centric information indices can be calculated by the same formulas on both vertex and edge graph partition based on graph (center path-based criteria) and center self-returning walk-based criteria. Moreover, edge-centric indices for multigraphs have different values from those calculated on the parent graph, the edge distance matrix of the multigraph being different from the edge distance matrix of the parent graph (Todeschini, 2009).

  ○ Radial centric information index

  $$^{v}\overline{I}_{C,R} = -\sum_{g=1}^{G} \frac{n_g}{A} \log_2 \frac{n_g}{A} \tag{9}$$

  where n_g is the number of graph vertices having the same atom eccentricity, that is the maximum distance from a vertex to any vertex in the graph, G the number of different vertex equivalence classes, and A the number of graph vertices (Todeschini, 2009).

  ○ Distance degree centric index

  $$^{v}\overline{I}_{C,\deg} = -\sum_{g=1}^{G} \frac{n_g}{A} \log_2 \frac{n_g}{A} \tag{10}$$

where n_g is the number of graph vertices having both the same atom eccentricity and the same vertex distance degree (the sum of all vertices from a vertex), G the number of different vertex equivalence classes, and A the number of graph vertices (Todeschini, 2009).

o Distance code centric index

$$^v\overline{I}_{C,code} = -\sum_{g=1}^{G} \frac{n_g}{A} \log_2 \frac{n_g}{A} \tag{11}$$

where n_g is the number of graph vertices contemporarily having the same eccentricity, the same vertex distance degree and the same vertex distance code (occurrence number of distances of different length from a vertex), G the number of different vertex equivalence classes, and A the number of graph vertices (Todeschini, 2009).

o Complete centric index

$$^v\overline{I}_{C,C} = -\sum_{g=1}^{G} \frac{n_g}{A} \log_2 \frac{n_g}{A} \tag{12}$$

where n_g is the number of graph vertices contemporarily having the same eccentricity, the same vertex distance degree, the same vertex distance code, but also distinct vertex than those defining the pseudo-center. G is the number of different vertex equivalence classes, and A is the number of graph vertices (Todeschini, 2009).

o Generalized radial centric information index

$$^v\overline{I}_{C,R}^{G} = -\sum_{g=1}^{G} \frac{n_g}{A} \log_2 \frac{n_g}{A} \tag{13}$$

where n_g is the number of graph vertices having the same average topological distance to the polycenter, G the number of different vertex equivalence classes, and A the number of graph vertices (Todeschini, 2009).

o Generalized distance degree centric index

$$^v\overline{I}_{C,deg}^{G} = -\sum_{g=1}^{G} \frac{n_g}{A} \log_2 \frac{n_g}{A} \tag{14}$$

where n_g is the number of graph vertices having both the same average topological distance to the polycenter and the same vertex distance degree, G the number of different vertex equivalence classes, and A the number of graph vertices (Todeschini, 2009).

○ Generalized distance code centric index

$$v\overline{I}_{C,code}^{G} = -\sum_{g=1}^{G} \frac{n_g}{A} \log_2 \frac{n_g}{A} \tag{15}$$

where n_g is the number of graph vertices having the same average topological distance to the polycenter, the same vertex degree and the same distance code, G is the number of different vertex equivalence classes, and A the number of graph vertices (Todeschini, 2009).

○ Generalized complete centric index

$$v\overline{I}_{C,C}^{G} = -\sum_{g=1}^{G} \frac{n_g}{A} \log_2 \frac{n_g}{A} \tag{16}$$

where n_g is the number of graph vertices contemporarily having the same average topological distance to the polycenter, the same vertex distance degree, the same vertex distance code but being distinct by the atoms defining the pseudocenter, that is, removing existing degeneracy of pseudocenter atoms. G is the number of different vertex equivalence classes, and A is the number of graph vertices (Todeschini, 2009).

○ Edge radial centric information index
○ Edge distance degree centric index
○ Edge distance code centric index
○ Edge complete centric index
○ Generalized edge radial centric information index
○ Generalized edge distance degree centric index
○ Generalized edge distance code centric index
○ Generalized edge complete centric index

The development of the new indices is justified by the need of sensitively assess the molecular characteristics of the compounds: the branching, the eccentricity, the symmetry, etc. Their calculation algorithm is also more and more complicated because series of compounds have to be more and more accurate studied; especially, if they prove efficiency as therapeutic agents. Nowadays, computer programs prove to be very efficient and rapid for obtaining these values.

ACKNOWLEDGMENT

This contribution is part of the research project "Fullerene-Based Double-Graphene Photovoltaics" (FG2PV), PED/123/2017, within the Romanian National Program "Exploratory Research Projects" by UEFISCDI Agency of Romania.

KEYWORDS

- **molecular graph**
- **topological index**

REFERENCES AND FURTHER READING

Balaban, A. T., & Harary, F., (1968). Chemical graphs V: Enumeration and proposed nomenclature of benzenoid catacondensed polycyclic aromatic hydrocarbons. *Tetrahedron, 24*, 2505–2506.

Balaban, A. T., (1979). Chemical graphs. Part 34. Five new topological indices for the branching of tree-like graphs. *Teor. Chem. Acta, 53*, 355.

Balaban, A. T., (1983). Topological indices based on topological distances in molecular graphs. *Pure Appl. Chem., 55*, 199.

Balaban, A. T., Bertelsen, S., & Basak, S. K., (1994). New centric topological indices for acyclic molecules (Trees) and substituents (Rooted-trees) and coding of rooted-trees. *MATCH, Comm. Math. Comp. Chem., 30*, 55–72.

Balaban, A., Chiriac, A., Moţoc, I., & Simon, Z., (1980). *Steric Fit in Quantitative Structure-Activity Relationships.* Springer Verlag, New York.

Basak, S. C., Gute, B. D., & Grunwald, G. D., (1999). A hierarchical approach to the development of QSAR models using topological, geometrical and quantum chemical parameters. In: Devillers, J., & Balaban, A. T., (eds.), *Topological Indices and Related Descriptors in QSAR and QSPR* (p. 692). Gordon and Breach Science Publishers.

Bonchev, D., & Balaban, A. T., (1993). Central vertices versus central rings in polycyclic systems. *Journal of Mathematical Chemistry, 14*, 287–304.

Bonchev, D., Balaban, A. T., & Mekenyan, O., (1980). Generalization of the graph center concept and derived topological centric indices. *Journal of Chemical Information and Computer Sciences, 20*(2), 106–113.

Cheng, Z., Zhang, Y., Zhou, C., Zhang, W., & Gao, S., (2009). Classification of 5-HT$_{1A}$ receptor ligands on the basis of their binding affinities by using PSO-AdaBoost-SVM. *Int. J. Mol. Sci., 10*(8), 3316–3337.

Devillers, J., & Balaban, A. T., (1999). *Topological Indices and Related Descriptors in QSAR and QSPR* (pp. 28–30). Gordon and Breach Science Publishers, Amsterdam.

Horvath, A. L., (1992). Studies in physical and theoretical chemistry 75. *Molecular Design. Chemical Structure Generation from the Properties of Pure Organic Compounds* (pp. 930–932). Elsevier Science Publishers, Amsterdam.

Hu, Q. N., Liang, Y. Z., & Fang, K. T., (2003). The matrix expression, topological index and atomic attribute of molecular topological structure. *Journal of Data Science, 1*, 361–389.

Merris, R., (2001). *Graph Theory* (pp. 35–37). John Wiley and Sons, Inc., New York.

Ray, S. K., Basak, S. C., Raychaudhury, C., Roy, A. B., & Ghosh, J. J., (1983). The utility of information content, structural information content, hydrophobicity and van der Waals volume in the design of barbiturates and tumor-inhibitory triazenes. *Arzneim. Forsch., 33*, 352.

Ray, S. K., Basak, S. C., Raychaudury, C., Roy, A. B., & Ghosh, J. J., (1982). A quantitative structure-activity relationship study of tumor-inhibitory triazenes using bonding information content and lipophilicity. *IRCS Med. Sci., 10*, 933.

Raychaudhury, C., Ray, S. K., Ghosh, J. J., Roy, A. B., & Basak, S. C., (1984). Discrimination of isomeric structures using information theoretic topological indices. *J. Comput. Chem., 5*, 581.

Todeschini, R., & Consonni, V., (2009). *Molecular Descriptors for Chemoinformatics*. Wiley-VCH Verlag GmbH & Co. KGaA; Weinheim.

U. S. Environmental Protection Agency, (2008). *Molecular Descriptors Guide* (p. 17). Description of the molecular descriptors appearing in the toxicity estimation software tool. Version 1.0.2.

Voiculetz, N., Balaban, A. T., Niculescu-Duvăz, I., & Simon, Z., (1991). *Modeling of Cancer Genesis and Prevention* (pp. 147–152). CRC Press, Inc.

CHAPTER 7

Characteristic Polynomial

LORENTZ JÄNTSCHI[1] and SORANA D. BOLBOACĂ[2]

[1]*Technical University of Cluj-Napoca, Romania*

[2]*Iuliu Hațieganu University of Medicine and Pharmacy Cluj-Napoca, Romania, E-mail: sbolboaca@gmail.com*

7.1 DEFINITION

A topological description of a molecule requires storing the adjacencies (the bonds) between the atoms and the identities (the atoms). If this problem is simplified at maximum by disregarding the bond and atom types, then the adjacencies are simply stored with 0 and 1 in the vertex adjacency matrix ([Ad]), and the identities are stored with 0 and 1 into the identity matrix ([Id]). The characteristic polynomial (ChP) is the natural construction of a polynomial in which the eigenvalues of the [Ad] are the roots of the ChP as it follows:

$$\lambda \text{ is an eigenvalue of } [Ad] \leftrightarrow \text{ it exists } [v] \neq 0$$
$$\text{eigenvector such that } \lambda \cdot [v] = [Ad] \cdot [v] \rightarrow$$
$$(\lambda \cdot [Id] - [Ad]) \cdot [v] = 0; \text{ since } v \neq 0 \rightarrow [\lambda \cdot Id - Ad]$$
$$\text{is singular} \rightarrow \det([\lambda \cdot Id - Ad]) = 0$$
$$\text{ChP} \overset{def}{=} |\lambda \cdot Id - Ad|$$

The characteristic polynomial is a polynomial in λ of degree the number of atoms.

Please note that this definition allows extensions. A natural extension is to store in the identity matrix (instead of unity) non-unity values accounting for the atom types, as well as to store in the adjacency matrix (instead of unity) non-unity values accounting for the bond types.

7.2 HISTORICAL ORIGIN(S)

First reports relating to the use of the characteristic polynomial in relation with the chemical structure appears shortly after the discovery of wave-based treatment of microscopic level in Hückel (1931). The Hückel's method of molecular orbitals it is actually the first extension of the Charact-poly definition. It uses the 'secular determinant,' the determinant of a matrix which is decomposed as [E·Id–Ad], standing with the energy of the system (E in the place of λ), for approximate treatment of π electron systems in organic molecules. In this approximate treatment of the Schrödinger's (Schrödinger, 1926) equation ($E\psi = \hat{H}\psi$), the wavefunction (ψ) of the system configuration is defined as a linear combination (c_i stands for unknown, to be determined, coefficients) of the π electrons (p_i, each assigned to an atom), $\psi = \Sigma_i c_i p_i$ and the components of the molecular Hamiltonian ($\hat{H}$) are identified based on the orthogonal states of the electrons ($<p_i|p_j> = \delta_{i,j}$; $\delta_{i,j} = 1$ when $i = j$ and $\delta_{i,j} = 0$ when $i \neq j$): $H_{i,j} = <p_i| \hat{H}|p_j>$ when $H_{i,i} = <p_i| \hat{H}|p_i> = \alpha$ (if $i = j$, the same for all atoms) and $H_{i,j} = <p_i| \hat{H}|p_j> = \beta$ if $[Ad]_{i,j} = 1$ and $H_{i,j} = <p_i| \hat{H}|p_j> = 0$ if $[Ad]_{i,j} = 0$. The roots of this extended version of Charact-poly are assigned to the individual electronic energies (ε_i) (see Coulson (1940, 1950) for further details).

Going in a different direction with the approximation of the wavefunction treatment, Hartree (1928a, b) and Fock (1930a, b) finds the same eigenvector-eigenvalue problem (§20 in Laplace, 1776; T1 in Cauchy, 1829) in the Slater's treatment (Slater, 1929; Hartree & Hartree, 1935). Here is the second extension of the Charact-poly, the eigenproblem (finding of eigenvalues and eigenvectors) being involved to any Hessian (Sylvester, 1880) matrix [A] ([Ad] → [A]).

The Charact-poly it is related with the matching polynomial (Godsil & Gutman, 1981), because both polynomials degenerates to same expression for forests (disjoint union of trees). Adapting (Godsil, 1995) for molecules, a k-matching in a molecule is a matching with exact k bonds between different atoms (each set containing a single edge is also an independent edge set; the empty set should be treated as a independent edge set with zero edges—this set is unique; also due to the constraint of connecting different atoms matching may involve no more than [n/2] bonds, where n is the number of atoms—see §3.1 & §3.3 in Diudea et al., 2001). It is possible to count the k-matches (Ramaraj & Balasubramanian, 1985), but nevertheless it is a hard problem (Curticapean, 2013), as well as to express the derived Z-counting polynomial (Hosoya, 1971) and matching polynomial (both are defined using $m(k)$ as the k-matching number of the selected molecule) (Table 7.1).

TABLE 7.1 Polynomials Derived from k-Matching

Z-counting polynomial	Matching polynomial (where n the number of atoms)
$\sum_{k\geq 0} m(k)\cdot \lambda^k$	$\sum_{k\geq 0} (-1)^k \cdot m(k)\cdot \lambda^{n-2k}$

7.3 NANO-SCIENTIFIC DEVELOPMENT(S)

There are many methods (algorithms) for calculation of the characteristic polynomial and its roots. A method with complexity of $O(n^4)$ extracts the coefficients one-by-one applying the Newton's identities (see Figure 7.1). Please note that the algorithm given below works only for the classical version of Charact-poly (ChP $\equiv |\lambda\cdot\text{Id-Ad}|$), where Trace($\cdot$) function sums the elements on the main diagonal.

```
Input data: adjacency matrix ([Ad])
[Bx] ← [Ad]
c₀ ← 1
For each k from 1 to n-1 do
        cₖ ← Trace([Bx])
        cₖ ← cₖ·(-1)/k
        [Bx] ← [Bx] - cₖ·[Id]
        [Bx] ← [Ad]×[Bx]
End for
cₙ ← Trace([Bx])
cₙ ← cₙ·(-1)/n
Output data: the series of the coefficients (cₖ)₀≤ₖ≤ₙ for Charact-poly, ChP = Σ₀≤ₖ≤ₙcᵢ·λⁿ⁻ᵏ
```

FIGURE 7.1 Tracing the coefficients of the Charact-poly from adjacencies.

When using the previous given algorithm, in order to avoid the lost of the precision with increasing of the number of atoms, someone should use the arbitrary-precision integer libraries for calculation (note that all coefficients of the Charact-poly are integers) of the arithmetic's given in the algorithm such as is bcmath (Morris & Cherry, 1975; Nelson, 1991).

It is possible to reduce the complexity of the calculation of Charact-poly by taking the advantage of its symmetry ($[Ad]_{i,j} = [Ad]_{j,i}$). Budde's method (Givens, 1957) is one alternative. Following (Rehman & Ipsen, 2011), the Budde's method requires first a tridiagonalization (Householder, 1958) of the adjacency matrix (let us call it Td) and is requested a number of significant (of the leading degrees) coefficients. Figure 7.2 provides the algorithm.

Input data: tridiagonalized matrix ([Td], as $(\alpha_i)_{1 \leq i \leq n}$ and $(\beta_i)_{2 \leq i \leq n}$)	
$c_0 \leftarrow 1$	$[Td] = \begin{bmatrix} \alpha_1 & \beta_2 & 0 & \cdots & 0 \\ \beta_2 & \alpha_1 & \ddots & \ddots & \vdots \\ 0 & \ddots & \ddots & \ddots & 0 \\ \vdots & \ddots & \ddots & \ddots & \beta_n \\ 0 & \cdots & 0 & \beta_n & \alpha_n \end{bmatrix}$

$c_{1,1} \leftarrow -\alpha_1$
$c_{1,2} \leftarrow c_{1,1} - \alpha_2$
$c_{2,2} \leftarrow \alpha_1 \cdot \alpha_2 - \beta_2 \cdot \beta_2$
For each i from 3 to k do
    $c_{1,i} \leftarrow c_{1,i} - \alpha_i$
    $c_{2,i} \leftarrow c_{2,i} - \alpha_i \cdot c_{1,i-1} - \beta_i \cdot \beta_i$
    For each j from 3 to i-1 do
        $c_{i,i} \leftarrow -\alpha_i \cdot c_{i-1,i-1} - \beta_i \cdot \beta_i \cdot c_{i-2,i-2}$
    End for
    $c_{i,i} \leftarrow -\alpha_i \cdot c_{i-1,i-1} - \beta_i \cdot \beta_i \cdot c_{i-2,i-2}$
End for
For each i from k+1 to n do
    $c_{1(i)} \leftarrow c_{1(i-1)} - \alpha_i$
    If k > 2 then
        $c_{2,i} \leftarrow c_{2,i-1} - \alpha_i \cdot c_{1,i-1} - \beta_i \cdot \beta_i$
        For each j from 3 to i-1 do
            $c_{i,i} \leftarrow -\alpha_i \cdot c_{i-1,i-1} - \beta_i \cdot \beta_i \cdot c_{i-2,i-2}$
        End for
    End if
End for
$c_{0,n} \leftarrow 1$
Return $(c_{j,n})_{0 \leq j \leq k}$

Output data: partial series of the coefficients, $(c_j)_{0 \leq j \leq k}$ for Charact-poly ChP $= \sum_{0 \leq i \leq n} c_i \cdot \lambda^{n-j}$

FIGURE 7.2   Budde's first coefficients of the Charact-poly from adjacencies.

The main inconvenient of the previous given method (Budde's method) is that it requires the tridiagonalization of the adjacency, which it means that a series of operations including divisions are involved and the resulted matrix no more contains only integers, and therefore is lost the feature to work with arbitrary-precision integers and to extract the exact values of the coefficients. The results may come only as floating point numbers and the precision strongly depends by the number of operations involved, therefore by the number of atoms (n).

On the other hand, the using of arbitrary-precision integer libraries for calculation in conjunction with the algorithm given in Figure 7.1; it is expected to increase the complexity of the calculation. Indeed, as was resulted from a study conducted on a series of fullerenes, the complexity becomes of order $O(n^4 \cdot \ln(n))$—see Table 7.2, where some additional information is provided too, containing the Total strain energy (from continuum elasticity) in eV (Tománek, 2014).

TABLE 7.2 Calculation Times of the Charact-poly on Fullerenes

Fullerene	Additional information	Calculation times(s)
	Molecular formula: C_{20} Molecular symmetry: I_h Total strain energy: 24.204 (the only one topology) Isomers: none	Run 1: 0 Run 2: 1 Average: 0.5 Estimated: 0.2
	Molecular formula: C_{30} Molecular symmetry: C_{2v} Total strain energy: 25.204 (smallest value among isomers) Isomers: 3	Run 1: 2 Run 2: 2 Average: 2.0 Estimated: 1.7
	Molecular formula: C_{40} Molecular symmetry: C_{2v} Total strain energy: 25.684 (smallest value among isomers) Isomers: 40	Run 1: 8 Run 2: 7 Average: 7.5 Estimated: 7.0
	Molecular formula: C_{50} Molecular symmetry: D_{5h} Total strain energy: 25.474 (smallest value among isomers) Isomers: 271	Run 1: 20 Run 2: 20 Average: 20.0 Estimated: 19.7
	Molecular formula: C_{60} Molecular symmetry: I_h Total strain energy: 24.849 (the only one topology) Isomers: none	Run 1: 43 Run 2: 48 Average: 45.5 Estimated: 45.6
	Molecular formula: C_{70} Molecular symmetry: D_{5h} Total strain energy: 26.486 (the only one topology) Isomers: none	Run 1: 88 Run 2: 95 Average: 91.5 Estimated: 91.7

TABLE 7.2 *(Continued)*

Fullerene	Additional information	Calculation times(s)
	Molecular formula: C_{80} Molecular symmetry: D_{5h} Total strain energy: 26.274 (smallest value among isomers) Isomers: 6	Run 1: 166 Run 2: 170 Average: 168.0 Estimated: 167.2
	Molecular formula: C_{90} Molecular symmetry: C_2 Total strain energy: 30.066 (smallest value among isomers) Isomers: 46	Run 1: 282 Run 2: 283 Average: 282.5 Estimated: 283.0
	Molecular formula: C_{100} Molecular symmetry: D_5 Total strain energy: 30.446 (smallest value among isomers) Isomers: 450	Run 1: 455 Run 2: 449 Average: 452.0 Estimated: 452.0

For reproducibility of the study, the calculation was conducted on a 2.33 Ghz dual core computer by running a single core tasked program (a PHP implementation) for the data and the results given in Table 7.2. The variability among execution times can be assigned to the multitasking operating system (which runs in background other tasks too) as well as to the CPU's cache memory (cached in 2 levels, $2 \times 2 \times 32$ kB in the first and, 4096 in the second).

The estimated times from Table 7.2 are from the best (among alternatives) fit, as is given in Table 7.3 (where independent variable was tenth part of the number of atoms, $x = n_c/10$, and the dependent variable was $y = $ time, in seconds).

As can be concluded from the values given in Table 7.3, all the models with intercept ($\hat{y} = a \cdot x^4 + b$; $\hat{y} = a \cdot x^5 + b$; $\hat{y} = a \cdot x^4 \cdot \ln(x) + b$) are susceptible to have the intercept (the b coefficient) not significantly different from zero– actually it is the expected result, since a fullerene with a zero size requires no calculations. Thus, is perfectly justified to try and use the models without intercept. Before to proceed, it is something else which it should keep our attention. Under assumption that exist intercept, then this should be seen as a small time required by the program implementing the algorithm for initializations and for displaying the results. But looking at the models, only one of them proposes a small value for that time, namely b = 0.26 s in the model $\hat{y} = a \cdot x^4 \cdot \ln(x) + b$, which it means that if it is a trustable model without intercept, this should be the one. Indeed, by conducting the analysis without intercept, the results sustain the hypothesis. The standard error of estimate (see) remains almost unchanged when the intercept is removed only for this model. Therefore, the best guess for the approximation of the complexity of the algorithm for the calculation of the Charact-poly with arbitrary-precision integers given in Figure 7.1 is of $O(n^4 \cdot \ln(n))$. For reproducibility of the study, the calculation was conducted on a 2.33 Ghz dual-core computer by running a single core tasked program (a PHP implementation) for the data and the results given in Table 7.2.

TABLE 7.3 Calculation Complexity of the Charact-poly on Fullerenes

Model	Coefficients and significances	Statistics	Remarks
$\hat{y} = a \cdot x^4 + b$	$a = 0.045$ $(t_a = 58)$; $b = -7.98$ $(t_b = 2.39)$	$r^2 = 0.99794$; see = 7.6	$p_b \approx 5\%$
$\hat{y} = a \cdot x^5 + b$	$a = 0.0045$ $(t_a = 58)$; $b = 7.45$ $(t_b = 2.20)$	$r^2 = 0.99765$; see = 8.1	$p_b \approx 6\%$
$\hat{y} = a \cdot x^4 \cdot \ln(x) + b$	$a = 0.02$ $(t_a = 1134)$; $b = 0.26$ $(t_b = 1.53)$	$r^2 = 0.99999$; see = 0.4	$p_b \approx 17\%$
$\hat{y} = a \cdot x^4$	$a = 0.044$ $(t_a = 59)$	$r^2 = 0.996$; see = 9.6	see below
$\hat{y} = a \cdot x^5$	$a = 0.0046$ $(t_a = 58)$	$r^2 = 0.996$; see = 9.9	
$\hat{y} = a \cdot x^4 \cdot \ln(x)$	$a = 0.01963$ $(t_a = 1346)$	$r^2 = 0.999996$; see = 0.4	

7.4 NANOCHEMICAL APPLICATION(S)

In classical molecular topology the atoms are considered indistinguishable and are represented as vertices, the bonds are considered unweighted and are represented as edges and the obtained molecular graphs are unweighted and unoriented. In this context, the set of the bonds (edges) is a subset of the Cartesian product of the set of the atoms (vertices) by itself and the molecules (graph) is defined as the collection of the set of vertices and of the set of edges (see Table 7.4).

There is something to consider when discuss calculations on molecular graphs. Thus, with the increasing of the simplification in the molecular graph representation (neglecting type of the atom, bond orders, geometry in the favor of topology) increases the degeneration of the whole pool of possible calculations (descriptors) on the graph structure – existing more and more molecules possessing same representation as molecular graph. This consequence is favorable for the problems seeking for similarities and is unfavorable for the problems seeking for dissimilarities.

TABLE 7.4 Classical Molecular Graph

Definition	Names (concepts)	Cardinality	Example
V: finite set	V: vertices (atoms)	$|V| = N$: number of vertices	$G = "A\text{-}B\text{-}C"$
$E \subseteq V \times V$	E: edges (bonds)	$|E| = M$: number of edges	$V = \{1(\leftrightarrow A), 2(\leftrightarrow B),$
$G = G(V, E)$	G: graph (molecule)	$\forall N, V \leftrightarrow \{1, 2,..., N\}$	$3(\leftrightarrow C)\}$
			$E = \{(1, 2), (2, 3)\}$

A necessary step to accomplish a better coverage of the similarity vs. dissimilarity dualism is to build and use a family of molecular descriptors, large enough to be able to provide answers for the all (by its individuals) when is feed with molecules datasets. On the natural way a such kind of family should possess a 'genetic code' – namely a series of variables of which values to (re)produce a (one by one) molecular descriptor, all descriptors being therefore obtained on same way (being breed in the family). All individuals of the family should be independent of the numbering of the atoms in molecule (should be molecular invariants).

Since these are all restrictions applying, may seem a simple construction, but it is in same time a complex one. First, is obviously that such construction—the family of molecular descriptors—can be build only with the help of the computer, because requires a great number of operations repeatedly (with different augments) applied on the same molecular structure. The molecular geometry should be considered too, and for this reason (of obtaining of the models for the molecular geometry) this subject will be continued in a later section.

In order to reflect the topology of a graph structure, three adjacency matrices can be built. If we store the full graph (each pair of vertices stored twice, in both ways) then the rectangular matrices reflects 1:1 the graph (these matrices are more convenient when we do matrix operations). The matrices of vertex adjacency and of edges adjacency are square matrices and the enumerating twice the edges is reflected in symmetry of the matrix relative to its main diagonal, which can be rebuild in the absence of the representation, by having only the lists of vertices and edges (see Table 7.4).

An extremely important problem in chemistry is to identify uniquely a chemical compound. If the visual identification (looking on the structure) seems simple, for compounds of large size this alternative is no more viable. The data of the structure of the compounds stored into the informational space may provide the answer to this problem. Together with the storing of the structure of the compound other issue is raised, namely the arbitrary in the numbering of the atoms. Namely for a chemical structure with N atoms stored as a (classical molecular) graph exists exactly N! possibilities of different numbering of the atoms. Unfortunately, storing the graphs as lists of edges (and eventually of vertices) does not provide a direct tool to check this arbitrary differentiation due to the numbering. The same situation applies on the adjacency matrices.

Therefore, seeking for graph invariants is perfectly justified: an invariant (graph invariant) does not depend on numbering. The adjacency matrix is not a graph invariant (and very simple examples may be created instantly to proof this). The ideal situation is that the invariant to be uniquely assigned to each (and any) structure, but this kind of invariants are very hard to be found.

A procedure to generate a no degenerated invariant is proposed by IUPAC as the international chemical identifier (InChI) which converts the chemical structure to a table of connectivity expressed as a unique and predictable series of characters (McNaught, 2006).

An important class of graph invariants are the graph polynomials. To this category belongs the characteristic polynomial, a graph invariant encoding important properties of the graph. Unfortunately, does not represent a bijective image of the graph, existing different graphs with same characteristic polynomial (cospectral graphs), and smallest cospectral graphs occurs for 5 vertices (Von Collatz & Sinogowitz, 1957). In order to count the cospectral graphs, one should compare A000088 (Sloane, 1996) and A082104 (Weisstein, 2003) integer sequences.

Let's take a chemical compound, namely hexamine (Pubchem CID: 4101). Hexamine ($C_6H_{12}N_4$) it uses in the production of powdery or liquid preparations of phenolic resins and phenolic resin molding compounds. It has been proposed that hexamethylenetetramine could work as a molecular building block for self-assembled molecular crystals (Markle, 2000). It has a cage-like structure similar to adamantine and its representation is given in Figure 7.3.

In Figure 7.3, the hydrogen atoms are represented with grey (and are not numbered), carbon with light blue, and nitrogen with blue.

Let's take the representation in a matrix form by its adjacency matrix by taking it conventionally without attached hydrogen atoms as well as by its distance matrix in two scenarios: topological and geometrical distances. Please note that this simple case of a molecule, but even here the geometrical distance is with a totally different meaning than the topological distance. The resulted matrices are given in the Figure 7.4.

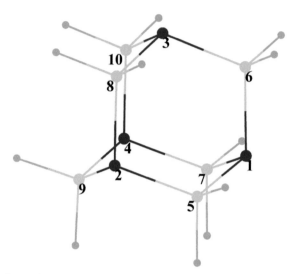

FIGURE 7.3   Hexamine.

Ad	1	2	3	4	5	6	7	8	9	10
1	0	0	0	0	1	1	1	0	0	0
2	0	0	0	0	1	0	0	1	1	0
3	0	0	0	0	0	1	0	1	0	1
4	0	0	0	0	0	0	1	0	1	1
5	1	1	0	0	0	0	0	0	0	0
6	1	0	1	0	0	0	0	0	0	0
7	1	0	0	1	0	0	0	0	0	0
8	0	1	1	0	0	0	0	0	0	0
9	0	1	0	1	0	0	0	0	0	0
10	0	0	1	1	0	0	0	0	0	0

Ad	1	2	3	4	5	6	7	8	9	10
1	0	0	0	0	$^{31}/_{46}$	$^{31}/_{46}$	$^{31}/_{46}$	0	0	0
2	0	0	0	0	$^{31}/_{46}$	0	0	$^{31}/_{46}$	$^{31}/_{46}$	0
3	0	0	0	0	0	$^{31}/_{46}$	0	$^{31}/_{46}$	0	$^{31}/_{46}$
4	0	0	0	0	0	0	$^{31}/_{46}$	0	$^{31}/_{46}$	$^{31}/_{46}$
5	$^{31}/_{46}$	$^{31}/_{46}$	0	0	0	0	0	0	0	0
6	$^{31}/_{46}$	0	$^{31}/_{46}$	0	0	0	0	0	0	0
7	$^{31}/_{46}$	0	0	$^{31}/_{46}$	0	0	0	0	0	0
8	0	$^{31}/_{46}$	$^{31}/_{46}$	0	0	0	0	0	0	0
9	0	$^{31}/_{46}$	0	$^{31}/_{46}$	0	0	0	0	0	0
10	0	0	$^{31}/_{46}$	$^{31}/_{46}$	0	0	0	0	0	0

bonds represented undistinguishable (with 1)	bonds represented from geometrical distances (inverse of the distance in Å)

FIGURE 7.4   Different adjacency matrices representing hexamine.

The unity (or identity) matrix stores 1 on the main diagonal and is easy to be extended to store a atomic property (such as something in relation with atomic mass, electronegativity, partial charge or even the number of attached hydrogen atoms, when also 0 is allowed) when the new matrices continues to have all non-null values on the main diagonal, but can be different from one and different one to each other depending now from the atom type. The result is exemplified in Figure 7.5 (where electronegativity is taken from Pauling scale and was divided by 4). The general idea when the weights was chosen in the identity matrices exemplified in Figure 7.5 is to have (almost everywhere) subunitary numbers, because when on these matrices is applied the procedure of calculation of the characteristic polynomial, then for large molecules numbers greater than 1 rapidly produces big numbers as coefficients as well as outcomes of the evaluation of the polynomial.

Ad	1	2	3	4	5	6	7	8	9	10
1	1	0	0	0	0	0	0	0	0	0
2	0	1	0	0	0	0	0	0	0	0
3	0	0	1	0	0	0	0	0	0	0
4	0	0	0	1	0	0	0	0	0	0
5	0	0	0	0	1	0	0	0	0	0
6	0	0	0	0	0	1	0	0	0	0
7	0	0	0	0	0	0	1	0	0	0
8	0	0	0	0	0	0	0	1	0	0
9	0	0	0	0	0	0	0	0	1	0
10	0	0	0	0	0	0	0	0	0	1

atoms represented with 1 (undistinguishable)

Ad	1	2	3	4	5	6	7	8	9	10
1	.75	0	0	0	0	0	0	0	0	0
2	0	.75	0	0	0	0	0	0	0	0
3	0	0	.75	0	0	0	0	0	0	0
4	0	0	0	.75	0	0	0	0	0	0
5	0	0	0	0	.625	0	0	0	0	0
6	0	0	0	0	0	.625	0	0	0	0
7	0	0	0	0	0	0	.625	0	0	0
8	0	0	0	0	0	0	0	.625	0	0
9	0	0	0	0	0	0	0	0	.625	0
10	0	0	0	0	0	0	0	0	0	.625

atoms represented by electronegativity (distinguishable)

FIGURE 7.5 Different identity matrices representing hexamine.

Based on the modified forms of the adjacency and identity matrices, the extension of the formula of the characteristic polynomial is immediate:

$$P_{\varphi,AP,MO}(\lambda) = P_{\varphi}(\lambda, G) = |\lambda \cdot Id(A_p) - Ad(M_O)|$$

where M_O is a certain metric operator (as were exemplified in Figure 7.4) and A_p is a certain atomic property (as were exemplified in Figure 7.5). For a single molecule it results a series of the polynomial formulas (given in the next for a clear reading as determinants) which can be evaluated for different values of the argument (X). The next figure exemplifies the calculation for hexamine.

The outcome of the regression analysis is a model with a certain explanatory power. This explanatory power is influenced by the number of the coefficients included in the model (n_c), as well as by the number of independent variables (n_d) used to explain the association when the model was feed with a certain number of molecules (m), and therefore the adjusted value (r^2_{adj}) of the correlation coefficient (r^2) provides a ordering of the explanatory powers:

$$r^2_{adj} = r^2 - (1 - r^2)\frac{n_d}{m - n_c}$$

	M_O from classical topology	M_O from geometry $(a = -\tfrac{31}{46})$
A_P from classical topology	$\begin{matrix} \lambda & 0 & 0 & 0 & -1 & -1 & -1 & 0 & 0 & 0 \\ 0 & \lambda & 0 & 0 & -1 & 0 & 0 & -1 & -1 & 0 \\ 0 & 0 & \lambda & 0 & 0 & -1 & 0 & -1 & 0 & -1 \\ 0 & 0 & 0 & \lambda & 0 & 0 & -1 & 0 & -1 & -1 \\ -1 & -1 & 0 & 0 & \lambda & 0 & 0 & 0 & 0 & 0 \\ -1 & 0 & -1 & 0 & 0 & \lambda & 0 & 0 & 0 & 0 \\ -1 & 0 & 0 & -1 & 0 & 0 & \lambda & 0 & 0 & 0 \\ 0 & -1 & -1 & 0 & 0 & 0 & 0 & \lambda & 0 & 0 \\ 0 & -1 & 0 & -1 & 0 & 0 & 0 & 0 & \lambda & 0 \\ 0 & 0 & -1 & -1 & 0 & 0 & 0 & 0 & 0 & \lambda \end{matrix}$	$\begin{matrix} \lambda & 0 & 0 & 0 & a & a & a & 0 & 0 & 0 \\ 0 & \lambda & 0 & 0 & a & 0 & 0 & a & a & 0 \\ 0 & 0 & \lambda & 0 & 0 & a & 0 & a & 0 & a \\ 0 & 0 & 0 & \lambda & 0 & 0 & a & 0 & a & a \\ a & a & 0 & 0 & \lambda & 0 & 0 & 0 & 0 & 0 \\ a & 0 & a & 0 & 0 & \lambda & 0 & 0 & 0 & 0 \\ a & 0 & 0 & a & 0 & 0 & \lambda & 0 & 0 & 0 \\ 0 & a & a & 0 & 0 & 0 & 0 & \lambda & 0 & 0 \\ 0 & a & 0 & a & 0 & 0 & 0 & 0 & \lambda & 0 \\ 0 & 0 & a & a & 0 & 0 & 0 & 0 & 0 & \lambda \end{matrix}$
A_P from electronegativities $(b = \tfrac{5}{8};\ c = \tfrac{3}{4})$	$\begin{matrix} c\cdot\lambda & 0 & 0 & 0 & 1 & 1 & 1 & 0 & 0 & 0 \\ 0 & c\cdot\lambda & 0 & 0 & 1 & 0 & 0 & 1 & 1 & 0 \\ 0 & 0 & c\cdot\lambda & 0 & 0 & 1 & 0 & 1 & 0 & 1 \\ 0 & 0 & 0 & c\cdot\lambda & 0 & 0 & 1 & 0 & 1 & 1 \\ 1 & 1 & 0 & 0 & b\cdot\lambda & 0 & 0 & 0 & 0 & 0 \\ 1 & 0 & 1 & 0 & 0 & b\cdot\lambda & 0 & 0 & 0 & 0 \\ 1 & 0 & 0 & 1 & 0 & 0 & b\cdot\lambda & 0 & 0 & 0 \\ 0 & 1 & 1 & 0 & 0 & 0 & 0 & b\cdot\lambda & 0 & 0 \\ 0 & 1 & 0 & 1 & 0 & 0 & 0 & 0 & b\cdot\lambda & 0 \\ 0 & 0 & 1 & 1 & 0 & 0 & 0 & 0 & 0 & b\cdot\lambda \end{matrix}$	$\begin{matrix} c\lambda & 0 & 0 & 0 & a & a & a & 0 & 0 & 0 \\ 0 & c\lambda & 0 & 0 & a & 0 & 0 & a & a & 0 \\ 0 & 0 & c\lambda & 0 & 0 & a & 0 & a & 0 & a \\ 0 & 0 & 0 & c\lambda & 0 & 0 & a & 0 & a & a \\ a & a & 0 & 0 & b\lambda & 0 & 0 & 0 & 0 & 0 \\ a & 0 & a & 0 & 0 & b\lambda & 0 & 0 & 0 & 0 \\ a & 0 & 0 & a & 0 & 0 & b\lambda & 0 & 0 & 0 \\ 0 & a & a & 0 & 0 & 0 & 0 & b\lambda & 0 & 0 \\ 0 & a & 0 & a & 0 & 0 & 0 & 0 & b\lambda & 0 \\ 0 & 0 & a & a & 0 & 0 & 0 & 0 & 0 & b\lambda \end{matrix}$

FIGURE 7.6 Different characteristic-like polynomials for the chemical structure of hexamine.

The use of the extended characteristic polynomial is exemplified on a series of 45 C_{20} fullerene congeners which were obtained by replacing the carbon atom with nitrogen and boron by a certain pattern which is illustrated in Figure 7.7 (S1 to S4 are shells; on each shell are atoms of same type).

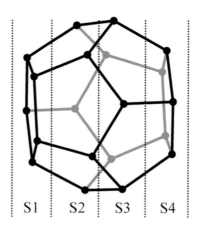

FIGURE 7.7 Pattern for generation of C_{20} fullerene congener structures.

Are 4^3 (64) possible arrangements of Carbon, Nitrogen and Boron in the sites defined by the shells S1 to S4 in Figure 7.7, but some of them defines identical molecules as long as the structure is free to move (and rotate). Due to this fact, there are only 45 different structures. The structures were drawn and stored in separate files. The geometries were build at HF 6-31G level of theory and a series of calculated properties were collected for them and are given in the Table 7.5 along with the file name (named accordingly to the design from Figure 7.7, where Homo: highest occupied molecular orbital energy, in eV; Lumo: lowest unoccupied molecular orbital energy, in eV; Pola: polarizability, in 10^{-30} m^3, DipM: dipole moment, in Debye).

TABLE 7.5 Selected Properties from HF 6-31G Calculations

Mol	Homo	Lumo	Pola	DipM	Mol	Homo	Lumo	Pola	DipM
bbbb	−0.2461	−0.0385	60.469	0.003541	cbnn	−0.3918	0.0903	55.059	8.237555
bbbn	−0.2546	−0.095	61.208	1.189393	ccbb	−0.2697	−0.0317	61.306	6.250461
bbcn	−0.2988	−0.0764	60.081	9.954801	ccbc	−0.3159	0.0574	59.911	0.713889
bbnb	−0.2959	−0.0608	60.272	1.381921	ccbn	−0.3237	0.074	57.661	5.714922
bbnn	−0.2747	−0.0603	58.602	7.083818	cccb	−0.2589	−0.0696	61.359	1.960172
bcbb	−0.2471	−0.0647	62.309	2.179006	cccc	−0.3843	0.1622	58.487	0.000832
bcbn	−0.3083	−0.0407	59.187	6.785335	cccn	−0.3122	0.1522	57.091	7.334984
bccb	−0.2036	−0.1014	62.714	0.002733	ccnb	−0.3265	0.0326	58.372	3.414014
bccn	−0.2994	−0.039	59.196	9.030429	ccnc	−0.3363	0.1803	56.784	1.457706
bcnb	−0.3023	−0.0469	59.881	5.572056	ccnn	−0.3565	0.1456	54.976	10.170182
bcnn	−0.2578	−0.1307	58.182	0.680023	cnbb	−0.3004	−0.087	59.378	3.2186
bnbn	−0.3545	−0.0214	54.813	6.939198	cnbn	−0.3617	0.1144	54.798	8.288544
bncn	−0.3513	−0.006	56.543	4.948867	cncb	−0.3145	−0.0319	59.111	3.170551
bnnb	−0.3466	−0.0157	57.038	0.008881	cncn	−0.3547	0.171	54.802	6.884452
bnnn	−0.3903	−0.0527	54.572	6.824007	cnnb	−0.3394	−0.0048	56.511	3.000425
cbbb	−0.2621	−0.0498	61.333	3.939499	cnnc	−0.334	0.1663	54.992	0.001746
cbbc	−0.2475	0.026	60.198	0.003295	cnnn	−0.3883	0.1159	53.051	9.415604
cbbn	−0.221	−0.0178	59.654	3.793689	nbbn	−0.2199	−0.0619	57.786	0.019205
cbcb	−0.2836	−0.006	61.228	0.128928	nbnn	−0.4057	0.0636	52.938	0.768984
cbcn	−0.3008	0.0269	58.345	8.600025	ncbn	−0.3241	0.0499	55.991	2.259238
cbnb	−0.3269	0.0075	58.474	0.04766	nccn	−0.3292	0.1694	54.967	0.003738
cbnc	−0.3749	0.1273	56.843	0.89975	ncnn	−0.3882	0.1361	52.906	1.953068
					nnnn	−0.454	0.0624	51.084	0.002512

The calculations for the extended characteristic polynomial were conducted diversifying the atomic property in 8 levels, as given in Table 7.6.

TABLE 7.6 Atomic Properties Included in the Extension of the Charact-poly

'A' – atomic mass (per 294.0)	'B' – cardinality (always 1)
'C' – charges (atomic electrostatic charge, ESP)	'D' – solid-state density (in kg/m³/30,000)
'E' – electronegativity (revised Pauling/4.00)	'F' – first ionization potential (in kJ/mol/1312.0)
'G' – melting point temperature (in K/3820.0)	'H' – attached hydrogen atoms (per 4)

The calculations for the extended characteristic polynomial were conducted diversifying the adjacency in 3 levels, as well as were used the distance matrix in place of the adjacency when the diversification were produced in 6 levels, as given in Table 7.7.

TABLE 7.7 Adjacency Weights Included in the Extension of the Charact-poly

On adjacencies	'g' – 0 or geometrical distance	't' – (0 or 1)	'c' – 0 or inverse of bond order
On distances	'G' – geometrical distance	'T' – topological distances	'C' – smallest sum of bond orders inverses

A FreePascal program were build to split the work in parallel depending on the number of processors available splitting the job by properties (no more than 8 parallel tasks can be produced). A huge file containing the descriptors names as well as calculated values of the polynomials for the series of the molecules is produced. The descriptors are named as described in Table 7.8.

It is expected that a diversification like the one considered here to produce degenerations, namely identical values of the descriptors for different descriptor names, which is the case in the dataset considered here. For instance, one degeneration is immediate, namely between the classical topological calculation when 1 is encoded in the adjacency matrix and the diversified calculation in which the inverse of the bond order is considered, because all bonds in the congener series are single bonds (see Figure 7.7). This is the reason for which, before regression analysis involving the properties from HF 6-31G calculations, a filtering program was designed to look for degenerations and to reduce the pool of descriptors eliminating them. Because also this program works in parallel, without channeling between tasks, may still contain few degenerations, and another program were designed to sort the data (acting in parallel, but all tasks are writing in a common shared output file) to clean the descriptors pool by repeated series of identical values.

TABLE 7.8 Names of the Descriptors Calculated Using the Extended Characteristic Polynomial

Variable	Description
$L_1L_2L_3L_4d_1d_2d_3d_4$	8 characters, first 4 being letters, last 4 being digits
$d_1d_2d_3d_4$	$\overline{d_1d_2d_3d_4}$ ranges from, 0000 to, 1000; is evaluated $P_{\varphi,L_2,L_3}(\pm\overline{d_1d_2d_3d_4}/1000)$
L_1	is 'I' when the evaluated polynomial value is unchanged ($f(x) = x$); is 'R' when reciprocal ($f(x) = 1/x$) value of the evaluated polynomial is calculated; is 'L' when logarithm ($f(x) = \ln(x)$) value of the evaluated polynomial is calculated;
L_2	Encodes the atomic property used to diversify the identity matrix (see Table 7.6)
L_3	Encodes the metric operator used to diversify the adjacency matrix (see Table 7.7)
L_4	Encoding for negative (N) or positive (P) argument of the polynomial

The next stage is to test the descriptors in regard to their ability to make simple linear regressions, but very different square deviations when the data are normalized have no meaning when a linear relation with a measured property is desired. Therefore, in a stage the values are normalized and compared the two distributions (of the property and of the descriptor) and the association is rejected for a departure with a probability of association less than 1% (this procedure should be called normalization) and in the last stage simple linear associations are obtained (when some descriptors are removed when possess more than 100 times or less than 100 times variance than the observed).

Table 7.9 contains the statistics of the descriptors generated, where stage refers the stage applied on the pool of descriptors.

A remark is immediate about the results given in Table 7.9: even if the design of the diversification admits the degeneracy (see Table 7.8) as well as the dataset is a simple pattern on which the congener series were constructed (see Figure 7.7) the level of degeneracy is very low (the descriptors pool size is reduced from 272,124 to 230,450 till the sorting stage included, which is about 84.7% of its initial size; a reduction to 40.7% in average for the last stage of simple linear regressions) and when is compared with other families of descriptors, such as MDF, for similar sample sizes the reduction is much less [in Jäntschi and Bolboacă (2006) on 40 compounds from 787,968 descriptors after a similar filtering remained 70,943 which is about 9%].

TABLE 7.9   Statistics from Preliminary Treatment of the Descriptors Pool

Stage	Number of descriptors	Remarks
Generating	272,124	Always using the defined configuration
Filtering	235,530	Degeneration depends on the complexity of the dataset
Sorting	230,450	Degeneration depends on the parallelization level
Normalization	140,447	For Homo
Normalization	147,071	For Lumo
Normalization	139,902	For Pola
Normalization	146,134	For DipM
Simple linear regression	114,936	For Homo
Simple linear regression	115,747	For Lumo
Simple linear regression	119,249	For Pola
Simple linear regression	93,299	For DipM

Regression was conducted with one and two dependent variables taking into account additive and/or multiplicative effects among them. The analysis produced more than one possible outcome for each case, and were selected the best candidate regressions with the highest explanatory power on the molecules subject to investigation for the property with which the model was feed. The equations are given in Table 7.10.

In all cases the best model selections were with both additive and multiplicative effects included ($\hat{Y} = a{\cdot}X_1 + b{\cdot}X_2 + c{\cdot}X_1{\cdot}X_2 + d$), and therefore someone can say that the association between the structure as can be described by the characteristic polynomial and the dipole moment, HOMO and LUMO energies and polarizability is hardly to be considered as being purely linear. Also in all cases linear models ($\hat{Y} = a{\cdot}X_1 + b{\cdot}X_2 + d$) were selected as the second best alternative in all cases in disfavor of the multiplicative effects models ($\hat{Y} = c{\cdot}X_1{\cdot}X_2 + d$), suggesting that however the linear component is the predominant one. Test results for significant differences among explanatory powers of the models (by using of Fisher Z transformation) are given in Table 7.11.

TABLE 7.10 Regressions with Highest Explanatory Power

No	Property	Model	Descriptors	Coefficients	r^2	r^2_{adj}
1	Dipole moment	$\hat{Y} = a \cdot X_1 + d$	$X_1 = REN0841$	$a = -0.1410\ (t = 5.74)$ $d = 3.742\ (t = 9.78)$	0.434	0.420
2	Dipole moment	$\hat{Y} = c \cdot X_1 \cdot X_2 + d$	$X_1 = RDtP0066$ $X_2 = IDtP0087$	$c = 4.221\ (t = 8.85)$ $d = 1.555\ (t = 4.04)$	0.651	0.635
3	Dipole moment	$\hat{Y} = a \cdot X_1 + b \cdot X_2 + d$	$X_1 = LFgN0609$ $X_2 = IDcP0908$	$a = 5.328 \times 10^1\ (t = 7.10)$ $b = 5.434 \times 10^1\ (t = 8.89)$ $d = 7.544\ (t = 15.1)$	0.689	0.675
4	Dipole moment	$\hat{Y} = a \cdot X_1 + b \cdot X_2 + c \cdot X_1 \cdot X_2 + d$	$X_1 = LFgN0612$ $X_2 = RDtN0065$	$a = -3.103 \times 10^1\ (t = 3.78)$ $b = -5.126 \times 10^1\ (t = 8.79)$ $c = -6.851 \times 10^2\ (t = 9.59)$ $d = 1.791\ (t = 4.54)$	0.725	0.711
5	HOMO energy	$\hat{Y} = a \cdot X_1 + d$	$X_1 = LFgP0454$	$a = -0.04027\ (t = 8.79)$ $d = -0.4028\ (t = 36.6)$	0.643	0.634
6	HOMO energy	$\hat{Y} = c \cdot X_1 \cdot X_2 + d$	$X_1 = IDTP0633$ $X_2 = IDTP0653$	$c = -7.434 \times 10^{-7}\ (t = 11.5)$ $d = -2.755 \times 10^{-1}\ (t = 50.6)$	0.758	0.747
7	HOMO energy	$\hat{Y} = a \cdot X_1 + b \cdot X_2 + d$	$X_1 = RGTN0155$ $X_2 = LBgN0874$	$a = -0.01407\ (t = 6.49)$ $b = -0.1827\ (t = 8.26)$ $d = -0.9522\ (t = 12.1)$	0.808	0.799
8	HOMO energy	$\hat{Y} = a \cdot X_1 + b \cdot X_2 + c \cdot X_1 \cdot X_2 + d$	$X_1 = RGCN0182$ $X_2 = LBgP0165$	$a = 1.060 \times 10^{-1}\ (t = 4.80)$ $b = 6.9852 \times 10^{-2}\ (t = 11.0)$ $c = 2.4080 \times 10^{-2}\ (t = 5.51)$ $d = -6.947 \times 10^{-1}\ (t = 19.7)$	0.830	0.822

TABLE 7.10 (Continued)

No	Property	Model	Descriptors	Coefficients	r^2	r^2_{adj}
9	LUMO energy	$\hat{Y} = a \cdot X_1 + d$	$X_1 = \text{LBgN0566}$	$a = 1.214\ (t = 8.79)$ $d = 0.1019\ (t = 8.33)$	0.643	0.634
10	LUMO energy	$\hat{Y} = c \cdot X_1 \cdot X_2 + d$	$X_1 = \text{IHGN0132}$ $X_2 = \text{IHGN0157}$	$c = 5.972 \times 10^{-11}\ (t = 13.7)$ $d = -5.854 \times 10^{-2}\ (t = 7.15)$	0.816	0.808
11	LUMO energy	$\hat{Y} = a \cdot X_1 + b \cdot X_2 + d$	$X_1 = \text{LGGN0094}$ $X_2 = \text{LBgN0584}$	$a = 7.458 \times 10^{-2}\ (t = 7.27)$ $b = 1.008\ (t = 10.0)$ $d = 6.230 \times 10^{-1}\ (t = 8.72)$	0.830	0.821
12	LUMO energy	$\hat{Y} = a \cdot X_1 + b \cdot X_2 + c \cdot X_1 \cdot X_2 + d$	$X_1 = \text{IHGN0150}$ $X_2 = \text{IHGN0131}$	$a = 4.098 \times 10^{-6}\ (t = 10.9)$ $b = 4.103 \times 10^{-6}\ (t = 11.6)$ $c = 6.936 \times 10^{-11}\ (t = 12.9)$ $d = -6.481 \times 10^{-2}\ (t = 7.97)$	0.840	0.832
13	Polarizability	$\hat{Y} = a \cdot X_1 + d$	$X_1 = \text{IATN0079}$	$a = -38.50\ (t = 18.9)$ $d = 500.4\ (t = 21.4)$	0.893	0.890
14	Polarizability	$\hat{Y} = c \cdot X_1 \cdot X_2 + d$	$X_1 = \text{LFTN0225}$ $X_2 = \text{LBGP0631}$	$c = 25.09\ (t = 24.9)$ $d = 71.64\ (t = 126)$	0.937	0.934
15	Polarizability	$\hat{Y} = a \cdot X_1 + b \cdot X_2 + d$	$X_1 = \text{LBgN0419}$ $X_2 = \text{LGGN0488}$	$a = -2.896\ (t = 26.1)$ $b = 0.6533\ (t = 12.7)$ $d = 37.82\ (t = 43.1)$	0.959	0.957
16	Polarizability	$\hat{Y} = a \cdot X_1 + b \cdot X_2 + c \cdot X_1 \cdot X_2 + d$	$X_1 = \text{LDGN0394}$ $X_2 = \text{LDGN0402}$	$a = 760.6\ (t = 14.9)$ $b = 735.4\ (t = 13.8)$ $c = -1.499\ (t = 5.76)$ $d = -58.45\ (t = 3.34)$	0.963	0.961

TABLE 7.11 Z values for Comparison of Explained Variances for Models from Table 7.10 $z(r^2_{adj}) = \mathrm{arctanh}(\sqrt{r^2_{adj}})$; $\sigma(r^2_{adj}) = 1/\sqrt{(45-3)}$; $z_{ij} = (z_i-z_j)/\sigma\sqrt{2}$; $z(2.5\%) = 1.96$

Dipole moment

z_{ij}	1	2	3	4
1				
2	9.78			
3	12.0	2.20		
4	14.1	4.37	2.17	

HOMO energy

z_{ij}	5	6	7	8
5				
6	6.82			
7	10.8	4.02		
8	12.9	6.09	2.07	

LUMO energy

z_{ij}	9	10	11	12
9				
10	11.6			
11	12.8	1.19		
12	13.9	2.27	1.07	

Polarizability

z_{ij}	13	14	15	16
13				
14	8.21			
15	15.0	6.77		
16	16.5	8.30	1.53	

Comparison of the explanatory powers of the models reveals only three cases in which the differences are not significant, namely for LUMO energy when multiplicative or additive effects are used to explain it, and when additive of both additive and multiplicative effects are used to explain it, and for polarizability when additive or both additive and multiplicative effects are used to explain it. In these cases, using of the larger models (with more coefficients) are not fully justified statistically based on what may come from a by chance association.

For the dipole moment (see Table 7.10) using of a model with multiplicative effects only in association with the structure of the compounds selects as a best pair a reciprocal of a characteristic polynomial value (RDtP0066) and a directly proportional one (IDtP0087) while using a model with additive effects only selects as a best pair a logarithm of a characteristic polynomial value (LFgN0609) and a directly proportional one (IDcP0908). Interesting is the fact that when a full additive and multiplicative effects model is used, it is kept the transformation of the descriptors from both partial effects models, being selected reciprocal (RDtN0065) and a logarithmic (LFgN0612) transformed descriptors, when not only the transformation is kept, it is kept also the polynomial formula ("Dt" in RDtP0066 and in RDtN0065; "Fg" in LFgN0609 and in LFgN0612). The only difference is at values in which the polynomials are evaluated (RDtN0065 evaluates the "Dt" polynomial in $-65/1000$ while RDtP0066 evaluates it in $-66/1000$; LFgN0612 evaluates the "Fg" polynomial in $-612/1000$ while LFgN0609 evaluates it in $-609/1000$). The model reveals an association of the dipole moment with the first ionization potential dependent on geometry and solid-state density dependent on topology, being able to explain about 71.1% of the variability by these factors.

Looking at HOMO energy (see Table 7.10) the additive effects model and the full multiplicative and additive effects model have the same composition of descriptors (RGTN0155 and LBgN0874 for additive effects only; RGCN0182 and LBgP0165 for the full effects one). While looking to the descriptors values in the original files in which were placed the evaluation results, one can found that actually RGCN0155 and RGTN0155 provides same series of values for the molecules on which the calculation were applied. Therefore, in this case, the additive effects model and the full effects model possess the same characteristic polynomials formulas ("GC" and "Bg") and the same operations ("R" and "L") on it, the polynomials being only evaluated in different points (RGCN0182 evaluates the "GC" polynomial in $-182/100$ while RGCN0155 evaluates it in $-155/1000$; LBgP0165 evaluates "Bg" polynomial in $165/1000$ while LBgN0874 evaluates it in

–874/1000). Since no change in the composition of the selected model is observed in this case of HOMO energy, it can be said that the multiplicative effects have a minor influence on the values of HOMO energy when this is related with the structure with two characteristic polynomials. The obtained model shows an association of HOMO energy with the melting points and the geometry of the molecules, being able to explain with them about 79.9% of the variability.

For the LUMO energy the situation is totally reversed than for the HOMO energy. The full model borrows its composition from the model with multiplicative effects only (IHGN0132 and IHGN0157 descriptors in multiplicative model; IHGN0131 and IHGN0150 descriptors in full model). Therefore since no change in the composition of the selected model is observed in this case of LUMO energy, it can be said that the additive effects have a minor influence on the values of LUMO energy when this is related with the structure with two characteristic polynomials. The obtained model shows that for LUMO energy the attached hydrogen atoms and the geometry of the molecule plays an important role, and the model is able to explain at least 80.8% of the total variability using these factors.

When looking at polarizability, all additive, multiplicative and full models keeps the same operation (logarithm, "L" letter at the beginning of the descriptors names) while the composition is subject to change from one model to another. It is also interesting that the full model selects the same polynomial for the both descriptors ("DG" in both LDGN0394 and LDGN0402) while the evaluation is in two closer points (-394/1000 and -402/1000). This fact suggests that the best model to be used is this model of full effects since is strongly related with the concept of polarization–a charge separation–which usually takes small values relative to the total charge of an atom or an molecule. The association between the solid-state densities as atomic properties ("D" letter) and polarizability as estimated property requires further studies designed to see more about. The model based on solid-state density and geometry of the molecule is able to explain about 96.1% of the total variability.

7.5 MULTI-/TRANS-DISCIPLINARY CONNECTION(S)

The other one connected polynomial with the Charact-poly is the Laplacian polynomial which use a modified form of the adjacency matrix ([Ad]), the Laplacian matrix ([La]), calculated as [La] = [Dg] – [Ad], where [Dg] simply counts on the main diagonal the number of atom's bonds (the rest of its elements are null; for convenience with the graph theory related concept were noted [Dg] – from vertex degree). The Laplacian polynomial is the Charact-poly of the Laplacian matrix:

$$\text{LaP} \overset{\text{def}}{=} \left| \lambda \cdot \text{Id} - \text{La} \right| = \left| \lambda \cdot \text{Id} - \text{Dg} + \text{Ad} \right|$$

The Laplacian matrix is often used in the analysis of electrical networks. The roots of the Laplacian polynomial uses too, under the name of Laplacian spectra.

7.6 OPEN ISSUES

Based on the conducted study the extension of the characteristic polynomial to take into account the type of the atom when is counted as a vertex in its classical approach and to take into account the type of the bond when is counted as a adjacency in its classical approach, as well as the alternative use of the distance matrix, computed by topologies as well as by geometries are fruitful extensions. The case study reveals that useful information may be bring out from the structure-property and structure-activity study when the extended characteristic polynomial is used. Some disappointments can be recorded as well, one of them being the relatively low (when compared with other family-based derived descriptors) as well as the inconvenience of the calculation of the polynomial for values of the argument outside of the [−1,1] interval, when results of the calculation goes outside of the precision of calculation for any reasonable sized molecule.

7.7 TERMS

The term secular function has been used for what is now called characteristic polynomial (in some literature the term secular function is still used). The term comes from the fact that the characteristic polynomial was used to calculate secular perturbations (on a time scale of a century, i.e., slow compared to annual motion) of planetary orbits, according to Lagrange's theory of oscillations.

Sachs graphs (Sachs, 1962) is a possible enumeration of what the characteristic polynomial counts as authors of Graovac et al. (1972) observed.

KEYWORDS

- **counts of random walks**
- **Eigenvalues**
- **Eigenvectors**
- **quantum chemistry**
- **structure-resonance theory**
- **topological theory of aromaticity**

REFERENCES AND FURTHER READING

Alan M., (2006). The IUPAC international chemical identifier, in Chapter I—A new standard for molecular informatics. *Chem. Int., 28*(6), 12–15.

Cauchy, A., (1829). On the equation with the help of which one determines the secular inequalities of the movements of the planets. *Exerc Math, 4,* 140–160.

Coulson, C. A., (1937). The evaluation of certain integrals occurring in studies of molecular structure. *Mathematical Proceedings of the Cambridge Philosophical Society, 33*(1), 104–110.

Coulson, C. A., (1940). On the calculation of the energy in unsaturated hydrocarbon molecules. *Mathematical Proceedings of the Cambridge Philosophical Society, 36*(2), 201–203.

Coulson, C. A., (1950). Notes on the secular determinant in molecular orbital theory. *Mathematical Proceedings of the Cambridge Philosophical Society, 46*(1), 202–205.

Curticapean, R., (2013). Counting matchings of size k Is ♯ W[1]-hard. *Lecture Notes in Computer Science, 7965,* 352–363.

Diudea, M. V., Gutman, I., & Jäntschi, L., (2001). *Molecular Topology.* New York: Nova Science.

Fock, V. A., (1930). "Self-consistent field" with exchange for sodium (In German). *Zeitschrift für Physik, 62*(11/12), 795–805.

Fock, V. A., (1930). Approximation method for solving the quantum mechanical many-body problem (In German). *Zeitschrift für Physik, 61*(1/2), 126–148.

Givens, W. B., (1957). The characteristic value-vector problem. *J. Assoc. Comput. Mach., 4,* 298–307.

Godsil, C. D., & Gutman, I., (1981). On the theory of the matching polynomial. *Journal of Graph Theory, 5*(2), 137–144.

Godsil, C. D., (1995). Algebraic matching theory. *The Electronic Journal of Combinatorics, 2*(R8), p. 14.

Graovac, A., Gutman, I., Trinajstić, N., & Živkovic, T., (1972). Graph theory and molecular orbitals: Application of Sachs theorem. *Theoretica Chimica Acta, 26,* 67–78.

Hartree, D. R., & Hartree, W., (1935). Self-consistent field, with exchange, for beryllium. *Proc. R. Soc. London, 150*(869), 9–33.

Hartree, D. R., (1928). The wave mechanics of an atom with a non-coulomb central field. Part I. Theory and methods. *Math. Proc. Cambridge, 24*(1), 89–110.

Hartree, D. R., (1928). The wave mechanics of an atom with a non-coulomb central field. Part II. Some results and discussion. *Math. Proc. Cambridge, 24*(1), 111–132.

Hosoya, H., (1971). Topological index. A newly proposed quantity characterizing the topological nature of structural isomers of saturated hydrocarbons. *Bull. Chem. Soc. Japan, 44*, 2332–2339.

Householder, A. S., (1958). Unitary triangularization of a nonsymmetric matrix. *Journal of the ACM, 5*(4), 339–342.

Huckel, E., (1931). Quantentheoretische beiträge zum benzolproblem. *Z. Phys., 70*, 204–286.

Jäntschi, L., & Bolboacă, S. D., (2006). Modeling the inhibitory activity on carbonic anhydrase IV of substituted thiadiazole- and thiadiazoline- disulfonamides: Integration of structure information. *Electronic Journal of Biomedicine, 2*, 22–33.

Laplace, P. S., (1776). Research on the integral calculus and the system of the world (In French). *Mem Acad Sci Paris, 2*, 47–179.

Markle, R. C., (2000). Molecular building blocks and development strategies for molecular nanotechnology. *Nanotechnology, 11*(2), 89–99.

Morris, R., & Cherry, L., (1975). *BC–Version 6 Unix, 6*[th] *Edition Unix BC Source Code.* Available at: http://minnie.tuhs.org/cgi-bin/utree.pl?file = V6/usr/source/s1/bc.y.

Nelson, P. A., (1991). *BC–an Arbitrary Precision Calculator Language, Documentation of Version 1.06.* Available at: https://www.gnu.org/software/bc/manual/html_mono/bc.html.

Ramaraj, R., & Balasubramanian, K., (1985). Computer generation of matching polynomials of chemical graphs and lattices. *J. Comput. Chem., 6*, 122–141.

Rehman, R., & Ipsen, I. C. F., (2011). *La Budde's Method for Computing Characteristic Polynomials.* arXiv:1104.3769 (URL: http://arxiv.org/format/1104.3769).

Sachs, H., (1962). Uber selbstkomplementiire graphen. *Publ. Math., 9*, 270–282.

Schrödinger, E., (1926). An undulatory theory of the mechanics of atoms and molecules. *Physical Review, 28*(6), 1049–1070.

Slater, J. C., (1929). The theory of complex spectra. *Phys. Rev., 34*(10), 1293–1295.

Sloane, N. J. A., (1996). *Number of Graphs on n Unlabeled Nodes.* OEIS(A000088), http://oeis.org/A000088.

Sylvester, J. J., (1880). On the theorem connected with Newton's rule for the discovery of imaginary roots of equations. *Messenger of Mathematics, 9*, 71–84.

Tománek, D., (2014). *Supplementary information (online) of: Guide through the Nanocarbon Jungle: Buckyballs, Nanotubes, Graphene, and Beyond. "C42 Isomers,"* URL: http://www.nanotube.msu.edu/fullerene/fullerene.php?C=42."

Von Collatz, L., & Sinogowitz, U., (1957). Spectra of finite graphs (In German). *Abhandlungen aus dem Mathematischen Seminar der Universität Hamburg, 21*(1), 63–77.

Weisstein, W. E., (2003). Nu*mber of Unique Characteristic Polynomials Among all Simple Undirected Graphs on n Nodes.* OEIS (A082104), http://oeis.org/A082104.

CHAPTER 8

Clar Structure

MARINA A. TUDORAN[1,2] and MIHAI V. PUTZ[1,2]

[1]*Laboratory of Structural and Computational Physical Chemistry for Nanosciences and QSAR, Biology-Chemistry Department, West University of Timisoara, Pestalozzi Street No. 44, Timisoara, RO-300115, Romania*

[2]*Laboratory of Renewable Energies-Photovoltaics, R&D National Institute for Electrochemistry and Condensed Matter, Dr. A. Paunescu Podeanu Str. No. 144, Timisoara, RO-300569, Romania, Tel: +40-256-592638, Fax: +40-256-592620, E-mail: mv_putz@yahoo.com or mihai.putz@e-uvt.ro*

8.1 DEFINITION

The Clar structure is defined as a structure, which presents the maximum number of benzenoid rings with inscribed circles with the property that there is no circle in adjacent rings. Another definition state that a Clar structure is represented by the valence structure having a maximum number of π-sextets which are nonadjacent.

8.2 HISTORICAL ORIGIN(S)

Even if the notion of the aromatic sextet was introduced for the first time in structural chemistry in 1925, Clar was the one who promotes, in 1958, the "aromatic sextet" from the single six-member rings to polycyclic benzenoid hydrocarbons (Randic, 2014). He formulated the Clar's rule on the great stability of systems with $6n$ π electrons (Clar, 1972): "*A fully benzenoid hydrocarbon is the one which present $6n$ π electrons, meaning that the Huckel's rule $(2n+4)n$ π electrons is not respected by the most stable hydrocarbons.*" A general definition (Clar, 1964) state that, for a benzenoid hydrocarbon, the Clar formula can be constructed by drawing circles (which represent the aromatic sextets) in some hexagons from the benzenoid system following three rules: Rule 1 states that neighboring hexagons most not contain circles, Rule 2 states that the arrangement of circles must be made in a way so that in the rest of the molecule one should be able to write Kekule structures and Rule 3 states that the final formula should contain the maximum number of circles (Gutman, 1985).

Clar classifies the benzenoids hydrocarbons in three types: fully benzenoid benzenoids which has only sextet rings and "empty" rings (e.g., cyclohexane), benzenoid systems with a single Clar structure which has rings with C = C bond besides aromatic sextet rings–this structures may or may not possess empty rings (e.g., naphthalene) and benzenoid systems with two or more Clar structure, where the "sextet" can migrate inside the molecule, and in general, only one formula is represented, the other possible cases being represented by an arrow for indicating the alternative locations for the aromatic sextet; e.g., coronene (Randic, 2014). Another classification can be made by considering the type of Clar representations. The first class is represented by benzenoid hydrocarbons with unique Clar formula (e.g., teropyrene): the Clar formula is possibly a proper representation of the chemical behavior and of the real electronic structure of the molecule. The second class contains benzenoid hydrocarbons with different Clar formula (e.g., dibenzo[a,e]pyrene): the actual electronic configuration for this type of molecule is represented by a resonance hybrid of the all possible Clar structures (Gutman, 1985).

Another aspect worth mentioning about the Clar theory is that the Clar formula can be seen as a shorted notation of a Kekule structures group (Clar & Zander, 1958; Clar et al., 1959), a Clar formula with n sextets being correspondent to 2^n Kekule structures. In a general manner, Clar's theory can be seen as a resonance theory which is based on a selected number of relevant Kekule structures, the irrelevant structures being neglected (Gutman, 1985).

8.3 NANO-SCIENTIFIC DEVELOPMENT(S)

The aromatic sextet theory proposed by Clar (1972) can be considered a method to estimate the degree of cyclic conjugation in benzenoid hydrocarbons in a semi-quantitative manner. In Clar model, the aromatic sextet formula is used to describe the π-electrons arrangement, the hexagon in which an aromatic sextet (a circle) is present is assuming to have a higher degree of cyclic conjugation, while the hexagons without circle are assuming to be less conjugated (Gutman & Ivanov-Petrovic, 1997). In order to construct the aromatic sextet, one should consider that the remaining parts of the molecule which do not present rings must have Kekule structures. Another condition is represented by the fact that adjacent hexagon does not possess aromatic sextets, but one should also consider, in the same time, the maximum possible number of rings which can be drawn with respecting the previous rules (Gutman, 1982; Gutman & Cyvin, 1988). For benzenoid hydrocarbons, Clar theory can be applied to describe, in a reliable way, the configuration of π-electrons and their physicochemical behavior (Gutman & Ivanov-Petrovic, 1997).

Another concept from the organic chemistry is represented by the Pauling ring bond order, defined as the sum of Pauling bond orders for the six carbon-carbon bonds which form the individual ring, and is usually used to promote the Clar theory of aromatic sextets from its qualitative character to a quantitative level (Randic, 2014). Studies in this area propose the idea that Pauling bond orders determine the average count of the carbon-carbon double bond in individual rings for all Kekule valence structures, and in this way is seen as a parallel for regularities of Clar sextet model. Figure 8.1 presents the Clar conformation and the Pauling ring bond order for several small polycyclic aromatic hydrocarbons (PAHs) molecules.

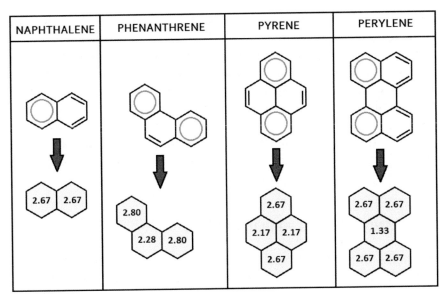

FIGURE 8.1   PAHs molecules, their Clar conformations, and Pauling ring bond order.

The parallelism between the relative magnitude of numerical Pauling ring bond order and the intuitive notion of aromatic sextets, empty rings, and migrating sextets create premises for interpreting numerical characterization of rings in benzenoid systems using Pauling bond orders, as a quantitative representation of the qualitative Clar theory of aromatic sextet. There are two approaches known to be most used to characterize the individual benzene rings in benzenoid hydrocarbons. For instance, one can consider the case where the π electrons are partitioned in carbon-carbon double bonds in individual Kekule valence structures to separate rings (Randic, 2004). Another approach assumes the Pauling bond order partition to individual benzene rings (Gutman et al., 2004). In his work, Randic (2014) propose an alternative approach, in which the contribution from the π electrons from the carbon-carbon double bond is assigned to both rings, instead of being partitioned between two rings as in the general case. This way, he gives an alternative interpretation of the numerical ring values in case of Clar aromatic sextets as the average of carbon-carbon double bond content of the ring which is taken over all Kekule valence structure (Randic, 2014).

A novel approach on Clar structures is represented by the work of Randic and Plavsic (2011) where they propose a new classification of these structures in three types: Clar Structural formulas, Algebraic Clar formulas, and Canonical Clar formulas. According to the definition, in a Clar structural formula, the aromatic π-sextets cannot be placed in adjacent benzene rings. On the other hands, for determining the algebraic Clar structures, one needs to count in each ring the number of times of a π-aromatic circle being inscribed in the set of all possible Clar structure of the respective molecule. Moving forward, a canonical Clar structure is defined as the Clar structure in which the non-adjacent benzenoid rings with the largest π-sextets ring values are the ones in which the ring are inscribed (the π-sextets). Their approach is based on the numerical characterization of Clar structures and can be used as an alternative method of local molecular features characterization for polycyclic conjugated hydrocarbons (Randic & Plavsic, 2011).

8.4 NANOCHEMICAL APPLICATION(S)

According to graph theory, a fullerene $\Gamma = (V,E,F)$ is defined as a three-regular plane graph which presents only pentagonal and hexagonal faces. For a given fullerene Γ, one can define a Kekule structure $K \subseteq E$ (also known as a perfect matching) as a set of edges with the property that each vertex is incident exactly with one edge in K (Hartung, 2013). The Clar number represents the maximum cardinality of a set which contains the independent benzene faces overall Kekule structures ("benzene faces" states for the faces which have exactly three of their bonding edges in K). For a given plane graph $\Omega = (V, E, F)$, one can define a vertex covering (C, A) as a set of edges A and faces C with the property that each vertex of Ω is incident with only one element from $C \cup A$. For a face f of Ω, one can say that an edge a (with $a \in A$) lies on f if both vertices of a are incident with f, and a exits in f if a and f share only one vertex. If the set A is empty, (C, A) is called a face-only vertex, or a perfect Clar structure (Fowler & Pisanski, 1994).

In his work, Hartung (2013) propose two theorems based on Clar theory. The first theorem states that a plane graph Ω with minimum degree 3 and a vertex covering (C, A) have the property that odd number of edges from A exit any face of odd degree which do not belong to, and even number of edges from A (or none) exit any face of even degree (Hartung, 2013). For a fullerene Γ, a Clar structure (C, A) is a vertex covering, with C consisting only of hexagonal faces and most of the two edges in A lie on any face of Γ. When the Clar structure (C, A) have the maximum number of faces in C, one will obtain the Clar number of fullerene. Starting from the fact that every vertex of Γ is incident with only one element from the Clar structure, every edge in A contains two vertices, and every face in C contains six vertices, will result in the formula (Hartung, 2013):

$$|C| = \frac{|V|}{6} - \frac{|A|}{3} \qquad (1)$$

For $|A| = 6$, one will obtain a complete Clar structure, given by the formula:

$$|C| = \frac{|V|}{6} - 2 \qquad (2)$$

In the same manner, theorem 2 states that, if a fullerene Γ has a complete structure (C, A), there is a pairing of the twelve pentagons with the property that each pair is connected by a single edge; the six edges between the pairs of pentagonal faces are the only edges in A (Hartung, 2013). Fullerenes with Clar number equal with $(|V|/6)-2$ are called by Ye and Zhang (Zhang & Ye, 2007) extremal fullerenes.

8.5 MULTI-/TRANS-DISCIPLINARY CONNECTION(S)

In one of their work, Zhang and Zang (Zhang, 1993, 1997; Zhang & Zhang, 1996, 2000) develop a mathematical object, called Zhang-Zhang polynomial, which is related to the Clar theory of aromatic sextets for benzenoid hydrocarbons and has the following form:

$$\zeta(x) = \zeta(B,x) = \sum_{k \geq 0} z(B,k) x^k \qquad (3)$$

with B as the benzenoid system, $z(B, k)$ as the polynomial coefficients which counts the number of ways in which k mutual isolated edges and hexagons can cover the vertices of B (Gutman et al., 2005). The Zhang-Zhang polynomial has several properties, depicted in Table 8.1.

TABLE 8.1   Properties of Zhang-Zhang Polynomial

Property	Definition
Property 1	There is an equality between the power of $\zeta(B, x)$ and the number of aromatic sextets in any of the Clar formulas (Randic, 2003, Hansen & Zheng, 1994), also known as the Clar number of B, noted with $Cl(B)$
Property 2	There is an equality between the coefficient $z(B, Cl)$ and the number of formulas of Clar aromatic sextet of B, noted with $C(B)$
Property 3	There is an equality between the coefficient $z(B, 1)$ and the number of the off-diagonal matrix elements with the highest value γ_1 in Herndon's approach of resonance theory (Gutman & Cyvin, 1989); there is also an equality between $z(B, 1)$ and the number of edges from the resonance graph, noted with $RG(B)$, which is associated with B (Randic, 1997)
Property 4	There is an equality between the coefficient $z(B, 0)$ and the number of Kekule structure, noted with $K(B)$, of the structure of B.

There is also a relation between the resonance energy (RE) and the Zhang-Zhang polynomial. By considering $B_1 \cup B_2$ as a molecular graph for two disconnected π electron systems, then will result in the equality (Gutman et al., 2005):

$$RE(B_1 \cup B_2) = RE(B_1) + RE(B_2) \qquad (4)$$

Knowing that:

$$K(B_1 \cup B_2) = K(B_1) \cdot K(B_2) \qquad (5)$$

and considering that resonance energy represent some function of the Kekule structure count, this will lead to the following formula (Wilcox, 1975; Swinborne-Sheldrake et al., 1975):

$$RE \approx a_0 \ln K \qquad (6)$$

where a_0 represents a constant depending on the RE type and the benzenoid system class.

Taking into consideration that Zhang-Zhang polynomial is following an analog relation of the form:

$$\zeta(B_1 \cup B_2, x) = \zeta(B_1, x) \cdot \zeta(B_2, x) \qquad (7)$$

the following approximation may be possible:

$$RE \approx a \ln \zeta(x) \qquad (8)$$

with a, the fitting parameter designed as a variable x.

If property 4 is considered and for $x = 0$, formula (7) may be seen as a special case of formula (8) (Gutman et al., 2005).

8.6   OPEN ISSUES

Even if the aromaticity of organic polybenzenoid hydrocarbons (PBHs) is well studied, there are few studies made on the aromaticity of their inorganic boron-nitride (BN) analogs, and the presence of Clar structures in this type of inorganic compounds is yet to be known. Worth mentioned are the works of Madura et al., (1998) and Shen et al., (2007) on the aromaticity of borazine and BN-naphthalene, their results determining that BN-acenes present more localized π-electron characteristics than their corresponding organic scenes. An investigation on the possible existence of the Clar structure in inorganic systems density was made by Wu and Zhu (2015) using density functional theory (DFT) calculations to determine the aromaticity and stability of a working set of several BN-[n]acenes ($n = 3$–10). The authors use different types of indices: electron localization function, noted with ELF, where π states for the π components; the bifurcation values of the electron localization function, noted with BV(ELF$_\pi$); the multicenter index of aromaticity, noted with MCI (Bultinck et al., 2005), which can be used to evaluate multifold aromaticity for both organic and inorganic species (Radenkovic et al., 2015) due to its ability to being split into σ- and π-contributions; and the six-center index, noted with SCI, which is used to correlate electron delocalization for six-membered rings. The resulted values for ELF$_\pi$ and SCI determined that BN analogs of PBHs present reduce aromaticity. However, isomers with more π-sextets are present higher thermodynamic stability; this fact suggesting that the Clar structure should exist in this type of structures (Wu & Zhu, 2015).

KEYWORDS

- **algebraic Clar formulas**
- **aromatic sextet**
- **canonical Clar formulas**
- **Clar structural formulas**
- **empty rings**
- **Pauling ring bond order**
- **vertex covering**

REFERENCES AND FURTHER READING

Bultinck, P., Ponec, R., & Van Damme, S., (2005). Multicenter bond indices as a new measure of aromaticity in polycyclic aromatic hydrocarbons. *Journal of Physical Organic Chemistry, 18*(8), 706–718.

Clar, E., & Zander, M., (1958). 1:12–2:3–10:11-Tribenzoperylene. *Journal of the Chemical Society,* 1861–1865.

Clar, E., (1972). *The Aromatic Sextet.* John Wiley & Sons, London.

Clar, E., Ironside, C. T., & Zander, M., (1959). The electronic interaction between benzenoid rings in condensed aromatic hydrocarbons. 1: 12–2: 3–4: 5–6: 7–8: 9–10: 11-hexabenzocoronene, 1: 2–3: 4–5: 6–10: 11-tetrabenzoanthanthrene, and 4: 5–6: 7–11: 12–13: 14-tetrabenzoperopyrene. *Journal of the Chemical Society,* 142–147.

Fowler, P. W., & Pisanski, T., (1994). Leapfrog transformations and polyhedra of Clar type. *Journal of the Chemical Society, Faraday Transactions, 90,* 2865–2871.

Gutman, I., & Cyvin, S. J., (1989). *Introduction to the Theory of Benzenoid Hydrocarbons* (Vol. 9, pp. 10). Springer, Berlin.

Gutman, I., & Ivanov-Petrovic, V., (1997). Clar theory and phenylenes. *Journal of Molecular Structure (Theochem), 389,* 227–232.

Gutman, I., (1985). Clar formulas and Kekule structures. *Match, 17,* 75–90.

Gutman, I., Gojak, S., & Furtula, B., (2005). Clar theory and resonance energy. *Chemical Physics Letters, 413,* 396–399.

Gutman, I., Morokawa, T., & Narita, S., (2004). On π-electron content of bonds and rings in benzenoid hydrocarbons. *Zeitschrift Fur Naturforschung, 59a,* 295–298.

Hansen, P., & Zheng, M., (1994). The Clar number of a benzenoid hydrocarbon and linear programming. *Journal of Mathematical Chemistry, 15*(1), 93–107.

Hartung, E., (2013). Fullerenes with complete Clar structure. *Discrete Applied Mathematics, 161,* 2952–2957.

Madura, I. D., Krygowski, T. M., & Cyranski, M. K., (1998). Structural aspects of the aromaticity of cyclic pi-electron systems with BN bonds. *Tetrahedron, 54*(49), 14913–14918.

Radenkovic, S., Tosovic, J., & Nikolic, J. D., (2015). Local aromaticity in naphtha-annelated fluoranthenes: Can the five-membered rings be more aromatic than the six-membered rings? *The Journal of Physical Chemistry A., 119*(20), 4972–4982.

Randic, M., & Plavsic, D., (2011). Algebraic Clar formulas–numerical representation of Clar structural formula. *Acta Chimica Slovenica, 58*, 448–457.

Randic, M., (1997). Resonance in catacondensed benzenoid hydrocarbons. *International Journal of Quantum Chemistry, 63*, 585–600.

Randic, M., (2004). Algebraic kekulé formulas for benzenoid hydrocarbons. *Journal of Chemical Information and Computer Sciences, 44*(2), 365–372.

Randic, M., (2014). Novel insight into Clar's aromatic π-sextets. *Chemical Physics Letters, 601*, 1–5.

Shen, W., Li, M., Li, Y., & Wang, S., (2007). Theoretical study of borazine and its derivatives. *Inorganica Chimica Acta, 360*, 619–624.

Swinborne-Sheldrake, R., Herndon, W. C., & Gutman, I., (1975). Kekulé structures and resonance energies of benzenoid hydrocarbons. *Tetrahedron Letters, 16*(10), 755–758.

Wilcox, C. F., (1975). Topological definition of resonance energy. *Croatica Chemica Acta, 47*, 87–94.

Wu, J., & Zhu, J., (2015). The Clar structure in inorganic BN analogs of polybenzenoid hydrocarbons: Does it exist or not? *A European Journal of Chemical Physics and Physical Chemistry, 16*, 3806–3813.

Zhang, H., & Zhang, F., (1996). The Clar covering polynomial of hexagonal systems. I. *Discrete Applied Mathematics, 69*, 147–167.

Zhang, H., & Zhang, F., (2000). The Clar covering polynomial of hexagonal systems. III. *Discrete Mathematics, 212*(3), 261–269.

Zhang, H., (1993). Clar covering polynomials of S. T-Isomers. *MATCH Communications in Mathematical and in Computer Chemistry, 29*, 189–197.

Zhang, H., (1997). The Clar covering polynomial of hexagonal systems with an application to chromatic polynomials. *Discrete Mathematics, 172*, 163–173.

CHAPTER 9

Connectivity Index

BOGDAN BUMBĂCILĂ[1] and MIHAI V. PUTZ[1,2]

[1]*Laboratory of Computational and Structural Physical Chemistry for Nanosciences and QSAR, Biology-Chemistry Department, Faculty of Chemistry, Biology, Geography at West University of Timișoara, Pestalozzi Street No.16, Timișoara, RO-300115, Romania*

[2]*Laboratory of Renewable Energies-Photovoltaics, R&D National Institute for Electrochemistry and Condensed Matter, Dr. A. Paunescu Podeanu Str. No. 144, RO-300569 Timișoara, Romania, Tel.: +40-256-592-638; Fax: +40-256-592-620, E-mail: mv_putz@yahoo.com or mihai.putz@e-uvt.ro*

9.1 DEFINITION

Connectivity indices are topological indices used in QSAR studies. They are molecular descriptors with mathematically obtained values from the characteristics of molecular graphs. Molecular connectivity indices are descriptors of the molecular accessibility. The first- and second-order connectivity indices represent molecular accessibility surface areas and volumes, and higher order indices represent magnitudes in higher dimensional spaces. The atom degrees are measures of the accessibility perimeter of the corresponding atom. The accessibility perimeter is computed here from the van der Waals and covalent radii of the atoms and the overlapping angle between the van der Waals circumferences of bonded atoms (Estrada, 2002).

The essence of molecular connectivity is the encoding of a molecular structure in a non-empirical way. It is not a measured characteristic, and it does not derive from a particular physical property, and it does not translate into one (Kier et al., 2002).

9.2   HISTORICAL ORIGINS

The first true connectivity index introduced in computational chemistry and QSAR/QSPR was Randić index–1975. Its name comes from Milan Randić (born 1930), an American Croatian physicist and theoretical chemist.

9.3   NANOCHEMICAL IMPLICATIONS

The Randić index, R, was proposed as a "branching degree" index, suitable for measuring the carbon atom branching of a saturated hydrocarbon.

$$R(G) = {}^{1}\chi = \sum_{i,j \in E} \frac{1}{\sqrt{\delta_i \delta_j}} \tag{1}$$

where δ_i, δ_j are the degree of vertices incident with the bond $i \sim j$.

It was experimentally determined that there is a good correlation between the Randić index and some physical-chemical properties of alkanes like chromatographic retention times, boiling points, enthalpies of formation (Yang, 2011).

The term of the sum $(\delta_i \delta_j)^{-1/2}$ is called *edge connectivity* and it characterizes the accessibility of the edges (the accessibility of a bond to encounter another bond in intermolecular interactions).

This connectivity index is known as a first order index. The 0 order Randić connectivity index is thus calculated with the following formula:

$$^{0}\chi = \sum_{i \in E} \frac{1}{\sqrt{\delta_i}} \tag{2}$$

Randić index was then generalized into a series of connectivity indices, where the sums were made over paths of different lengths. The generalized connectivity index χ of length h is then calculated as:

$$^{h}\chi = \sum \frac{1}{\sqrt{\delta_i \delta_j \delta_k \dots \delta_h}} \tag{3}$$

Calculations for χ of different lengths are made bellow, using the graph (Figure 9.1):

$$^{0}\chi = (1)^{-1/2} + (1)^{-1/2} + (1)^{-1/2} + (4)^{-1/2} + (2)^{-1/2} + (3)^{-1/2} + (1)^{-1/2} + (1)^{-1/2}$$
$$^{0}\chi = 8.37$$

$$^1\chi = \left(4\times1\right)^{-1/2} + \left(4\times1\right)^{-1/2} + \left(4\times1\right)^{-1/2} + \left(4\times2\right)^{-1/2} + \left(2\times3\right)^{-1/2} + \left(3\times1\right)^{-1/2}$$
$$+ \left(3\times3\right)^{-1/2} + \left(3\times1\right)^{-1/2} + \left(3\times1\right)^{-1/2}$$
$$^1\chi = 4.33$$

$$^2\chi = \left(1\times4\times1\right)^{-1/2} + \left(1\times4\times1\right)^{-1/2} + \left(1\times4\times1\right)^{-1/2} + \left(1\times4\times2\right)^{-1/2} + \left(1\times4\times2\right)^{-1/2}$$
$$+ \left(1\times4\times2\right)^{-1/2} + \left(4\times2\times3\right)^{-1/2} + \left(2\times3\times1\right)^{-1/2} + \left(2\times3\times3\right)^{-1/2} + \left(1\times3\times1\right)^{-1/2}$$
$$+ \left(3\times3\times1\right)^{-1/2} + \left(3\times3\times1\right)^{-1/2} + \left(1\times3\times1\right)^{-1/2}$$
$$^2\chi = 4.31$$

Other connectivity indices used today in QSAR studies are Wiener Index, Zagreb group Indices, Hosoya Index, Balaban Index, and Atom Bond Connectivity Index.

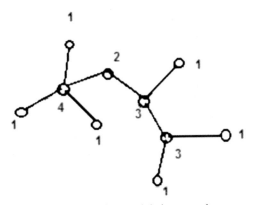

FIGURE 9.1 Molecular graph with vertices and their vertex degrees.

The Atom Bond Index can be defined as

$$ABC = \sum_{i \sim j} \sqrt{\frac{2\left(\delta_i + \delta_j - 2\right)}{\delta_i \delta_j}} \tag{4}$$

where δ_i and δ_j are the degrees of two vertices, i and j which are not necessarily adjacent (Furtula et al., 2009).

The Wiener index is a topological index which is showing the branching of a molecular graph, but today it is used to assess the centricity of a graph (Rouvray et al., 2002). The Wiener index is named after Harry Wiener, who introduced it in 1947 as the "path number"–the sum of the lengths of the shortest paths between all pairs of vertices in the molecular graph represented by non-hydrogen atoms (Wiener, 1947)

$$W = 1/2 \sum_{i=1}^{n} \sum_{j=1}^{n} d_{ij} \qquad (5)$$

where d_{ij} is the distance matrix–meaning the shortest path between a pair of vertices (i and j)–the smallest number of edges connecting them (Mohar et al., 1988).

It was shown that the Wiener index can be well correlated with the boiling points of alkanes molecules (Stiel et al., 1962). Later work in QSAR showed that it is also correlated with other physical-chemical characteristics including their density, surface tension and viscosity of their liquid phase and the van der Waals surface of the alkane molecule (Rouvray et al., 1976, Gutman et al., 1995).

The Zagreb Indices have been introduced in 1972, by Nenad Trinajstić and Ivan Gutman.

They are expressed as:

$$M_1 = \sum_{i \in V} \delta_i^2 \qquad (6)$$

$$M_2 = \sum_{i,j \in E} \delta_i \delta_j \qquad (7)$$

Here, M_1 is called as the First Zagreb Index or the Gutman Index and M_2 the Second Zagreb Index. δ_i is the degree of the vertex i (the number of its neighbors) and the term $\delta_i \delta_j$ is the weight of the edge defined by vertexes i and j (Gutman et al., 1972).

The Balaban index J is a connectivity index defined for a graph with n nodes and m edges by the following equation:

$$J = \frac{m}{\gamma + 1} \sum_{i=1}^{n} \sum_{j=1}^{n} \left(D_i D_j \right)^{-\frac{1}{2}} \qquad (8)$$

where m is the total number of edges in the graph, n–the total number of vertices, $\gamma = (m-n+1)$–the cyclomatic number of the molecular graph, D_i and D_j are distance sums calculated as the sums over the rows or columns of the topological distance matrix of the molecule, D (Balaban, 1982). The term $(D_i D_j)$ calculated from the distance matrix has to be taken into consideration only for adjacent vertices i and j.

An example is given in Figure 9.2.

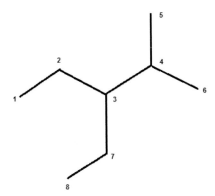

FIGURE 9.2   Molecular graph with numbered vertices.

The distance matrix of the graph is:

Atom	1	2	3	4	5	6	7	8	Distance sum
1	0	1	2	3	4	4	3	4	21
2	1	0	1	2	3	3	2	3	15
3	2	1	0	1	2	2	1	2	11
4	3	2	1	0	1	1	2	3	13
5	4	3	2	1	0	2	3	4	19
6	4	3	2	1	2	0	3	4	19
7	3	2	1	2	3	3	0	1	15
8	4	3	2	3	4	4	1	0	21

$m = 7$ (edges), $n = 8$ (vertices), $\gamma = 7/\left[(7-8+1)+1\right] = 7$

$J = 7\times[(21\times15)^{-1/2} + (15\times11)^{-1/2} + (11\times13)^{-1/2} + (13\times19)^{-1/2}$
$+ (13\times19)^{-1/2} + (11\times15)^{-1/2} + (15\times21)^{-1/2}]$

$J = 7\times(0.056 + 0.078 + 0.084 + 0.064 + 0.064 + 0.078 + 0.056) = 3.36$

The first order molecular connectivity index (Randić index of length 1) was largely used in QSAR/QSPR studies. It has been discovered that this particular index has good correlations with the molecular surface area. It is a simple and accurate measure of the molecular surface for various classes of molecules, and it correlates well with most of the molecular surface-dependent properties and processes.

Kier and Hall developed a valence molecular connectivity index, derived from the non-hydrogen part of the molecule. So this leads to a first classification of the molecular connectivity indices: simple molecular connectivity indices and valence molecular connectivity indices. In the valence approximation non-hydrogen atoms are described by their atomic valence δ^v values, calculated from the electron configuration:

$$\delta^v = \frac{Z^v - h}{Z - Z^v - 1} \qquad (9)$$

Z^v is the number of valence electrons in the atom, Z–the atomic number, h–the number of hydrogen atoms bound to the same atom (Kier et al., 2002).

Thus, the valence connectivity index will be assessed with the following formula:

$$^1\chi = \sum \frac{1}{\sqrt{\left(\delta_i^v \delta_j^v\right)}} \qquad (10)$$

The second classification in both the simple and valence molecular connectivity indices can be made. Especially for molecular connectivity indices with orders/lengths higher than 2, it is necessary to specify the subclass of the index: path indices, cluster indices, path/cluster indices and chain indices $^n\chi$, $n = 0,1.../\chi_C/\chi_{PC}/\chi_{CH}$. These subclasses are defined by the type of the structural subunit they are describing. In fact, the order of path indices can go from 0 up to the number of bonds between non-hydrogen atoms. The lowest order/length for cluster indices is 3 but orders higher than 4 don't have much chemical sense. The lowest order for path-cluster indices is 4 and lengths higher than 6 don't have too much chemical significance. The lowest order for chain indices is 3 and it goes up to the size of the largest ring in the molecule.

The valence counterparts of all four subclasses of higher length indices can be defined by analogy described above for the first-order valence molecular connectivity index (Karcher et al., 1990).

The path indices can also be divided into 2 subgroups: the first containing the indices of lengths 0,1 and 2, which is best describing the global molecular properties like size, surface, volume and the second subgroup, containing all the higher lengths indices, which are more refined in describing local structural properties and possible interactions between the molecule in cause and another one.

The main property of cluster type indices is that all bonds are connected to the common central atom, in a star-like structure. The third order cluster molecular connectivity index is the first of the cluster type indices where three bonds are joined with the common central atom—for example, t-butane.

$$^3\chi_C = \sum \frac{1}{\sqrt{\left(\delta_i \delta_j \delta_k \delta_l\right)}} \qquad (11)$$

where i, j, k, l correspond to the individual non-hydrogen atoms that form the subgraph.

The fourth order path-cluster molecular connectivity index is the first member of the path-cluster indices, and it refers to subgraphs consisting of four adjacent bonds between non-hydrogen atoms, from which three of them are joined to the same non-hydrogen atom (for example, i-pentane).

$$^4\chi_{PC} = \sum \frac{1}{\sqrt{\left(\delta_i \delta_j \delta_k \delta_l \delta_m\right)}} \tag{12}$$

The cluster and path-cluster indices are describing local structural properties, for example, the branching degree in a molecule. They are highly sensitive to changes in branching and their value rapidly increases with the degree of branching. For example, the fourth order path-cluster molecular connectivity index is accurate in describing activities or physical-chemical properties of branches alkyl-alcohols and alkyl-chlorides and also the substitution pattern on the benzene ring, so it can describe well properties of polysubstituted benzenes. Its value rapidly increases with the degree of substitution on the benzene ring or with the proximity of substitution in benzene-substituted isomers (Karcher et al., 1996).

The chain type molecular connectivity indices are describing the rings in a molecule and the substitution patterns of those rings. They were found to be useful in accurately describing the chromatographic behavior of substituted benzenes (Karcher et al., 1990).

9.4 TRANS-DISCIPLINARY CONNECTIONS

The most successful for describing a molecule in QSAR/QSPR studies of all the topological indices are the molecular connectivity indices because of their wide applications in various areas of science: physics, chemistry, biology, pharmacology and drug design, environmental sciences. This statement is sustained by the fact that these indices are based on sound chemical, topological and mathematical/geometrical grounds, and they were developed with the idea of parallelism with important physical-chemical properties like boiling points, mobility on chromatography columns, enthalpies of formation and molecular surface areas. Comparative studies proved that the novel molecular indices are performing better than Wiener and Balaban Numbers and this can be explained by the fact that in the old indices there is contained too much structural information which can obscure the significant factors for a particular chemical property (Karcher et al., 1990).

One of the main goals of Chemistry of polymers is the synthesis of new molecules having defined properties. The use of topological and connectivity indices as structural descriptors of their macromolecules could be important for an optimal polymer design (Camarda et al., 1999).

The Atom Bond Connectivity Index (ABC) can be a good predictor for the thermochemical data of alkanes, like the heats of formation. The ABC index is an example of how a topological index can be used in reproducing and predicting experimental data, according to Gutman et al., (2012).

One study, in 2011, conducted by Shobha, J. et al., proved that there can be found a good correlation between the minimum inhibitory concentrations of some phenols with bacterial growth inhibitory activity (over *Porphyromonas gingivalis, Streptococcus sobrinus,* and *Selemonas artemidis*) and Randić connectivity indices of 0-4th degree (length). Calculations of the indices were made using DRAGON software (Shobha et al., 2011).

In 1999, Estrada et al., demonstrated the linear independence of the edge-connectivity index to other topological indices by performing an analysis on octane isomers. The edge-connectivity index did not produce linear correlations with any of the approx. 40 topological indices studied. This index produced the best single-variable quantitative structure-property relationship models for some physic-chemical properties of the studied octanes. It was concluded that the edge connectivity index is an independent index containing important structural information to be used in QSAR/QSPR studies (Estrada et al., 1999).

In 1982, Sabljić and co-workers studied the second-order valence molecular connectivity indices, which were found to correlate extremely well with the bioconcentration factors in fish, obtained from the flowing water method, for various halocarbons (hydrocarbons, benzenes, biphenyls, and diphenyl oxides). Bioconcentration factor is the concentration of a chemical in an organism divided by the concentration in water and it is one of the most important indicators for the fate of chemicals in the environment. Present methods for a preliminary estimation of the concentration of hazardous chemicals in biological systems are based on empirical parameters, like water solubility (WS) and octanol-water partition coefficient (K_{o-w}). The accuracy of these methods is very low and the experimental determination of empirical parameters can be very costly and time-consuming. Thus, a practical and efficient way for the prediction of bioconcentration factors of hazardous chemicals has been proposed. Molecular connectivity indices, based on molecular topology (number and types of atoms and chemical bonds), are purely non-empirical data and their calculation is quite simple. Several well known and extensively used pesticides (DDT, DDD, DDE, heptachlor, and dieldrin) have been chosen to test the predictive ability of the equation describing the parabolic relationship between second-order valence molecular connectivity indices and bioconcentration factors (Sabljić et al., 1982).

9.5 OPEN ISSUES

Novel connectivity indices like the eccentric connectivity index (Sharma et al., 1997) and sum-connectivity index (Zhou et al., 2009) are today developed, especially for improving the low degeneracy of the already known indices, for a higher sensitivity toward branching, for the presence of a heteroatom, and also for better correlations between these calculated indices and the laboratory determined physical-chemical properties of compounds, so that the estimations would be more accurate.

For example, in 2000, Li et al., have proposed a novel improved connectivity index. A correct choice of the index in a QSAR/QSPR study is very important. Only when the index is accurately reflecting the structure of the molecules, it can direct the molecule design. Two parameters are very important in QSAR/QSPR studies: topology indices and quantum chemistry characteristics. These are obtained at different levels of assessing the expression of the molecular structure. Topological indices are obtained from molecular ichnography, and they are calculated only based on the framework of a molecule, but some non-binding forces are also important in QSAR/QSPR studies. Quantum chemistry indices are calculated from the molecular configuration at the quantum level, which can reveal all the electronic information of a molecule, such as the electron distribution probabilities, energy, polarity, and electrical charge. Still, the quantum chemistry method has some limitations: it cannot describe the volume effects of a molecule when interacting with another molecule, and these effects are very important in QSAR/QSPR when the molecule has therapeutic properties and a mechanism of action explained by the process of binding to an *in-vivo* receptor. The authors of the study concluded that the methods are complementary to each other.

The improved formula is:

$$\delta_i' = \frac{Z_i^v (\sigma_i - h_i)}{N^2} \qquad \text{(9-bis)}$$

where δ_i' is the delta value of atom i, Z_i^v is the number of valence electrons of atom i, h_i the number of hydrogen atoms forming σ bonds with the atom, σ_i is the number of the bonding electrons, and N is the main quantum number of atom i. The σ' value can be expressed as the ability of nuclear attraction of valence electrons, excluding those forming σ bonds with hydrogen atoms. δ_i' value is used not only to distinguish different atoms, but also can reflect some characteristic of atoms.

The calculation showed that the new delta can be applied not only to carbon atoms, but to other types of atoms (Li et al., 2000).

ACKNOWLEDGMENT

This contribution is part of the research project "Fullerene-Based Double-Graphene Photovoltaics" (FG2PV), PED/123/2017, within the Romanian National program "Exploratory Research Projects" by UEFISCDI Agency of Romania.

KEYWORDS

- **Balaban connectivity index**
- **Randić connectivity index**
- **zero/higher-order connectivity index**

REFERENCES AND FURTHER READING

Balaban, A. T., (1982). Distance connectivity index. *Chem. Phys. Lett., 89*, 399–404.
Camarda, K. V., & Maranas, C. D., (1999). Optimization in polymer design using connectivity indices. *Industrial & Engineering Chemistry Research, 38*(5), 1884–1892.
Devillers, J., & Balaban, A. T., (1999). *Topological Indices and Related Descriptors in QSAR and QSPR* (pp. 117–119). Amsterdam, Netherlands: Gordon and Breach.
Estrada, E., & Rodriguez, L., (1999). Edge-connectivity indices in QSPR/QSAR studies. 1. Comparison to other topological indices in QSPR studies. *J. Chem. Inf. Comput. Sci., 39*(6), pp. 1037–1104.
Estrada, E., (2002). Physicochemical interpretation of molecular connectivity indices. *J. Phys. Chem. A., 106*(39), pp. 9085–9091.
Furtula, B., Graovac, A., & Vukičević, D., (2009). Atom-bond connectivity index of trees. *Discrete Appl. Math., 157*, 2828–2835.
Gutman, I., & Körtvélyesi, T., (1995). Wiener indices and molecular surfaces. *Zeitschrift für Naturforschung, 50a*, 669–671.

Gutman, I., & Trinajstić, N., (1972). Graph theory and molecular orbitals. Total π-electron energy of alternant hydrocarbons. *Chemical Physics Letters, 17*(4), 535–538.

Gutman, I., Tošović, J., Radenković, S., & Marković, S., (2012). On atom-bond connectivity index and its chemical applicability. *Indian Journal of Chemistry, 51A*, 690–694.

Karcher, W., & Devillers, J., (1990). *Practical Applications of Quantitative Structure-Activity Relationships (QSAR) in Environmental Chemistry and Toxicology.* Kluwer Academic Publishers. Dordrecht, Netherlands.

Karcher, W., & Karabunarliev, S., (1996). The use of computer-based structure-activity relationships in the risk assessment of industrial chemicals. *J. Chem. Inf. Comput. Sci., 36*, 672–677.

Kier, L. B., & Hall, L. W., (2002). The meaning of molecular connectivity: A bimolecular accessibility model. *Croatica Chemica Acta, 75*(2), 371–382.

Li, X., Yu, Q., & Zhu, L., (2000). An improved molecular connectivity index. *Science in China, Series B, 43*(3), 288–294.

Mohar, B., & Pisanski, T., (1988). How to compute the Wiener index of a graph. *Journal of Mathematical Chemistry, 2*(3), 267–277.

Rouvray, D. H., & Crafford, B. C., (1976). The dependence of physical-chemical properties on topological factors. *South African Journal of Science, 72*, 47.

Rouvray, D. H., & King, B. R., (2002). *Topology in Chemistry: Discrete Mathematics of Molecules* (pp. 16–37). Horwood Publishing.

Sabljić, A., & Protić, M., (1982). Molecular connectivity: A novel method for prediction of bioconcentration factor of hazardous chemicals. *Chemical-Biological Interactions, 42*(3), pp. 301–310.

Sharma, V., Goswami, R., & Madan, A. K., (1997). Eccentric connectivity index: A novel highly discriminating topological descriptor for structure-property and structure-activity studies. *Journal of Chemical Information and Computer Sciences, 37*(2), 273–282.

Shobha, J., Yadav, M., Paradkar, L., Anuraj, N. S., & Sharma, S., (2011). QSAR studies on bacterial growth by using connectivity type topological indices. *Oxidation Communications, 34*(3), 640–649.

Stiel, L. I., & Thodos, G., (1962). The normal boiling points and critical constants of saturated aliphatic hydrocarbons. *AIChE Journal, 8*(4), 527–529.

Wiener, H., (1947). Structural determination of paraffin boiling points. *Journal of the American Chemical Society, 1*(69), 17–20.

Yang, Y., & Lu, L., (2011). The Randić index and the diameter of graphs. *Discrete Mathematics, 311*, 1333–1343.

Zhou, B., & Trinajstić, N., (2009). On a novel connectivity index. *Journal of Mathematical Chemistry, 46*, 1252–1270.

CHAPTER 10

Counting Polynomials

LORENTZ JÄNTSCHI[1] and SORANA D. BOLBOACĂ[2]

[1]*Technical University of Cluj-Napoca, Romania*

[2]*Iuliu Haţieganu University of Medicine and Pharmacy Cluj-Napoca, Romania, E-mail: sbolboaca@gmail.com*

10.1   DEFINITION

A topological description of a molecule requires the storing of the adjacencies (the bonds) between the atoms and the identities (the atoms). By disregarding the bond and atom types, if this problem is simplified at maximum, then adjacencies are simply stored with 0 and 1 in the vertex adjacency matrix ([Ad]) and the identities are stored with 0 and 1 into the identity matrix ([Id]). Even more, any square matrix derived from adjacencies by keeping its symmetry (relative to the main diagonal, $Ad_{i,j} = Ad_{j,i}$) or not carries a series of structural features of the originating molecule. The counting polynomial (CoP) is a construction of a polynomial in which the values in the originating matrix (let's label this matrix [Tm]) are expressed in a polynomial function in which the coefficient of each monomial count the occurrences of the value used to express the degree of the monomial (for instance $7x^8$ for [Tm] express that are exactly 7 occurrences of the value of 8 in [Tm]):

$$CoP \overset{def}{=} \sum_{k \geq 0} Count(Tm_{i,j} = k) \cdot x^k$$, where [Tm] is any topological matrix

It is a convenience that the counting polynomial to be constructed for matrices containing integer (or natural) numbers.

10.2   HISTORICAL ORIGIN(S)

The concept of a counting polynomial in chemistry was first introduced by Polya (1936), however its uses appeared later (Hosoya et al., 1973).

Although the counting polynomials have no direct chemical interpretation, one might be able to use it as a device for coding, sorting, or classifying graphs [adapted from Hosoya (1988), where it is actually discussing about distance counting polynomials, but works for all others as well]. A particular case of Count-poly is Hosoya polynomial, when is calculated on the distance matrix (Gutman et al., 2001).

An generalized definition of the counting polynomials is given by Diudea et al. (2007). A counting polynomial $P(G,x)$ is a description of a graph property $P(G)$, in terms of a sequence of numbers, so that the exponents express the extent of its partitions while the coefficients are related to the frequency of the occurrence of partitions.

10.3   NANO-SCIENTIFIC DEVELOPMENT(S)

It is much easier to compute a counting polynomial than a characteristic one since it is calculated directly on the matrix with which is feed. For exemplification of the calculation, let's take a molecular graph (C_{20}) and calculate some typical matrices on adjacency, distance, and unsymmetrical Szeged (see Figure 10.1). Please note that the numbering is not relevant to the formula of the counting polynomial; in other words, it is an invariant relative to the numbering of the atoms. Even further, the adjacency matrix is actually the distance matrix, which contains all zeros and ones from a distance, and everything else is zero.

As can be seen in Figure 10.1, counting polynomial on distances always contains separable polynomial on adjacencies (as the monomials of zero and one degree). Also, in the monomial of degree, one is double of the number of bonds (actually the number of bonds counted from both directions of the ending atoms), and the free term is always the number of the atoms.

The complexity of the calculation for counting polynomial is always $O(n^2)$. If the matrix is symmetrical, it is required to count the frequencies for $n(n-1)/2$ atom pairs (upper part relative to the main diagonal for instance), to twice the obtained numbers (lower part relative to the main diagonal contains exactly the same numbers), and to add the free term which is always n (multiplied with x^0 to formally describe the number of the atoms). If the matrix is unsymmetrical, then it is required to count the frequencies for $n(n-1)$ atom pairs (excepting the zero's from the main diagonal).

The complexity of the calculation for the Count-poly comes actually from the complexity of the calculation of the matrices. The following algorithms (see Figure 10.2) are helpful for calculating the distance and unsymmetrical Szeged matrix from the adjacency matrix.

Distance matrix (Di)

0	1	1	2	2	2	3	3	2	3	4	1	2	3	2	3	4	5	4	3
1	0	2	2	1	3	4	3	3	4	5	2	2	3	1	2	3	4	3	2
1	2	0	1	2	1	2	2	2	3	3	2	3	4	3	4	5	4	3	3
2	2	1	0	1	2	2	1	3	4	3	3	4	5	3	3	4	3	2	2
2	1	2	1	0	3	3	2	4	5	4	3	3	4	2	2	3	3	2	1
2	3	1	2	3	0	1	2	1	2	2	2	3	3	4	5	4	3	3	4
3	4	2	2	3	1	0	1	2	2	1	3	4	3	5	4	4	3	3	2
3	3	2	1	2	2	1	0	3	3	2	4	5	4	4	3	3	2	1	2
2	3	2	3	4	1	2	3	0	1	2	1	2	2	3	4	3	4	5	4
3	4	3	4	5	2	2	3	1	0	1	2	2	1	3	3	2	3	4	5
4	5	3	3	4	2	1	2	2	1	0	3	3	2	4	3	3	2	3	4
1	2	2	3	3	2	3	4	1	2	3	0	1	2	2	3	3	4	5	4
2	2	3	4	3	3	4	5	2	2	3	1	0	1	1	2	2	3	4	3
3	3	4	5	4	3	3	4	2	1	2	2	1	0	2	2	1	2	3	3
2	1	3	3	2	4	5	4	3	3	4	2	1	2	0	1	2	2	3	3
3	3	4	5	4	3	3	4	2	1	2	2	1	0	2	2	1	2	3	3
2	1	3	3	2	4	5	4	3	4	3	3	3	2	2	1	0	1	1	2
3	2	4	3	2	5	4	3	4	3	3	3	2	2	3	2	1	0	1	2
4	3	5	4	3	4	3	3	3	2	2	3	2	1	2	1	2	1	0	1
5	4	4	3	3	3	2	2	3	2	1	4	3	2	3	2	1	0	1	2
4	3	3	2	2	3	2	1	4	3	2	5	4	3	3	2	2	1	0	1
3	2	3	2	1	4	3	2	5	4	3	4	3	4	3	3	2	1	2	2	1	0

Unsymmetrical Szeged matrix (USzp)

0	8	8	8	8	8	8	8	8	8	8	8	8	8	8	8	10	8	8	
8	0	8	8	8	8	8	8	8	10	8	8	8	8	8	8	8	8	8	
8	8	0	8	8	8	8	8	8	8	8	8	8	8	8	10	8	8	8	
8	8	8	0	8	8	8	8	10	8	8	8	8	8	8	8	8	8	8	
8	8	8	8	0	8	8	8	8	8	8	8	8	8	10	8	8	8	8	
8	8	8	8	8	0	8	8	8	8	8	8	8	10	8	8	8	8	8	
8	8	8	8	8	8	0	8	8	8	8	8	10	8	8	8	8	8	8	
8	8	8	8	8	8	8	0	8	8	8	10	8	8	8	8	8	8	8	
8	8	8	8	10	8	8	8	0	8	8	8	8	8	8	8	8	8	8	
8	10	8	8	8	8	8	8	8	0	8	8	8	8	8	8	8	8	8	
8	8	8	8	8	8	8	8	8	8	0	8	8	8	8	8	8	8	8	10
8	8	8	10	8	8	8	8	8	8	8	0	8	8	8	8	8	8	8	
8	8	8	8	8	8	10	8	8	8	8	8	0	8	8	8	8	8	8	
8	8	8	8	8	8	8	8	8	8	8	8	8	0	8	8	8	8	8	
8	8	8	8	8	8	8	10	8	8	8	8	8	8	0	8	8	8	8	
8	8	8	8	8	8	8	8	8	8	8	10	8	8	8	0	8	8	8	
8	8	8	8	8	8	8	8	8	8	8	8	8	8	8	8	0	8	8	
8	8	10	8	8	8	8	8	8	8	8	8	8	8	8	8	8	0	8	
8	8	8	8	8	8	8	8	8	8	10	8	8	8	8	8	8	8	0	8
10	8	8	8	8	8	8	8	8	8	8	8	8	8	8	8	8	8	8	0

Counting polynomials

CoP([Ad],C_{20}) =	$60x^1 + 20x^0$
CoP([Di],C_{20}) =	$20x^5 + 60x^4 + 120x^3 + 120x^2 + 60x^1 + 20x^0$
CoP([USzP],C_{20}) =	$20x^{10} + 360x^8 + 20x^0$

FIGURE 10.1   Some counting polynomials for C_{20}.

As can be seen, the previously given algorithms have a complexity of $O(n^3)$ for both calculating the distance matrix and the unsymmetrical Szeged matrix. The algorithm for the distance matrix derived from (Floyd, 1962).

10.4 NANOCHEMICAL APPLICATION(S)

Count-poly is very useful in discriminating among similar structures. To exemplify, let's take the isomers of C_{32} fullerene. There are exactly six isomers (see Goedgebeur, 2012). In the Table 10.1, the six isomers of the C_{32} fullerene are given along with their InChi strings as well as the calculated Count-poly's.

Input data: ÷ adjacency matrix ([Ad]) ÷ number of atoms (n) ÷ number of bonds (m)	Input data: ÷ distance matrix ([Di]) ÷ number of atoms (n)

For each i from 1 to n do
    $Di_{i,i} \leftarrow 0$
    For each j from 1 to i-1 do
      If $Ad_{i,j} = 0$ then
          $Di_{i,j} \leftarrow m + 1$
          $Di_{j,i} \leftarrow m + 1$
      else
          $Di_{i,j} \leftarrow 1$
          $Di_{j,i} \leftarrow 1$
      End if
    End for
End for
For each k from 1 to n do
    For each i from 1 to n do
      For each j from 1 to n do
        If $Di_{i,k} + Di_{k,j} < Di_{i,j}$ then
          $Di_{i,j} \leftarrow Di_{i,k} + Di_{k,j}$
          $Di_{j,i} \leftarrow Di_{i,j}$
        End if
      End for
    End for
End for
	For each i from 1 to n do
	    For each j from 1 to n do
	      $USzp_{i,j} \leftarrow 0$
	    End for
	End for
	For each i from 1 to n do
	    For each j from 1 to n do
	      If i = j then
	        Continue
	      End if
	      For each k from 1 to n do
	        If $Di_{i,k} < Di_{j,k}$ then
	          $USzp_{i,j} \leftarrow USzp_{i,j} + 1$
	        End if
	      End for
	    End for
	End for

Output data: distance matrix ([Di])	Output data: unsymmetrical Szeged matrix ([USzp])

FIGURE 10.2 Szeged unsymmetrical from distances which from adjacencies.

As can be seen in Table 10.1, actually are no identical formulas for any different C_{32} isomers for any of the Count-poly given (on Distance and on unsymmetrical Szeged matrices). Thus, the discriminating power of a Count-poly formula is high.

10.5 MULTI-/TRANS-DISCIPLINARY CONNECTION(S)

The other one polynomial connected with the Count-poly is the chromatic polynomial (Birkhoff, 1912), generalized later in the Tutte polynomial (Tutte, 1954; Crapo, 1969).

TABLE 10.1 Counting Polynomials for C_{32} Fullerene Isomers

Structure	InChi strings and polynomials (on [Di] and [USzp])
	1S/C32/c1–3–9–13–5(1)17–19–7(1)15–11(3)23–20–8 –2–4–10–14–6(2)18(20)28–29(21(9)23)25(13)31–27 (17)30(26(14)32(28)31)22(10)24(19)12(4)16(8)15 $CoP([Di]) = 2x^7 + 64x^6 + 180x^5 + 230x^4 + 228x^3 + 192x^2 + 96x^1 + 32x^0$ $CoP([USzp]) = 6x^{17} + 54x^{16} + 160x^{15} + 330x^{14} + 224x^{13} + 148x^{12} + 40x^{11} + 24x^{10} + 6x^9 + 32x^0$
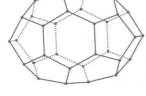	1S/C32/c1–5–6–2–10–17(5)21–13(1)29–23–15–4–7–8 –3–11(19(7)23)26–25(29)9(1)18(6)22–14(2)32–24 (16(3)31(22)26)20(8)12(4)28(27(10)32)30(15)21 $CoP([Di]) = 4x^7 + 64x^6 + 180x^5 + 228x^4 + 228x^3 + 192x^2 + 96x^1 + 32x^0$ $CoP([USzp]) = 4x^{17} + 88x^{16} + 144x^{15} + 336x^{14} + 176x^{13} + 164x^{12} + 40x^{11} + 24x^{10} + 12x^9 + 4x^8 + 32x^0$
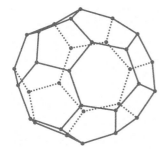	1S/C32/c1–2–14–5–7–15–4–3(13(1)31(14)15)21–19 (1)22–6–8–17–11–12–18–10–9(16(6)32(17)18)23(22) 20(2)24(5)27(10)30(12)29(7)26(4)28(11)25(8)21 $CoP([Di]) = 8x^7 + 60x^6 + 180x^5 + 228x^4 + 228x^3 + 192x^2 + 96x^1 + 32x^0$ $CoP([USzp]) = 80x^{16} + 168x^{15} + 336x^{14} + 168x^{13} + 168x^{12} + 36x^{11} + 12x^{10} + 12x^9 + 12x^8 + 32x^0$
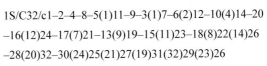	1S/C32/c1–2–4–8–5(1)11–9–3(1)7–6(2)12–10(4)14–20 –16(12)24–17(7)21–13(9)19–15(11)23–18(8)22(14)26 –28(20)32–30(24)25(21)27(19)31(32)29(23)26 $CoP([Di]) = 64x^6 + 180x^5 + 232x^4 + 228x^3 + 192x^2 + 96x^1 + 32x^0$ $CoP([USzp]) = 6x^{17} + 14x^{16} + 172x^{15} + 346x^{14} + 266x^{13} + 138x^{12} + 38x^{11} + 6x^{10} + 6x^9 + 32x^0$

TABLE 10.1   *(Continued)*

Structure	InChi strings and polynomials (on [Di] and [USzp])
	1S/C32/c1–13–2–20–6–15–3(19(1)20)27–8–16(15)10 –24–12–18–9(23(8)24)28–7–17(18)11–22–5(26(2)32 (29(6)10)30(11)12)14(13)4(21(7)22)25(1)31(27)28 CoP([Di]) = $68x^6 + 180x^5 + 228x^4 + 228x^3 + 192x^2 + 96x^1 + 32x^0$ CoP([USzp]) = $72x^{16} + 246x^{15} + 162x^{14} + 278x^{13} + 126x^{12} + 78x^{11} + 24x^{10} + 6x^9 + 32x^0$
	1S/C32/c1–7–2–10–15(1)27–24–13–5–8–4–12–18(5)29 (27)21–9(1)16–3(7)11–17(2)30–26–14(6(8)19(13)31 (30)22(10)24)20(4)32(25(11)26)28(16)23(12)21 CoP([Di]) = $2x^7 + 60x^6 + 180x^5 + 234x^4 + 228x^3 + 192x^2 + 96x^1 + 32x^0$ CoP([USzp]) = $32x^{16} + 174x^{15} + 318x^{14} + 270x^{13} + 144x^{12} + 48x^{11} + 6x^9 + 32x^0$

Hosoya (1971) also introduced *Z*-counting polynomial, which is related by its name to the Count-poly, but by its definition to the Charact-poly.

Sextet polynomial (Ohkami et al., 1981) and the Omega polynomial (Diudea, 2006) are also different sorts of counting polynomials.

The independence polynomial (Gutman & Hosoya, 1990) counts the selections of k-independent vertices of G. The other related graph polynomials are the king, domino, and star polynomials (Motoyama & Hosoya, 1977; Farrell & De Matas, 1988).

If one counts the sets of mutually adjacent vertices instead of the sets of independent vertices, then the clique polynomial is obtained (Hoede & Li, 1994).

10.6   OPEN ISSUES

It is a convenience that the counting polynomial to be constructed for matrices containing integer (or natural) numbers.

KEYWORDS

- **counts of random walks**
- **eigenvalues**
- **eigenvectors**
- **quantum chemistry**
- **structure-resonance theory**
- **topological theory of aromaticity**

REFERENCES AND FURTHER READING

Birkhoff, G. D., (1912). A determinant formula for the number of ways of coloring a map. *Ann. of Math.*, *14*(1–4), 42–46.

Crapo, H. H., (1969). The Tutte polynomial. *Aequationes Mathematicae, 3*(3), 211–229.

Diudea, M. V., (2006). Omega polynomial. *Carpathian J. Math., 22*(1/2), 43–47.

Diudea, M. V., Vizitiu, A. E., & Janežič, D., (2007). Cluj and related polynomials applied in correlating studies. *Journal of Chemical Information and Modeling, 47*(3), 864–874.

Farrell, E. J., & De Matas, C., (1988). On star polynomials, graphical partitions and reconstruction. *Int. J. Math. Math. Sci., 11*, 87–94.

Floyd, R. W., (1962). Algorithm 97: Shortest path. *Commun. ACM, 5*(6), 345–345.

Goedgebeur, J., (2012). *Number of Fullerenes with 2n Vertices (or Carbon Atoms). OEIS(A007894).* http://oeis.org/A007894.

Gutman, I., & Hosoya, H., (1990). Molecular graphs with equal Z-counting and independence polynomials. *Z. Naturforsch. A: Phys. Sci., 45*, 645–648.

Gutman, I., Klavžar, S., Petkovšek, M., & Žigert, P., (2001). On Hosoya polynomials of benzenoid graphs. *MATCH Commun. Math. Comput. Chem., 43*, 49–66.

Hoede, C., & Li, X. L., (1994). Clique polynomials and independent set polynomials of graphs. *Discrete Math., 125*, 219–228.

Hosoya, H., (1971). Topological index. A newly proposed quantity characterizing the topological nature of structural isomers of saturated hydrocarbons. *Bull. Chem. Soc. Japan, 44*, 2332–2339.

Hosoya, H., (1988). On some counting polynomials in chemistry. *Discrete Applied Mathematics, 19*, 239–257.

Hosoya, H., Murakami, M., & Gotoh, M., (1973). Distance polynomial and characterization of a graph. *Natl. Sci. Rept. Ochanomiau Univ., 24*, 27–34.

Motoyama, A., & Hosoya, H., (1977). King and domino polynomials for polyomino graphs. *J. Math. Phys., 18*, 1485–1490.

Ohkami, N., Motoyama, A., Yamaguchi, T., & Hosoya, H., (1981). Mathematical properties of the set of the Kekule patterns and the sextet polynomial for polycyclic aromatic hydrocarbons. *Tetrahedron, 37*, 1113–1122.

Pólya, G., (1936). Combinatorial number determinations for groups, graphs and chemical compounds (In German). *Acta Math., 68*, 145–253.

Tutte, W. T., (1954). A contribution to the theory of chromatic polynomials. *Canad. J. Math., 6,* 80–91.

CHAPTER 11

Cubic Silicon Carbide Thin Films Deposition

MATTEO BOSI[1], MARCO NEGRI[1,2], and GIOVANNI ATTOLINI[1]

[1]IMEM-CNR Institute, Parco Area delle Scienze 37A 43124 PARMA, Italy, E-mail: matteo.bosi@imem.cnr.it

[2]Department of Mathematical, Physical and Computer Sciences, Parma University, Viale delle Scienze 7/A, I-43124 Parma, Italy

11.1 DEFINITION

Cubic silicon carbide (3C-SiC) is a wide-bandgap material with characteristics that make it an ideal candidate for applications in high power electronic devices working at high frequency, high temperature, harsh environments, and for micro electro mechanical systems (MEMS) (Ferro, 2015).

It is biocompatible, so it can be used for MEMS working in the human body (Saddow, 2012; Coletti, 2007). Its high breakdown electric field strength (1–3 MV/cm) make it an ideal material for high power applications (Gerhardt, 2011).

3C-SiC is usually heteroepitaxially grown on low-cost, large-sized silicon substrates, that gives the possibility to fabricate MEMS using the well-developed Si microfabrication technology (Bosi, 2013).

Due to the large lattice mismatch of about 20% and to the difference in thermal expansion coefficients, SiC/Si heteroepitaxy is still problematic, and issues such as poor crystal quality, surface roughness, residual stress, and wafer bending still prevent the widespread realization of 3C-SiC devices and applications (Nishino, 1983; Ferro, 2015).

11.2 HISTORICAL ORIGIN(S)

SiC synthesis has been studied since the 60s using the chemical vapor deposition (CVD) technique (Jennings et al., 1966; Campbell and Chu, 1966), using $SiCl_4$ and CCl_4 as precursors. The experiments resulted in the deposition of 6H on 6H Si-face Lely platelets at temperatures of over 1700°C. An in-situ hydrogen etching was performed prior to growth. Later a process was reported using $SiCl_4$ and C_6H_{12} using a vertical cold-wall reactor (Von Muench and Pfaffeneder, 1976). All these works, however, were related to hexagonal polytypes.

The first successful experiments on 3C-SiC/Si deposition were carried out in 1983 by S. Nishino, introducing the so-called carbonization layer between the Si substrate and the SiC layer (Nishino et al., 1983). This process permitted to obtain a thick SiC (up to 34 μm thick) on large area substrates.

11.3   NANO-SCIENTIFIC DEVELOPMENT(S)

The growth of SiC thin films via homoepitaxy is a usually performed in hot wall VPE reactors (see also, Vapor Phase Epitaxy (VPE)) using silicon and carbon containing precursors, transported to the substrate by using a carrier gas, typically hydrogen.

Since nowadays there are no 3C-SiC substrates available, the growth is carried out heteroepitaxially on silicon substrates, exploiting their low cost and large area. However, the main drawback for 3C-SiC/Si deposition is the large lattice mismatch, of about 20%, between substrate and film and the difference in thermal expansion coefficients. The use of different deposition techniques, precursors, temperature ranges, and growth recipes permitted to limit this issue. But, despite all these efforts, poor crystal quality, high surface roughness, residual stress, and wafer bending still prevent the widespread realization of 3C-SiC devices and applications. Its mainly because the residual stress can induce a significant bow to the substrate that limits photolithography processing and device fabrication (Ferro, 2015).

The first breakthrough in 3C-SiC/Si heteroepitaxy has been achieved by Nishino et al., (1983), with the development of the so-called "carbonization process": the C-precursor is introduced at temperatures lower than film growth to convert Si surface into SiC. The carbonization usually happens at temperatures of about 1000–1100°C. The temperature is then increased, and a thick 3C-SiC layer is grown at a higher temperature from 1200 to 1400°C.

The precursors used for the deposition of 3C-SiC are usually silane and propane, but recently ethane and ethylene were also employed as C-containing species, and chlorine-based precursors were adopted. The advantage of adding chlorinated species to the gas phase is to avoid homogeneous nucleation and the formation of Si-clusters in the gas phase. Precursors such as

HCl, trichlorosilane, and methyl trichlorosilane were successfully employed for 3C-SiC growth, with the advantage of achieving growth rates up to 100 μm/h (La Via et at. 2007).

High crystallinity 3C-SiC layers with a flat surface and low defect density are deposited at temperatures close to the silicon melting point (1414°C) usually between 1350 and 1380°C. Lower deposition temperatures (1000–1200°C) are also employed but the resulting film, even if it is suitable for MEMS realization, generally has poorer crystal quality and rough surfaces and may not be suitable for applications that require a very flat and perfect surface.

The main requirement of 3C-SiC/Si epitaxy is the minimization of the residual stress and stress gradients inside the epitaxial layer, that can also block photolithography and microfabrication processes. The stress originates from both lattice and thermal mismatch and may cause large macroscopic bending of the substrate.

11.4 NANOCHEMICAL APPLICATION(S)

3C-SiC films have great potential for applications in power devices and MEMS for harsh environments, including biosensing operation.

The increasing demand for high-voltage, high-power electronic devices and the limits faced by the traditional silicon technology promoted the research and development of new power electronic devices based on wide bandgap semiconductors such as GaN and SiC.

The availability of GaN and 4H-SiC substrates fostered the development of power devices based on these materials, but the progress is still limited by the high cost of substrates themselves. The possibility to realize reliable 3C-SiC power devices on silicon has gathered a lot of attention in recent years, and there are a lot of efforts devoted to the development of technology based on 3C-SiC. 4H-SiC it is still considered the best candidate for high power-high frequencies applications due to its high breakdown field, electron drift velocity, and mobility. However, 3C-SiC has properties that make it a good candidate for power devices, such as a smaller bandgap that reduce the electric field strength necessary to achieve channel inversion inside metal oxide semiconductor field-effect transistors (MOSFET) and a smaller trap density at the SiO_2/3C-SiC interface. Its hole mobility is much higher than 4H polytype, and it has been shown that defects such as stacking faults do not propagate during device operation in 3C like in 4H.

SiC excellent mechanical, chemical, and electrical properties make it an excellent structural material for micro-electro mechanical systems (MEMS) and nano electro mechanical systems (NEMS) applications, thanks also to the compatibility with Si micromachining techniques.

The main interest in 3C-SiC MEMS is for harsh environments devices and sensors, where Si devices experience their limits. In applications where the device faces high temperature (up to 600°C), mechanical wear, high radiation, high oxidation, and harsh chemicals SiC is considered the material of choice.

Moreover, the high Young modulus of SiC (~400 GPa) enables fabrication of mechanical resonators that can operate up to the GHz range.

SiC biocompatibility with organic tissues makes it attractive for the realization of devices to be used in biological media such as neural probes or blood sensors (Saddow, 2012).

Due to the presence of strain given by the lattice and thermal mismatch between 3C-SiC and Si, stress and stress gradients may bend the microstructures or introduce unwanted and unpredictable forces to the system that may alter the behavior of the device or change the predicted resonant frequencies.

Strategies adopted to minimize the strained content of 3C-SiC/Si include the optimization of the carbonization process and the deposition of a buffer layer to reduce the strain due to lattice mismatch (Bosi, 2013). However, the presence of thermal mismatch cannot be avoided and could be limited only by reducing the growth temperature or by cooling the layer after the growth with a long ramp.

11.5 MULTI-/TRANS-DISCIPLINARY CONNECTION(S)

In its amorphous form, SiC is a material of choice for cardiovascular applications, and it was successfully used to coat heart stents for *in-vivo* clinical trials. Its superior tribological and mechanical properties and the hydroxyapatite-like osseointegration make it an excellent candidate material for medical prosthetic implants (Li and Wang, 2015). Despite these applications, the cytotoxicity of SiC particles remains unclear Chapter 14 in Gerhardt, 2011).

The properties of amorphous SiC fostered the research on the biocompatibility of other SiC polytypes, in particular, 3C-SiC.

Coletti et al., (2007) investigated the electrochemical potential and the surface energy based interactions between a semiconductor and a biological cell, demonstrating a superior cytocompatibility of 3C-SiC films:

single-crystal SiC was proven to be biocompatible and capable of directly interfacing cells without the need for surface functionalization. SiC is significantly better than Si as a substrate for cell culture, with lower toxicity and enhanced cell proliferation.

Thanks to these results, 3C-SiC has found interesting applications in biomedical devices, such as neural interfaces for Brain Machine Interfaces (BMI) (Saddow, 2012).

Recently it has been observed that several intrinsic defects of SiC polytypes are associated with electron spin and may be used as a quantum bit for quantum computing and spintronics (Castelletto, 2015). The source was identified as a carbon antisite – vacancy pair, and it can be created by electron irradiation and annealing of ultrapure SiC. These intrinsic defects were initially observed on 4H and 6H SiC polytypes, but recently they were also demonstrated on 3C-SiC. Hexagonal polytypes show single photon emission rate in the Mcounts s^{-1} regime, and this is considered one of the brightest single photon sources at room temperature. The possibility to obtain single photon sources in SiC, combined with MEMS engineering to realize waveguides and photon cavities tuned to wavelengths from 1.25 to 1.6 µm pave the way for possible applications in nonlinear optics and quantum information science.

11.6 OPEN ISSUES

The main issues that are limiting 3C-SiC exploitation for MEMS and MOSFET technologies are related to the high strain content in the hetero-epitaxial layers.

The high lattice and thermal mismatch cannot be completely avoided in the epitaxial process and, generally, the higher is the layer crystal quality, the higher the strain. The strain may cause a severe bending of the wafer that in the worst cases may also break.

The wafer bow causes severe problems in technological processes such as photolithography, masking, and surface polishing that in some case may be completely impossible. In MEMS and NEMS fabrication the residual strain content of the layer can deform, bend or break structures such as membranes, beams, and cantilever.

A lot of effort is being directed towards alternative methods to relieve the strain, including customized carbonization processes and optimization of the growth method.

KEYWORDS

- **epitaxy**
- **vapor phase epitaxy**
- **strain**
- **micro electro mechanical systems**

REFERENCES AND FURTHER READING

Bosi M., Attolini, G., Negri, M., Frigeri, C., Buffagni, E., Ferrari, C., et al., (2013). Optimization of a buffer layer for cubic silicon carbide growth on silicon substrates. *J. Cryst. Growth, 383*, 84.

Campbell, R. L., & Chu, T. L., J., (1966). Epitaxial growth of silicon carbide by the thermal reduction technique. *J. Electrochemical Society, 113*, 827.

Castelletto, S., Rosa, L., & Johnson, B. C., (2015). In: Saddow, S. E., & La, V. F., (eds.), *Silicon Carbide for Novel Quantum Technology Devices, Chapter 8 in Advanced Silicon Carbide Devices and Processing.* ISBN 978–953–51–2168–8, Intech, doi: 10.5772/61166.

Coletti, C., Jaroszeski, M. J., Pallaoro, A., Hoff, A. M., Iannotta, S., & Saddow, S. E., (2007). Biocompatibility and wettability of crystalline SiC and Si surfaces. *Engineering in Medicine and Biology Society, 29th Annual International Conference of the IEEE,* p. 5849.

Ferro, G., (2015). 3C-SiC Heteroepitaxial growth on silicon: The quest for holy grail. *Critical Reviews in Solid State and Materials Sciences, 40*, 56.

Gerhardt, R., (2011). *Properties and Applications of Silicon Carbide.* ISBN 978-953-307-201-2, InTech. doi: 10.5772/615.

Jennings, V. J., Sommers, A., & Chang, H. C., (1966). The epitaxial growth of silicon carbide. *J. Electrochemical Society, 113*(7), 728.

La Via, F., Leone, S., Mauceri, M., Pistone, G., Condorelli, G., Abbondanza, G., et a., (2007). Very high growth rate epitaxy processes with chlorine addition. *Mater. Sci. Forum, 556/557*, 157.

Li, X., & Wang, X., (2015). Micro/nanoscale mechanical and tribological characterization of SiC for orthopedic applications. *J. Biomed. Mater. Res. B. Appl. Biomater., 72*, 353.

Nishino, S., Powell, A., & Will, H., (1983). Production of large-area single-crystal wafers of cubic SiC for semiconductor devices. *Appl. Phys. Lett., 42*, 460.

Saddow, S. E., (2012). *Silicon Carbide Biotechnology: A Biocompatible Semiconductor for Advanced Biomedical Devices and Applications.* Waltham, MA: Elsevier. eBook ISBN: 9780128030059; Hardcover ISBN: 9780128029930.

Von Muench, W., & Pfaffeneder, I., (1976). Epitaxial deposition of silicon carbide from silicon tetrachloride and hexane. *Thin. Solid Films, 31*, 39.

CHAPTER 12

Degree Distance

ASMA HAMZEH and ALI IRANMANESH

Department of Mathematics, Faculty of Mathematical Sciences, Tarbiat Modares University, P.O. Box: 14115-137, Tehran, Iran, Tel: +982182883493, Fax: +982188006544, E-mail: iranmanesh@modares.ac.ir

12.1 DEFINITION

Let G be a simple connected graph with vertex set $V(G)$. For $u,v \in V(G)$, let $d_G(u,v)$ be the distance between u and v in G. For $u \in V(G)$, let $d_G(u)$ be the degree of u in G, and let $D_G(u)$ be the sum of distances between u and all vertices of G i.e., $D_G(u) = \sum_{v \in V(G)} d_G(u,v)$. The degree distance of G is defined as:

$$D'(G) = \sum_{u \in V(G)} d_G(u) D_G(u).$$

It is a useful molecular descriptor. Earlier as noted, this graph invariant appeared to be part of the molecular topological index (or Schultz index), which may be expressed as $D'(G) + \sum_{u \in V(G)} d_G(u)^2$, where the latter part $\sum_{u \in V(G)} d_G(u)^2$ is known as the first Zagreb index. Thus, the degree distance is also called the true Schultz index in the chemical literature.

12.2 HISTORICAL ORIGIN(S)

The degree distance seems to have been considered first in connection with certain chemical applications by Dobrynin and Kochetova (1994) and at the same time by Gutman (1994), who named it the Schultz index. This name was eventually accepted by most other authors (Dobrynin, 1999;

Schultz et al., 2000; Zhou, 2006). In fact, there are several names (degree distance, MTI index, Schultz index, and Zagreb index) associated with this index or closely related indices. The degree distance may be considered a weighted version of the Wiener index (Dobrynin et al., 2001) the Wiener index being the sum of all distances between vertices of G.

12.3 NANO-SCIENTIFIC DEVELOPMENT(S)

For chemists, the use of computing tools became an obligation in order to manipulate molecular information that were, during the last years, numerically stocked on computers in databases with huge quantities. Moreover, the multiplication of exploitable data by chemists resulted in an obligation of numeration, in order to be able to stock, visualize and treat the same data easily, which gave birth to the molecular describers that are often manipulated by quantitative structure-activity relationships (QSAR) or of quantitative structure-property relationships (QSPR). During the study of QSAR/QSPR, we study relations between structure and the activity of a component or a molecule. The numeral value which is linked to the chemical construction with regards to the related chemical's composition to the biological activity, chemical reactivity, as well as the numerous physical characteristics represents, the topological index/ molecular structure descriptor. In this respect, the innumerable descriptors of the structure have been suggested in spite of not analyzing their correlation with the numerous physical characteristics, chemical reactivity, or biological activity in several cases. This can be observed more specifically in the case of the multitudinal molecular-graph-based descriptors of the structure.

A topological index of a graph is a real number related to the graph; it does not depend on labeling or pictorial representation of a graph. In theoretical chemistry, molecular structure descriptors (also called topological indices) are used for modeling physicochemical, pharmacologic, toxicologic, biological and other properties of chemical compounds (Gutman et al. 1986).

In connection with certain investigations in mathematical chemistry, authors in 1994 introduced a graph-theoretical descriptor for characterizing alkanes by an integer, namely degree distance.

12.4 NANOCHEMICAL APPLICATION(S)

Mathematical chemistry is a branch of theoretical chemistry for discussion and prediction of the molecular structure using mathematical methods without necessarily referring to quantum mechanics. Chemical graph theory is a branch of mathematical chemistry which applies graph theory to mathematical modeling of chemical phenomena. This theory had an important effect on the development of the chemical sciences.

A topological index is a numeric quantity from the structural graph of a molecule. Usage of topological indices in chemistry began in 1947 when chemist Harold Wiener developed the most widely known topological descriptor, the Wiener index, and used it to determine physical properties of types of alkanes known as paraffin.

Diudea and his co-authors were the first scientist considered topological indices of nanostructures into account. In some research paper, he and his team computed the Wiener index of armchair and nanotubes. In this section, we bring some results of degree distance, for chemical graphs and nanostructures.

The bar polyhex graph is composed of exclusively of hexagonal rings that are face bounded by six-membered cycles in the plane. Any two rings have either one common edge (and are then said to be adjacent) or have no common vertices (Figure 12.1).

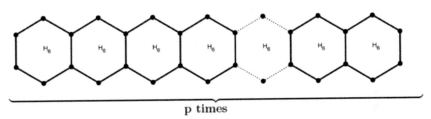

p times

FIGURE 12.1   The bar Polyhex graph with p hexagons: $G_p = H_6 | \dots | H_6$.

As we can see, n the order of G_p, m the size of G_p and D. The diameter of G_p, can be expressed by $p(p \geq 1)$ the number of hexagons in the bar polyhex graph G_p like the following:

$$n = 4p + 2; \quad m = 5p + 1; \quad D = 2p + 1.$$

In Kordala et al. (2014), the authors gave the results of degree distance index of bar-polyhex graph as follows:

Let be G_p bar-polyhex graph with p number of hexagons ($p \geq 1$), the degree distance index of this graph:

$$D'(G_p) = \frac{80p^3 + 144p^2 + 94p + 6}{3}.$$

In Eliasi et al. (2008), the authors obtained degree distance of $G: = TUV$ $C_6[2p; q]$ that denotes an arbitrary armchair polyhex nanotube in terms of its circumference $2p$ and their length q, see Figure 12.2.

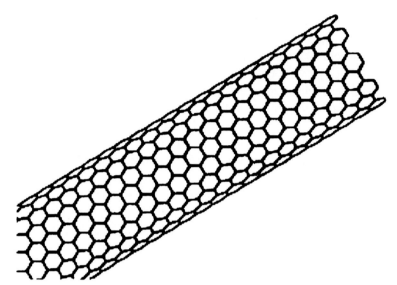

FIGURE 12.2   An armchair polyhex nanotube. Reprinted from Diudea, M. V.; Stefu, M.; Parv, B.; John, P. E. Wiener Index of Armchair Polyhex Nanotubes. Croat. Chem. Acta 2004, 77, 111-115. Open Access.

The degree distance of $G: = TUV$ $C_6[2p; q]$ nanotubes is given by:

Case 1: p is even

$$D'(G) = \begin{cases} \dfrac{p}{6}[-48p^2q + 72p^2q^2 + 3(-1)^{q+1} + 3 - 8q^3 - 12q^2 + 6q^4 + 8q] & \text{if } p > q \\[2ex] \dfrac{-p^2}{3}[-18q^2p + 3p^3 - 6p - 12p^2q - 12q^3 + 12q + 4p^2 - 4 + 12pq + 12q^2] & \text{if } p \le q. \end{cases}$$

Case 2: p is odd

$$D'(G) = \begin{cases} \dfrac{p}{6}[72p^2q^2 + 6q^4 - 12q^2 - 3 + 3(-1)^q - 48p^2q - 8q^3 + 8q] & \text{if } p > q \\[2ex] \dfrac{-p}{3}[-12p^3q - 12pq^3 - 18q^2p^2 + 3 + 12pq - 6p^2 + 3p^4 + 4p^3 - 4p + 12p^2q + 12pq^2] & \text{if } p \le q. \end{cases}$$

Suppose L_h and Z_h denote the molecular graphs of linear polyacene and zig-zag hexagonal belt with h hexagons. In Ashrafi et al. (2011), the authors gave the degree distance index of L_h and Z_h.

The degree distance index of L_h and Z_h are computed as follows:

$$D'(L_h) = \frac{8h(2h+1)(5h+2)}{3} + 24h^2 + 26h + 2,$$

$$D'(Z_h) = 20h^3 + 24h^2 + 4h.$$

12.5   MULTI-/TRANS-DISCIPLINARY CONNECTION(S)

Topological indices and graph invariants based on the distances between the vertices of a graph are widely used in theoretical chemistry for establishing relations between the structure and the properties of molecules. They give correlations with physical, chemical, and thermodynamic parameters of chemical compounds. One such topological index is the degree distance, which was introduced by Dobrynin, Kochetova, and Gutman as a weighted version of the Wiener index.

KEYWORDS

- **degree distance**
- **Schultz index**
- **Wiener index**

REFERENCES AND FURTHER READING

Ashrafi, A. R., Hamzeh, A., & Hossein-Zadeh, S., (2011). Computing Zagreb, hyper wiener and degree-distance indices of four new sums of graphs. *Carpathian J. Math., 27,* 153–164.
Dobrynin, A. A., & Kochetova, A. A., (1994). Degree distance of a graph: A degree analog of the Wiener index. *J. Chem. Inf. Comput. Sci., 34,* 1082–1086.
Dobrynin, A. A., (1999). Explicit relation between the Wiener index and the Schultz index of catacondensed benzenoid graphs. *Croat. Chem. Acta, 72,* 869–874.
Dobrynin, A. A., Entringer, R., & Gutman, I., (2001). Wiener index of trees: Theory and applications. *Acta Appl. Math., 66,* 211–249.
Eliasi, M., & Salehi, N., (2008). Schultz index of armchair polyhex nanotubes. *Int. J. Mol. Sci., 9,* 2016–2026.

Gutman, I., & Polansky, O. E., (1986). *Mathematical Concepts in Organic Chemistry* (3rd edn.). Publisher: Springer-Verlag, Berlin. 212 pp. ISBN 9783642709845

Gutman, I., (1994). Selected properties of the Schultz molecular topological index. *J. Chem. Inf. Comput. Sci., 34*, 1087–1089.

Kordala, A., El Marraki, M., & Essalih, M., (2014). The Wiener, the hyper-wiener and the degree distance indices of the bar polyhex graph. *Applied Mathematical Sciences, 8*, 8771–8779.

Schultz, H. P., & Schultz, T. P., (2000). Topological organic chemistry. 12. Whole-molecule Schultz topological indices of alkanes. *J. Chem. Inf. Comput. Sci., 40*, 107–112.

Zhou, B., (2006). Bounds for the Schultz molecular topological index. *MATCH Commun. Math. Comput. Chem., 56*, 189–194.

CHAPTER 13

Distance Algorithm in Chemical Graphs

YASER ALIZADEH[1] and ALI IRANMANESH[2]

[1]*Department of Mathematics, Hakim Sabzevari University, Sabzevar, Iran, E-mail: y.alizadeh@hsu.ac.ir*

[2]*Department of Pure Mathematics, Faculty of Mathematical Sciences, Tarbiat Modares University, P.O. Box: 14115-137, Tehran, Iran, Tel: +982182883447; Fax: +982182883493, E-mail: iranmanesh@modares.ac.ir*

13.1 DEFINITION

The core of computer science is algorithms. An algorithm is a set of instructions for solving a problem. The word "algorithm" is a distortion of al-Khwarizmi, a Persian mathematician who wrote an influential treatise about algebraic methods. A topological index of a molecular graph G is a real number derived from the structure of graph that is invariant under automorphisms of G. Such indices based on the distances in the graph are used, for example, in biological activities or physic-chemical properties of alkenes which are correlated with their chemical structure. Some algorithms were proposed to compute topological indices based on distance. These are the so-called shortest path algorithms, and in general, solve even more complicated problems where edges are allowed to carry weights. Some of the well-known algorithms to compute the Wiener index is the Floyd-Warshall algorithm (Pisanski-Mohar, 1988). Also, an algorithm was proposed by Alizadeh and Iranmanesh to compute the distance based topological indices.

13.2 HISTORICAL ORIGIN(S)

Many methods and algorithms, for computing the topological indices of a graph, were proposed. In a series of papers, the algorithms for calculating some topological indices, based on distance in a graph, for a special class of graphs were presented (Aringhieri et al., 2001; Cai & Zhou, 2008; Cash et al., 2002; Chepoi & Klavzar, 1997).

The presented algorithms in Aringhieri et al., (2001) and Klavzar et al., (2000) are linear, and a linear algorithm is the best algorithm for performing a computation or solving a problem, but these algorithms are not applicable for any graph (just for benzenoid systems or trees).

13.2.1   FLOYD-WARSHAL ALGORITHM

Let G has vertex set $V(G) = \{v_1, \ldots v_n\}$. Distance between two vertices v_i and v_j, $d(v_i, v_j)$ is defined as length of shortest path connecting them. The distance matrix $D(G)$ is defined as $D[G] = [d_{ij}]$ where $d_{ij} = d(v_i, v_j)$, $1 \leq i,j \leq n$.

It is clear that $W(G) = \dfrac{1}{2}\sum\limits_{i=1}^{n}\sum\limits_{j=1}^{n} d(i,j) = \sum\limits_{1 \leq i < j \leq n} d(i,j)$

Floyd-Warshal is an algorithm for computing the distance matrix of a graph.

Its input is the adjacency matrix $A(G) = [a\{ij\}]$, where $a_{ij} = 1$, if vertices i and j are adjacent and $a_{ij} = 0$ otherwise.

The steps of algorithm are as follows:

for vertices i and j, $1 \leq i \leq j \leq n$, put $d[i,j]: = a[i,j]$ if i is adjacent to j or i $= j$ otherwise if i is not adjacent to J, $d[i,j]: = n$.
for each vertex s in $V(G)$ do
    for each vertex t in $V(G)$–s which $d(s, t) < n$ do
      for each vertex u in $V(G)$–s do
        if $d(s,t) + d(t,u) < d(s,u)$ then put $d(s,u): = d(s,t) + d(t,u)$.

The complexity time of the algorithm is $O(n)^2$. The Wiener index equals to half of sum of elements of the distance matrix. Then $O(n)^2$ added to time complexity. The algorithm is easy to program. Below is a GAP program to compute the Wiener index according to the Floyd-Warshal algorithm.

```
#Input A[i,j];
for i in [1..n] do
   for j in [1..n] do
      if a[i][j] = 1 or i = j then d[i][j]: = a[i][j];
         else d[i][j]: = n;
         fi;
```

```
            od;
    od;
    for s in [1..n] do
        for t in Difference([1..n],[s]) do
            if d[s][t]<n then
                for u in [1..n] do
                    if d[s][t]+d[t][u]<d[s][u] then
                        d[s][u]: = d[s][t]+d[t][u];
                    fi;
                od;
            fi;
        od;
    od;
    W: = 0;
    for i in [1..n-1] do
        for j in [i+1..n]] do
            w: = w+d[i][j];
        od;
    od;
    w; # The Wiener index of graph
```

13.2.2 MOHAR–PISANSKI ALGORITHM

In this section, an algorithm in order to computing the Wiener index of a tree is given. In Mohar & Pisanski (1998), the authors improved the algorithm of Canfield, Robinson, and Rouvray for compute the Wiener index of an arbitrary tree in linear time.

The input of the algorithm of Mohar-Pisanski is a rooted tree. This means a vertex $v_0 \in V(T)$ is distinguished and called the root of T. All other vertices are indexed as $v_1, v_2, ... v_{n-1}$ in such way that each v_i $(i \geq 1)$ is adjacent to exactly one vertex among $v_1, v_2, ... v_{i-1}$. This unique neighbor is denoted by T_i and called the predecessor of v_i. A rooted tree is illustrated in Figure 13.1.

The complexity time of the Mohar-Pisanski algorithm is linear time, and the algorithm has the following steps:

- Input rooted tree as array T_i $i = 1,2...n-1$,
- For each vertex v_i compute in degree $ind(v_i) = |\{j \,|v_i = T_j\}|$,
- Form a queue of all leaves of T, the vertices with $ind(v_i) = 0$,

- Delete all leaves one after another and compute $w(e)$ for the corresponding deleted edge,
- compute $W(T)$ by $W(T) = \sum_{e \in E(T)} w(e)$.

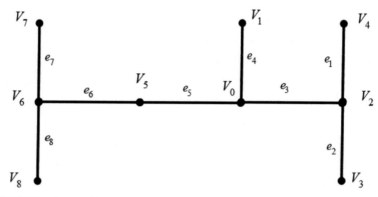

FIGURE 13.1   A rooted tree.

Each step of the above algorithm takes at most linear time. Therefore the complexity time of the algorithm is $O(n)$. In the following, a GAP program (The GAP Group, 2015) is presented for the presented algorithm in Müller et al. (1987) called matrix multiplication (MM) algorithm.

```
for i in [1..n] do
    for j in [1..n] do
        if (D[i,j] = 0) and i <> j) then D[i,j] : = n;
    od;
od;
s: = 1;
while s<n do
    for j in [1..n] do
        for i in [1..n] do
            if D[i,j] = 1 then
                for w in [1..n] do
                    if D[W,i]+s < D[w,j] then D[w,j] : = D[w,i] + s;
                od;
        od;
    s: = s+1;
    od;
od;
```

```
W : = O;
for i in [1...n] do
    for j in [1..n] do
        W : = W + D[i,j];
    od;
od;
MM : = W # The Wiener index;
```

13.2.3 ALIZADEH–IRANMANESH ALGORITHM

The set of vertices that their distance to vertex u is equal to t, $t \geq 0$ is denoted by $D_{u,t}$ and consider $D_{u,0} = u$. The distance between the vertex u and its adjacent vertices is equal to 1, therefore, $D_{u,1} = N(u)$. It is easy to see that for each vertex u of the graph G, $V(G) = \bigcup_{t \geq 0} D_{u,t}$. For each vertex $v \in D_{u,t}$, $t \geq 1$, the distance between u and vertex $w \in N(v)$ is equal to $t-1$ or t or $t+1$ and for each vertex $w \in D_{u,t+1}$, there exist a vertex $v \in D_{u,t}$ such that $w \in N(v)$.

Hence $D_{u,t+1} = \bigcup_{v \in D_{u,t}} N(v) - \{D_{u,t} \cup D_{u,t-1}\}$ $t \geq 1$.

In the follows, the topological indices such as the Wiener, the reverse Wiener, the Szeged and the eccentric connectivity indices can be expressed in terms of the set $D_{u,t}$.

By obtaining the $D_{u,t}$ for each vertex $u \in V(G)$ and the following relations, distances between all vertices can be computed.

The following algorithm was proposed by Alizadeh and Iranmanesh (Iranmanesh & Alizadeh 2008; Iranmanesh et al., 2009) for computing the sets $D_{u,t}$. The input of the algorithm is the set of neighbors of each vertex of G.

For each vertex u, perform the following steps:

1. Put $D_{u,0} = u$, $D_{u,1} = N(u)$, $t := 1$, $l := 1$,

2. Put S as a list:
   $S[w] := 1$ if $w = u$ or $w \in N(u)$, otherwise $S[w] := 0$
3. While $l \neq 0$ do
   for each vertex $v \in D_{u,t}$ do
      If $S[v] := 0$ then add v to $D_{u,t+1}$ and then $S[v] := 1$.
      If $D_{u,t+1} = \phi$ then put $l := 0$

$t: = t + 1.$

Since in this algorithm every edge will be explored two times then the complexity times of the algorithm is $O(mn)$.

In the following, a GAP program is presented to compute the sets $D_{i,t}$. For an arbitrary graph.

```
# Input "N" the set of neighbors of each vertex
D: = [];
for i in [1..n] do
    S: = [];
    for j in [1..n] do
        S[j]: = 0;
    od;
    D[i]: = [];
    D[i][1]: = N[i];
    for j in N[i] do
        S[j]: = 1;
    od;
    S[i]: = 1;
    s: = 1;
    t: = 1;
    while s<>0 do
        D[i][t+1]: = [];
        for j in D[i][t] do
            for x in N[j] do
                if S[x]<>1 then Add(D[i][t+1],x); fi;
                S[x]: = 1;
            od;
        od;
        if D[i][t+1] = [] then s: = 0; fi;
        t: = t+1;
    od;
od;
```

13.3 NANO-SCIENTIFIC DEVELOPMENT(S)

In some papers by using the mentioned algorithms, topological indices based on distance in graph such as Wiener index, Szeged index, edge Wiener index, eccentric connectivity index were computed for some nanostructure nano-tubes and fullerenes. For example, these topological indices were obtained for C_{12k+5} fullerenes in Alizadeh et al., (2012), for coronene/circumcoronene in Alizadeh & Klavzar (2013), and for $HAC_5C_7[p,q]$ and $HAC_5C_6C_7[p,q]$ in Iranmanesh & Alizadeh (2013).

13.4 NANOCHEMICAL APPLICATION(S)

Let G be a connected graph, then the distance between vertices u and v is denoted by d(u,v). The diameter of G is the maximum distance between its vertices and denoted diam(G). The eccentricity ecc(u) of u is the maximum distance between u and any other vertex of G. The distance between edges g $= u_1v_1$ and $f = u_2v_2$ is defined as:

$$d_e(g,f) = min\{d(u_1,u_2),d(u_1,v_2),d(v_1,u_2),d(v_1,v_2)\}+1.$$

Equivalently, this is the distance between the vertices g and f in the line graph of G. The degree of u will be denoted *deg(u)*. For a simple connected graph G, four more applicable topological indices based on distance The Wiener index $W(G)$, the edge Wiener index $W_e(G)$, the eccentric connectivity index $\xi(G)$, the Szeged index $Sz(G)$ are defined as:

$$W(G) = \sum_{\{u,v\}\subseteq V(G)} d(u,v),$$

$$W_e(G) = \sum_{\{f,g\}\subseteq E(G)} d_e(f,g),$$

$$RW(G) = \frac{1}{2}n(n-1)\text{diam}(G)-W(G),$$

$$\zeta(G) = \sum_{u\in V(G)} ecc(u)\deg(u),$$

$$Sz(G) = \sum_{uv\in E(G)} n_u n_v$$

where n_u (resp. n_v) is the number of vertices that are closer to vertex u (resp. v) than vertex v (resp. u).

Using the Alizadeh-Iranmanesh algorithm the topological indices of an important family of nanostructure coronene/circumcoronene were computed (for more information see, Alizadeh & Klavzar, 2013). The structure of coronene/circumcoronene H_n is shown in Figure 13.2. The parameter n defines the number of layers.

$$W\left(H_n\right) = \frac{164}{5}n^5 - 6n^3 + \frac{1}{5}n$$

$$W_e\left(H_n\right) = \frac{369}{5}n^5 - \frac{123}{2}n^4 + 15n^3 - \frac{3}{2}n^2 + \frac{6}{5}n$$

$$RW\left(H_n\right) = \frac{196}{5}n^5 - 18n^4 - 6n^3 + 3n^2 - \frac{1}{5}n$$

$$\xi(H_n) = 60n^3 - 33n^2 + 9n$$

$$Sz\left(H_n\right) = 54n^6 - \frac{3}{2}n^4 + \frac{3}{2}n^2$$

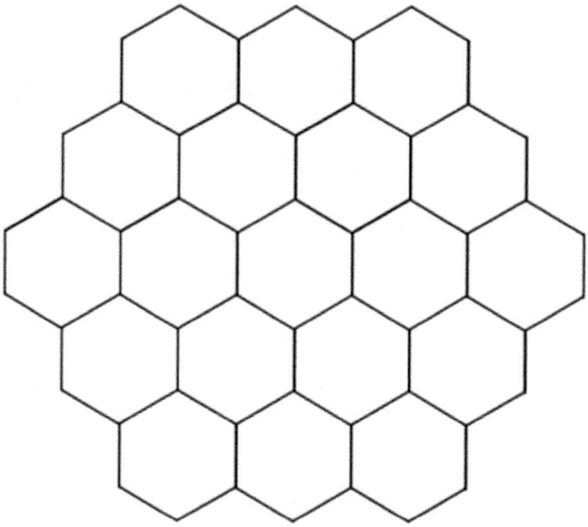

FIGURE 13.2   Coronene/circumcoronene H_3. Reprinted from Alizadeh & Klavzar, 2013. Open Access.

13.5   MULTI-/TRANS-DISCIPLINARY CONNECTION(S)

The algorithms proposed in order to computing distance based topological indices such as Alizadeh-Iranmanesh algorithm can be used to characterize some mathematical properties of graphs such as diameter, radius, center, and distance matrix eccentricity of each vertex.

KEYWORDS

- **distance algorithm**
- **distance matrix**
- **molecular graph**
- **topological index**

REFERENCES AND FURTHER READING

Alizadeh, Y., & Klavzar, S., (2013). Interpolation method and topological indices: 2-Parametric families of graphs. *MATCH Commun. Math. Comput. Chem., 69*, 523–534.

Alizadeh, Y., Iranmanesh, A., & Klavzar, S., (2012). Interpolation method and topological indices: The case of fullerenes C_{12k+4}. *MATCH Commun. Math. Comput. Chem., 68*, 303–310.

Aringhieri, R., Hansen, P., & Malucelli, F., (2001). A linear algorithm for the hyper-Wiener index of chemical trees. *J. Chem. Inf. Comput. Sci., 41*, 958–963.

Cai, X., & Zhou, B., (2008). Reverse Wiener index of connected graphs. *MATCH Commun. Math. Comput. Chem., 60*, 95–105.

Cash, G., Klavzar, S., & Petkovsek, M., (2002). Three methods for calculation of the hyper-Wiener index of molecular graphs. *J. Chem. Inf. Comput. Sci., 42*, 571–576.

Chepoi, V., & Klavzar, S., (1997). The Wiener index and the Szeged index of benzenoid systems in linear time. *J. Chem. Inf. Comput. Sci., 37*, 752–755.

Iranmanesh, A., & Alizadeh, Y., (2008). Computing Wiener index of $HAC_5C_7[p,q]$ nanotube by GAP program. *Iranian Journal of Mathematical Sciences and Informatics, 3*, 1–12.

Iranmanesh, A., & Alizadeh, Y., (2013). Eccentric connectivity index of $HAC_5C_7[p,q]$ and $HAC_5C_6C_7[p,q]$ nanotubes. *MATCH Commun. Math. Comput. Chem., 69*, 175–182.

Iranmanesh, A., Alizadeh, Y., & Mirzaie, S., (2009). Computing Wiener polynomial, Wiener index, hyper Wiener index of C_{80} fullerene by GAP program. *Fullerenes, Nanotubes and Carbon Nanostructures, 17*, 560–566.

Klavzar, S., Zigert, P., & Gutman, I., (2000). An algorithm for the calculation of the hyper-Wiener index of benzenoid hydrocarbons. *Computers and Chemistry, 24*, 229–233.

Mohar, B., & Pisanski, T., (1998). How to compute the Wiener index of graph. *J. Math. Chem., 2*, 267–277.

Müller, W. R., Szymanski, K., Knop, J. V., & Trinajstic, N., (1987). An algorithm for construction of the molecular distance matrix. *J. Comput. Chem., 8*, 170–173.

The GAP Group, (1992). GAP, groups, algorithms and programming. Lehrstuhl De fur Mathematik, RWTH: Achen.

CHAPTER 14

Eccentric Distance Sum

MAHDIEH AZARI[1] and ALI IRANMANESH[2]

[1]*Department of Mathematics, Kazerun Branch, Islamic Azad University, P. O. Box: 73135-168, Kazerun, Iran*

[2]*Department of Pure Mathematics, Faculty of Mathematical Sciences, Tarbiat Modares University, P. O. Box: 14115-137, Tehran, Iran, Tel: +982182883447; Fax: +982182883493; E-mail: iranmanesh@modares.ac.ir*

14.1 DEFINITION

Let G be a simple connected graph with vertex set $V(G)$. Let $\varepsilon_G(u)$ denote the eccentricity of the vertex u in G, which is the maximum distance between u and any other vertex of G, and let $D_G(u)$ denote the distance sum of the vertex u in G, which is the sum of distances between u and all other vertices of G. The eccentric distance sum (EDS) of G is denoted by $\xi^{ds}(G)$ and defined as the summation of the product of the eccentricity and distance sum of each vertex of G,

$$\xi^{ds}(G) = \sum_{u \in V(G)} \varepsilon_G(u)D_G(u).$$

The eccentric distance sum can be defined alternatively as

$$\xi^{ds}(G) = \sum_{\{u,v\} \subseteq V(G)} (\varepsilon_G(u) + \varepsilon_G(v))d_G(u,v),$$

where $d_G(u, v)$ denotes the distance between the vertices u and v in G and the summation are taken over all unordered pairs of vertices $\{u,v\}$ of G.

14.2 HISTORICAL ORIGIN(S)

Chemical graphs, particularly molecular graphs, are models of molecules in which atoms are represented by vertices and chemical bonds by edges of a graph. A graph invariant (also known as a topological index or structural descriptor) is any function calculated on a molecular graph irrespective of the labeling of its vertices. The chemical information derived through graph invariants has been found to be useful in chemical documentation, isomer discrimination, structure-property/activity correlations, and pharmaceutical drug design (Basak et al., 1991). Hundreds of topological indices have been introduced and studied, starting with the seminal work by Wiener (1947a, b) in which he used the sum of all shortest-path distances of a (molecular) graph for modeling physical properties of alkanes. Structure-activity/property relationships (SAR/SPRs) are models that relate structural aspects of a molecule to its physicochemical properties. The inherent problem in the development of a suitable correlation between chemical structures and physical properties can be attributed to the nonquantitative nature of chemical structures. Graph theory can be employed through the translation of chemical structures into a characteristic polynomial, matrix, sequence or numerical graph invariants (Balaban et al., 1983; Bonchev et al., 1986; Basak et al., 1991; Katritzky & Gordeeva, 1993; Estrada & Ramirez, 1996). Since the structure of an analog depends on the connectivity of its constituent atoms, the numerical graph invariants derived from information based on connectivity can reveal structural or substructural information of a molecule. Molecular topology as represented by the connectivity of the atoms can relate physical properties and biological activity with the analogs (Gupta et al., 2002a).

Several topological indices based on graph theoretical notion of eccentricity have been recently proposed and used in QSAR and QSPR studies by mathematicians, chemists, and biologists. Namely, eccentric connectivity index (Sharma et al., 1997; Alizadeh et al., 2013; Azari et al., 2016), eccentric distance sum (Gupta et al., 2002b), total eccentricity (Dankelmann et al., 2004), average eccentricity (Dankelmann et al., 2004), and augmented and super augmented eccentric connectivity index (Bajaj et al., 2005, 2006).

The eccentric distance sum was introduced by Gupta et al. (2002b) as a novel eccentricity-based graph invariant in 2002. Gupta et al. (2002b) investigated the structure-activity relationship of the eccentric distance sum with regard to the anti-HIV activity of dihydroseselins. The values of the eccentric distance sum of each analog in the data set were computed and active range identified. Subsequently, biological activity was assigned to each analog in the data set, which was then compared with the reported anti-HIV activity of dihydroseselin analogs. Surprisingly, the accuracy of prediction was found to be more than 88% with regard to anti-HIV activity. Gupta et al. (2002b) also investigated the quantitative structure-property relationship of the eccentric distance sum with regard to various physical properties of diverse nature, for data sets consisting of primary amines, secondary amines, and alcohols. Values of the eccentric distance sum of all the compounds in various data sets were computed and the resultant data subjected to regression analysis. The mathematical models along with statistical analysis for various data sets and physical properties involved were tabulated in Gupta et al. (2002b). Excellent correlations were obtained using the eccentric distance sum in all six data sets employed in Gupta et al. (2002b). Correlation percentages ranging from 93% to more than 99% were obtained in data sets using eccentric distance sum. The average errors were also on the lower side (from 0.17% to 11.84%) indicating higher correlation abilities of the novel graph invariant. Gupta et al. (2002b) concluded that the overall results with regard to structure–activity and quantitative structure–property studies using the eccentric distance sum were better than the corresponding values obtained using the Wiener index.

14.3 NANO-SCIENTIFIC DEVELOPMENT(S)

The eccentric distance sum has attracted a lot of interest in the last several years both in mathematical and chemical research communities, and numerous results and structural properties of the EDS were established. The graph theoretical concepts and notations not defined in this essay can be found in the standard books of graph theory such as West (1996).

Rodríguez (2005) computed the EDS of distance-regular hypergraphs in terms of its intersection array. Moreover, using the alternating polynomials and the Laplacian polynomials, he obtained upper bounds on the EDS of hypergraphs.

Yu et al. (2011) characterized the extremal unicyclic graphs among n-vertex unicyclic graphs with given girth having the minimal and second minimal EDS, respectively. In addition, they characterized the extremal trees with the minimal and second minimal EDS in the class of trees with a given diameter. In particular, Yu et al. (2011) proved that for an n-vertex tree T, $\xi^{ds}(T) \geq 4n^2 - 9n + 5$, with equality holding if and only if T is the n-vertex star S_n, and for an n-vertex unicyclic graph G, $\xi^{ds}(G) \geq 4n^2 - 9n + 1$, with equality holding if and only if G is the graph obtained by adding an edge between two pendent vertices of n-vertex star.

Hua et al. (2011) presented a short and unified proof of Yu et al.'s (2011) results on the EDS of trees and unicyclic graphs by treating the lower bound for EDS of a more general graph class, namely, cacti, than trees and unicyclic graphs.

Ilić et al. (2011) introduced a graph transformation that increased the EDS and proved that the path P_n is the unique extremal trees with n vertices having maximum EDS. Various lower and upper bounds for the EDS in terms of other graph invariants including the Wiener index, degree distance (Dobrynin & Kochetova, 1994; Gutman, 1994), eccentric connectivity index, independence number, connectivity, matching number, chromatic number, and clique number were also established in Ilić et al. (2011). In addition, they presented explicit formulae for the EDS of the Cartesian product, applied to some graphs of chemical interest (like C_4-nanotubes and C_4-nanotori).

Zhang and Li (2011) determined the n-vertex trees with the maximum, second-maximum, third-maximum, and fourth-maximum EDS, respectively, for $n \geq 8$. They also characterized the extremal unicyclic graphs on n vertices with the maximal, second maximal, and third maximal EDS, respectively.

Hua et al. (2012) gave some new lower and upper bounds for EDS in terms of other graph invariants. They also presented two Nordhaus-Gaddum-type results for EDS. Moreover, for a given nontrivial connected graph, they gave explicit formulae for EDS of its double graph and extended double cover, respectively. For all possible k values, they also characterized the graphs with the minimum EDS within all connected graphs on n vertices with k cut edges, and all graphs on n vertices with edge-connectivity k, respectively.

Li et al. (2012) characterized the extremal trees among n-vertex conjugated trees (trees with a perfect matching) having the minimal and second minimal EDS. They also identified the trees with the minimal and second minimal EDS among the n-vertex trees with matching number m, and characterized the extremal tree with the second minimal EDS among the n-vertex trees of a given diameter. This work led to finding the trees with the third and fourth minimal EDS among the n-vertex trees.

Songhori (2012) computed the EDS of Volkmann tree. She also computed this index for vertex-transitive graphs and non-vertex transitive graphs whose automorphism group has exactly two orbits.

Geng et al. (2013) determined the tree among n-vertex trees with domination number γ having the minimal EDS and identified the tree among n-vertex trees with domination number γ satisfying the relation $n = k\gamma$ having the maximal EDS, for $k = 2, 3, \frac{n}{3}, \frac{n}{2}$. Sharp upper and lower bounds on the EDS among the n-vertex trees with k leaves, and the trees among the n-vertex trees with a given bipartition having the minimal, second minimal, and third minimal EDS were also determined in Geng et al. (2013).

Azari and Iranmanesh (2013) presented explicit formulae for computing the EDS of the most important graph operations such as the Cartesian product, join, composition, disjunction, symmetric difference, cluster, and corona product of graphs. They also applied the obtained results to compute the EDS for some important classes of molecular graphs and nanostructures by specializing components in graph operations.

Tavakoli et al. (2013) computed the EDS of the hierarchical product of graphs and applied their results to compute the EDS of octanitrocubane, bridge graphs, and bridge-cycle graphs.

Eskendar and Vumar (2013) calculated the EDS of the generalized hierarchical product of graphs.

Hemmasi et al. (2014) presented an algorithm for computing the EDS of any simple connected graph. According to this algorithm and using the GAP program, they wrote a program to compute this index. They tested the algorithm to calculate the EDS of C_{12n+4} and $C_{12(2n+1)}$ fullerenes.

Qu and Yu (2014) characterized the chain hexagonal cactus with the minimal and maximal EDS among all chain hexagonal cacti of length n, respectively. Moreover, they presented exact formulae for the EDS of two types of hexagonal cacti.

Mukungunugwa and Mukwembi (2014) determined the asymptotic sharp upper bounds for EDS of graphs according to its order and minimal degree. Their result extended a result of Ilić et al. (2011), and that of Zhang & Li (2011).

Gao et al. (2014) determined the EDS of r-coronagraphs of fan graph, wheel graph, gear fan graph, and gear wheel graph.

Using the Tutte-Berge formula and the method of graph transformation, An et al. (2015) could obtain sharp lower bound for the EDS of graphs with a given matching number, and determine the extremal graphs.

Miao et al. (2015) characterized the trees having the maximal EDS among n-vertex trees with maximum degree Δ and among those with domination number 3. The trees having the maximal or minimal EDS among n-vertex trees with independence number α and the trees having the maximal EDS among n-vertex trees with matching number m were also determined in Miao et al. (2015).

Li et al. (2015) determined sharp lower bound on the EDS in the class of all connected bipartite graphs with a given matching number m, the minimum EDS was realized only by the complete bipartite graph $K_{q,n-q}$. They also characterized the extremal graph with the minimum EDS in the class of all the n-vertex connected bipartite graphs of odd diameter. All the extremal graphs having the minimum EDS in the class of all connected n-vertex bipartite graphs with a given vertex connectivity were also identified in Li et al. (2015).

Venkatakrishnan and Sathiamoorthy (2015) gave an explicit formula to calculate the EDS of the join of k graphs.

Wang and Kang (2016) compared the degree distance and EDS for some graph families.

Some derivative indices of the eccentric distance sum such as the adjacent eccentric distance sum (Sardana & Madan, 2002, 2003) and super augmented eccentric distance sum connectivity indices (Gupta et al., 2012) have recently been introduced. These indices were found to exhibit high sensitivity towards the presence and relative position of heteroatoms; therefore their mathematical properties have been studied too.

14.4   NANOCHEMICAL APPLICATION(S)

The eccentric distance sum offers a vast potential for structure-activity/ property relationships. Gupta et al. (2002b) showed that this graph invariant displays high discriminating power with respect to both biological activity and physical properties. Comparatively, the EDS exhibited much better correlation and lesser average errors than the Wiener index. The excellent prediction of the physical properties by EDS can be attributed to probable contribution of distance sum in addition to eccentricity. The physical properties are significantly responsible for the biological activity of a chemical compound.

14.5   MULTI-/TRANS-DISCIPLINARY CONNECTION(S)

It has been shown by Gupta et al. (2002b) that, the eccentric distance sum can provide valuable leads for the development of safe and potent thera-peutic agents of diverse nature. Though both EDS and Wiener index showed almost the same predictability of anti-HIV activity of dihydroseselins, but EDS exhibited far superior discriminating power and correlating ability with regard to physical properties.

14.6   OPEN ISSUES

There are many open questions for further study on EDS. For example, one could try to compute the values of the EDS for various classes of graphs, especially for dendrimers, nanotubes, nonotori, carbon nanocones, and other classes of nanostructures. One may study the relations of the EDS with other distance-based invariants. It would be interesting to investigate the graph which is extremal with respect to EDS among all connected graphs on n vertices and k cut vertices. It is also interesting, but a little difficult, to solve the following open problems proposed by Li et al. (2015).

- **Problem 1.** How to determine the graph with minimum EDS among n-vertex bipartite graphs with a given even diameter?
- **Problem 2.** How to determine the graph with minimum EDS among n-vertex bipartite graphs with a given radius?
- **Problem 3.** How to determine the graph with minimum EDS among n-vertex bipartite graphs with a given edge-connectivity?

KEYWORDS

- **anti-HIV activity**
- **graph invariant**
- **quantitative structure-activity relationship**
- **quantitative structure-property relationship**
- **Wiener index**

REFERENCES AND FURTHER READING

Alizadeh, Y., Azari, M., & Došlić, T. (2013). Computing the eccentricity-related invariants of single-defect carbon nanocones. *J. Comput. Theor. Nanosci., 10*(6), 1297–1300.

An, M., Sun, M., Liu, Y., & Meng, X. (2015). Eccentric distance sum of graphs with a given matching number. *J. Tianjin Univ. Sci. Tech., 30*(2), 75–77.

Azari, M., & Iranmanesh, A. (2013). Computing the eccentric-distance sum for graph operations. *Discrete Appl. Math., 161*(18), 2827–2840.

Azari, M., Iranmanesh, A., & Diudea, M. V. (2017). A topological study of some molecular graphs by vertex-eccentricity-based descriptors. *Studia Univ. Babes Bolyai. Chem., 62*(1), 129–142.

Bajaj, S., Sambi, S. S., & Madan, A. K. (2005). Topological models for prediction of anti-HIV activity of acylthiocarbamates. *Bioorg. Med. Chem., 13*, 3263–3268.

Bajaj, S., Sambi, S. S., Gupta, S., & Madan, A. K. (2006). Model for prediction of anti-HIV activity of 2-pyridinone derivatives using novel topological descriptor. *QSAR Comb. Sci., 25*, 813–823.

Balaban, A. T., Motoc, I., Bonchev, D., & Mekennyan, O. (1983). Graph invariant indices for structure-property correlation. *Top. Curr. Chem., 114*, 21–55.

Basak, S. C., Niemi, G. J., & Veith, G. D. (1991). Predicting properties of molecules using graph invariants. *J. Math. Chem.*, *7*, 243–272.

Bonchev, D., Mekenyan, O., & Balaban, A. T. (1986). Algorithms for coding chemical compounds. In: Trinajstic, N. (ed.), *Mathematical and Computational Concepts in Chemistry* (pp. 34–47). Ellis Horwood: Chichester, England.

Dankelmann, P., Goddard, W., & Swart, C. S. (2004). The average eccentricity of a graph and its subgraphs. *Util. Math.*, *65*, 41–51.

Dobrynin, A., & Kochetova, A. A. (1994). Degree distance of a graph: A degree analog of the Wiener index. *J. Chem. Inf. Comput. Sci.*, *34*, 1082–1086.

Eskender, B., & Vumar, E. (2013). Eccentric connectivity index and eccentric distance sum of some graph operations. *Trans. Comb.*, *2*(1), 103–111.

Estrada, E., & Ramirez, A. (1996). Edge adjacency relationships and molecular topographic graph invariants: Definition and QSAR applications. *J. Chem. Inf. Comput. Sci.*, *36*, 837–843.

Gao, Y., Liang, L., & Gao, W. (2014). Degree distance and eccentric distance sum of certain special molecular graphs. *Sch. Acad. J. Biosci.*, *2*(7), 454–458.

Geng, X. Y., Li, S. C., & Zhang, M. (2013). External values on the eccentric distance sum of trees. *Discrete Appl. Math.*, *161*(16/17), 2427–2439.

Gupta, M., Gupta, S., Dureja, H., & Madan, A. K. (2012). Superaugmented eccentric distance sum connectivity indices: Novel highly discriminating topological descriptors for QSAR/QSPR. *Chem. Biol. Drug. Des.*, *79*, 38–52.

Gupta, S., Singh, M., & Madan, A. K. (2002a). Application of graph theory: Relationship of eccentric connectivity index and Wiener's index with anti-inflammatory activity. *J. Math. Anal. Appl.*, *266*(2), 259–268.

Gupta, S., Singh, M., & Madan, A. K. (2002b). Eccentric distance sum: A novel graph invariant for predicting biological and physical properties. *J. Math. Anal. Appl.*, *275*, 386–401.

Gutman, I. (1994). Selected properties of the Schultz molecular topological index. *J. Chem. Inf. Comput. Sci.*, *34*, 1087–1089.

Hemmasi, M., Iranmanesh, A., & Tehranian, A. (2014). Computing eccentric distance sum for an infinite family of fullerenes. *MATCH Commun. Math. Comput. Chem.*, *71*, 417–424.

Hua, H., Xu, K., & Wen, S. (2011). A short and unified proof of Yu et al.'s two results on the eccentric distance sum. *J. Math. Anal. Appl.*, *382*, 364–366.

Hua, H., Zhang, S., & Xu, K. (2012). Further results on the eccentric distance sum. *Discrete Appl. Math.*, *160*(1/2), 170–180.

Ilić, A., Yu, G. H., & Feng, L. H. (2011). On the eccentric distance sum of graphs. *J. Math. Anal. Appl.*, *381*, 590–600.

Katritzky, A. R., & Gordeeva, E. V. (1993). Traditional graph invariant indices versus electronic, geometric and combined molecular graph invariants in QSAR/QSPR research. *J. Chem. Inf. Comput. Sci.*, *33*, 835–857.

Li, S. C., Wu, Y. Y., & Sun, L. L. (2015). On the minimum eccentric distance sum of bipartite graphs with some given parameters. *J. Math. Anal. Appl.*, *430*, 1149–1162.

Li, S. C., Zhang, M., Yu, G. H., & Feng, L. H. (2012). On the external values of the eccentric distance sum of trees. *J. Math. Anal. Appl.*, *390*(1), 99–112.

Miao, L. Y., Cao, Q. Q., Cui, N., & Pang, S. Y. (2015). On the external values of the eccentric distance sum of trees. *Discrete Appl. Math.*, *186*, 199–206.

Mukungunugwa, V., & Mukwembi, S. (2014). On eccentric distance sum and minimum degree. *Discrete Appl. Math.*, *175*, 55–61.

Qu, H., & Yu, G. (2014). Chain hexagonal cacti with the external eccentric distance sum. *Scientific World J.,* 897918.

Rodríguez, J. A. (2005). On the Wiener index and the eccentric distance sum of hypergraphs. *MATCH Commun. Math. Comput. Chem.*, *54*, 209–220.

Sardana, S., & Madan, A. K. (2002). Predicting anti-HIV activity of TIBO derivatives: A computational approach using a novel topological descriptor. *J. Mol. Model.*, *8*, 258–265.

Sardana, S., & Madan, A. K. (2003). Relationship of Wiener's index and adjacent eccentric distance sum index with nitroxide free radicals and their precursors as modifiers against oxidative damage. *J. Mol. Struct. (Theochem), 624*, 53–59.

Sharma, V., Goswami, R., & Madan, A. K. (1997). Eccentric connectivity index: A novel highly discriminating topological descriptor for structure-property and structure-activity studies. *J. Chem. Inf. Comput. Sci.*, *37*, 273–282.

Songhori, M. (2012). A note on eccentric distance sum. *J. Math. NanoSci.*, *2*(1), 37–41.

Tavakoli, M., Rahbarnia, F., & Ashrafi, A. R. (2013). Distribution of some graph invariants over hierarchical product of graphs. *Applied Math. Comput.*, *220*, 405–413.

Venkatakrishnan, Y. B., & Sathiamoorthy, G. (2015). Eccentricity based topological indices of the joint of *k* graphs. *Int. J. Pure Appl. Math.*, *101*(3), 381–385.

Wang, H., & Kang, L. (2016). Further properties on the degree distance of graphs. *J. Comb. Optim.*, *31*(1), 427–446.

West, D. B., (1996). *Introduction to Graph Theory* (pp. 512). 1st ed.; Prentice Hall: Upper Saddle River, USA.

Wiener, H. (1947a). Structural determination of paraffin boiling points. *J. Am. Chem. Soc., 69*, 17–20.

Wiener, H. (1947b). Correlation of heats of isomerization and differences in heats of vaporization of isomers among the paraffin hydrocarbons. *J. Am. Chem. Soc., 69*(11), 2636–2638.

Yu, G. H., Feng, L. H., & Ilić, A. (2011). On the eccentric distance sum of trees and unicyclic graphs. *J. Math. Anal. Appl.*, *375*(1), 99–107.

Zhang, J., & Li, J. (2011). On the maximal eccentric distance sums of graphs. *ISRN Appl. Math.,* 421456.

CHAPTER 15

Electrodeposition

MIRELA I. IORGA

National Institute for Research and Development in Electrochemistry and Condensed Matter, Department of Applied Electrochemistry, 300569 Timisoara, Romania

15.1 DEFINITION

The metals electrodeposition is an extremely complex process, since beside electrochemical steps–as adsorption and mass transfer, are involved processes as complexes formation and dissociation equilibrium, and incorporation of impurities and/or additives deliberately introduced in the electrochemical bath. The deposition reaction represents the reaction of the charged particles at the interface between a metal electrode (solid phase) and the solution (liquid phase), and the two types of charged particles that can pass through the interface are metal ion and electron.

15.2 HISTORICAL ORIGIN(S)

The beginnings of electroplating (the process used in electroplating is called electrodeposition) of precious metals go back in time to 1800. Modern electrochemistry was invented by the Italian chemist and university professor Luigi Valentino Brugnatelli in 1805. He is considered the pioneer in utilizing gold in the electroplating process, by Alessandro Volta's invention (1800)–the Voltaic pile–to facilitate the electrodeposition (Hunt, 1973). Consequently, this method used by Brugnatelli implying voltaic electricity enabled him to experiment with various metallic-plating solutions. By 1805, his process was refined to plate a fine layer of gold over large silver metals (Garcia & Burleigh, 2013).

Brugnatelli wrote to the *Belgian Journal of Physics and Chemistry*: "I have lately gilt in a complete manner two large silver medals, by bringing them into communication by means of a steel wire, with a negative pole of a voltaic pile, and keeping them one after the other immersed in ammoniuret of gold newly made and well saturated." His letter was reprinted later in Great Britain.

Unfortunately, any of Brugnatelli's important work was not published in the scientific journals of his day, due to some reasons of Napoleon's French Academy of Sciences, the leading scientific body of Europe. Only a few close people knew about his discovery, beyond the Italy borders his scientific work was widely unknown. His inventions didn't become industrial used for the next thirty years (Hunt, 1973).

By 1839, other scientists from Great Britain and Russia had independently invented similar metal deposition processes (as Brugnateli's), for electroplating of copper in printing press plates (Dufour, 2006).

In Russia, Boris Jacobi developed electroplating, electrotyping and galvanoplastic sculpture (Hunt, 1973). He was not only the person who discovered galvanoplastics, but he developed electrotyping and galvanoplastic sculpture. These came quickly into fashion in Russia. Here scientists as Peter Bagration, Heinrich Lenz, and Vladimir Odoyevsky (a science fiction author) have contributed to this technology development (Salmond, 2004).

For coating objects in gold, in 1800s to 1845 period, two main commercial processes were used. The first one was to introduce some objects in a diluted gold chloride solution (water gilding), and allowing that a very thin flash of gold be deposited onto inexpensive objects. The second was applied for objects with high durability and value, the primary technique for a thick, durable gold plate over a surface, was a dangerous process involving mercury amalgam and gold leaf (fire gilding).

Then, by the Birmingham Jewelry Quarter: "It was a Birmingham doctor, John Wright, who first showed that items could be electroplated by immersing them in a tank of silver held in solution, through which an electric current was passed." He finds that potassium cyanide was an appropriate electrolyte for gold and silver electroplating.

Meanwhile, other scientists were carrying on similar work. In 1840 occurred several patents for electroplating processes. First, the electroplating process was patented by cousins Henry and George Richard Elkington. The patent rights for John Wright's process were bought by the Elkington's. They held a monopoly on electroplating for many years due to their patent for an inexpensive method of electroplating.

In Birmingham, the Elkington's founded the electroplating industry, and from there this industry spread around the world (Thomas, 1991; Beauchamp, 1997). The earliest electrical generator used by the Elkington's in an industrial process (1844) was the Woolrich Electrical Generator, now at Birmingham Science Museum (Hunt, 1973).

Struggling to commercialize their patented processes, they met resistance from the traditional Sheffield plate manufacturers. Because they were unable to license their patents, later they would build their factories in which plated objects such as silverware, frames of glasses, gifts, and many other low-cost items were produced. They had such a success in adapting their processes that soon they will dominate the decorative metals industry from their region.

The gold and silver electroplating process spread very quickly from Great Britain throughout the Europe and the United States of America. For example, in France, such electroplated decorative objects were readily accepted by upper society to sense their richness and the new fashion tendencies.

The next new wonder in economic jewelry dated from 1857 when electroplating was first applied to costume jewelry (Garcia & Burleigh, 2013).

In Russia, cathedral domes, icons, and religious statues were being successfully plated by large-scale gold plating. In the 19th century, Russia mentioned the use of electroplating for the massive galvanoplastic sculptures in Saint Petersburg at St. Isaac's Cathedral. Another example is at the tallest Orthodox Church, the Cathedral of Christ the Saviour from Moscow, the dome was electroplated with gold (Salmond, 2004). The electroplating baths and the equipment were made suitable to permit the plating process of several larger objects. Subsequently, the old techniques such as mercury amalgam gold gilding and water gilding are displaced by the electroplating process (Hunt, 1973).

As knowledge of electrochemistry and electroplating process spread up, by the 1850s, other types of non-decorative metal plating occur, for example, electroplating processes for bright nickel, brass, tin, and zinc– adapted for commercial purposes (McDonald & Hunt, 1982). Many of these were utilized for specific manufacturing and engineering applications.

The first modern electroplating plant which began its production in 1876 was the Norddeutsche Affinerie in Hamburg (Stelter & Bombach, 2004).

Then, except some technical improvements in direct current (DC) power supplies, the period from 1870 to 1940 was characterized only by small improvements in manufacturing processes, anodic principles, and plating bath formulas. The extension of electroplating processes in other industries did not produce any significant scientific developments since the occurrence, in the middle of 1940s, of the electronics industry.

In the late 19th century, a great stimulus for the plating industry was the development of electric generators. Then, with higher currents, occurs the possibility to process in bulk, automotive items corrosion protected, with a better aspect, and with enhanced wearing properties. Also, by the same technique could be treated metal machine components or hardware.

Further developments and refinements were determined by the two World Wars and by the aviation industry. The following processes were spread in industry: bronze alloy plating, hard chromium plating, sulfamate nickel plating, and several other plating methods. The equipment used for the plating processes developed from the tar-lined wooden tanks, operated manually, to automatized equipment, which could work over thousands of kilograms per hour of parts (McDonald & Hunt, 1982).

In the late 1940s the heavy gold plating for electronic components was rediscover (Garcia & Burleigh, 2013). On large-scale commercial use, in the middle of 1950's, the traditional cyanide-based baths were displaced by the newer and safer plating baths which had in their composition acidic formulas.

Then, since the environmental protection must be taken into consideration, in the 1970s many regulatory laws for wastewater emissions and disposal setting the direction for the electroplating industry for the next 30 years were done. Improvements in chemical formulas and specialized hardware allowed for the rapid and continuous plating of wire, metal strips, semiconductors, and complex metal shapes.

After analyzing the emergence and development of metals electrochemical deposition processes from aqueous solutions, and of main use areas for metal electrochemical deposition, was found a wide variety of domains in which the electrodeposition processes have emerged lately.

Nowadays due to the chemical development and to a higher understanding of their basic electrochemical principles, more sophisticated plating bath formulas were achieved. In this regard, greater control of the layer thickness, the working parameters, and the performance of the electroplated items was made. These innovative developments in chemistry have enabled better plating speed, throwing power and high-quality, reliable plated finishes.

The electroplating of materials such as platinum, ruthenium, and osmium are now finding broader usages on electronic connectors, circuit boards, and contacts. In electronics, osmium, platinum, and ruthenium were electroplated and were used in circuit boards, contacts, or electronic connectors.

The opinion of more and more experts is that the telecommunication industry will increase in high dependence on the new and innovative electroplating technology. The electronics industry and the need to support the development of their underlying infrastructure will continue to make improvements in the electroplating industry, all over the world.

Future progress in waveform technologies for DC power supplies may lead to even greater achievements for the electroplating and metal finishing industry. Besides, safer "closed loop" manufacturing processes and waste-water recycling will continue to diminish the exposure of workers to harmful chemical substances and toxic byproducts.

The electroplating processes implied in all types of plating application (decorative or technical) will find new areas in the emerging global markets.

One of the American physicist Richard Feynman's (1918–1988) first projects was to develop technology for electroplating metal onto plastic. Feynman developed the original idea into a successful invention (Feynman, 1985).

Nowadays metal coating applications range from jewelry and automotive parts to some of the most critical technologies used in spacecraft and modern weaponry. For example, the Mars Orbiter, the Keck Observatory, the TOW missile, and the latest infrared sensor and countermeasures devices wouldn't work without plating.

15.3   NANO-SCIENTIFIC DEVELOPMENT(S)

15.3.1   FUNDAMENTAL THEORETICAL ISSUES INVOLVED IN THE ELECTROCHEMICAL DEPOSITION

The metals electrodeposition is an extremely complex process, besides electrochemical steps, as adsorption and mass transfer, are involved complexes formation and dissociation equilibrium, and incorporation of impurities and/or additives deliberately introduced in the electrochemical bath. The resulting surface, microscopically detectable (i.e., deposit appearance) depends on either the primary and secondary distribution of the current (or current density), and specific current densities for the parallel electrochemical reaction taking place on the electrode surface.

In general, the metals are electrolytically deposited from complex or simple salt solutions: sulfates, chlorides or nitrates. In this case, the overall cathodic overall reaction is the discharging of the hydrated metal ions, followed by their subsequent incorporation in the crystal lattice of the deposit formed at the cathode.

$$M^{z+} \cdot xH_2O + ze^- = [M] + xH_2O \qquad (1)$$

Very frequent this is the cathodic deposition of metals from their complexes ion solutions. For this type of solutions, the overall cathodic reaction will be as follows:

$$MA_x^{z-x} + ze^- = [M] + xA^- \qquad (2)$$

where z is the metal valence in the MA_x^{z-x} complex.

Electrochemical deposition of metals and alloys involves metal ions reduction from aqueous electrolytes, organic electrolytes and salt melts.

In the case of aqueous solutions metal ions M^{z+} reduction is represented by the reaction:

$$M_{sol}^{z+} + ze^- \rightarrow M_{crist} \qquad (3)$$

Two different processes can achieve the electrochemical deposition:

- an electrodeposition process, where an external source provides z electrons (e^-);
- a deposition process without current (autocatalytic), where the electron source is the reducing agent from solution (in this case an external source is not involved).

These two processes, electrodeposition and deposition without current, in fact, consists the electrochemical deposition. In both cases, the interest is focused on the metal electrode in contact with the aqueous ion solution. The deposition process is the reaction of the charged particles at the interface between a metal electrode (solid phase) and the solution (liquid phase); the two types of charged particles that can pass through the interface are metal ion and electron.

15.3.2 METAL-SOLUTION INTERFACE: WHERE DEPOSITION PROCESS PROGRESS

Metal-solution interface involves two components: metal and aqueous ionic solution. To understand phenomena occurring on the interface is necessary to have a basic knowledge about the metal structure and electric properties, molecular structure of water and structure and properties of ionic solutions (Schlesinger & Paunovic, 2000).

When a metal is contacted with an electrolyte solution, the mutual charging of phases occurs for the following reasons (Vaszilcsin, 1995):

- charged particles (ions or electrons) passing through the interface;
- solvent dipole orientation at the interface;
- between positive and negative charge center on the metal surface, there is a disparity.

In the following issue, one will consider the processes occurring when the two phases are in contact. The structure of the interface between the metal ion solutions (Paunovic, 1998) is shown in Figure 15.1. The interface appears as a region between two phases, with a different composition from that inside of each of them (i.e., by metal, respectively solution mass).

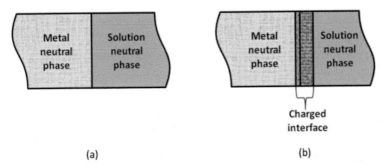

(a)   (b)

FIGURE 15.1   Two phases in contact: **(a)** t = 0, when contacts; **(b)** at equilibrium (Paunovic, 1998).

The electric field at metal–solution interface reach very high values (for example, 10^6, 10^7 V/cm). The importance of metal–electrolyte solution interface understanding is that the electrodeposition processes take place in this narrow region, where there is an appreciable electric field that can be controlled by an external voltage (Paunovic, 1998; Schlesinger & Paunovic, 2000; Vaszilcsin, 1995).

15.3.3 METAL–SOLUTION INTERFACE FORMATION: INTERFACE CHARGING

When a metal, M, is immersed in an aqueous solution containing M^{z+} ions, (e.g., MA salt), between the two phases, metal and solution, will be an exchange of ions. Some M^{z+} ions from crystal lattice will pass into solution, while other ions from the solution will penetrate into the crystal lattice (Schlesinger & Paunovic, 2000; Trasatti, 1981).

It is assumed that there are such conditions as M^{z+} ions that are leaving the crystal lattice will exceed the number entering in it. Thus, there is an excess of electrons on the metal surface, and the metal is negative charged q_M^- (metal charge per area unit). In response to metal charge at the interface, a charge rearrangement occurs in the solution phase from the interface. The metal negative charged attract M^+ ions positively charged from solution and reject A^{z+} ions negative charged from solution. The result of this is a surplus of positive ions, M^+, in solution, near the metal interface.

In this case, the solution at the interface has an opposite and equal charge, q_s^+ (charge per interface solution area unit). The positive charge of interface solution results in a rate decreasing of M^+ ions which leave the crystal lattice (due to the irrejection) and increasing the speed at which ions enter the crystal lattice.

After a while, between metal M and its ions from solution a dynamic equilibrium will be established, given by Eq. (4):

$$M^{z+} + ze^- \leftrightarrow M \tag{4}$$

where z represents the number of electrons.

The reaction from left to right is electron consumer and is called reduction, and the opposite, from right to left, generates electrons and is called oxidation. At the moment of dynamic equilibrium establish, the same number of M^+ ions, $\bar{n}$, enter, and the same amount of M^+ ions are leaving the crystal lattice, $\vec{n}$, as shown in Figure 15.2 and Eq. (5).

$$\bar{n} = \vec{n} \tag{5}$$

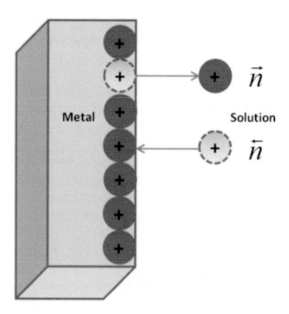

FIGURE 15.2 Metal–solution interface formation; steady state: $\bar{n} = \vec{n}$ (Paunovic, 1998).

When the equilibrium is established, the interface region is neutral:

$$q_M = -q_S \qquad (6)$$

15.3.4   ELECTRICAL DOUBLE LAYER

The next question is how the charge excess is distributed to the interface, in both the metal side and in the solution (Paunovic, 1998).

The contact region between the two phases in which the component particles (ions, molecules, or atoms) distribution differs from the inside contact phases, is called an *electrical double layer*. The term of the double layer has the meaning of charged interface, because not always such an interface can be equated with two layers of electrical charges (Vaszilcsin, 1995). Knowledge of electrical double layer structure is crucial to understand how the electrode processes kinetics is influenced.

15.3.5   ELECTRODE POTENTIAL

Electrode characteristic properties manifest metal–electrolyte solution interface, generated by interactions resulted in specific local changes. These interactions determine the qualitative transformations of the metal surface and the adjacent electrolyte layer (Oniciu & Constantinescu, 1982).

Adsorption of charged or neutral particles at the electrode surface and electrode reactions rate is conditioned by the size and the profile of potential difference between metal and electrolyte solution in contact with it. The origin and nature of this potential difference are essential in electrode kinetics, electrochemical thermodynamics, and electrochemical adsorption understanding.

By definition the electric potential in a point of a material system is the work done in the transport of a positive unit charge from that point to infinite; it is assumed that during the transport, the unit charge doesn't influence the electric field.

If the charge must pass material environments with variable densities, the definition of the potential is complicated due to electrical and chemical properties of a charged particle (electron or ion) crossing the interface, assuming the simultaneously overcoming of chemical and electrostatic forces (Oniciu & Constantinescu, 1982; Vaszilcsin, 1995).

15.4 NANOCHEMICAL APPLICATION(S)

15.4.1 ELECTRODEPOSITION MECHANISM: NUCLEATION AND CRYSTAL GROWTH IN METAL LATTICE

A metal cathodic deposition is a simultaneously discharging and crystallization process, and since crystallization takes place under the influence of electricity, was named electrocrystallization. It is a phase transformation under the electric field and has many common features with normal crystallization (Facsko, 1969).

The metal electrocrystallization theory, ideally–on a smooth surface, was developed in 1931 by Volmer and Erdey-Gruz and relies on mechanism model of occurrence of a new phase from a supersaturated phase, where the energy used with germs formation is lower as germ's dimensions are smaller. The slowest stage of the process was considered the bi- or tri-dimensional crystal germs nucleation on the electrode surface. The bi-dimensional germ is formed by a mono-atomic layer, on which another layer is spread, etc., thus developing the tri-dimensional germ.

The electrocrystallization process shows two successive stages: deposition of the hydrated ion within the metal crystal lattice and crystallization–three-dimensional crystal development. In turn, the deposition process consists of ion's dehydration, followed by its reduction and subsequent incorporation into the metal lattice. It is assumed that by ion incorporation in metal lattice the electrode reaction is finalized.

The metal ion in solution goes through a route that can be sketched as follows: the ion comes from the volume phase of the solution to the interface by transport. At the interface take place the deformation of the hydrating shell and progressively lose of this, passing into an adsorbed ion, partially discharged, further transfer reaction moving in an adsorbed atom, and finally included in the metal lattice. The occurrence of adsorbed ion is demonstrated indirectly through activation energy calculations, showing that the direct formation of the neutral superficial atom is improbable, because it would require too high activation heat compared to the experimentally determined value.

The electrodeposition (electrocrystallization) mechanism involves as a first step the reduction of a cation to the substrate surface (aided by the application of a current or voltage) to produce the adsorbed atom and its migration to the surface at an energetically favorable point. Atoms that will continue to agglomerates together will form the nucleus of a new phase. The core growth parallel and/or perpendicular to the surface, of course, on the surface more nuclei can shape and develop.

Since the whole electrode surface was coated with at least one layer, it is easier to deposit on the same metal substrate rather than on a different metal substrate. The deposition process takes place quickly on the same metallic substrate than on a different substrate. Apparently, the first layers formation determines the deposit's structure and adhesion (Brett et al., 1998; Facsko, 1969; Paunovic, 1998).

Hydrated ion interaction with the electrode surface depends on where it occurs. Although a plan of a crystal lattice, the metallic electrode surface is not perfect and may contain a lot of defects: steps, corners, vacancies in steps and gaps, as shown in Figure 15.3.

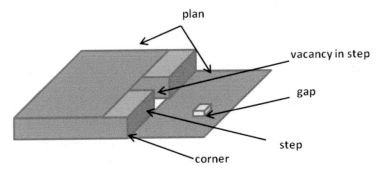

FIGURE 15.3   Different types of defects on metallic electrode surface (Brett et al., 1998; Facsko, 1969; Paunovic, 1998).

Following research on the growth mechanism of already formed crystals was found that cation discharging do not take place on the entire facet of crystal growing, but only in some active parts of it, as presented in Figure 15.4.

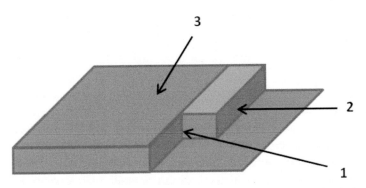

FIGURE 15.4   Scheme of metal crystal growth: 1 – the most active sector; 2 – the less active sector; 3–the least active sector (Brett et al., 1998; Facsko, 1969; Paunovic, 1998).

Crystal growth takes place mainly in the more active sector (1), as atoms resulting from the discharge of cations are attracted to a greater number of atoms from crystal lattice (I, II, and III) and the deposition energy at this point is minimal. After completing the most active sector, the atoms begin to fixate on the less active sector (2) producing two more active areas. This mode of growth continues until metal atoms completely cover the entire crystal facet. Since then, the atoms in the crystal lattice can be placed in completed facet corners (3), where atoms in the lattice exert a higher attraction, and the covering process of the crystal facet is restarted; the deposition takes places in successive layers (Firoiu, 1983).

In the case of more complex crystalline systems, the formation work of a two-dimensional germ on crystal different facets, in general, is not equivalent. Thus, those facets will grow preferentially, where structure follows the solution supersaturation degree (Facsko, 1969).

In the case of electrocrystallization, the role of solution supersaturation is played by the supplementary potential increasing at the metal–solution interface, i.e., the overpotential. The germs initiation will require higher crystallization overpotential, equivalent to a high supersaturation, and growth of existing crystalline formations – a low crystallization overpotential, equal to a low supersaturation (Oniciu & Constantinescu, 1982). In electrocrystallization, the overpotential (and therefore supersaturation) must overcome the forces that tend to destroy the germs before they reach critical dimensions which give them stability.

Since the electrodeposition is an electrode process, the reaction rate is determined by the electrode applied voltage. Thus, the deposition rate and therefore the deposit structure are affected: a small overpotential means more time to form a perfect crystalline structure deposit (Brett et al., 1998; Koryta et al., 1993; Pletcher & Walsh, 2000).

The crystal particle size is determined by the ratio of the nucleation rate and already formed crystal growth rate. The higher number of crystal germs is created in a time unit, the crystal grain size of the metal deposit will be lower.

The probability of nucleation increases with cathodic polarization. All factors (deposited metal ions concentration, current density, surface-active substances in the solution, etc.) that increase the cathodic polarization will contribute to the finely crystallized layers (Firoiu, 1983).

In crystal evolution when the step has reached the electrode limits, increasing does not stop because support crystals real deposition surfaces have defects that may initiate new actions. In this regard, it may be mentioned shearing dislocations, those progress by edge rotation (helical motion).

When more dislocations interact, the collision of two steps in the opposite helical development may determine double helix crystal growth (Antropov, 2001; Hine, 1985; Oniciu & Constantinescu, 1982; Pletcher & Walsh, 2000).

During electrocrystallization process, crystal facets with different crystallographic indices were characterized by various growth rates; distinct deposition rates on several crystal facets have a seemingly paradoxical consequence, namely: facets with high growth rates tend to disappear since those characterized by low growth rates tend to develop.

The texture occurrence can be explained as follows: at the beginning of the electrolysis, on the cathode surface germs randomly-oriented are formed. As electrolysis progresses, germs oriented perpendicular with maximum growth rate will occupy the entire electrode surface, whereas after a while germs with oriented parallel will stop growing (Firoiu, 1983).

Conventionally admitting the cubic crystal where crystals do not have a well-defined orientation – lack of texture – a situation which does not occur again in practice. If one of crystal facets is on a parallel plan to the electrode surface, and the remaining facets are oriented differently, an incomplete texture is obtained. When most of the crystals are oriented the same way, the texture is complete. The textures achieved in practice range from incomplete to complete texture.

Sometimes complex and spectacular crystalline developments are met, such as dendrites and needle crystals, whose formation has practical implications. Thus, in the case of electrochemical power sources, dendrites can penetrate and damage the electrodes separating membranes causing shorting source, and in the case of metallic powder deposition, the deposit quality will be affected (Oniciu & Constantinescu, 1982).

In the case of metal electrodeposition, several types of crystal growth occur, such as pyramid, layers, blocks, ridge, cubic layers, spiral, needles, dendrite (tree), and texture.

After Bockris, block and dendrite types shall be regarded as modifications of the pyramid and layer types. If a polycrystalline structure growth after a systematic orientation, on the cathode surface may be are peat macroscopic pattern or texture (Bockris, 1972).

Different crystal growth forms were highlighted by recent research conducted on copper or silver electrodeposition processes (Kondo & Murakami, 2004; Polk et al., 2004). The electrodeposition morphology may be affected by some factors, such as solution composition, including pH and impurities, operating temperature, electrolyte flow rate, cathode potential and overpotential, current density and substrate (Antropov, 2001; Hine, 1985; Oniciu & Constantinescu, 1982; Pletcher & Walsh, 2000).

15.5 MULTI-/TRANS-DISCIPLINARY CONNECTION(S)

Understanding nucleation and surface crystalline germs growth, the mono- and multi-layers formation, as uniform deposit forming, are based on the knowledge of condensed matter physics and surface chemistry-physics (Schlesinger & Paunovic, 2000).

Metal electrochemical deposition is used for base metal surface protection (more active) against corrosion by coating the surfaces with more stable metals or alloys (galvanotechnics) or giving the products a decorative aspect (galavnoplastics). In this regard, the most critical areas that use metals electrochemical deposition are getting replicas and reproductions of works of art, manufacture bands, laminated pipes, printed circuits, etc.

Very new areas are applications in automotive and aerospace industries; by electrodeposition, pieces with complicated shapes, which cannot be machined, can be made.

Electrochemical deposition of metals from aqueous solutions form the basis of hydroelectrometalurgical processes, for example, metal extraction from ore (electrowinning) and further purification by electrolysis (electrorefining). From the metals produced and refined by hydroelectrometalurgical processes, one can mention copper, nickel, zinc, cadmium, tin, lead, silver, gold, manganese, etc. Hydroelectrometalurgy can get technical purity metals, and metals can be recovered profitably from poor ores. Electrodeposition also can be used as a preconcentration technique for trace analysis.

Another area of utmost importance for the whole society today is the treatment of wastewater containing metal ions. As increasingly require legislative provisions that have found that conventional remediation methods are no longer sufficient, and in particular are not economic. This fact determined the emergence of more modern methods for metal ions removal. With the increasing interest in environmental protection, the electrodeposition method has imposed higher use for removing metallic ions pollutants from effluents and process flows. In these processes, the ions are deposited on a support electrode structure to be subsequently recycled or removed. It will prevent the dissipation of toxic materials in soil, from the landfill waste from industrial processes.

Thus, some methods are carried out by a direct recovery in metallic form, while in other electrodeposition is followed by the chemical or anodic dissolution of the deposit in a small volume of electrolyte. The methods can also be applied for the separation of metal ions from dilute solutions, at a concentration in the range $1 \div 1000 \ mg/dm^3$.

15.6 OPEN ISSUES

These can be understood and interpreted-based on a variety of mass and surface analytical methods and techniques, studying electrical, magnetic and physical properties of metals and alloys (Schlesinger & Paunovic, 2000).

The nature of the obtained deposit, and the conditions of its formation while applying a constant current (or constant voltage) depends not only on the metal type but, to a large extent on the solution composition and the existent impurities in the solution. The presence of surface-active substances and various oxidants (e.g., dissolved oxygen) affects the metal electrodeposition kinetics. The solution purity degree and the nature of the impurities can affect the crystal growth, the number of crystallization centers which occur per time unit per cathode surface unit, polarization at a given current density, the variation thereof in time, etc.

Recent research (Haatja et al., 2003; Schwartz et al., 2004; Mohd & Pletcher, 2005) attempt to establish patterns to predict the morphological evolution of electrodeposition processes according to process parameters, electrolyte composition and the species to be deposited.

In conclusion, depending on the current density, the cation concentration in solution, the temperature, the electrolyte agitation, and the surface-active substances presence, crystal growth and passivation processes have a different rate. These features will determine the size of crystals from the cathodic deposit (Antropov, 2001; Brett et al., 1998; Firoiu, 1983; Golumbioschi, 1995).

ACKNOWLEDGMENT

This contribution is part of the Nucleus-Programme under the project "Deca-Nano-Graphenic Semiconductor: From Photoactive Structure to The Integrated Quantum Information Transport," PN-18-36-02-01/2018, funded by the Romanian National Authority for Scientific Research and Innovation (ANCSI).

KEYWORDS

- **electrochemistry**
- **electrocrystallization**
- **electrodeposition**
- **metal deposition**

REFERENCES AND FURTHER READING

Antropov, L. I., (2001). *Theoretical Electrochemistry*. University Press of the Pacific, Honolulu, Hawaii.

Beauchamp, K. G., (1997). *Exhibiting Electricity IET*, p. 90.

Bockris, J. O. M., (1972). *Electrochemistry of Cleaner Environments*. Plenum Press, New York.

Brett, C. M. A., & Brett, A. M. O., (1998). *Electrochemistry: Principles, Methods, and Applications*. Oxford University Press Inc., New York.

Dufour, J., (2006). *An Introduction to Metallurgy* (5th edn.), Cameron.

Facsko, G., (1969). *Tehnologie Electrochimică* (Eng.: Electrochemical Technology). Technical Publishing, Bucharest, Romania.

Feynman, R., (1985). *Surely You're Joking, Mr. Feynman!*. In Chapter 6, The chief research chemist of the metaplast corporation.

Firoiu, C., (1983). *Tehnologia proceselor electrochimice* (Eng.: Technology of Electrochemical Processes). Didactic & Pedagogical Publisher, Bucharest, Romania.

Garcia, J. C., & Burleigh, T. D., (2013). *The Beginnings of Gold Electroplating*. The electrochemical society interface.

Golumbioschi, F., (1995). *Tehnologia proceselor electrochimice–Curs*. (Eng.: Technology of Electrochemical Processes - Lectures) Technical University, Timisoara, Romania.

Haatja, M., Srolovitz, D., & Bocarsly, A. B., (2003). Morphological stability during electrodeposition. I. Steady states and stability analysis. *Journal of the Electrochemical Society*, *150*(10), C699–C706.

Hine, F., (1985). *Electrode Processes and Electrochemical Engineering*. Plenum Press, New York.

Hunt, L. B., (1973). The early history of gold plating. A tangled tale of disputed priorities. *Gold Bulletin, 6*(1), pp. 16–27.

Kondo, K., & Murakami, H., (2004). Crystal growth of electrolytic Cu foil. *Journal of the Electrochemical Society*, *151*(7), C514–C518.

Koryta, J., Dvorak, J., & Kavan, L., (1993). *Principles of Electrochemistry*. John Wiley & Sons Ltd., New York.

McDonald, D., & Hunt, L. B., (1982). *A History of Platinum and its Allied Metals*. Johnson Matthey, Hatton Garden, London, EC1.

Mohd, Y., & Pletcher, D., (2005). The influence of deposition conditions and dopant ions on the structure, activity, and stability of lead dioxide anode coatings. *Journal of the Electrochemical Society*, *152*(6), D97–D102.

Oniciu, L., & Constantinescu, E., (1982). *Electrochimie şi coroziune* (Eng.: Electrochemistry and Corrosion). Didactic & Pedagogical Publisher, Bucharest, Romania.

Paunovic, M., (1998). *Fundamentals of Electrochemical Deposition*. John Wiley & Sons Inc., New York.

Pletcher, D., & Walsh, F. C., (2000). *Industrial Electrochemistry*. Chapman and Hall, London.

Polk, B., Bernard, M., Kasianowicz, J., Misakian, M., & Gaitan, M., (2004). Microelectroplating silver on sharp edges toward the fabrication of solid-state nanopores. *Journal of the Electrochemical Society, 151*(9), C559–C566.

Salmond, W., (2004). *Tradition in Transition, Russian Icons in the Age of the Romanovs*. Washington DC, Hillward Museum.

Schlesinger, M., & Paunovic, M., (2000). *Modern Electroplating*. John Wiley & Sons Inc., New York.

Schwartz, M., Myung, N. V., & Nobe, K., (2004). Electrodeposition of iron group-rare earth alloys from aqueous media. *Journal of the Electrochemical Society, 151*(7), C468–C477.

Stelter, M., & Bombach, H., (2004). Process optimization in copper electrorefining. *Advanced Engineering Materials, 6(7), 558.*

Thomas, J. M., (1991). *Michael Faraday and the Royal Institution: The Genius of Man and Place*. Bristol: Hilger.

Trasatti, S., (1981). *Electrodes of Conductive Metallic Oxides, Part B*. Elsevier Scientific Publishing Company, Amsterdam.

Vaszilcsin, N., (1995). *Electrochimie–Curs* (Eng.: Electrochemistry-Lectures). University Politehnica Timişoara, Romania.

CHAPTER 16

f-Graph: Bounded Vertex Degree

LOUIS V. QUINTAS and EDGAR G. DUCASSE

Mathematics Department, Pace University, New York, NY 10038, USA,
E-mail: lvquintas@gmail.com, egducasse@pace.edu

16.1 DEFINITION

A graph with no vertex of degree greater than a constant f is said to be of *bounded vertex degree f* and in an appropriate context is referred to simply as an *f-graph.*

16.2 ORIGINS AND DEVELOPMENT:

16.2.1 *GRAPHS WITH BOUNDED VERTEX DEGREE (F-GRAPHS)*

A graph with no vertex of degree greater than a specified non-negative integer f is said to be of *bounded vertex degree f* and in an appropriate context is referred to simply as an *f-graph.*

In Figure 16.1, all graphs on five vertices (that is, of *order* 5) are partitioned by *size* (number of edges) and f class. In the figure, all graphs above an f curve constitute the set of f-graphs of order 5. The only 0-graph is the *empty graph,* in this case, $5K_1$, the 1-graphs of order 5 consists of $5K_1$, K_2 U $3K_1$, and $2K_2$ U K_1. There are eleven 2-graphs of order 5, twenty-three 3-graphs of order 5, and thirty-four 4-graphs of order 5.

In the study of f-graphs, not all graphs of a given order need to be considered. For example, in Balinska et al., (2014) the graphs studied are all the f-forests of order n, where an *f-forest* is a union of f-trees. Another example of a restricted class of f-graphs consists of all 4-graphs of order n having no cycle of length greater than or less than 6.

The investigation of f-graphs as a self-standing subject came about naturally with the recurring appearance of f-graphs in applications. For example, in many chemical, electrical, communication, and transportation studies the underlying graph model is an *f*-graph (Balaban, 1976; Deo, 1974; Doyle & Snell, 1984; Roberts, 1976, 1993). In nanochemistry, the bounded valence of atoms forces molecular models to make use of *f*-graphs (for example, carbon having valence 4 requires $f = 4$). In electrical, communication, and transportation models the number of edges incident to a vertex is naturally bounded by physical or monetary constraints. The bound on the number of vertices in such models, although present, is not as restricted as the bound on the vertex degrees (Balińska & Quintas, 2006).

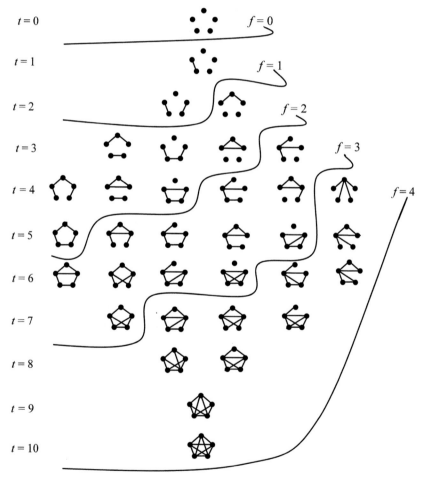

FIGURE 16.1 All graphs of order $n = 5$ partitioned into size (*t*) classes, $0 \leq t \leq 10$, and *f* classes, $0 \leq f \leq 4$.

Prior to the 1960s, research concerning graphs with a bounded degree was robust producing many significant results in the physical sciences and mathematical areas such as probability and statistics. However, due to very little communication among researchers in different disciplines, diverse notations and terminology emerged. This communication gap was still being worked on in 1975 (see Forward in Balaban, 1976). It was inevitable that the common thread of graphs with a bounded degree would extend to the explicit study of *f*-graphs. This, in turn, would result in a common link between the various areas of application.

One example of the preceding was the work of Manfred Gordon and John W. Kennedy (Good, 1963, 1962; Gordon & Kennedy, 1973; Gordon & Ross-Murphy, 1975) being done in England in the 1970s on polymerization of molecules. Here the evolution of graph models of chemical entities with no valence greater than *f* was being studied. Prior to this, in their study of random graphs, Paul Erdős and Alfréd Rényi (1960), studied the evolution of graphs with no restriction on the bound of the vertex degrees as the number of vertices went unbounded. Although the preceding was not applicable to many applications, this clearly suggested that a focused study of the evolution and structure of graphs with a bounded degree was needed. Many papers with *f*-graphs in this context have appeared (Gross et al., 1988; Kennedy, 1981, 1983; Kennedy & Quintas, 1989). Note that for applications the interest is in the number of vertices getting large but not necessarily going to infinity. This has been explicitly explored in Balińska & Quintas (2006).

Today *f*-graphs are studied from many different points of view both theoretically and in applications (Balińska et al., 2014; Cha et al., 2015, 2015; DuCasse & Quintas, 2011; DuCasse et al., 2012). Namely, their properties and structure are being studied, such as their graph invariants (topological indices), their algebraic properties (automorphism groups, eigenvalues) and their probabilistic/statistical characteristics (random *f*-graphs, vertex degree, and state distributions).

16.3 HISTORICAL COMMENT

Although there is no specific time at which the beginning of the study of *f*-graphs, as a unified subject can be identified. The book (Balińska & Quintas, 2006) is the first attempt to treat such a study as a unified topic.

16.4   OPEN ISSUES

16.4.1   *STRUCTURE PROBLEMS*

The structure of f-graphs is of basic interest. For large n the number of vertices of degree less than f is negligible relative to n. However, for small to moderate n the vertices of degree less than f constitute an integral part of the structure of an f-graph. The study of the structure of f-graphs has been dealt with via the detailed investigation of *edge maximal f-graphs*, EM f-graphs, (see Chapter 3 in Balińska & Quintas, 2006).

In Kennedy & Quintas (1989), a vertex of degree strictly less than f in an f-graph is called *orexic*. A basic result states:

If G is an EM f-graph of order n *with* $m \geq 1$ orexic vertices, then the orexic vertices induce a complete subgraph of order m in G with $m \leq f$.

The detailed study of an EM f-graph G shows how the induced orexic subgraph lies in G (Balińska & Quintas, 2006).

From Figure 16.1, the three EM 2-graphs of order 5 are easily seen to be the 5-cycle C_5, $C_3 \cup K_2$, and $C_4 \cup K_1$. Note that C_5 has no orexic vertices and the orexic vertices in the latter two graphs induce K_2 and K_1, respectively. For the EM 3-graphs, one obtains the two graphs shown in Figure 16.2. The only EM 4-graph is K_5.

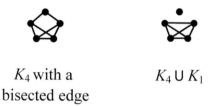

K_4 with a
bisected edge

$K_4 \cup K_1$

FIGURE 16.2   The two EM 3-graphs of order 5.

There are other subsets of graphs of order 5 that are of interest, for example, from Figure 16.2, it is easy to locate the nine 3-forests.

Enumeration results for various classes of EM f-graphs are given in Balińska & Quintas (2006). In particular, there are formulas for the number of unlabeled EM 2-graphs and the number of unlabeled EM (n–2)-graphs having order n and m orexic vertices, respectively.

Problem. Determine formulas for the number of EM f-graphs having order n, m orexic vertices, and with $3 \leq f \leq n$–3.

16.4.2 DISTANCE/SIMILARITY PROBLEMS

The *distance* between two graphs G and H is defined as the least number of deletions and insertions of edges in G needed to obtain H and is denoted $d(G, H)$ (see Cha et al., 2015). Note that $d(G, H) = d(H, G)$ and that $d(G, H)$ is undefined if G and H do not have the same order. Furthermore, if G and H are *f*-graphs, then it is also required that all intermediate graphs in going from G to H be *f*-graphs.

For *f*-graphs G and H that represent chemical structures, $d(G, H)$ is a natural definition for the *similarity* of G and H (see DuCasse & Quintas, 2011 and references therein). For example, for an alkane C_nH_{2n+2}, represented by its carbon skeleton as a 4-tree of order n, the distance from a given alkane G to alkanes H such that $d(G, H)$ is maximum yields alkanes most dissimilar to G.

Note that deleting an edge from a 4-tree of order n results in a 4-forest of order n made up of two 4-trees whose orders n_1 and n_2 sum to n. Such a 4-forest can be interpreted as an alkane mixture made up of two types of alkanes having n_1 and n_2 carbons, respectively (DuCasse & Quintas, 2011). Thus, a transition from an alkane to one of its isomers can be viewed as a transition involving passing through various alkane mixtures.

A small example to illustrate this is the graph $F(6, 4)$ whose vertex set consists of all the unlabeled 4-forests of order 6 with vertices G and H being adjacent if and only if G and H differ by exactly one edge (see Figure 16.3).

The basic theorem used in the investigating of these concepts are:

Let G and H be two graphs of order n and I a maximum size unlabeled subgraph common to both G and H. Then,

$$d(G, H) = |E(G)| + |E(H)| - 2|E(I)|.$$

There is a variety of open problems in this context that can be obtained by restricting the class of *f*-graphs being considered. Using $F(6, 4)$ with the above alkane interpretation one can say that the five 6-carbon hexane isomers, represented by the five 4-trees on the bottom level of $F(6, 4)$, are each maximally dissimilar to the methane mixture $6K_1$, because each isomer is at distance 5 from $6K_1$. Among the five hexane isomers, the isomer most dissimilar to the 6-chain is the isomer represented by the tree shown in Figure 16.4, which is at distance 4 from the 6-chain, and this is the only hexane isomer with this property.

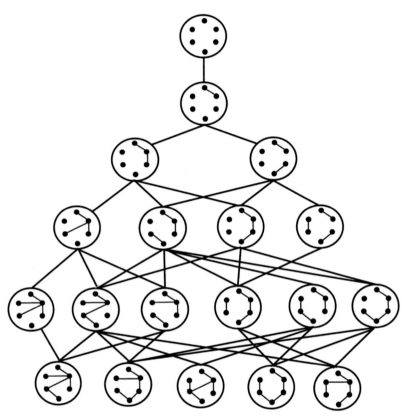

FIGURE 16.3   The graph $F(6, 4)$.

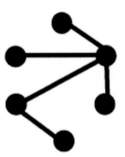

FIGURE 16.4   The only hexane isomer most dissimilar to the 6-chain.

Further note that the 6-chain alkane is most dissimilar to the pentane/ methane mixture represented, in $F(6, 4)$, by the 4-star union K_1 and the methane mixture $6K_1$ because they are each at distance 5 from the 6-chain.

In general, the following is an open problem.

Problem. Let *G* be a 4-forest of order *n*. Among the set of all 4-forests of order *n*, what is the set of 4-forests of order *n* most dissimilar to *G*? That is, what 4-forests are most distant from *G*?

KEYWORDS

- **bounded vertex degree**
- ***f*-graph**
- **vertex degree**

REFERENCES

Balaban, A. T., (1976). *Chemical Applications of Graph Theory*. Academic Press, New York.

Balińska, K. T., & Quintas, L. V., (2006). *Random Graphs with a bounded degree*. Publishing House of the Poznań University of Technology, Poznań.

Balińska, K. T., Zwierzyński, K. T., Quintas, L. V., & DuCasse, E. G., (2014). Domination in graphs whose vertices are forests with a bounded degree . *Graph theory notes of New York LXVII. The Mathematical Association of America*, 24–33.

Cha, S. H., DuCasse, E. G., Quintas, L. V., & Shor, J. P., (2015). Graphs whose vertices are forests with a bounded degree : Planarity. *Bulletin of the Institute of Combinatorics and its Applications*. In press.

Cha, S. H., DuCasse, E. G., Quintas, L. V., Kravette, K., & Mendoza, D. M., (2015). A graph model related to chemistry: Conversion of methane to higher alkanes. *International J. Chemical Modeling, 7*(1), 17–29.

Deo, N., (1974). *Graph Theory with Applications to Engineering and Computer Science*. Prentice-Hall, Englewood Cliffs, New Jersey.

Doyle, P. G., & Snell, J. L., (1984). *Random Walks on Electric Networks* (p. 22). Carus Mathematical Monographs, Mathematical Association of America.

DuCasse, E. G., & Quintas, L. V., (2011). A graph model related to chemistry. *International J. Chemical Modeling, 3*(4), 449–454. Also appears in book: *Advances in Chemical Modeling* (2012) (Vol. 3). *Nova Science Publications, Inc.*

DuCasse, E. G., Quintas, L. V., & Zimmler, M. O., (2012). Graphs whose vertices are forests with a bounded degree : Traceability. *Bulletin of the Institute of Combinatorics and its Applications, 66*, 65–71.

Erdős, P., & Rényi, A., (1960). On the evolution of random graphs. *Publ. Math. Inst. Hungarian Acad. Sci., 5*, 17–61.

Good, I. J., (1963). Cascade theory and the molecular weight averages of the sol fraction. *Proc. Roy. Soc. A., 272*, 54–59.

Gordon, M., & Kennedy, J. W., (1973). The graph-like state of matter: LCGI Scheme for the thermodynamics of alkanes and the theory of inductive inference. *J.C.S. Faraday II, 69*, 484–504.

Gordon, M., & Ross-Murphy, S. B., (1975). The structure and properties of molecular trees and networks, *Pure Applied Chem., 43*, 1–26.

Gordon, M., (1962). Good's theory of cascade processes applied to the statistics of polymer distributions. *Proc. Roy. Soc. A., 268*, 240–259.

Gross, R., Kennedy, J. W., Quintas, L. V., & Yarmush, M. L., (1988). Antigensis: A cascade-theoretical analysis of the size distribution of antigen-antibody complexes. *Discrete Applied Math., 19*, 177–194.

Kennedy, J. W., & Quintas, L. V., (1989). Probability models for random *f*-graphs, combinatorial mathematics (New York, 1985). *Ann. N.Y. Acad. Sci., 555*, 248–261.

Kennedy, J. W., (1981). Statistical mechanics and large random graphs. In: Hippe, Z., (ed.), *Data Processing in Chemistry* (pp. 113–132). Elsevier, Amsterdam.

Kennedy, J. W., (1983). The random-graph-like state of matter. In: *Computer Applications in Chemistry* (pp. 151–178). Elsevier, Amsterdam.

Roberts, F. S., (1976). *Discrete Mathematical Models with Applications to Social, Biological and Environmental Problems*. Prentice-Hall, Englewood Cliffs, New Jersey.

Roberts, F. S., (1993). New directions in graph theory (with the emphasis on the role of applications), Quo Vadis, graph theory? (Fairbanks, Alaska, 1990), Elsevier, Amsterdam. *Ann. Discrete Math., 55*, 13–43.

CHAPTER 17

Fries Structure

MARINA A. TUDORAN[1,2] and MIHAI V. PUTZ[1,2]

[1]*Laboratory of Structural and Computational Physical Chemistry for Nanosciences and QSAR, Biology-Chemistry Department, West University of Timisoara, Pestalozzi Street No. 44, Timisoara, RO-300115, Romania*

[2]*Laboratory of Renewable Energies-Photovoltaics, R&D National Institute for Electrochemistry and Condensed Matter, Dr. A. Paunescu Podeanu Str. No. 144, Timisoara, RO-300569, Romania, Tel: +40-256-592638, Fax: +40-256-592620, E-mail: mv_putz@yahoo.com or mihai.putz@e-uvt.ro*

17.1 DEFINITION

A Fries structure is defined as a Kekule structure with maximum numbers of double bonds, with the property that most double bonds are common to two rings. In other words, these types of structures can be obtained in the same manner as Clar structures (inscribing circles inside benzenoid rings which preset six π-electrons), with the difference that in the Fries structure case the circles can share an edge.

17.2 HISTORICAL ORIGIN(S)

Benzenoid hydrocarbon is one of the most important π-electron systems, being used in analyzing different theories about chemical and/or physicochemical behavior in relation with aromaticity and electron structure (Ciesielski et al., 2010; Minkin et al., 1994; Pascal, 2006; Schleyer, 2001, 2005; Randić, 2003). Starting from this considerate, benzene-like Kekule electron structures were studied and characterized by different mean, e.g., in his work, Fries try to explain the chemical stability (Fries, 1927) by making a distinction between the resonance contributor which maximizes the number of rings for these structures. He proposes a model similar with the one proposed by Clar, inscribing circles inside benzenoid rings which preset six π-electrons, the so-called "Fries structures," with the difference that in this case, the circles may share an edge. Fries structures have an interesting characteristic: one may obtain few Clar structures with the maximum number of separated sextets (Figure 17.1), by respecting additional rules, such as the separation between the isolated formal double bonds.

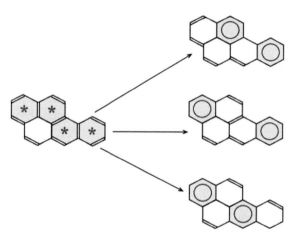

FIGURE 17.1   Three Clar structures derived from the Fries structure of Benzo[a]pyrene.

Studies in this area using descriptors of aromaticity, such as NCIS (Schleyer et al., 1996; Chen et al., 2005) or HOMA (Krygowski & Cyrański, 2001), lead to the conclusion that the most efficient pi-electron delocalization is given by the separated sextets in Clar structures (Balaban & Klein, 2009; Ciesielski et al., 2009). Due to the fact that for larger benzenoid hydrocarbon with low symmetry is difficult to determine the Clar structure with the maximum number of isolated sextets, Fries structures can be used as an initial step in this type of studies (Ciesielski et al., 2010).

17.3   NANO-SCIENTIFIC DEVELOPMENT(S)

In the organic chemistry, Clar sextets (Clar, 1972) and Fries structures (Fries, 1927) have a major contribution, being used in qualitative studies for understanding the π-electronic structure and the stability of benzenoids and other compounds with similar structure, both of them being based on the

Kekule structure (or, on the perfect matching of the graph G) (Fowler et al., 2012). In the same context, Fries (1927) determine an association between the stability and the existence of a Kekule structure which achieves a large number of benzenoid hexagons (in other words, the hexagons containing three double bonds which are not necessary switchable). Based on this considerate, the Fries number $F(G)$ can be defined as the cardinality of the maximum set of the hexagons of benzenoid over all Kekule structures (Fowler et al., 2012).

On the other hand, the Fries canonical structure can be used in aromaticity studies, and can be calculated according to Ciesielski and his coworkers (2010) from the matrix K using a recurrence function, namely Fries structure generating function (FGF), which follows the relation:

$$FGF(A,n) = P_n : \begin{cases} FGF(A,1) = P_1 = A \circ A^{-1} \\ FGF(A,i+1) = P_{i+1} = P_i \circ P_i^{-1} \end{cases} \tag{1}$$

where A represents the adjacency matrix, P represents the matrix for Pauling bond order, "$\circ$" states for the Hardman product, and n represents the number of recurrence steps (Horn & Johnson, 1985, 1994).

Moving forward with the itineration, one can obtain the general formula for obtaining the matrix $K(F)$ of the Fries structure for any benzenoid hydrocarbons as follow (Ciesielski et al., 2010):

$$\lim_{n \to \infty} FGF(A,n) = \lim_{n \to \infty} P_n = K(F) \tag{2}$$

This way, the authors (Ciesielski et al., 2010) obtained an algorithm which can be successfully applied even on large non-symmetrical benzenoid systems.

17.4 NANOCHEMICAL APPLICATION(S)

In literature, a benzenoid (graphene patches, hexagonal patches, graphite patches or benzenoid hydrocarbons) of the form $\Gamma = (V, E, F)$ can be defined as a plane graph having one distinguished face – the outside face and all the other faces hexagonal, and with the property that all vertices have degree 2 or 3 and all vertices with degree 2 are bonded to the outside face. The boundary to the outside face of the benzenoid Γ is referring as the boundary. Moving forward, the boundary of a hexagonal face of Γ is defined as the set of edges and vertices shared with the outside face. Boundary faces are the hexagonal faces which present non-empty boundaries. For a given boundary face f, its boundary is represented by the one (or many) simple path which are called the boundary segments of Γ. One can define H as the benzenoids class with the property that all boundary segments present odd length and the boundary is an elementary circuit. On the other hand, a fullerene $\Gamma = (V, E, F)$ is defined as a trivalent plane graph which has pentagonal and hexagonal faces. A Kekule structure (also known as a perfect matching) of a fullerene or a benzenoid Γ, with $K \subseteq E$, is defined as a set of edges with the property that each vertex is incident with exactly one edge in K. For a given Kekule structure on Γ, there are 0, 1, 2 or 3 of its bounding edges of a face of Γ which can be found in Γ. One can denote by $B_i(K)$ the set of faces which have exactly i edges in K. The faces in $B_0(K)$ are known as the void faces of K, while the faces in $B_3(K)$ are known as the benzene faces of K (Randić, 1977). Moving forward, one can define the Fries number of a fullerene or of a benzenoid Γ, denoted by $\phi(\Gamma)$ as the maximum number of benzene faces over all the Kekule structures which are possible for Γ. A Fries set is represented by a set of benzene faces $\phi(\Gamma)$ in some Kekule structure in Γ (Graver et al., 2013). In their work, Graver et al. (2013) propose a coloring method in order to determine the Fries number of a benzenoid. For $\Gamma \in H$ they make the following choices: red as the class of faces which do not lie on the boundary; blue as the largest of the remaining color classes; yellow as the third color class, and starting from these, they propose two theorems associated with the Fries number. The first one states that, for a benzenoid $\Gamma = (V, E, F)$ which admits a Kekule structure K, and with K_b as the number of edges of K on the boundary of Γ, the following expression is valid:

$$|B_3(K)| = \frac{|V|}{3} - \frac{|B_1(K)| + 2|B_2(K)| + |K_b|}{3} \tag{3}$$

and when $\Gamma \in H$:

$$\phi(\Gamma) \leq \frac{|V|}{3} - \frac{l_B + l_Y}{3} \tag{4}$$

The second theorem considers that, for a benzenoid in H, the unique Fries set is denoted by $B \cup Y$ representing the union of the two color classes which appear on the boundary (Graver et al., 2013), and is given by:

$$\phi(\Gamma) = \frac{|V|}{3} - \frac{l_B + l_Y}{3} \tag{5}$$

In order for the equality to be valid, there are several conditions which needs to be fulfilled, such as K_b should be equal with the summation of l_B and l_Y, and $B_1(K)$ should be equal with $B_2(K)$, both being equal with 0 (Graver et al., 2013).

17.5 MULTI-/TRANS-DISCIPLINARY CONNECTION(S)

In literature, a fullerene is known a molecule which contains only carbon atoms, with applications in biotechnology (Bosi et al., 2003), medical science (Faraji & Wipf, 2009), electronic and optic engineering (Venturini et al., 2002; Eckert et al., 2000). On the other hand, any molecule can be represented by a molecular graph, with vertices as atoms and edges as bonds between atoms (Salami & Ahmadi, 2015).

For a given fullerene C_n with n the number of carbon atoms, one can define a graph $\Gamma_n = (V, E, F)$, having $|V| = n$ vertices, $|E| = 3n/2$ edges as the carbon-carbon bonds and $|F| = 2 + n/2$ faces (both the pentagonal and the hexagonal ones) (Ye & Zhang, 2009). For a graph $G\langle V, E\rangle$, one can define a matching M (Doslic, 2007; Shiu & Zhang, 2008; Wang et al., 2007) as a set of pair of nonadjacent edges, and when every vertex from $V(G)$ is incident whit an edge from M, then is called a perfect matching (Kardos et al., 2009). For an organic compound, a Kekule structure represents a perfect matching of its molecular graph. Considering a fullerene C_n, one can define a Kekule structure $K \subseteq E$ as a perfect matching of its associated graph molecule Γ_n with edges which correspond to double bonds, meaning that every hexagonal face can have 0, 1, 2, or all 3 edges with double bonds in K. For a given Kekule structure, the hexagonal face which has exactly three double bond edges in it is called benzene-like (or full face) of K.

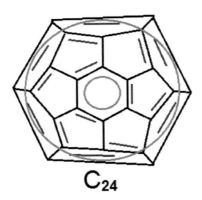

C_{24}

FIGURE 17.2 Fries and Kekule structure for fullerene C_{24}.

The Fries number of a fullerene C_n can be defined as the maximum number of the full faces over all possible perfect matching of Γ_n (Fries, 1972), and is strongly correlated with the isomers of the fullerene. In order to

find the Fries number of a fullerene, one can use the binary linear programming problem, with the form:

- **Objective Function** (Eq. 6): is counting the number of benzene-like
  hexagonal faces in the Kekule structure:

$$(P_1): \text{maximize} \sum_{k=1}^{m} y_k \tag{6}$$

- **Constrain I** (Eq. 7): assure that every vertex is incident only with a
  double bond edge and represent a condition so that no two double
  bond edges have simultaneous a vertex in common:

$$s.t. \sum_{j \in N(i)} x_{ji} = 1, \ i = 1, 2, \ldots, n; \tag{7}$$

- **Constrain II** (Eq. 8): assure that in a Kekule structure which respects
  the *CONSTRAIN I*, the k the hexagonal faces is of the benzene-like
  type if and only if all the three double bond edges are sitting in a
  perfect matching:

$$\sum_{i \in h_k} \sum_{j \in N(i) \cap h_k} x_{ij} - 6 y_k \geq 0 \ \ k = 1, 2, \ldots, m; \tag{8}$$

- **Constrain III** (Eq. 9): assure the symmetric properties of the double
  bond edges, meaning that e_{ij} is a double bond edge if and only if e_{ji} is
  also a double bond edge:

$$x_{ij} = x_{ji} \ \ i = 1, 2, \ldots, n; \tag{9}$$

- **Constrain IV** (Eq. 10): contain the condition for variable x_{ij} and y_k:

$$\begin{cases} x_{ij} \in \{0,1\} \ \ i = 1, 2, \ldots, n; \ j \in N(i) \, j > i; \\ y_k \in \{0,1\} \ \ k = 1, 2, \ldots, m; \end{cases} \tag{10}$$

All the above points will results the *Theorem 1* which states that there is
a one-to-one correspondence between the Fries number and its associated
Kekule structure, for a fullerene of the type C_n and the optimal objective
value and optimal solution determined for the binary integer linear programming problem (P_1). A simplified model of (P_1) will have the following form:

$$(P_2): \max \sum_{k=1}^{m} y_k \tag{11}$$

$$s.t. \sum_{j \in N(i) \, j < i} x_{ij} + \sum_{j \in N(i) \, j > i} x_{ji} = 1, \ i = 1, 2, \ldots, n; \tag{12}$$

$$\sum_{i\in h_k}\sum_{j\in N(i)\cap h_k\,j>i} x_{ij}-3y_k\geq 0 \quad k=1,2,\dots,m; \tag{13}$$

$$\begin{cases} x_{ij}\in\{0,1\} & i=1,2,\dots,n;\ j\in N(i)\,j>i;\\ y_k\in\{0,1\} & k=1,2,\dots,m; \end{cases} \tag{14}$$

The linear binary integer mathematical model (P_2) can be formulated as a quadratic programming problem using a simple polynomial transformation, and after some mathematical artifice one will obtain the equivalent concave quadratic programming reformulation for determining the Fries number for a fullerene (Salami & Ahmadi, 2015):

$$(P_5):\max \sum_{k=1}^{m}(1-M)y_k-\sum_{i\in V}\sum_{j\in N(i)\,j>i} Mx_{ji}+\sum_{k=1}^{m}My_k^2+\sum_{i\in V}\sum_{j\in N(i)\,j>i} Mx_{ij}^2 \tag{15}$$

$$s.t. \quad \sum_{j\in N(i)\,j<i} x_{ij}+\sum_{j\in N(i)\,j>i} x_{ji}=1,\ i=1,2,\dots,n; \tag{16}$$

$$\sum_{i\in h_k}\sum_{j\in N(i)\cap h_k\,j>i} x_{ij}-3y_k\geq 0 \quad k=1,2,\dots,m; \tag{17}$$

$$\begin{cases} 0\leq x_{ij}\leq 1 & i=1,2,\dots,n;\ j\in N(i)\,j>i;\\ 0\leq y_k\leq 1 & k=1,2,\dots,m; \end{cases} \tag{18}$$

The notations used in algorithm are: *indices*, such as i,j as vertices of fullerene graph Γ_n, with $i,j=1,2,\dots,n$; and k for the hexagonal face of Γ_n, with $k=1,2,\dots,m$; *parameters*, h_k as the set of k the vertices of the hexagon (the k the row of $H_{m\times 6}$); e_{ij} as the edge which connects the vertex i with the vertex j, $j\in N(i)$; $N(i)$ as the set of vertices which are adjacent to vertex i; $A_{n\times n}$ as the adjacency matrix of Γ_n; $H_{m\times 6}$ as the hexagonal-node matrix; variables x_{ij} and y_k which states for bond and hexagons, respectively, in a Kekule structure, and are calculated as:

$$x_{ij}=\begin{cases}1 & \text{if } e_{ij}\text{ represent a bond in a Kekule structure}\\ 0 & \text{otherwise}\end{cases} \tag{19}$$

$$y_k=\begin{cases}1 & \text{if } k^{th}\text{ hexagon is a benzene-like in a Kekule structure}\\ 0 & \text{otherwise}\end{cases} \tag{20}$$

The algorithm proposed by Salami and Ahmadi was successfully tested on different fullerene isomers, and the Fries number of C_{24} fullerene (see Figure 17.2) was obtained the value 2. Here, the red line depicts the double bonds which define the perfect matching of C_{24} fullerene, and the green circle depicts the benzene-like faces (Salami & Ahmadi, 2015).

17.6   OPEN ISSUES

Even if the Fries structures are studied and applied in different areas of chemistry, a general method to determine the Fries number for fullerene class has yet to be discovered. In their work, Fowler and his co-workers (2012) re-examine the procedure proposed by Ciesielski (2010), the CKC (Ciesielski-Krygowski-Cyrański) algorithm, as the way in which Fowler is referring to, showing that when applied to small benzenoid molecules (in this case, the smallest has seven hexagons) do not give in all cases canonical Fries structure.

The authors (Fowler et al., 2012) starts from the CKC algorithm, defined as follows: *"Giving a benzenoid with n vertices, one can construct the adjacency matrix n × n-dimensional of the form $A = P_0$, with $a_{uv} = 1$ if the vertex u is connected with the vertex v, and 0 otherwise. Then, the matrix P_k (for k = 1, 2 …) is defined as"*:

$$P_k = P_{k-1} \circ P_{k-1}^{-1} \tag{21}$$

The CKC procedure presents some properties. First states that if the molecular graph is non-singular, in other words, if A has an inverse, then is possible to make the specific construction; the process will be stopped if a singular matrix P_k appear at any later stage. Another property is referring to the adjacency matrix A and its inverse A^{-1} which are symmetric, leading to P_1 to be also symmetric; this property is transferred to each subsequent iterate P_k, which will have the entries in the position u, v equal to the entries from the position v, u, meaning that the matrix P_k is transformed into itself under all automorphism of the graph. In addition, P_1 and all subsequent P_k present the property that for non-bonded pairs u, v there is no occurrence of non-zero entries, and all matrices P_k have zero entries as the diagonal elements.

The authors also employ in their study the Fries structure construction starting from the Hardman product, in $h \neq 0$ case. A possible method for finding $F(G)$ propose an exponential algorithm with 2^h choices, derived from the fact that for each hexagon with six entries ½ there are two possible ways of the formal double bonds assignments inside that face (Fowler et al., 2012). Table 17.1 presents the benzenoids maximum values of h and the worst case for the exponential algorithm. One can deduce that only for

2 of the 2^h possible assignment of the double bonds inside the ambiguous hexagon one can obtain a correct canonical Fries structure. When tested for a general set of large benzenoids, the CKC algorithm is impractical if the exponential step is not avoided.

TABLE 17.1 Values Obtained for the CKC Algorithm Applied to a Set of Benzenoids (Fowler et al., 2012)

H (number of hexagonal faces)	1	2	3	4	5	6	7	8	9	10	11	12	13
h_{max} (the largest number of faces with entries ½ for all edges appearing in the convergent structures)	1	0	1	2	1	2	3	3	4	4	4	5	7

The authors conclude that there are two ways in which limitation of the CKC algorithm might occur: first, the algorithm can determine a perfect matching with unique characteristics, having fewer that $F(G)$ simultaneous benzenoid hexagons; and second, there may be cases when the algorithm will lead to a Hadamard product matrix which generates 2^h choices of perfect matchings ($h \neq 0$), but neither of these choices realizes the Fries number for that specific benzenoid (Fowler et al., 2012).

KEYWORDS

- **Clar sextets**
- **Fries number**
- **Fries structure**
- **π-electron systems**

REFERENCES AND FURTHER READING

Balaban, A. T., & Klein, D. J., (2009). Claromatic carbon nanostructures. *The Journal of Physical Chemistry C., 113*, 19123–19133.

Bosi, S., DaRos, T., Spalluto, G., & Prato, M., (2003). Fullerene derivatives: An attractive tool for biological applications. *European Journal of Medicinal Chemistry, 38*, 913–923.

Chen, Z., Wannere, C. S., Corminboeuf, C., Puchta, R., & Schleyer, P. V. R., (2005). Nucleus-independent chemical shifts (NICS) as an aromaticity criterion. *Chemical Reviews, 105*, 3842–3888.

Ciesielski, A., Krygowski, T. M., & Cyrański, M. K., (2010). How to find the Fries structures for benzenoid hydrocarbons. *Symmetry, 2*, 1390–1400.

Ciesielski, A., Krygowski, T. M., Cyrański, M. K., Dobrowolski, M. A., & Aihara, J. I., (2009). Graph topological approach to magnetic properties of benzenoid hydrocarbons. *Physical Chemistry Chemical Physics, 11*, 11447–11445.

Clar, E., (1972). *The Aromatic Sextet*. Wiley: New York.

Doslic, T., (2007). Fullerene graphs with exponentially many perfect matchings. *Journal of Mathematical Chemistry, 41*, 183–192.

Eckert, J. F., Nicoud, J. F., Nierengarten, J. F., Liu, S. G., Echegoyen, L., Barigelletti, L., et al., (2000). Fullerene-oligophenylenevinylene hybrids: Synthesis, electronic properties, and incorporation in photovoltaic devices. *Journal of the American Chemical Society, 122*, 7467–7479.

Faraji, A. H., & Wipf, P., (2009). Nanoparticles in cellular drug delivery. *Bioorganic & Medicinal Chemistry, 17*, 2950–2962.

Fowler, P. W., Myrvold, W., & Bird, W. H., (2012). Counterexamples to a proposed algorithm for Fries structures of benzenoids. *Journal of Mathematical Chemistry, 50*, 2408–2426.

Fries, K., (1927). Über bicyclische verbindungen und ihren vergleich mit dem naphtalin. III. Mitteilung. (Eng.: On bicyclic compounds and their comparison with the naphthalene. III. Communication). *Justus Liebigs Annalen der Chemie, 454*, 121–324.

Graver, J. E., Hartung, E. J., & Souid, A. Y., (2013). Clar and fries numbers for benzenoids. *Journal of Mathematical Chemistry, 51*, 1981–1989.

Horn, R. A., & Johnson, C. R., (1985). *Matrix Analysis*. Cambridge University Press: Cambridge, UK.

Horn, R. A., & Johnson, C. R., (1994). *Topics in Matrix Analysis*. Cambridge University Press: Cambridge, UK.

Kardos, F., Kral, D., Miskuf, J., & Sereni, J., (2009). Fullerene graphs have exponentially many perfect matchings. *Journal of Mathematical Chemistry, 46*, 443–447.

Krygowski, T. M., & Cyrański, M. K., (2001). Structural aspects of aromaticity. *Chemical Reviews, 101*, 1385–1419.

Minkin, V. I., Glukhovtsev, M. N., & Simkin, B. Y., (1994). *Aromaticity and Antiaromaticity*. J. Wiley: New York, USA.

Pascal, R. A., Jr., (2006). Twisted acenes. *Chemical Reviews, 106*, 4809–4819.

Randic, M., (1977). A graph theoretical approach to conjugation and resonance energies of hydrocarbons. *Tetrahedron, 33*, 1843–2016.

Randić, M., (2003). Aromaticity of polycyclic conjugated hydrocarbons. *Chemical Reviews, 103*, 3449–3605.

Salami, M., & Ahmadi, M. B., (2015). A mathematical programming model for computing the fries number of a fullerene. *Applied Mathematical Modeling, 39*(18), 5473–5479.

Schleyer, P. V. R., (2001). Special issue of the journal devoted to aromaticity. *Chemical Reviews, 101*, 1115–1566.

Schleyer, P. V. R., (2005). Special issue of the journal devoted to delocalization sigma and pi. *Chemical Reviews, 105*, 3433–3948.

Schleyer, P. V. R., Maerker, C., Dransfeld, A., Jiao, H. J., & Hommes, N. J. R. V., (1996). Nucleus-independent chemical shifts: A simple and efficient aromaticity probe. *Journal of the American Chemical Society, 118,* 6317–6318.

Shiu, W. C., & Zhang, H., (2008). A complete characterization for k-resonant Klein-bottle polyhexes. *Journal of Mathematical Chemistry, 43,* 45–59.

Venturini, J., Koudoumas, E., Couris, S., Janot, J. M., Seta, P., Mathis, C., & Leach, S., (2002). Optical limiting and nonlinear optical absorption properties of C_{60}-polystyrene star polymer films: C_{60} concentration dependence. *Journal of Materials Chemistry, 12,* 2071–2076.

Wang, W. H., Chang, A., & Lu, D. Q., (2007). Unicyclic graphs possessing Kekulé structures with minimal energy. *Journal of Mathematical Chemistry, 42,* 311–320.

Ye, D., & Zhang, H., (2009). External fullerene graphs with the maximum Clar number. *Discrete Applied Mathematics, 157,* 3152–3173.

CHAPTER 18

Fullerenes

FRANCO CATALDO[1,2]

[1]*Actinium Chemical Research Institute, Via Casilina 1626A,
00133 Rome, Italy*

[2]*Università degli Studi della Tuscia, Via S. Camillo de Lellis, Viterbo,
Italy, Tel: +39 06-94368230, Fax: +39 06-94368230,
E-mail: franco.cataldo@fastwebnet.it, cataldo.franco@fastwebnet.it*

18.1 DEFINITION

Fullerenes are a family of molecules composed exclusively by elemental carbon. In the fullerene molecules, the carbon atoms are sp^2 hybridized and closed in a polyhedral cage structure (Kroto et al., 1991). The sp^2 hybridization is typical of the carbon atoms disposed in a graphene sheet (a single layer of carbon atoms in graphite). In a graphene sheet, the sp^2 hybridized carbon atoms are arranged in a series of condensed hexagonal rings without strain. The introduction of a single pentagonal ring in such a graphene plane causes a curvature in the plane (Cataldo, 2002). With 12 pentagons, the entire graphene sheet undergoes a closure into a fullerenic polyhedron. In this situation, the sp^2 hybridized carbon atoms undergo a considerable strain and distortions that the pure aromatic character of a graphene plane is lost. Thus, the π electrons which are delocalized over the entire graphene plane become localized and weakly cross-conjugated in a fullerene cage (Haddon et al., 1986).

According to Euler's formula, $v - e + f = 2$ (with v = vertex, e = edge, f = face) and from that formula it is possible to demonstrate that just 12 pentagons are need to close a polyhedral cage with almost any number of hexagons. Thus, a correct definition of fullerenes is as follows: molecules made by sp^2 hybridized carbon atoms closed in polyhedral cages which possess 12 pentagons and an arbitrary number of hexagons, such that the number of carbon atoms (n) is given by $n = 20 + 2h$ where h is the number of hexagons in each structure (Albert et al., 2010). Another key rule linked to fullerene stability is the isolated pentagons rule. According to this rule, only the fullerene polyhedral cages where the 12 pentagon rings are fully surrounded by hexagons are stable. This is the reason why C_{60} is the smallest fullerene isolated and produced in macroscopic quantity. The lower C_{60} homologs can be conceived only with at least two adjacent pentagons and are not stable and isolable (Kroto et al., 1991). Two adjacent unsaturated pentagons produce a pentalene structure which is extremely reactive edge in a fullerene cage and destabilizes the cage structure (Cataldo, 1999).

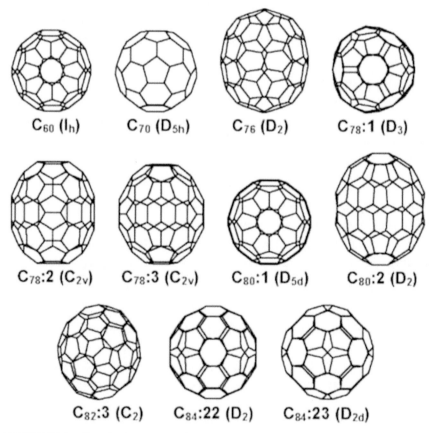

FIGURE 18.1   Chemical structures of stable fullerenes.

18.2   HISTORICAL ORIGINS

A lively account about the discovery of fullerenes can be found in the book of Bagott (1996). Another lively account can be found in an excellent paper written by Kroto (1992), where the link with polyynes can be found (Kroto et al., 1993). If the discovery of fullerenes belongs to Kroto, Smalley, and Curl as well as the full characterization of C_{60} and C_{70} (Kroto et al., 1991), the merit for the macroscopic production of C_{60} and C_{70} must be attributed to Kratschmer and colleagues who used the electric arc between two graphite electrodes under He atmosphere to produce a carbon soot containing about 3–5% by weight of C_{60} and C_{70} (Krätschmer et al., 1990; Krätschmer 2014). The fullerenes can be extracted from the carbon soot produced from the arc using organic solvents like benzene or toluene, and then the fullerenes can be separated and isolated in pure form using liquid chromatography (Hirsch et al., 2005; Taylor 1999). Fullerenes can be produced also in large quantities in certain types of laminar flames starting from benzene combustion. Benzene combustion yields carbon soot from where C_{60} and C_{70} can be recovered by solvent extraction (Howard et al., 1991). The industrial production of fullerenes is already started in Japan some year ago (Murayama 2005).

18.3   NANO-SCIENTIFIC DEVELOPMENTS

The chemistry and physics of fullerenes is now fully developed, and complete reviews can be found for example in the books of Hirsch et al., (2005) and Taylor (1999). However, a number of reviews are available, and there is a journal "Fullerenes Nanotubes and Carbon Nanostructures" published by Taylor & Francis Group, dedicated to the chemistry and physics of fullerenes and related carbon nanostructures and carbon materials.

Being polyolefins, fullerenes undergo a number of addition reactions. Limiting our discussion to C_{60}, it possesses 30 weakly conjugated double bonds, halogens, hydrogen, amines, ozone, and many other molecules can be added to the double bonds (Taylor, 1999; Hirsch et al., 2005; Cataldo and Iglesias-Groth, 2015). Fullerenes form Diels-Alder adducts with dienes, thus acting as a dienophile (Cataldo et al., 2015) and forms adducts also with transition metals (Garcia-Hernandez et al., 2015).

Fullerenes are soluble in a number of different hydrocarbon and halogenated solvents (Hirsch et al., 2005) but they are soluble also in fatty alcohols

(Heymann, 1996) and in fatty acids esters (Cataldo and Braun, 2007; Cataldo and Da Ros, 2008).

Endohedral fullerenes were prepared by inserting, for example, a metal atom inside the fullerene cage.

18.4   NANOCHEMICAL APPLICATIONS

C_{60} fullerene but also the higher homologs are powerful free radical acceptors (Tumanskii et al., 2006) and for these reason were intensively studied as potential drugs for example as tumor growth inhibitors (Cataldo and Da Ros, 2008) or other purposes like antiviral activity (Hirsch, 2010) or antibiotic activity (Kai et al., 2003). Fullerenes in their crystalline powder form can be considered non-toxic (Andrievsky et al., 2005). This has paved the way for fullerene application in many different fields ranging from cosmetics (Benn, 2011) for its antimicrobial, antioxidant, anti-aging properties toward skin (Takada et al., 2006) to additive for fuels (Morris et al., 1997) or potential new explosives (Cataldo et al., 2013). Other emerging application fields of fullerenes regard their use as electron acceptors in solar cells assembly (Sabirov, 2013). In this specific application, the fullerene adduct with acenes or other substituents seem the most promising (He et al., 2010). Further emerging fields of fullerenes applications are reviewed elsewhere (Langa et al., 2007)

18.5   MULTI-/TRANS-DISCIPLINARY CONNECTIONS

Since the early discovery of fullerenes, Kroto (1992) and Kroto et al., (1993) made the prediction that fullerenes may exist in space. Indeed this prediction turned out to be correct when using the infrared spectra collected by the Spitzer orbiting telescope it was possible to detect the spectral signature of fullerenes in a young planetary nebula called TC-1 (Cami et al., 2010). Later fullerenes were found in different space environments ranging from the circumstellar to the interstellar medium to the extragalactic planetary nebulae (Garcia-Hernandez et al., 2011). Although fullerenes are not naturally-occurring in the Earth, they are widespread in the Universe and at present C_{60} and C_{70} are the largest organic molecules known to exist in the space to date.

Since fullerenes are highly reactive with atomic hydrogen, it is thought that the hydrogenated fullerenes, called fulleranes, should be present in space (Cataldo and Iglesias-Groth, 2010) and indeed some preliminary hint about the possible detection of fulleranes were reported (Zhang and Kwok 2013). There is also interest in searching fullerenes adducts with metals and other molecules (Dunk et al., 2013).

18.6 OPEN ISSUES

One aspect of fullerenes regards their topological modeling (Cataldo et al., 2011; Schwerdtfeger et al., 2015), for example, in relation to their stability, giant fullerenes, endohedral fullerenes, and onion-like carbon (concentric fullerenes). Another important open issue regards the exact mechanism of fullerenes formation in different conditions. Fullerenes are formed in relatively high yields in a carbon arc only under helium with hydrogen having an inhibiting effect on fullerene formation but then fullerenes are formed in laminar flames where there is no helium at all. This aspect is also important in explaining the fullerene formation in space in different conditions.

KEYWORDS

- **developments**
- **fullerenes**
- **history**
- **structure**

REFERENCES AND FURTHER READING

Albert, V. V., Chancey, R. T., Oddershede, L. B., Harris, F. E., & Sabin, J. R., (2010). Fragmentation of fullerenes. In: Sattler, K. D., (ed.), *Handbook of Nanophysics: Clusters and Fullerenes* (pp. 1–13). CRC Press: Boca Raton, Florida, USA, Chapter 26.

Andrievsky, G., Klochkov, V., & Derevyanchenko, L., (2005). Is the C_{60} fullerene molecule toxic?!. *Fullerenes, Nanotubes, and Carbon Nanostructures, 13*, 363–376.

Bagott, J., (1996). *Perfect Symmetry, the Accidental Discovery of Buckminsterfullerene.* Oxford University Press, Oxford, UK.

Benn, T. M., Westerhoff, P., & Herckes, P., (2011). Detection of fullerenes (C_{60} and C_{70}) in commercial cosmetics. *Environmental Pollution, 159*, 1334–1342.

Cami, J., Bernard-Salas, J., Peeters, E., & Malek, S. E., (2010). Detection of C_{60} and C_{70} in a young planetary nebula. *Science, 329*, 1180–1182.

Cataldo, F., & Braun, T., (2007). The solubility of C_{60} fullerene in long chain fatty acids esters. *Fullerenes, Nanotubes, and Carbon Nonstructures, 15*, 331–339.

Cataldo, F., & Da Ros, T., (2008). *Medicinal Chemistry and Pharmacological Potential of Fullerenes and Carbon Nanotubes.* Springer Science & Business Media, Dordrecht.

Cataldo, F., & Iglesias-Groth, S., (2010). *Fulleranes: The Hydrogenated Fullerenes.* Springer Science & Business Media, Dordrecht.

Cataldo, F., & Iglesias-Groth, S., (2015). A differential scanning calorimetric (DSC) study on heavy ozonized C_{60} fullerene. *Fullerenes, Nanotubes and Carbon Nanostructures, 23*, 253–258.

Cataldo, F., (1999). On the isolated pentagon rule: Could fullerenes exist violating this rule? *Fullerenes, Nanotubes and Carbon Nanostructures, 7*, 289–295.

Cataldo, F., (2002). The impact of a fullerene-like concept in carbon black science. *Carbon, 40*, 157–162.

Cataldo, F., García-Hernández, D. A., & Manchado, A., (2015). Chemical thermodynamics applied to the Diels–Alder reaction of C_{60} fullerene with polyacenes. *Fullerenes, Nanotubes and Carbon Nanostructures, 23*, 760–768.

Cataldo, F., Graovac, A., & Ori, O., (2011). *The Mathematics and Topology of Fullerenes.* Springer Science & Business Media, Dordrecht.

Cataldo, F., Ursini, O., & Angelini, G., (2013). Synthesis and explosive decomposition of polynitro [60] fullerene. *Carbon, 62*, 413–421.

Dunk, P. W., Adjizian, J. J., Kaiser, N. K., Quinn, J. P., Blakney, G. T., Ewels, C. P., & Kroto, H. W., (2013). Metallofullerene and fullerene formation from condensing carbon gas under conditions of stellar outflows and implication to stardust. *Proceedings of the National Academy of Sciences, 110*, 18081–18086.

Garcia-Hernandez, D. A., Cataldo, F., & Manchado, A., (2015). About the iron carbonyl complex with C_{60} and C_{70} fullerene: [Fe (CO) $_4$ ($\eta 2C_{60}$)] and [Fe (CO) $_4$ ($\eta 2C_{70}$)]. *Fullerenes, Nanotubes and Carbon Nanostructures.* Published online, doi: 10.1080/1536383X.2015.1125343.

García-Hernández, D. A., Iglesias-Groth, S., Acosta-Pulido, J. A., Manchado, A., García-Lario, P., Stanghellini, L., & Cataldo, F., (2011). The formation of fullerenes: Clues from new C_{60}, C_{70}, and (possible) planar C_{24} detections in Magellanic cloud planetary nebulae. *The Astrophysical Journal Letters, 737*, L30–L35.

Haddon, R. C., Brus, L. E., & Raghavachari, K., (1986). Rehybridization and π-orbital alignment: The key to the existence of spheroidal carbon clusters. *Chemical Physics Letters, 131*, 165–169.

He, Y., Chen, H. Y., Hou, J., & Li, Y., (2010). Indene– C_{60} bisadduct: A new acceptor for high-performance polymer solar cells. *Journal of the American Chemical Society, 132*, 1377–1382.

Heymann, D., (1996). Solubility of fullerenes C_{60} and C_{70} in seven normal alcohols and their deduced solubility in water. *Fullerenes, Nanotubes and Carbon Nanostructures, 4*, 509–515.

Hirsch, A., & Brettreich, M., (2005). *Fullerene. Chemistry and Reactions.* Wiley-VCH, Weinheim.

Hirsch, A., (2010). The era of carbon allotropes. *Nature Materials, 9*, 868–871.

Howard, J. B., McKinnon, J. T., Makarovsky, Y., Lafleur, A. L., & Johnson, M. E., (1991). Fullerenes C_{60} and C_{70} in flames. *Nature, 352*, 139–141.

Kai, Y., Komazawa, Y., Miyajima, A., Miyata, N., & Yamakoshi, Y., (2003). [60] Fullerene as a novel photoinduced antibiotic. *Fullerenes, Nanotubes and Carbon Nanostructures, 11*, 79–87.

Krätschmer, W., (2014). Fullerenes, carbon chains, and interstellar matter. *Fullerenes, Nanotubes and Carbon Nanostructures, 22*, 23–34.

Krätschmer, W., Lamb, L. D., Fostiropoulos, K., & Huffman, D. R., (1990). Solid C_{60}: A new form of carbon. *Nature, 347*, 354–358.

Kroto, H. W., (1992). C_{60}: Buckminsterfullerene, the celestial sphere that fell to earth. *Angewandte Chemie International Edition in English, 31*, 111–129.

Kroto, H. W., Allaf, A. W., & Balm, S. P., (1991). C_{60}: Buckminsterfullerene. *Chemical Reviews, 91*, 1213–1235.

Kroto, H. W., Walton, D. R. M., Jones, D. E. H., & Haddon, R. C., (1993). Polyynes and the formation of fullerenes. *Philosophical Transactions of the Royal Society of London A: Mathematical, Physical and Engineering Sciences, 343*, 103–112.

Langa, F., & Nierengarten, J. F., (2007). *Fullerenes: Principles and Applications*. Royal Society of Chemistry, London.

Loutfy, R. O., Lowe, T. P., Moravsky, A. P., & Katagiri, S., (2002). Commercial production of fullerenes and carbon nanotubes. In: *Perspectives of Fullerene Nanotechnology* (pp. 35–46). Springer Netherlands.

Morris, R. E., Pruitt-Mentle, D., Black, B. H., & Malhotra, R., (1997). The impact of [60] fullerene on jet fuel stability. *Petroleum Science and Technology, 15*, 381–396.

Murayama, H., Tomonoh, S., Alford, J. M., & Karpuk, M. E., (2005). Fullerene production in tons and more: From science to industry. *Fullerenes, Nanotubes and Carbon Nanostructures, 12*, 1–9.

Ross, R. B., Cardona, C. M., Guldi, D. M., Sankaranarayanan, S. G., Reese, M. O., Kopidakis, N., & Holloway, B. C., (2009). Endohedral fullerenes for organic photovoltaic devices. *Nature Materials, 8*, 208–212.

Sabirov, D. S., (2013). Anisotropy of polarizability of fullerene higher adducts for assessing the efficiency of their use in organic solar cells. *The Journal of Physical Chemistry C., 117*, 9148–9153.

Schwerdtfeger, P., Wirz, L. N., & Avery, J., (2015). The topology of fullerenes. *Wiley Interdisciplinary Reviews: Computational Molecular Science, 5*, 96–145.

Takada, H., Mimura, H., Xiao, L., Islam, R. M., Matsubayashi, K., Ito, S., & Miwa, N., (2006). Innovative anti-oxidant: Fullerene (INCI#:7587) is a "radical sponge" on the skin. Its high level of safety, stability, and potential as premier anti-aging and whitening cosmetic ingredient. *Fullerenes, Nanotubes, and Carbon Nonstructures, 14*, 335–341.

Taylor, R., (1999). *Lecture Notes on Fullerene Chemistry: A Handbook for Chemists*. Imperial College Press, London.

Tumanskii, B. L., & Kalina, O., (2006). *Radical Reactions of Fullerenes and Their Derivatives*. Springer Science & Business Media, Dordrecht.

Zhang, Y., & Kwok, S., (2013). On the detections of C60 and derivatives in circumstellar environments. *Earth, Planets and Space, 65*, 1069–1081.

CHAPTER 19

Geometric-Arithmetic Index

MAHDIEH AZARI[1] and ALI IRANMANESH[2]

[1]*Department of Mathematics, Kazerun Branch, Islamic Azad University, P.O. Box: 73135-168, Kazerun, Iran*

[2]*Department of Pure Mathematics, Faculty of Mathematical Sciences, Tarbiat Modares University, P. O. Box: 14115-137, Tehran, Iran, Tel: +982182883447; Fax: +982182883493; E-mail: iranmanesh@modares.ac.ir*

19.1 DEFINITION

Consider a simple graph G with the edge set $E(G)$. Let d_u denote the degree of the vertex u in G which is the number of all first neighbors of u in G, and let $uv \in E(G)$ be the edge connecting the vertices u and v of G. The geometric–arithmetic index of G is defined as

$$GA = GA(G) = \sum_{uv \in E(G)} \frac{\sqrt{d_u d_v}}{(d_u + d_v)/2} = \sum_{uv \in E(G)} \frac{2\sqrt{d_u d_v}}{d_u + d_v}, \tag{1}$$

where the summation goes over all edges uv of G.

19.2 HISTORICAL ORIGIN(S)

Mathematical chemistry is a branch of theoretical chemistry for discussion and prediction of molecular structures using mathematical methods without necessarily referring to quantum mechanics. Chemical graph theory is a branch of mathematical chemistry which applies graph theory to mathematical modeling of chemical phenomena (Gutman & Polansky, 1986; Trinajstić, 1992; Todeschini & Consonni, 2000). Chemical graphs, particularly molecular graphs, are models of molecules in which atoms are represented by vertices and chemical bonds by edges of a graph. Physico-chemical or biological properties of molecules can be predicted by using the information encoded in the molecular graphs, eventually translated in the adjacency or connectivity matrix associated to these graphs. This paradigm is achieved by considering various graph theoretical invariants of molecular graphs (also known as topological indices or structural descriptors/measures) and evaluating how strongly are they correlated with various molecular properties. In this way, chemical graph theory plays an important role in the mathematical foundation of QSAR and QSPR research (Diudea, 2001). A topological index is any function calculated on a molecular graph (connected graph with maximum vertex degree at most 4, whose graphical representation may resemble a structural formula of a molecule), irrespective of the labeling of its vertices. Hundreds of topological indices have been introduced and studied, starting with the seminal work by Wiener (1947a, b) in which he used the sum of all shortest-path distances of a (molecular) graph for modeling physical properties of alkanes. Topological indices can be divided into various categories such as graph entropy (Bonchev, 1983; Dehmer, 2008; Dehmer et al., 2009; Dehmer & Mowshowitz, 2011) representing information-theoretic indices, eigenvalue-based measures (Randić et al., 2001; Estrada, 2002; Dehmer et al., 2012b), distance-based measures (Balaban, 1982; Khalifeh et al., 2009; De Corato et al., 2012; Putz et al., 2013; Alizadeh et al., 2013, 2014; Xu et al., 2014; Azari & Iranmanesh, 2013a, 2014, 2015b, c, Iranmanesh & Azari, 2015a), symmetry-based descriptors (Todeschini & Consonni, 2000), and degree-based invariants (Vukičević & Gašperov, 2010; Furtula et al., 2013; Gutman & Tošović, 2013; Gutman, 2013; Zhong & Xu, 2014; Azari & Iranmanesh, 2013b, 2015a; Azari, 2014; Falahati-Nezhad et al., 2014; Iranmanesh & Azari, 2015b).

Topological indices based on end-vertex-degrees of edges have been used over 40 years. Among them, several indices are recognized to be useful tools in QSPR/QSAR studies. Some oldest and most thoroughly investigated are the first Zagreb index M_1 (Gutman & Trinajstić, 1972), the second Zagreb index M_2 (Gutman et al., 1975), the Randić connectivity index χ (Randić, 1975), the harmonic index H (Fajtlowicz, 1987), and the atom-bond connectivity index ABC (Estrada et al., 1998).

Probably, the best-known such descriptor is the Randić connectivity index which is suitable for measuring the extent of branching of the carbon-atom skeleton of saturated hydrocarbons (Randić, 1975). There are a large number of papers and some books dealing with this molecular descriptor. For example, see (Li & Gutman, 2006; Gutman & Furtula, 2008; Li & Shi, 2008), and the references cited therein.

During years of research, scientists were trying to improve the predictive power of the Randić connectivity index. This led to the definition of a number of modifications and new topological indices similar to the original Randić index.

The *GA* index was first introduced in a paper by Vukičević and Furtula (2009) published in the *Journal of Mathematical Chemistry* as one of the successors of the Randić connectivity index. This index was named as geometric-arithmetic index, because as it can be seen from Eq. (1), it consists of a geometrical mean of end-vertex degrees of an edge *uv*, $\sqrt{d_u d_v}$, as numerator and arithmetic mean of end-vertex degrees of the edge *uv*, $(d_u + d_v)/2$, as denominator. Vukičević and Furtula (2009) showed that for physicochemical properties such as entropy, enthalpy of vaporization, standard enthalpy of vaporization, enthalpy of formation, and acentric factor, the predictive power of *GA* index is somewhat better than predictive power of the Randić connectivity index. Vukičević and Furtula (2009) also analyzed some basic mathematical properties of *GA* index. They gave lower and upper bounds for the *GA* index in the case of simple connected graphs and trees as well as in the case of chemical graphs and chemical trees (or molecular trees are trees with maximum degree at most four used to model carbon skeletons of acyclic hydrocarbons), and identified the trees with the minimum and the maximum *GA* indices, which are the star and the path, respectively. Thus, the path is the unique molecular tree with the maximum *GA* index.

19.3 NANO-SCIENTIFIC DEVELOPMENT(S)

Although *GA* index has been introduced just 10 years ago, a large number of scientific investigations were reported on this index whose origin is from chemistry, and are claimed to have chemical applications. The graph theoretical concepts and notations not defined in this essay can be found in the standard books of graph theory such as West (1996).

Soon after introducing *GA* by Vukičević and Furtula (2009), and Fath-Tabar et al. (2010) generalized its definition by introducing a new class of topological indices, based on some properties of vertices of graph. These indices are called general geometric-arithmetic indices ($GA_{general}$) and defined for a simple graph *G* as (Fath-Tabar et al., 2010)

$$GA_{general} = GA_{general}(G) = \sum_{uv \in E(G)} \frac{\sqrt{Q_u Q_v}}{(Q_u + Q_v)/2} = \sum_{uv \in E(G)} \frac{2\sqrt{Q_u Q_v}}{Q_u + Q_v}, \qquad (2)$$

where Q_u is some quantity that in a unique manner can be associated with the vertex u of the graph G and the summation goes over all edges uv of G. The name of this class of indices is evident from their definition. Namely, indices belonging to this group are calculated as the ratio of geometric and arithmetic means of some properties of adjacent vertices u and v (vertices u and v connected by an edge). Five members of GA group topological indices have been put forward up to now. The geometric-arithmetic index GA is the first member of this group which is obtained from Eq. (2) by setting Q_u to be the degree d_u of the vertex u in the graph G. So, in the literature, GA is also called the first geometric-arithmetic index and denoted by GA_1. The second member of this class, denoted by GA_2, was considered by Fath-Tabar et al. (2010) by setting Q_u to be the number n_u of vertices of G lying closer to the vertex u than to the vertex v for the edge uv of the graph G. The third member of this class, denoted by GA_3, was considered by Zhou et al. (2009) by setting Q_u to be the number m_u of edges of G lying closer to the vertex u than to the vertex v for the edge uv of the graph G. The fourth member of this class, denoted by GA_4, was proposed by Ghorbani and Khaki (2010) by setting Q_u to be the eccentricity ε_u of the vertex u in the graph G which is the largest distance between u and any other vertex of G. The fifth member of this class, denoted by GA_5, was introduced by Graovac et al. (2011) by setting Q_u to be the quantity S_u defined as the sum of degrees of all neighbors of the vertex u in the graph G.

Another generalization of GA index called the ordinary generalized geometric-arithmetic index was proposed by Eliasi and Iranmanesh (2011). This index is defined for a graph G as

$$OGA_k(G) = \sum_{uv \in E(G)} \left[\frac{\sqrt{d_u d_v}}{(d_u + d_v)/2} \right]^k = \sum_{uv \in E(G)} \frac{(4d_u d_v)^{\frac{k}{2}}}{(d_u + d_v)^k},$$

where k is an arbitrary positive real number. Some basic mathematical properties of this index were investigated in Eliasi & Iranmanesh (2011).

The majority of people is familiar with the well-known arithmetic and geometric means. However, the number of means is practically infinite, and a general formula for their computation is given by

$$M_p = M_p(x_1, x_2, ..., x_n) = \left(\frac{1}{n} \sum_{i=1}^{n} (x_i)^p \right)^{1/p}. \tag{3}$$

It can be easily seen that arithmetic mean is attained when $p = 1$, and geometric mean when $p = 0$. Taking into mind Eq. (3), it is natural asking ourselves if the ratio between geometric and arithmetic means, in the case of *GA* index, is the right choice. In order to find the answer of this question, Došlić et al. (2011) performed an exhaustive computer research. They considered a general class of indices based on means as follows:

$$Z_{p,q}(G) = \sum_{uv \in E(G)} \frac{M_p(d_u, d_v)}{M_q(d_u, d_v)}.$$

Note that $GA \equiv Z_{0,1}$ and $GA \equiv Z_{-1,0}$. Došlić et al. (2011) examined the correlations between the $Z_{p,q}$ indices for $-2 \le p, q \le 2$ with *GA* index for several sets of trees. It was found that, in all cases, the $Z_{p,q}$ index is well correlated with *GA*; in most cases, these correlations are linear or almost linear. The obtained results indicated that the ratio between geometric and arithmetic means in the definition of the *GA* index is not a necessary choice and other $Z_{p,q}$ indices may be used in chemical researches instead of it. However, there is no scientific justification to introduce any further structure descriptor of this kind, because the *GA* index gathers the same information on observed molecule as other $Z_{p,q}$ indices.

The edge version of the geometric-arithmetic index $GA_e(G)$ of a given graph *G* was introduced by Mahmiani et al. (2012) to be the *GA* index of the line graph *L(G)* of *G*, and the total version of the geometric-arithmetic index $GA_t(G)$ of *G* was proposed by Mahmiani and Khormali (2013) to be the *GA* index of the total graph *T(G)* of *G*, where *L(G)* is the graph whose vertices correspond to the edges of *G* with two vertices being adjacent if and only if the corresponding edges in *G* have a vertex in common, and where *T(G)* is the graph whose vertex set is the union of the vertex set and the edge set of *G* and two vertices of *T(G)* being adjacent if and only if the corresponding elements of *G* are adjacent or incident.

A descriptor is called degenerate if it possesses the same value for more than one graph. Dehmer et al. (2012a) evaluated the uniqueness, discrimination power or degeneracy of some degree, distance, and eigenvalue-based descriptors for investigating graphs holistically. In fact, the uniqueness of structural descriptors has been investigated in mathematical chemistry and related disciplines for discriminating the structure of isomeric structures and other chemical networks. A highly discriminating graph measure is desirable for analyzing graphs; hence, measuring the degree of its degeneracy is important for understanding its properties, limits, and quality. For the degree-based descriptors such as *GA*, it is not surprising that these measures have only little discrimination power, as many graphs can be realized by identical degree sequences. This effect is even stronger if the cardinality of the underlying graph set increases. In order to tackle the question of what kind of degeneracy the indices possess, Dehmer et al. (2012a) plotted their characteristic value distributions. For a graph class, they used the class of exhaustively generated non-isomorphic, connected and unweighted graphs with 9 vertices. In the case of GA, it was found that by using this index there exist many degenerate graphs possessing quite similar index values where the hull of the distributions forms a Gaussian curve.

A thread in a graph G is any maximal connected subgraph induced by a set of vertices of degree 2 in G. A string in G is a subgraph induced by a thread and the vertices adjacent to it. A graph G consists of s strings if it can be represented as a union of s strings so that any two strings have at most two vertices in common. Došlić et al. (2012) computed GA index for all graphs that consist of at most three strings. It was concluded that among the graphs with at most three strings the GA index assumes rational values only on cycles and on the trivial string K_2. In addition, the GA index is quite discriminative on the considered class of graphs; given the number of edges, the graph type can be in most cases reconstructed from the index value. So, the GA index measures both diversity of edge types and a disparity of their end-vertices.

In graph theory, a regular graph is a graph in which every vertex has the same degree. A graph which is not regular is said to be irregular. In order to establish "how irregular" a given graph is, several irregularity measures, denoted here by $irr(G)$, have been proposed. Each of such measures must satisfy the condition $irr(G) \geq 0$ and $irr(G) = 0$ if the graph G is regular. In chemical applications, quantifying the irregularity of molecular graphs and of biomolecular networks seems to be of marginal importance. Gutman et al. (2014) proposed an irregularity measure by means of the GA index. Since, for a connected graph G with m edges, $GA(G) \leq m$ with equality if and only if G is regular (Vukičević & Furtula, 2009), so

$$irr_{GA}(G) = 1 - \frac{GA(G)}{m}.$$

can be viewed as a measure of graph irregularity (Gutman et al., 2014).

A chemical reaction can be represented as the transformation of the chemical (molecular) graph representing the reaction's substrate into another chemical graph representing the product. The type of chemical reaction where two substrates combine to form a single product (combination reaction) motivated Ramakrishnan et al. (2013) to study the effect of topological indices, particularly GA, when a bridge is introduced between the respective vertices (of degree i, $i = 1,2,3$) of two copies of the same graph. The authors claimed that the graph obtained in this manner may or may not represent a stable chemical compound in reality, but it is the interest of the chemist to check the stability of the so obtained structure of the resulting graph. Ramakrishnan et al. (2013) also presented an algorithm to compute the distance matrix of the resultant graph obtained after each iteration and thereby tabulated various topological indices including *GA*.

The bounds of a topological index are important information of a molecular graph in the sense that they establish the approximate range of the index in terms of molecular structural parameters. Computing upper and lower bounds on *GA* index has been subject of many papers.

Yuan et al. (2010) established some bounds for the *GA* index using some graph invariants. In particular, they found a sharp lower bound for the *GA* index of triangle-free graphs with n vertices and m edges. In addition, they proposed sharp lower and upper bounds for the molecular graph G with $n \geq 4$ vertices and $m \in [n-1, 2n]$ edges and determined the n-vertex molecular trees with first, second, and third minimum *GA* indices as well as the second and the third maximum *GA* indices.

Das (2010) presented a sharp lower bound for the *GA* index of a simple connected graph with m edges, in terms of the number of edges, maximum vertex degree, and minimum vertex degree. He also obtained several upper and lower bounds on *GA* index of a simple connected graph in terms of maximum and minimum vertex degree, minimum non-pendent vertex degree, number of pendent vertices, number of non-pendent edges, second Zagreb index, and degree sequence of the graph, and characterized graphs for which these bounds are best possible. The Nordhaus-Gaddum-type results for the *GA* index of a graph and its connected complement were also given in Das (2010).

Das and Trinajstić (2010) compared the *GA* index and the *ABC* index for chemical trees and molecular graphs. Besides chemical trees and molecular graphs, general graphs were treated. It was also shown that (Das & Trinajstić, 2010), the *GA* index is greater than the *ABC* index for the difference between the maximum and minimum degree, less than or equal to three. Comparison between these two indices, in the case of general trees and general graphs, remained an open problem.

Mogharrab and Fath-Tabar (2011) obtained lower and upper bounds for the *GA* index of a simple n-vertex graph with no isolated vertex and m edges, in terms of the first Zagreb index M_1 and the second Zagreb index M_2. They also presented exact formulae for computing the *GA* index of the Cartesian product of two paths which is called the rectangular grid, and the Cartesian product of a path and a cycle which is known as a G_4-nanotube.

Das et al. (2011b) proposed further upper and lower bounds for the *GA* index of a simple connected graph in terms of some structural graph parameters and characterized graphs for which these bounds are best possible. Moreover, they discussed the effect on *GA* of inserting an edge into a graph.

Du et al. (2011) determined the *n*-vertex (molecular) trees with the second and the third for $n \geq 7$, the fourth and the fifth for $n \geq 10$, and the sixth for $n \geq 11$ maximum *GA* indices, unicyclic (molecular) graphs with the first for $n \geq 3$, the second and the third for $n \geq 5$, the fourth for $n \geq 7$, and the fifth and the sixth for $n \geq 9$ maximum *GA* indices, and bicyclic (molecular) graphs with the first for $n \geq 4$, the second and the third for $n \geq 6$, and the fourth, the fifth and the sixth for $n \geq 8$ maximum *GA* indices.

Das and Trinajstić (2012) compared *GA* and GA_2 indices for chemical trees and starlike trees. Comparison between these indices, in the case of general graphs, remained an open problem.

Lokesha et al. (2013) presented some upper and lower bounds on *GA* index of a simple connected graph in terms of the number of edges, number of pendant vertices, maximum vertex degree, minimum non-pendant vertex degree, and second Zagreb index.

Divnić et al. (2014) found extremal graphs in the class of *n*-vertex simple connected graphs with minimum degree δ for which *GA* attains its minimum value, or gave a lower bound. They showed that when δ or *n* is even, the extremal graphs are regular graphs of degree δ. The result was obtained only in the case $\delta \geq \lceil \delta_0 \rceil$, with $\delta_0 = q_0(n-1)$, where $q_0 \approx 0.088$ is the unique positive root of the equation $q\sqrt{q} + q + 3\sqrt{q} - 1 = 0$, and the case $\delta < \lceil \delta_0 \rceil$ was left as an open problem.

Cruz et al. (2014) found the extremal values of the *GA* index over the set of Kragujevac trees with the central vertex of fixed degree.

Rodríguez and Sigarreta (2015a) obtained new inequalities involving *GA* index and characterized graphs extremal with respect to them. In particular, they could improve some previous inequalities and related *GA* to some well-known topological indices such as the first and second Zagreb indices, Randić index, harmonic index, sum connectivity index (Zhou & Trinajstić, 2009), and modified Narumi-Katayama index (Todeschini & Consonni, 2010).

Rodríguez and Sigarreta (2015b) also studied *GA* index from an algebraic point of view. Their main tool for this study was an appropriate matrix that is a modification of the classical adjacency matrix involving the degrees of the vertices. Besides, they used the known sum-connectivity matrix in order to obtain more algebraic properties of *GA*.

Raza et al. (2016) proved that *GA* index is greater than *ABC* index for line graphs of molecular graphs, for general graphs in which the difference between maximum degree Δ and minimum degree δ ($\delta > 1$) is less than or equal to $(2\delta-1)^2$, and for some families of trees; giving partial solution to an open problem proposed in Das & Trinajstić (2010). The complete solution of the said problem was left as a task for future.

Since the discovery of buckminsterfullerene (Kroto et al., 1985) and latter of nanotubes (Iijima, 1991), the investigation of nanomolecules, both by experimental and theoretical chemists, has been intensively conducted. Nowadays, there is a vast number of papers and several books dealing with these molecules. Theoreticians examined many structures of fullerenes, dendrimers, nanotubes, nanotori, etc., expecting to be synthesized in the future (Diudea & Nagy, 2007).

In recent years, there has been considerable interest in the general problem of determining topological indices of nanostructures. Diudea (1995, 2002) was the first chemist considered this problem. There exist a large number of papers on computing *GA* index of chemical graphs and nanostructures.

Ghorbani and Jalali (2009) presented exact formulae for computing *GA* index of the two-dimensional lattice of the $TUC_4C_8(S)$ graph, $TUC_4C_8(S)$ nanotube, and $TC_4C_8(S)$ nanotorus. Shabani and Ashrafi (Shabani & Ashrafi, 2010) computed *GA* index of $TUC_4C_8(R)$ nanotube, $TC_4C_8(R)$ nanotorus, and polyhex nanotorus. Iranmanesh and Zeraatkar (2010, 2011) computed *GA* index of $TUAC_6[p,q]$ nanotube, $TUZC_6[p,q]$ nanotube, $HC_5C_7[p,q]$ nanotube, $SC_5C_7[p,q]$ nanotube, $HAC_5C_7[p,q]$ nanotube, and $HAC_5C_6C_7[p,q]$ nanotube. Ghorbani et al. (2011) computed *GA* index of V-Phenylenic nanotube and nanotorus. Ashrafi and Shabani (2010) studied *GA* index of cabon nanocones. Shabani et al. (2010) presented a group theoretical method for computing *GA* index of graphs and applied this method to compute *GA* index of two classes of dendrimers namely the all-aromatic dendrimers and Wang's helicene-based dendrimers. Ghorbani and Ghazi (2010) computed *GA* index of polyomino chains of k cycles and triangular benzenoid graphs. Xiao et al. (2010) computed *GA* index of benzenoid systems and phenylenes and established a simple relation between the *GA* index of a phenylene and the corresponding hexagonal squeeze. Yarahmadi (2011) gave the expressions for computing the *GA* index of hexagonal systems and phenylenes and presented a new method for describing a hexagonal system by relating a simple graph to it. Fath-Tabar et al. (2011) studied *GA* index under several graph products and applied the results to compute *GA* index of some molecular graphs and nanostructures. Asadpour (2012) computed *GA* index for an infinite family of nanostructures derived from a class of composite graphs called bridge graphs. Cruz et al. (2013) studied the maximal and minimal values of *GA* index over the class of all hexagonal systems with a fixed number of hexagons. Farahani (2013) computed the *GA* index of polycyclic aromatic hydrocarbons (PAHs). Nikmehr et al. (2014, 2015) computed *GA* index of $VAC_5C_6C_7[p,q]$ nanotube, *H*-phenylenic nanotube, *H*-naphthylenic nanotube, and *H*-anthracenic nanotube. Maadanshekaf and Moradi (2014) computed *GA* index of PAMAM dendrimers and polypropylenimin octaamin dendrimers. Raut (2014) determined *GA* index of some members of isomers of decane with the molecular formula $(C_{10}H_{22})$ namely *n*-decane, 3,4,4-trimethyl heptanes, and 2,4-dimethyl-4-ethyl hexane. Imran and Hayat (2014) computed *GA* index of Aztec diamond. Hayat and Imran (2014) determined *GA* index for oxide and chain silicate Networks. Imran et al. (2014) extended this study to interconnection networks and derived analytical closed results of *GA* index for butterfly and Benes networks. Bača et al. (2015) studied a carbon nanotube network that is motivated by the molecular structure of a regular hexagonal lattice and determined *GA* index of this class of networks. Baig et al. (2015) studied the *GA* index of dominating oxide network (DOX), dominating silicate network (DSL), and regular triangulene oxide network (RTOX). Soleimani et al. (2015) gave exact formulae for computing *GA* index of linear [n]-tetracene, V-tetracenic nanotube, *H*-tetracenic nanotube, and tetracenic nanotorus. Sridhara et al. (2015) computed *GA* index of graphene. Farahani (2015) obtained *GA* index for a type of benzenoid system called jagged-rectangle. Azari and Falahati-Nezhad (2015, 2017) computed *GA* index of splice and link of graphs and applied their results to compute the *GA* index of an infinite family of dendrimers. It is worth mentioning that fullerene graphs are 3-regular, so *GA* index of these graphs is equal to the number of their edges. The *GA* index of some other families of nanotubes and nanostar dendrimers were computed in Hosseinzadeh & Ghorbani (2009); Chen & Liu (2010); Husin et al. (2013, 2015).

19.4 NANOCHEMICAL APPLICATION(S)

The reason for introducing a new topological index is to gain prediction of some properties of molecules somewhat better than obtained by already presented indices. Thus, a test study of predictive power of a new index must be done. As a standard for testing new topological descriptors, the properties of octanes are commonly used. Octanes are particularly convenient for such studies, because the number of their structural isomers (18) is large enough to make statistical inferences reliable, and because experimental data are available for all isomers. Sixteen physicochemical properties of octanes can be found at www.moleculardescriptors.eu. Vukičević and Furtula (2009) compared the predictive power of *GA* index with that of the Randić connectivity index using the following physicochemical properties of octanes: boiling point (*BP*), entropy (*S*), enthalpy of vaporization (*HVAP*), standard enthalpy of vaporization (*DHVAP*), enthalpy of formation (*HFORM*), and acentric factor (*AcenFac*). The motivation for choosing just these physicochemical properties is that both *GA* and χ indices give relatively good linear correlations, i.e., correlation coefficients are larger than 0.8. The results, taken from Vukičević & Furtula (2009), are reported in Table 19.1.

TABLE 19.1 Correlation Coefficients for *GA* and Randić Connectivity Indices and Some Physico-Chemical Properties of Octanes[*]

	Correlation coefficient (R)		1–RQR (%)
	GA index	χ index	
BP	0.823	0.821	0.562
S	0.912	0.906	2.942
HV AP	0.941	0.936	4.152
DHV AP	0.966	0.958	9.005
HFORM	0.858	0.850	2.494
AcenFac	0.912	0.904	4.051

[*]*RQR* is the ratio of quadratic mean of residuals. Reprinted from Das et al., (2011a). Open Access.

A superficial glance on *R*'s does not justify the introduction of the *GA* index because (even though the *GA* gives better correlation coefficients than χ) the differences between them are not significant. However, the predictive power of the *GA* index compared with the Randić connectivity index is reasonably better and that can be seen from the ratio of quadratic mean of residuals (*RQR*) (Das et al., 2011a),

$$RQR = \frac{\sqrt{\sum_{i=1}^{n}(a.GA_i + b - Exp_i)^2}}{\sqrt{\sum_{i=1}^{n}(a'.\chi_i + b' - Exp_i)^2}} .$$

Vukučević & Furtula (2009) observed that only for boiling point (*BP*), *GA* index does not give better results than the Randić connectivity index. In all other cases, the prediction power of *GA* index is at least for 2.5% better than the prediction power of the Randić connectivity index. The greatest improvement in prediction with *GA* index comparing to Randić connectivity index is obtained in the case of standard enthalpy of vaporization (more than

9%). So, using this test, Vukičević and Furtula (2009) concluded that *GA* index should be considered as a predictive tool for QSPR/QSAR researches.

Benzenoid hydrocarbons (*B*) belong to the most important polycyclic aromatic compounds. They consist of fused benzene rings. Their characteristic physicochemical properties, especially their thermal stability, were subject to intensive research. Benzenoid hydrocarbons containing no internal carbons atoms (carbons atoms belonging to three sic-membered rings) are said to be "catacondensed." Whereas nowadays only ca. 1000 benzenoid hydrocarbons are known, the number of possible benzenoid hydrocarbons is unimaginatively large. For instance, the number of possible benzenoid hydrocarbons with 35 benzene rings is 5851000265625801806530 (Vöge et al., 2002). Therefore, the modeling of their physicochemical properties is very important in order to predict properties of currently unknown species.

Das et al. (2011a) modeled the heat of formation of 25 benzenoid hydrocarbons using the *GA* index. The data set, collected from Das et al. (2011a), is given in Table 19.2. The correlation graphic between the *GA* indices and the heat of formation of the 25 benzenoid hydrocarbons from Table 19.2, taken from Das et al. (2011a), is shown in Figure 19.1. It is evident from this graphic that there exists a good linear correlation between *GA* index and the heat of formation of benzenoid hydrocarbons. The respective correlation coefficient is equal to 0.972. This observation also leads to the conclusion that *GA* index deserves to be used in chemical applications.

TABLE 19.2 Heat of Formation and the *GA* Index of Some Benzenoid Hydrocarbons

Name	n	ΔH_f (g) (kJ/mol)	GA
Benzene	6	82.9	6.000
Naphthalene	10	150.6	10.919
Anthracene	14	227.7	15.838
Phenanthrene	14	207.1	15.879
Pyrene	16	225.7	18.838
Benzo[*a*]anthracene	18	291.0	20.798
Benzo[*c*]phenanthrene	18	302.4	20.838
Chrysene	18	262.8	20.838
Napthacene	18	291.4	20.758
Triphenylene	18	269.8	20.879
Benzo[*a*]pyrene	20	301.0	23.798
Benzo[*e*]pyrene	20	304.0	23.838
Perylene	20	324.0	23.838

TABLE 19.2 (*Continued*)

Name	n	ΔH_f (g) (kJ/mol)	GA
Benzo[*b*]chrysene	22	346.0	25.758
Benzo[*c*]chrysene	22	334.0	25.798
Benzo[*g*]chrysene	22	333.0	25.838
Benzo[*a*]napthacene	22	359.0	25.717
Dibenzo[*a,c*]anthracene	22	345.0	25.798
Dibenzo[*a,h*]antharacene	22	343.0	25.758
Dibenzo[*a,j*]antharcene	22	343.0	25.758
Dibenzo[*b,g*]phenanthrene	22	347.0	25.758
Dibenzo[*c,g*]phenanthrene	22	335.0	25.798
Pentacene	22	374.5	25.677
Pentaphene	22	359.0	25.717
Picene	22	334.0	25.798

Reprinted from Das et al., 2011a. Open Access.

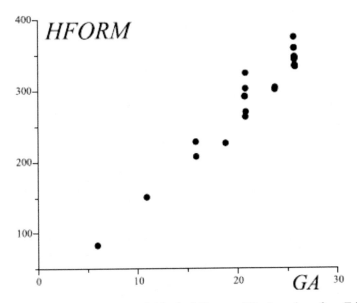

FIGURE 19.1 Heat of formation vs. *GA* for the 25 benzenoid hydrocarbons from Table 19.2. Reprinted from Das et al., (2011a). Open Access.

However, it is well-known that the heat of formation roughly depends on the number of atoms in the molecule, and therefore the correlation shown in Figure 19.1 may, in fact, look unrealistically good. In order to overcome this problem, the examination of correlation between the heat of formation and *GA* index should be limited to isomers (Das et al., 2011a). Among experimental results given in Table 19.2, there are all twelve catacondensed benzenoid hydrocarbons with five benzene rings (i.e., 22 carbon atoms). The correlation between the *GA* index and the heat of formation for the 12 catacondensed benzenoid hydrocarbons with five benzene rings is shown in Figure 19.2. The correlation coefficient is −0.939. It should be noted from Figure 19.2 that two outliers exist. By inspecting the data set of 12 catacondensed benzenoid hydrocarbons with five benzene rings, Das et al. (2011a) determined which molecules correspond to these two outliers. They found that these molecules are the only two branched catacondensed benzenoid hydrocarbons in this data set, benzo[g]chrysene and dibenz[a,c]anthracene. This observation showed that for the modeling of the heat of formation of benzenoid hydrocarbons, other structural details should be incorporated besides the *GA* index.

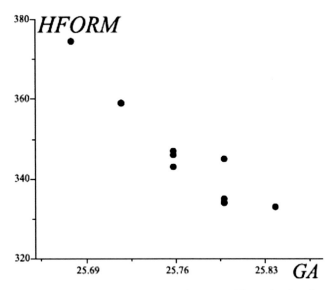

FIGURE 19.2   Correlation between *GA* index and the heat of formation for all catacondensed benzenoid hydrocarbons with 5 hexagons (i.e., with 22 carbon atoms). Reprinted from Das et al., (2011a). Open Access.

Dehmer et al. (2012a) studied the correlation ability of some degree, distance, and eigenvalue-based topological indices by calculating the linear correlation between them and depicted the results as correlation networks. They used two classes of graphs; the class of generated undirected, connected and nonisomorphic graphs with 9 vertices and the class of chemical graphs with 21 vertices, and chose different thresholds for the correlation coefficient, resulting in different networks. Their results showed that, among the indices that are highly correlated by using these graph classes, the *GA* index was highly correlated with other considered indices that belong to different categories, e.g., degree-based and eigenvalue-based, etc.

19.5 MULTI-/TRANS-DISCIPLINARY CONNECTION(S)

As explained before, most of the results on *GA* index have been achieved by using mathematical; particularly graph/group theoretical, methods. Some mathematical software such as GAP, MATLAB, MATHEMATICA, etc. has also been applied in solving computational problems on *GA* index. As far as we know, the applications of *GA* index in other fields of science have not yet been reported in the literature.

19.6 OPEN ISSUES

In order to reduce the arbitrariness in the production of novel topological indices, several criteria are put forward that such a molecular structure descriptor should be required to satisfy. One of these criteria is always that a topological index "should change gradually with gradual change in (molecular) structure." This property may be called the smoothness of the topological index in question. In order to quantify this concept, two measures thereof have been introduced (Furtula et al., 2013), called "structure sensitivity" (*SS*) and "abruptness" (*Abr*). These are defined as follows (Furtula et al., 2013):

Let *G* be a molecular graph and let *TI* be the topological index considered. The set $\Gamma(G)$ consists of all connected graphs obtained from *G* by replacing one of its edges by another edge. Intuitively, the elements of $\Gamma(G)$ are those graphs that are structurally most similar to *G*. Then

$$SS(TI,G) = \frac{1}{|\Gamma(G)|} \sum_{\gamma \in \Gamma(G)} \left| \frac{TI(G) - TI(\gamma)}{TI(G)} \right|, \tag{4}$$

$$Abr(TI,G) = \max_{\gamma \in \Gamma(G)} \left| \frac{TI(G) - TI(\gamma)}{TI(G)} \right|. \tag{5}$$

According to Eq. (4), *SS* is the average relative sensitivity of *TI* to small changes in the structure of the graph *G*. According to Eq. (5), *Abr* shows how much a small structural change may cause a jump-wise (large) change in the considered topological index. From a practical point of view, the best is if the structure sensitivity is as large as possible, but the abruptness is as small as possible.

Furtula et al. (2013) examined the smoothness of 12 vertex-degree-based topological indices, including *GA* index, recently. The test was done for the set of all trees with $n = 6,7,...,13$ vertices. Computational details and the other calculated values for *SS* and *Abr* can be found in Furtula et al. (2013). The degree-based topological indices with the greatest structure sensitivity were found to be the augmented Zagreb index *AZI*, recently invented in Furtula et al. (2010), and the second Zagreb index M_2. Thus, at least in the case of trees, the degree-based topological indices with best structure sensitivity are the augmented and the second Zagreb indices, and these appear to be superior to the other examined indices such as *GA*. As a sort of unpleasant surprise, the very same indices were found to have the greatest abruptness. The results also showed that the Randić connectivity index and the *GA* index have almost identical structure sensitivities. For instance, for $n = 12$, they calculated $SS(\chi) = 0.026$ and $SS(GA) = 0.024$. Or for $n = 8$, $SS(\chi) = 0.0484$ and $SS(GA) = 0.0474$. These findings shed doubts (at least in the case of trees) as to whether the introduction of the *GA* index was fully justified.

Another comparative test was performed in Gutman & Tošović (2013) of how well a class of 20 vertex-degree-based topological indices, including *GA* index, are correlated with two simple physicochemical parameters of octane isomers. These parameters were chosen to be the standard heats of formation (representative for thermo-chemical properties) and the normal boiling points (representative for intermolecular, van-der-Waals-type, interactions). Gutman & Tošović (2013) found that the correlation ability of many of these indices is either rather weak or nil. The *GA* index although having reasonably good correlation abilities (more than 0.8), the only vertex-degree-based topological index that had correlation coefficients over 0.9 and could successfully pass the test was the augmented Zagreb index. Among the examined vertex-degree-based molecular structure-descriptor, the second-best index appeared to be *ABC* index. Consequently, these indices should be preferred to other considered vertex-degree-based indices like *GA* in designing quantitative structure-property relations.

Another comparative test performed by (Gutman et al., 2014) on the set of all isomeric octanes showed that for the standard enthalpy of formation, the predictive ability of *GA* index is slightly lower than the predictive ability of some other examined degree-based indices such as *AZI*, *ABC*, and the recently-introduced reduced reciprocal Randić connectivity index *RRR* (Manso et al., 2012).

KEYWORDS

- **chemical graph theory**
- **Furtula**
- **molecular graph**
- **topological index**
- **Vukičević**

REFERENCES AND FURTHER READING

Alizadeh, Y., Azari, M., & Došlić, T., (2013). Computing the eccentricity-related invariants of single-defect carbon nanocones. *J. Comput. Theor. Nanosci., 10*(6), 1297–1300.

Alizadeh, Y., Iranmanesh, A., Došlić, T., & Azari, M., (2014). The edge Wiener index of suspensions, bottlenecks, and thorny graphs. *Glas. Mat. Ser. III., 49*(69), 1–12.

Asadpour, J., (2012). Computing some topological indices of nanostructures of bridge graph. *Dig. J. Nanomater. Bios., 7*(1), 19–22.

Ashrafi, A. R., & Shababi, H., (2010). GA index and Zagreb indices of nanocones. *Optoelectron. Adv. Mater.-Rapid Commun., 4*(11), 1874–1876.

Azari, M., & Falahati-Nezhad, F., (2015). On vertex-degree-based invariants of link of graphs with application in dendrimers. *J. Comput. Theor. Nanosci., 12*(12), 2611–5616.

Azari, M., & Falahati-Nezhad, F., (2017). Splice graphs and their vertex-degree-based invariants. *Iranian J. Math. Chem., 8*(1), 61–70.

Azari, M., & Iranmanesh, A., (2013a). Computing the eccentric-distance sum for graph operations. *Discrete Appl. Math., 161*(18), 2827–2840.

Azari, M., & Iranmanesh, A., (2013b). Chemical graphs constructed from rooted product and their Zagreb indices. *MATCH Commun. Math. Comput. Chem., 70*, 901–919.

Azari, M., & Iranmanesh, A., (2014). Harary index of some nanostructures. *MATCH Commun. Math. Comput. Chem., 71*, 373–382.

Azari, M., & Iranmanesh, A., (2015a). Some inequalities for the multiplicative sum Zagreb index of graph operations. *J. Math. Inequal., 9*(3), 727–738.

Azari, M., & Iranmanesh, A., (2015b). Edge-Wiener type invariants of splices and links of graphs. *Politehn. Univ. Bucharest Sci. Bull. Ser. A Appl. Math. Phys., 77*(3), 143–154.

Azari, M., & Iranmanesh, A., (2015c). Clusters and various versions of Wiener-type invariants. *Kragujevac J. Math., 39*(2), 155–171.

Azari, M., (2014). Sharp lower bounds on the Narumi-Katayama index of graph operations. *Appl. Math. Comput., 239C*, 409–421.

Bača, M., Horváthová, J., Mokrišová, M., Semaničová-Feňovčíková, A., & Suhányiova, A., (2015). On topological indices of carbon nanotube network. *Can. J. Chem., 93*, 1–4.

Baig, A. Q., Imran, M., & Ali, H., (2015). On topological indices of polyoxide, polysilicate, DOX, and DSL networks. *Can. J. Chem., 93*(7), 730–739.

Balaban, A. T., (1982). Highly discriminating distance-based topological index. *Chem. Phys. Lett., 89*, 399–404.

Bonchev, D., (1983). *Information Theoretic Indices for Characterization of Chemical Structures*. Research Studies Press: Chichester, England.

Chen, S., & Liu, W., (2010). The geometric–arithmetic index of nanotubes. *J. Comput. Theor. Nanosci., 7*, 1993–1995.

Cruz, R., Giraldo, H., & Rada, J., (2013). External values of vertex–degree topological indices over hexagonal systems. *MATCH Commun. Math. Comput. Chem., 70*, 501–512.

Cruz, R., Gutman, I., & Rada, J., (2014). Topological indices of Kragujevac trees. *Proyecciones, 33*(4), 471–482.

Das, K. C., & Trinajstić, N., (2010). Comparison between first geometric–arithmetic index and atom–bond connectivity index. *Chem. Phys. Lett., 497*, 149–151.

Das, K. C., & Trinajstić, N., (2012). Comparison between geometric-arithmetic indices. *Croat. Chem. Acta, 85*(3), 353–357.

Das, K. C., (2010). On geometric–arithmetic index of graphs. *MATCH Commun. Math. Comput. Chem., 64*, 619–630.

Das, K. C., Gutman, I., & Furtula, B., (2011a). Survey on geometric–arithmetic indices of graphs. *MATCH Commun. Math. Comput. Chem., 65*, 595–644.

Das, K. C., Gutman, I., & Furtula, B., (2011b). On first geometric–arithmetic index of graphs. *Discrete Appl. Math., 159*, 2030–2037.

De Corato, M., Benedek, G., Ori, O., & Putz, M. V., (2012). Topological study of Schwarzitic junctions in 1D lattices. *Int. J. Chem. Model., 4*(2/3), 105–113.

Dehmer, M., & Mowshowitz, A., (2011). A history of graph entropy measures. *Inf. Sci., 1*, 57–78.

Dehmer, M., (2008). Information processing in complex networks: Graph entropy and information functional. *Appl. Math. Comput., 201*, 82–94.

Dehmer, M., Grabner, M., & Furtula, B., (2012a). Structural discrimination of networks by using distance, degree and eigenvalue-based measures. *PLoS One, 7*(7), e38564.

Dehmer, M., Sivakumar, L., & Varmuza, K., (2012b). Uniquely discriminating molecular structures using novel eigenvalue-based descriptors. *MATCH Commun. Math. Comput. Chem., 67*, 147–172.

Dehmer, M., Varmuza, K., Borgert, S., & Emmert-Streib, F., (2009). On entropy-based molecular descriptors: Statistical analysis of real and synthetic chemical structures. *J. Chem. Inf. Model., 49*, 1655–1663.

Diudea, M. V., & Nagy, C. L., (2007). *Periodic Nanostructures*. Springer: Dordrecht, Netherlands.

Diudea, M. V., (1995). Wiener index of dendrimers. *MATCH Commun. Math. Comput. Chem., 32*, 71–83.

Diudea, M. V., & Nagy, C. L., (2007). *Periodic Nanostructures* (pp. 212). 1st ed.; Springer: Dordrecht, Netherlands.

Diudea, M. V., (2002). Hosoya polynomial in tori. *MATCH Commun. Math. Comput. Chem., 45*, 109–122.

Divnić, T., Milivojević, M., & Pavlović, L., (2014). Extremal graphs for the geometric–arithmetic index with given minimum degree. *Discrete Appl. Math., 162*, 386–390.

Došlić, T., Furtula, B., Graovac, A., Gutman, I., Moradi, S., & Yarahmadi, Z., (2011). On vertex-degree-based molecular structure descriptors. *MATCH Commun. Math. Comput. Chem., 66*, 613–626.

Došlić, T., Loghman, A., & Badakhshian, L., (2012). Computing topological indices by pulling a few strings. *MATCH Commun. Math. Comput. Chem., 67*, 173–190.

Du, Z., Zhou, B., & Trinajstić, N., (2011). On geometric-arithmetic indices of (molecular) trees, unicyclic graphs and bicyclic graphs. *MATCH Commun. Math. Comput. Chem., 66,* 681–697.

Eliasi, M., & Iranmanesh, A., (2011). On ordinary generalized geometric–arithmetic index. *Applied Math. Lett., 24,* 582–587.

Estrada, E., (2002). Characterization of the folding degree of proteins. *Bioinformatics, 18,* 697–704.

Estrada, E., Torres, L., Rodriguez, L., & Gutman, I., (1998). An atom-bond connectivity index: Modeling the enthalpy of formation of alkanes. *Indian J. Chem., 37A,* 849–855.

Fajtlowicz, S., (1987). On conjectures on Graffiti-II. *Congr. Numer., 60,* 187–197.

Falahati-Nezhad, F., Iranmanesh, A., Tehranian, A., & Azari, M., (2014). Strict lower bounds on the multiplicative Zagreb indices of graph operations. *Ars. Combin., 117,* 399–409.

Farahani, M. R., (2013). Some connectivity indices of polycyclic aromatic hydrocarbons (PAHs). *Adv. Mat. Corrosion, 1,* 65–69.

Farahani, M. R., (2015). On atom bond connectivity and geometric-arithmetic indices of a benzenoid system. *International Journal of Engineering and Technology Research, 3*(2), 1–4.

Fath-Tabar, G. H., Hossein-Zadeh, S., & Hamzeh, A., (2011). On the first geometric-arithmetic index of product graphs. *Util. Math., 86,* 279–288.

Fath-Tabar, G., Furtula, B., & Gutman, I., (2010). A new geometric–arithmetic index. *J. Math. Chem., 47,* 477–486.

Furtula, B., Graovac, A., & Vukičević, D., (2010). Augmented Zagreb index. *J. Math. Chem., 48*(2), 370–380.

Furtula, B., Gutman, I., & Dehmer, M., (2013). On structure-sensitivity of degree-based topological indices. *Appl. Math. Comput., 219,* 8973–8978.

Ghorbani, M., & Ghazi, M., (2010). Computing some topological indices of triangular benzenoid. *Dig. J. Nanomater. Bios., 5*(4), 1107–1111.

Ghorbani, M., & Jalali, M., (2009). Computing a new topological index of nanostructures. *Dig. J. Nanomater. Bios., 4,* 681–685.

Ghorbani, M., & Khaki, A., (2010). A note on the fourth version of geometric-arithmetic index. *Optoelectron. Adv. Mater.-Rapid Commun., 4*(12), 2212–2215.

Ghorbani, M., Mesgarani, H., & Shakeraneh, H., (2011). Computing *GA* index and *ABC* index of V–phenylenic nanotube. *Optoelectron. Adv. Mater.-Rapid Commun., 5*(3), 324–326.

Graovac, A., Ghorbani, M., & Hosseinzadeh, M. A., (2011). Computing fifth geometric-arithmetic index for nanostar dendrimers. *J. Math. Nanosci., 1*(1), 32–42.

Gutman, I., & Furtula, B., (2008). *Recent Results in the Theory of Randić Index.* University Kragujevac: Kragujevac, Serbia.

Gutman, I., & Polansky, O. E., (1986). *Mathematical Concepts in Organic Chemistry* (pp. 214). 1st ed.; Springer: Berlin, Germany

Gutman, I., & Tošović, J., (2013). Testing the quality of molecular structure descriptors. Vertex-degree-based topological indices. *J. Serb. Chem. Soc., 78,* 805–810.

Gutman, I., & Trinajstić, N., (1972). Graph theory and molecular orbitals. Total π–electron energy of alternant hydrocarbons. *Chem. Phys. Lett., 17,* 535–538.

Gutman, I., Ruščić, B., Trinajstić, N., & Wilcox, C. F., (1975). Graph theory and molecular orbitals. XII. Acyclic polyenes. *J. Chem. Phys., 62,* 3399–3405.

Gutman, I., Furtula, B., & Elphick, C., (2014). Three new/old vertex–degree–based topological indices. *MATCH Commun. Math. Comput. Chem., 72,* 617–632.

Hayat, S., & Imran, M., (2014). Computation of topological indices of certain networks. *Appl. Math. Comput., 240*, 213–228.

Hosseinzadeh, M. A., & Ghorbani, M., (2009). GA index of nanostar dendrimers. *J. Optoelec. Adv. Mater., 11*(11), 1671–1674.

Husin, N. M., Hasni, R., & Arif, N. E., (2013). Atom-bond connectivity and geometric arithmetic indices of dendrimer nanostars. *Austral. J. Basic Appl. Sci., 7*(9), 10–14.

Husin, N. M., Hasni, R., & Arif, N. E., (2015). Atom-bond connectivity and geometric arithmetic indices of certain dendrimer nanostars. *J. Comput. Theor. Nanosci., 12*(2), 204–207.

Iijima, S., (1991). Helical microtubules of graphitic carbon. *Nature, 354*, 56–58.

Imran, M., & Hayat, S., (2014). On computation of topological indices of Aztec diamonds. *Sci. Int. (Lahore), 26*(4), 1407–1412.

Imran, M., Hayat, S., & Mailk, M. Y. H., (2014). On topological indices of certain interconnection networks. *Appl. Math. Comput., 244*, 936–951.

Iranmanesh, A., & Azari, M., (2015a). Edge-Wiener descriptors in chemical graph theory: A survey. *Curr. Org. Chem., 19*(3), 219–239.

Iranmanesh, A., & Azari, M., (2015b). The first and second Zagreb indices of several interesting classes of chemical graphs and nanostructures. In: Putz, M. V., & Ori, O., (eds.), *Exotic Properties of Carbon Nanomatter* (Vol. 8, pp. 153–183). Springer: Dordrecht, Netherlands (Chapter 7).

Iranmanesh, A., & Zeraatkar, M., (2010). Computing Ga index for some nanotubes. *Optoelectron. Adv. Mater.-Rapid Commun., 4*(11), 1852–1855.

Iranmanesh, A., & Zeraatkar, M., (2011). Computing Ga index of $HAC_5C_7[p,q]$ and $HAC_5C_6C_7[p,q]$ nanotubes. *Optoelectron. Adv. Mater.-Rapid Commun., 5*(7), 790–792.

Khalifeh, M. H., Yousefi-Azari, H., Ashrafi, A. R., & Wagner, S. G., (2009). Some new results on distance-based graph invariants. *European J. Combin., 30*(5), 1149–1163.

Kroto, H. W., Heath, J. R., O'Brien, S. C., Curl, R. F., & Smalley, R. E., (1985). C_{60}: buckminsterfullerene. *Nature, 318*, 162–163.

Li, X., & Gutman, I., (2006). *Mathematical Aspects of Randić–Type Molecular Structure Descriptors*. Univ. Kragujevac: Kragujevac, Serbia.

Li, X., & Shi, Y., (2008). A survey on the Randić index. *MATCH Commun. Math. Comput. Chem., 59*, 127–156.

Lokesha, V., Shwetha, S. B., Ranjini, P. S., Cangul, I. N., & Cevik, A. S., (2013). New bounds for Randic and GA indices. *J. Inequal. Appl.,* 180.

Madanshekaf, A., & Moradi, M., (2014). The first geometric–arithmetic index of some nanostar dendrimers. *Iranian J. Math. Chem., 5*, 1–6.

Mahmiani, A., & Khormali, O., (2013). On the total version of geometric-arithmetic index. *Iranian J. Math. Chem., 4*(1), 21–26.

Mahmiani, A., Khormali, O., & Iranmanesh, A., (2012). On the edge version of geometric-arithmetic index. *Dig. J. Nanomater. Bios., 7*(2), 411–414.

Manso, F. C. G., Júnior, H. S., Bruns, R. E., Rubira, A. F., & Muniz, E. C., (2012). Development of a new topological index for the prediction of normal boiling point temperatures of hydrocarbons: The *Fi* index. *J. Mol. Liquids, 165*, 125–132.

Mogharrab, M., & Fath, T. G. H., (2011). Some bounds on GA_1 index of graphs. *MATCH Commun. Math. Comput. Chem., 65*, 33–38.

Nikmehr, M. J., Soleimani, N., & Veylaki, M., (2014). Topological indices based end-vertex degrees of edges on nanotubes. *Proceedings of IAM, 3*(1), 89–97.

Nikmehr, M. J., Soleimani, N., & Veylaki, M., (2015). Computing some topological indices of carbon nanotubes. *Proceedings of IAM, 4*(1), 20–25.

Putz, M. V., Ori, O., Cataldo, F., & Putz, A. M., (2013). Parabolic reactivity "coloring" molecular topology: Application to carcinogenic PAHs. *Curr. Org. Chem., 17*(23), 2816–2830.

Ramakrishnan, S., Senbagamalar, J., & Baskar, B. J., (2013). Topological indices of molecular graphs under specific chemical reactions. *Int. J. Comput. Algorithm, 2*, 224–234.

Randić, M., (1975). On characterization of molecular branching. *J. Am. Chem. Soc., 97*, 609–6615.

Randić, M., Vračko, M., & Novič, M., (2001). Eigenvalues as molecular descriptors. In: Diudea, M. V., (ed.), *QSPR/QSAR Studies by Molecular Descriptors* (pp. 93–120). Nova: New York, USA.

Raut, N. K., (2014). Degree based topological indices of isomers of organic compounds. *International Journal of Scientific and Research Publications, 4*(8), 4.

Raza, Z., Ali, A., & Bhatti, A. A., (2016). More on comparison between first geometric-arithmetic index and atom-bond connectivity index. *Miskolc. Math. Notes., 17*(1), 561–570.

Rodíguez, J. M., & Sigarreta, J. M., (2015a). On the geometric-arithmetic index. *MATCH Commun. Math. Comput. Chem., 74*, 103–120.

Rodíguez, J. M., & Sigarreta, J. M., (2015b). Spectral study of the geometric-arithmetic index. *MATCH Commun. Math. Comput. Chem., 74*, 121–135.

Shababi, H., & Ashrafi, A. R., (2010). Computing the GA index of nanotubes and nanotori. *Optoelectron. Adv. Mater.-Rapid Commun., 4*(11), 1860–1862.

Shababi, H., Ashrafi, A. R., & Gutman, I., (2010). Geometric-arithmetic index: An algebraic approach. *Studia Univ. Babes Bolyai Chem., 55*(4), 107–112.

Soleimani, N., Nikmehr, M. J., & Agha, T. H., (2015). Computation of the different topological indices of nanostructures. *J. Natn. Sci. Foundation Sri Lanka, 43*(2), 127–133.

Sridhara, G., Rajesh, K. M. R., & Indumathi, R. S., (2015). Computation of topological indices of graphene. *J. Nanomater., 969348.*

Todeschini, R., & Consonni, V., (2000). *Handbook of Molecular Descriptors* (Vol. 11, pp. 668). Wiley-VCH: Weinheim, Germany.

Todeschini, R., & Consonni, V., (2010). New local vertex invariants and molecular descriptors based on functions of the vertex degrees. *MATCH Commun. Math. Comput. Chem., 64*, 359–372.

Trinajstić, N., (1992) *Chemical Graph Theory* (pp. 352). 2nd ed.; CRC Press: Boca Raton, FL.

Vöge, M., Guttmann, A. J., & Jensen, I., (2002). On the number of benzenoid hydrocarbons. *J. Chem. Inf. Comput. Sci., 42*, 456–466.

Vukičević, D., & Furtula, B., (2009). Topological index based on the ratios of geometrical and arithmetical means of end–vertex degrees of edges. *J. Math. Chem., 46*, 1369–1376.

Vukičević, D., & Gašperov, M., (2010). Bond additive modeling 1. Adriatic indices. *Croat. Chem. Acta, 83*(3), 243–260.

West, D. B., (1996). *Introduction to Graph Theory.* Prentice Hall: Upper Saddle River, USA.

Wiener, H., (1947a). Structural determination of paraffin boiling points. *J. Am. Chem. Soc., 69*, 17–20.

Wiener, H., (1947b). Correlation of heats of isomerization and differences in heats of vaporization of isomers among the paraffin hydrocarbons. *J. Am. Chem. Soc., 69*(11), 2636–2638.

Xiao, L., Chen, S., Guo, Z., & Chen, Q., (2010). The geometric-arithmetic index of benzenoid systems and phenylenes. *Int. J. Contemp. Math. Sciences, 5*(45), 2225–2230.

Xu, K., Liu, M., Das, K. C., Gutman, I., & Furtula, B., (2014). A survey on graphs extremal with respect to distance-based topological indices. *MATCH Commun. Math. Comput. Chem.*, *71*, 461–508.

Yarahmadi, Z., (2011). A note on the first geometric-arithmetic index of hexagonal systems and phenylenes. *Iranian J. Math. Chem.*, *2*(2), 101–108.

Yuan, Y., Zhou, B., & Trinajstić, N., (2010). On geometric–arithmetic index. *J. Math. Chem.*, *47*, 833–841.

Zhong, L., & Xu, K., (2014). Inequalities between vertex-degree-based topological indices. *MATCH Commun. Math. Comput. Chem.*, *71*, 627–642.

Zhou, B., & Trinajstić, N., (2009). On a novel connectivity index. *J. Math. Chem.*, *46*, 1252–1270.

Zhou, B., Gutman, I., Furtula, B., & Du, Z., (2009). On two types of geometric-arithmetic index. *Chem. Phys. Lett.*, *482*, 153–155.

CHAPTER 20

GAP Programming

YASER ALIZADEH[1] and ALI IRANMANESH[2]

[1]Department of Mathematics, Hakim Sabzevari University, Sabzevar, Iran, E-mail: y.alizadeh@hsu.ac.ir

[2]Department of Pure Mathematics, Faculty of Mathematical Sciences, Tarbiat Modares University, P.O. Box: 14115-137, Tehran, Iran, Tel: +982182883447; Fax: +982182883493; E-mail: iranmanesh@modares.ac.ir

20.1 DEFINITION

Groups, Algorithms, and Programming (GAP) (The GAP Group, 2015) is a system for computational discrete algebra, with particular emphasis on Computational Group Theory. GAP provides a programming language, a library of thousands of functions implementing algebraic algorithms written in the GAP language as well as large data libraries of algebraic objects (see also the overview and the description of the mathematical capabilities). GAP is used in research and teaching for studying groups and their representations, rings, vector spaces, algebras, combinatorial structures, graph theory, etc. You can study and easily modify or extend it for your special use.

20.2 HISTORICAL ORIGIN(S)

GAP stands for Groups, Algorithms, and Programming. The name was chosen to reflect the aim of the system.

GAP was started as a joint 'Diplom' project of four students whose names are hidden in the GAP banner: Johannes Meier, Alice Niemeyer, Werner Nickel, and Martin Schönert. While all four have meanwhile finished their 'Diplom;' further students are presently extending the library in the course of their work for the 'Diplom.'

The first version of GAP was operational by the end of 1986. The system was first presented at the Oberwolfach meeting on computational group theory in May 1988. Officially, Version 2.4 was the first to be given away from Aachen starting in December 1988. The strong interest in this version, in spite of its still rather small collection of group theoretical routines, as well as constructive criticism by many colleagues, confirmed our belief in the general design principles of the system. Nevertheless over three years have gone by until now in April 1992 version 3.1 is released.

Although fortunately, GAP is not yet part of history, we have kept a few documents marking steps of its development. In particular, the prefaces of the manuals of the various releases sketch the respective progress. We present here links to (relevant extracts from) these prefaces as well as to some 'interim report' that appeared in the GAP Forum (for recent developments see the page on 'Updates').

1. Preface for GAP 2.4, the first publically released version of GAP (JN, December 1988)
2. Preface for GAP 3.1 (JN, March 1991)
3. From the preface for GAP 3.2 (JN, January 1993)
4. Preface for GAP 3.4.3, the last GAP 3 release from Aachen (JN, June 1994)
5. Interim Report GAP, 1997, me and all that (JN, June 1995)
6. Interim Report Organizing GAP (JN, December 1995)
7. Interim Report A European summer of CGT '97 (JN, July 1996)
8. Interim Report WWW pages and refereeing (SL + JN, September 1996)
9. From the preface for GAP 3.4.4 (SL, April 1997)
10. Interim Report Move of GAP addresses (JN, June 1997)
11. Forum letter Contributions to GAP (CW, July 1997)
12. From the preface for GAP 4.B.1, the first beta release of GAP 4 (JN, July 1997)
13. From the preface for GAP 4.1 (SL, July 1999)
14. Preface for GAP 4.3 (SL, May 2002)
15. Preface for GAP 4.4 (AH, March 2004)
16. Preface for GAP 4.5 (June 2012)

20.3 NANO-SCIENTIFIC DEVELOPMENT(S)

A topological index of a molecular graph is a real number derived from the structure of the graph such that isomorphic graphs have the same topological index. The study of topological indices based on distance in a graph in biological sciences, physical chemistry and QSAR and QSPR studies started from 1947 when Harold Wiener introduced the Wiener index to establish the relationships between physico-chemistry properties of alkenes and the structures of their molecular graphs. Wiener index, Szeged index, edge Wiener index, eccentric connectivity index are some of the well-known topological indices. In a series of papers, topological indices were computed for some nanostructure nanotubes and fullerenes. For example, these topological indices were obtained for C_{12k+4} fullerenes in Alizadeh et al., (2012) for coronene/circumcoronene in Alizadeh & Klavzar (2013), for benzenoid hydrocarbon and phenylenes in Gutman & Klavzar (1996). In some other papers by using the GAP program, the topological indices were calculated for some nanotubes and fullerenes such as C_{80} fullerene $HAC_5C_7[p,q]$ and $HAC_5C_6C_7[p,q]$ nanotubes (see Iranmanesh et al., 2009; Iranmanesh & Alizadeh, 2013).

20.4 NANOCHEMICAL APPLICATION(S)

The chemical graph theory is a branch of mathematical chemistry that is mostly concerned with finding topological indices of chemical graphs that correlate well with certain physicochemical properties of the corresponding molecules. The GAP program is used to finding topological indices of molecular graphs. The following GAP program (Iranmanesh et al., 2009) is an example of its nanochemical applications. By this program topological indices as Wiener polynomial, Wiener index and hyper Wiener index of IPR isomers of C_{80} fullerene. The structure of the IPR Isomer of C_{80} fullerene is shown in Figure 20.1.

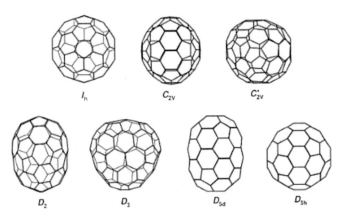

FIGURE 20.1 Isomers of C_{80} fullerene. Reprinted from Iranmanesh et al., (2011). Open Access.

```
md: = 1; v: = []; D: = [];
for i in [1..n] do
      D[i]: = []; u: = [i]; D[i][1]: = N[i];
      v[i]: = Size(N[i]);
      u: = Union(u,D[i][1]);
      r: = 1; t: = 1;
while r<>0 do
      D[i][t+1]: = [];
      for j in D[i][t] do
         for m in Difference (N[j],u) do
            AddSet(D[i][t+1],m);
         od;
      od;
      u: = Union(u,D[i][t+1]);
      if D[i][t+1] = [] then r: = 0;
      fi;
      t: = t+1;
      od;
      md: = MaximumList([md,Size(D[i])]);
   od;
   p: = [];
   for i in [1..md] do
      p[i]: = 0;
   od;
   for t in [1..md] do
      for i in [1..n] do
         p[t]: = p[t]+Size(D[i][t]);
      od;
   od;
p: = p/2;#(p is the set the wiener polynomial's coefficients of C_{80} fullerene)
```

20.4   MULTI-/TRANS-DISCIPLINARY CONNECTION(S)

GAP has a package, GRAP package that is used in graph theory. GRAP is a package for computing with graphs and groups. GRAPE is primarily designed for the construction and analysis of finite graphs related to groups, designs, and geometries. Special emphasis is placed on the determination of regularity properties and subgraph structure.

TABLE 20.1 The Wiener Polynomial, the Wiener and the Hyper Wiener Indices of IPR Isomers of C_{80} Fullerene

IPR Isomers of C_{80} fullerene	Wiener polynomial $W(C_{80}, x)$	Wiener index $W(G)$	Hyper Wiener index $WW(G)$
D5d	$120x + 240x^2 + 330x^3 + 420x^4 + 450x^5 + 480x^6 + 440x^7 + 350x^8 + 250x^9 + 70x^{10} + 10x^{11}$	17340	64530
D2	$120x + 240x^2 + 330x^3 + 420x^4 + 450x^5 + 478x^6 + 438x^7 + 348x^8 + 248x^9 + 86x^{10} + 2x^{11}$	17352	64622
C2v	$120x + 240x^2 + 330x^3 + 420x^4 + 450x^5 + 472x^6 + 433x^7 + 343x^8 + 245x^9 + 97x^{10} + 10x^{11}$	17412	65174
D3	$120x + 240x^2 + 330x^3 + 420x^4 + 450x^5 + 474x^6 + 435x^7 + 351x^8 + 249x^9 + 91x^{10}$	17368	64750
C2v	$120x + 240x^2 + 330x^3 + 420x^4 + 450x^5 + 466x^6 + 429x^7 + 342x^8 + 245x^9 + 104x^{10} + 14x^{11}$	17454	65549
D5h	$120x + 240x^2 + 330x^3 + 420x^4 + 450x^5 + 460x^6 + 425x^7 + 340x^8 + 245x^9 + 110x^{10} + 20x^{11}$	17500	65965
Ih	$120x + 240x^2 + 330x^3 + 420x^4 + 450x^5 + 450x^6 + 420x^7 + 330x^8 + 240x^9 + 120x^{10} + 40x^{11}$	17600	66900

Reprinted with permission from Iranmanesh et al., (2009). © Taylor & Francis.

KEYWORDS

- **fullerenes**
- **GAP programming**
- **nanotubes**

REFERENCES AND FURTHER READING

Alizadeh, Y., & Klavzar, S., (2013). Interpolation method and topological indices: 2-Parametric families of graphs. *MATCH Commun. Math. Comput. Chem., 69*, 523–534.

Alizadeh, Y., Iranmanesh, A., & Klavzar, S., (2012). Interpolation method and topological indices: The case of fullerenes C_{12k+4}, *MATCH Commun. Math. Comput. Chem., 68*, 303–310.

Iranmanesh, A., & Alizadeh, Y., (2013). Eccentric connectivity index of $HAC_5C_7[p,q]$ and $HAC_5C_6C_7[p,q]$ nanotubes. *MATCH Commun. Math. Comput. Chem., 69*, 175–182.

Iranmanesh, A., Alizadeh, Y., & Mirzaie, S., (2009). Computing wiener polynomial, Wiener index, hyper Wiener index of C_{80} fullerene by GAP program. Fullerenes. *Nanotubes and Carbon Nanostructures, 17*, 560–566.

The GAP Group, (2015). *GAP - Groups, Algorithms, and Programming*, Version 4.7.8.

CHAPTER 21

Kekulé Structure

MARINA A. TUDORAN[1] and MIHAI V. PUTZ[1,2]

[1]*Laboratory of Renewable Energies-Photovoltaics, R&D National Institute for Electrochemistry and Condensed Matter, Dr. A. Paunescu Podeanu Str. No. 144, Timisoara, RO-300569, Romania*

[2]*Laboratory of Structural and Computational Physical Chemistry for Nanosciences and QSAR, Biology-Chemistry Department, West University of Timisoara, Pestalozzi Street No. 44, Timisoara, RO-300115, Romania, Tel: +40-256-592638, Fax: +40-256-592620, E-mail: mv_putz@yahoo.com or mihai.putz@e-uvt.ro*

21.1 DEFINITION

A general definition states that the Kekulé structure is a Lewis structure in which the lines are used to represent the bonded electron pairs from the covalent bonds. From the organic chemistry perspective, the structural formula of benzene represented as a hexagonal ring with alternate single and double bonds between the carbon atoms, every carbon atom being tetravalent and incident with exactly one double bond is also referred as a Kekulé structure.

21.2 HISTORICAL ORIGIN(S)

The Kekulé structure concept was first stated by August Kekulé in 1865, when he proposed the structural formula of benzene as a cyclic molecule with the form of a regular hexagon (Wizinger-Aust et al., 1966). Because this formula implies that the carbon atom may have the valence 3, even though the known value of valence was 4, Kekulé inserts, in addition, three carbon-carbon bonds, this way resulting in the so-called the Kekulé structure of benzene. In other words, in general, a Kekulé structure has alternate single and double carbon-carbon bonds. Recently, studies in this area revealed that the number of Kekulé structures (notated with K) have many application in the theoretical chemistry of benzenoid hydrocarbons, presented below (Cyvln & Gutman, 1986).

1. For a benzenoid hydrocarbon C_nH_m, one can determine the total π electron energy by using a linear function of K with the formula:

$$E = \left(0.201n - 0.049m + 0.043K \cdot 0.795^{n-m}\right) \cdot E\{\text{benzen}\} \qquad (1)$$

2. For calculating the resonance energy (RE) one can use the formula:

$$RE = 114.3 \ln K \left[kJ \cdot mol^{-1} \right] \qquad (2)$$

3. For calculating the rate constant from the reaction of benzenoid hydrocarbon and maleic anhydride, after the formula:

$$k = 0.002 \left(K_P / K_R \right)^{16} \left[dm^3 \cdot mol^{-1} \cdot s^{-1} \right] \qquad (3)$$

with K_P is the number Kekulé structures of the reactant and K_R is the number Kekulé structures of the product.

4. For calculating the interatomic distance between the atoms n and m which form a double bond in a benzenoid hydrocarbon K_{nm}, whit an error of 1 [pm] (Herndon, 1974; Herndon & Parkanyi, 1976), using the formula:

$$d_{nm} = 146.5 - 13.0 \left(K_{nm} / K \right) \left[pm \right] \qquad (4)$$

with K as the total number of the Kekulé structure for the respective compound.

21.3 NANO-SCIENTIFIC DEVELOPMENT(S)

Determining of all benzenoid hydrocarbons which possess K Kekulé structures (Cyvin & Gutman, 1986a) with $0 < K < 9$ was in scientists attention since decades (Gutman, 1982, 1983), being implied that it is a finite number, knowing that benzene graph is the only one with $K = 2$, naphthalene is the only one with $K = 3$, anthracene is the only one with $K = 4$ and both tetracene and phenanthrene has $K = 5$. From literature it is known in that benzenoid systems with nine Kekulé structure is infinite, as the case $K = 9$ is referring to disconnected benzenoids (Cyvin et al., 1985) in which two naphthalene units (each one of them having three Kekulé structures) are behaving independently and may create connections with random large junctions of hexagons with fixed bonds (Cyvin & Gutman, 1985). Balaban and Haray

(1968) propose a classification of benzenoid systems. The first category is represented by catacondensed and includes: acene (linear, noted with "a"), mirror-symmetrical (unbranched, noted with "m"), center-symmetrical (unbranched, noted with "c"), unsymmetrical (unbranched, noted with "u") and branched (noted with "b"). The second category is represented by pericondensed and includes: non-Kekuléan (with no fixed bonds, noted with "n"), Kekulean with fixed bonds (essentially disconnected, noted with "e"), and non-Kekulean (noted with "o").

By definition, a graph is represented by a set of vertices and a set of edges (pairs of vertices) with vertices represented by points and edges by-line, having many applications in chemistry (Bonchev & Rouvrav 1991, 1992). A molecule C can be represented as a graph G by replacing atoms with vertices and bonds with edges. For a graph G, a set of disjoint edges which covers all its vertices is called a perfect matching. Another method of obtaining a perfect matching assumes choosing all the edges of G which correspond to the double edges of C, this way being created a correspondence between perfect matching and a Kekulé structure. The reverse process is also valid; namely, a Kekulé structure of C can be created by assigning double bonds to edges from the perfect matching of G (Hansen & Zheng, 1995).

In the chemistry field, there are different studies on enumerating Kekulé structures for conjugated molecules (Dionova-Jerman-Blaiii & Trinajistic, 1982). In case that a single Kekulé structure needs to be determined, worth mention is the method proposed by Kearsley (1993), which is a heuristic method and is not general applicable. Starting from these considerate, Hansen and Zheng (1995) propose a new method based on the Edmond Matching Algorithm (*EMA*). The algorithm starts with M a matching of G (empty, or not) and is proposing to find, if possible, M augmenting path P. The set M has the form $(M - M \cap P) \cup (P - M)$ and is constructed by adding the edges of P which are not belonging to M and deleting the edges which are common with P. What is obtained represents a matching which has one more edge compared to M, and the process is repeated as long as there are augmenting paths. At the end of the process, the maximum matching M is obtained, which become a perfect matching if covers all vertices of G. In other words, one can say that *EMA* appears to be a fast method for determining the existence of a perfect matching which presents an available computer code and also can be applied when only some vertices need to be covered. The enumeration problem from the graph theory perspective was also in the attention of Cyvin and Gutman, which develop, in their work (Cyvin & Gutman, 1986b), two theorems.

Theorem 1 states that a normal benzenoid having h hexagons possess $h + 1$ Kekulé structures, with $L(h)$ as the only benzenoid with h hexagons which respect the relation $K = h + 1$. From here one can deduce the corollary: considering $h_{max}(K)$, defined as the maximum value of h which can occur for a given number of K, its value will be given by the formula:

$$h_{max}(K) = K - 1 \tag{5}$$

Theorem 2 states that for B and B' defined as two normal benzenoids with the property that B' is generated from B by hexagon adding, the following relation is respected:

$$K(B') > K(B) \tag{6}$$

The second theorem demonstrates that in the case of normal (Kekulean) benzenoids, the number of Kekulé structure increase with adding a hexagon.

According to the graph theory, a 1-factor of a graph G is defined as a spanning subgraph of G, which presents only components with to vertices. The Kekulé structure has the equivalent on the 1 factor for the case when the carbon-atom skeleton of an unsaturated conjugated hydrocarbon is represented in a graph manner. Because of this association, one can approach some chemical investigations using mathematical techniques (Harary, 1969). Considering a molecular graph, with A the adjacency matrix, one can develop three theorems (Cyvln & Gutman, 1986).

Theorem 1 is referring to alternant hydrocarbons for which are valid the following relation:

$$per(A) = K^2 \tag{7}$$

with *per* as the permanent.

The **theorem 2** state that, for benzenoid hydrocarbons which present n carbon atoms, is valid the following relation (Cyvln & Gutman, 1986):

$$\det(A) = (-1)^{n/2} K^2 \tag{8}$$

with *det* as the determinant.

The next one, **theorem 3**, is referring to a molecular graph G with x_n and x_m two adjacent vertices and y_{nm} the edge connecting them for which is valid the following relation (Cyvln & Gutman, 1986):

$$K\{G\} = K\{G - y_{nm}\} + K\{G - y_n - y_m\} \tag{9}$$

And last, **theorem 4**, states that for a molecular graph G with x_n and x_m two adjacent vertices, x_n with graph-theoretical valence 1 (knowing as primary), the following relation is valid (Cyvln & Gutman, 1986):

$$K\{G\} = K\{G - y_n - y_m\} \qquad (10)$$

Using this theorem, one can calculate the number of Kekulé structure in an efficient way.

21.4 NANOCHEMICAL APPLICATION(S)

In chemistry, the conjugated circuits can be defined within individual Kekulé valence structures for conjugated hydrocarbons, and are of two types: *4n* (designed as R_n and Q_n) and *4n* + 2 (with a major contribution to molecular stability) (Randic, 1976, 1977, 2003). Regarding the Kekulé valence structures, Fries (1927) formulated an empirical rule for characterizing the most important ones as the structures with the largest number of fused benzene rings with the same Kekulé formula as the benzene (Vukicevic & Randic, 2005). The term of "innate degree of freedom" for a Kekulé structure for both non-benzenoid and benzenoid hydrocarbon is defined as the smallest possible number of carbon-carbon double bonds, which can determine the localization of the all remaining bonds after they are set. Studies (Figure 21.1) made on the degree of freedom (*df*) index determine that a larger value indicates Kekulé valence structures with larger numbers of π aromatic sextets present larger *df* values point, while a low value indicates weak Kekulé structures with long-range order.

FIGURE 21.1 The algorithm for finding Kekulé structures.

Applying to C_{60}, by finding the value of the degree of freedom one can determine which one of its Kekulé structures are important and which are not (Vukicevic & Randic, 2005).

The algorithm used in finding the *df* value for C_{60} molecule is schematically presented in Figure 21.1. The results are presented in Table 21.1, the 125,000 Kekulé valence structures being divided into six classes based on the value of their degree of freedom (Vukicevic & Randic, 2005). The most important Kekulé structure (which is also the Fries structure of buckminsterfullerene) is represented by the Kekulé valence structure with *df* = 10, showing high symmetry and being the structure in which all the 60 π electrons can be equally divided by assigning three π electrons to each benzene ring (Vukicevic & Randic, 2005).

TABLE 21.1   The Degree of Freedom for the Kekulé Valence Structure of C_{60}

Total number of Kekulé structure	Distinct Kekulé structure	Degree of freedom
1	1	10
80	2	9
2073	33	8
4060	47	7
3116	39	6
31500	36	5

The carbon nanotubes (*CNT*) were discovered by Iijima in 1991, a C_{70} fullerene tiny tube with a length of several microns and a diameter of about 1.5 nm (Iijima, 1991). Because they present sp^2 carbons, i.e., each carbon participates in two single bonds and one double bond, *CTN* is considered a "benzenoid system," and their stability may be determined, like in case of benzenoids, by determining the Kekulé structure count (*K*) (Hall, 1991). Known as the numbers of ways in which the double bonds can be positioned in the hexagons network, the *K* value for nanotubes (tubulenes) represents a complex problem, and enumeration methods were proposed by different scientists, such as Klein et al., (1993) who proposed a general solution for the π-electron model applied to all possible buckytubes, or Kirby (1993) who succeed in determining the π-electron energy of nanotubes. Saches et al., (1996) determined that the number of Kekulé structures for *CNT* appears to be situated much less in "zigzag" than in "armchair" type. In their work, Lukovits et al., (2003) determined that the extension of tubulene in vertical direction determine a high number of Kekulé structures than the extension in horizontal direction. The value of K_N^T was calculated with the formula:

$$K_N^T = 1'T^{N-1}1 \tag{11}$$

where T^{N-1} represents the transfer matrix for tubulene, 1 represents a column vector (all elements are equal with 1), 1 represents the transpose of 1, and N represents the number of vertical naphthalene layers. The authors also determine that the value of resonance energy increases with increasing values of N (Lukovits et al., 2003).

21.5 MULTI-/TRANS-DISCIPLINARY CONNECTION(S)

Benzenoid hydrocarbons are defined as polycyclic conjugated molecules with six-membered rings, having each pair of carbon being connected by a single or double bond (Vukicevic & Zigert, 2008). When the double bonds are situated in a single ring, they represent the π electrons, whose distribution is represented by Kekulé structures. In mathematical literature, they are known as benzenoid graphs, defined as 2-connected subgraphs of the hexagonal lattice with the property that every bounded face is a hexagon. A benzenoid graph B in which all of its vertices lie on its perimeter is called catacondensed, and if not is called peri-condensed. The edge from B which does not belong to its perimeter is called inner edge, and a hexagon which contains only one inner edge is called terminal hexagon or a leaf. For a benzenoid graph B, the set of all of its hexagon is noted with $H(B)$, and the set of the terminal hexagons with $L(B)$. One can define a matching for a graph B as a set of pairwise independent edge, and a perfect matching or 1-factor, if the set covers all vertices of B. Studies in this area, revealed that there is a correspondence one-to-one between the 1-factor of benzenoid graph and Kekulé structure (Cyvin & Gutman, 1989).

The algebraic Kekulé structure was introduced by Randic (2004) for benzenoids graphs (also known as Randic structures), which for a catacondensed case can be defined as: to each hexagon, an integer named Randic hexagon number was assigned, representing that 1-factor contribution of π electrons on the hexagon. The Randic hexagon number can take one of the two values:

- 2: if the π electrons of an edge in the 1-factor are on the perimeter of B;
- 1: if the π electrons of an edge in the 1-factor are not on the perimeter of B, the electrons being shared by two hexagons.

For a hexagon H from B, one can obtain the Randic hexagon by summing up all contribution of π electrons of double bonds around H (see Figure 21.2).

Giving a catacondensed benzenoid graph B and a function for assessing the Randic hexagon number $\rho:H(B) \to \{0,1,2,3,4,5,6\}$, and $L \in L(B)$, the labeling function of the form $f_{B,L}: H(B) \to \{0,1\}$ can be defined as:

$$f_{B,L}(L) = \begin{cases} 1, & \rho(L) > 4 \\ 0, & \text{otherwise} \end{cases} \tag{12}$$

and for $H \in H(B) \backslash \{L\}$ will results:

$$f_{B,L}(L) = \rho(H) \tag{13}$$

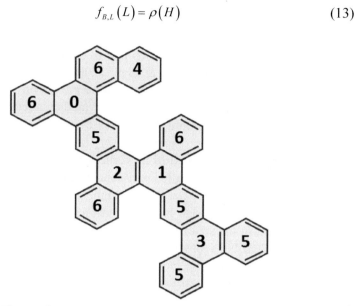

FIGURE 21.2　The Randic numbers for a given benzenoid structure (Vukicevic & Zigert, 2008).

There are two theorems which can be applied to the enumeration problem. Theorem 1 is referring to the labeling function $f_{B,L}(L)$, which uniquely determines the 1-factor of B for B a finite catacondensed benzenoid graph, except the benzene with a terminal hexagon L (Vukicevic & Zigert, 2008). Theorem 2 states that, by giving a labeling function $f_{B,L}$, one can determine if an inner edge of B belongs to the 1-factor or not, with B a finite catacondensed benzenoid graph, except the benzene with a terminal hexagon L. Also, for a catacondensed benzenoid graph, a binary code of length $|H\ (B)|$ can be assigned to the algebraic Kekulé structure of B by using the labeling function $f_{B,L}(L)$ (Vukicevic & Zigert, 2008).

21.6　OPEN ISSUES

After years of studies, scientists determine that there are a series of problems which arise in the Kekulé theory for benzene. First is represented by the *low reactivity of benzene*, meaning that benzene does not react with alkenes and bromine water, as it should. Another problem is referring to *the carbon-carbon bond length*, more precisely, the symmetrical characteristic of a molecule in Kekulé form with all six carbon-carbon bonds with the same length, even if in general the single and the double bond have different length. Also, the *hydrogenation enthalpies* are in question, due to the fact that based on the actual chemical name of the Kekulé structure, cyclohexa-1,3,5-triene, and based on the fact that cyclohexene (with 1 double bond) has the values of its hydration enthalpy equal with 120 kJ/mol, benzene (with 3 double bonds) should have a hydrogenation enthalpy equal with 360 kJ/mol while the experimentally measured value is 152 kJ/mol, less than the expected value.

KEYWORDS

- **benzenoid hydrocarbons**
- **buckminsterfullerene**
- **degree of freedom**
- **Kekulé structure**

REFERENCES AND FURTHER READING

Balaban, A. T., & Harary, F., (1968). Chemical graphs-V: Enumeration and proposed nomenclature of benzenoid catacondensed polycyclic aromatic hydrocarbons. *Tetrahedron, 24*(6), 2505.

Bonchev, D., & Rouvrav, D. H., (1991). *Chemical Graph Theory: Introduction and Fundamentals*. Abacus Press, Philadelphia.

Bonchev, D., & Rouvray, D. H., (1992). *Chemical Graph Theory: Reactivity and Kinetics*. Abacus Press: Philadelphia.

Cyvin, S. J., & Gutman, I., (1985). Topological properties of benzenoid systems–XXXIX–The number of Kekulé structures of benzenoid hydrocarbons containing a chain of hexagons. *Journal of the Serbian Chemical Society, 50*, 443–450.

Cyvin, S. J., & Gutman, I., (1986b). Number of Kekulé structures as a function of the number of hexagons in benzenoid hydrocarbons. *Zeitschrift für Naturforschung, 41a*, 1079–1086.

Cyvin, S. J., Cyvin, B. N., & Gutman, I., (1985). Number of Kekulé structures of five-tier strips. *Zeitschrift für Naturforschung, 40a*, 1253–1261.

Cyvln, S. J., & Gutman, I., (1986a). Kekulé structures and their symmetry properties. *Computers & Mathematics with Applications, 12B*(3/4), 859–876.

Dzonova-Jerman-Blazic, B., & Trinajistic, N., (1982). Computer-aided enumeration and generation of the Kekulé structures in conjugated hydrocarbons. *Computers & Chemistry, 6*(3), 121–132.

Fries, K., (1927). Über bicyclische Verbindungen und ihren Vergleich mit dem Naphtalin. III. Mitteilung. (Eng.: On bicyclic compounds and their comparison with the naphthalene. III. Communication). *Liebig Ann. Chem., 454*, 121–324.

Gutman, I., (1983). Topological properties of benzenoid systems. XXI. Theorems, conjectures, unsolved problems. *Croatica Chemica Acta, 56*, 365–374.

Hall, G. G., (1991). Aromaticity measured by Kekulé structures. *International Journal of Quantum Chemistry, 39*, 605–613.

Hansen, P., & Zheng, M., (1995). Assigning a Kekulé structure to a conjugated molecule. *Computers & Chemistry, 19*(1), 21–26.

Harary, E., (1969). *Graph Theory*. Addison-Wesley, Reading.

Herndon, W. C., & Parkanyi, C., (1976). π bond orders and bond lengths. *Journal of Chemical Education, 53*, 689–692.

Herndon, W. C., (1974). Resonance theory. VI. Bond orders. *Journal of the American Chemical Society, 96*, 7605–7614.

Iijima, S., (1991). Helical microtubules of graphitic carbon. *Nature, 354*, 56–58.

Kearsley, S. K., (1993). A quick robust method for assigning a Kekulé structure. *Computers & Chemistry, 17*, 1–10.

Kekulé Von Stradonitz, (Friedrich), (2008). *Complete Dictionary of Scientific Biography*. Encyclopedia.com, http://www.encyclopedia.com/doc/ 1G2-2830902274.html (3 September 2015).

Kirby, E. C., (1993). Cylindrical and toroidal polyhex structures. *Croatica Chemica Acta, 66*, 13–26.

Klein, D. J., & Randic, M. J., (1987). Innate degree of freedom of a graph. *Journal of Computational Chemistry, 8*, 516–521.

Klein, D. J., Seitz, W. A., & Schmalz, T. G., (1993). Symmetry of infinite tubular polymers: Application to buckytubes. *Journal of Physical Chemistry, 97*, 1231–1236.

Lukovits, I., Graovac, A., Kalman, E., Kaptay, G., Nagy, P., Nikolic, S., Sytchev, J., & Trinajstic, N., (2003). Nanotubes: Number of Kekulé structures and aromaticity. *Journal of Chemical Information and Computer Sciences, 43*, 609–614.

Randic, M., (1976). Conjugated circuits and resonance energies of benzenoid hydrocarbons. *Chemical Physics Letters, 38*, 68–70.

Randic, M., (1977). Aromaticity and conjugation. *Journal of the American Chemical Society, 99*, 444–450.

Randic, M., (2003). Aromaticity of polycyclic conjugated hydrocarbons. *Chem. Rev., 103*, 3449–3606.

Sachs, H., Hansen, P., & Zheng, M., (1996). Kekulé count in tubular hydrocarbons. *Communications in Mathematical and in Computer Chemistry (MATCH), 33*, 169–241.

Vukicevic, D., & Randic, M., (2005). On Kekulé structures of buckminsterfullerene. *Chemical Physics Letters, 401*, 446–450.

Vukicevic, D., & Zigert, P., (2008). Binary coding of algebraic Kekule structures of catacondensed benzenoid graphs. *Applied Mathematics Letters, 21*, 712–716.

Wizinger-Aust, R., Gillis, J. B., Helferich, B., & Wurster, C., (1966). *Kekulé und Seine Benzolformel* (Eng.: *Kekulé and his benzene formula.*). Verlag Chemie, Weinheim.

CHAPTER 22

Layered Double Hydroxide

OCTAVIAN D. PAVEL, ADRIANA URDĂ, and IOAN-CEZAR MARCU

University of Bucharest, Faculty of Chemistry, Department of Organic Chemistry, Biochemistry and Catalysis, 4-12 Regina Elisabeta Av., 030018, Bucharest, Romania, Tel: +4 021 3051464, Fax: +4 021 3159249, E-mail: ioancezar.marcu@chimie.unibuc.ro

22.1 DEFINITION

Layered double hydroxides (LDHs) compounds are anionic clays having the general formula $M^{2+}_{1-x}M^{3+}_x(OH)_2A^{n-}_{x/n}\cdot mH_2O$, where M^{2+} and M^{3+} are bivalent and trivalent cations, respectively, with ionic radii close to that of Mg^{2+}, A^{n-} are compensating anions and x varies in the range 0.2–0.4 which corresponds to a M^{2+}/M^{3+} molar ratio between 1.5 and 4. The parent material of this class is the naturally occurring mineral hydrotalcite with $M^{2+} = Mg^{2+}$, $M^{3+} = Al^{3+}$, $A^{n-} = CO_3^{2-}$, $x = 0.25$ and $m = 4$. This is why the LDHs are also known as hydrotalcite-like materials. The cations hexa-coordinated to hydroxyl groups form brucite-like sheets which stack to create a layered structure. The anions A^{n-} are placed in the inter-layer space in order to compensate the positive charge brought by the M^{3+} cations partially replacing M^{2+} cations in the layers. Water is also present in the interlayer space. Hydroxyl groups from the layers are oriented toward the interlayer space and can be hydrogen bonded to the interlayer anions and water molecules. More than two different cations can enter simultaneously the brucite-like sheets where they are homogeneously distributed and intimately mixed together. Also, a large variety of inorganic and organic anions can be intercalated between the brucite-like layers. Due to their structure, together with their compositional flexibility, the LDHs possess highly versatile physicochemical properties leading to various applications in catalysis, drug delivery, and environmental remediation.

22.2 HISTORICAL ORIGIN(S)

The LDHs belong to anionic clays, which themselves are part of phyllites family (from the Greek "phyllon" meaning leaf). Clay minerals and clays are very common compounds on the earth surface, so they constitute the main component of soils and sedimentary rocks (Del Hoyo, 2007).

The history of LDH materials begins with the discovery of the representative solid, namely the hydrotalcite, in Sweden around 1842 (Cavani et al., 1991). The stoichiometry of hydrotalcite, $[Mg_6Al_2(OH)_{16}]CO_3 \cdot 4H_2O$, was first correctly determined by Manasse in 1915 claiming that carbonate ions were essential for this type of structure. Meanwhile, many compounds having the LDH structure were synthesized by Feitknecht and Gerber (1942). They were named double sheet structures in which it was supposed that a layer of one cation's hydroxide is alternated with a layer of the second cation hydroxide. In the 1960s, the basal structure of LDHs was determined after X-ray diffraction studies on mineral samples carried out by Allmann (1968) and Taylor (1969) the confusion created by Feitknecht and Gerber being thus eliminated. They showed that two different cations are located in a layer while the anions with water molecules are located in the interlayer space. Utilization of these materials as catalysts started in 1924 when the Ni-Al-LDH-based system was shown to have good catalytic activity in the hydrogenation reaction (Zelinski & Kommarewsky, 1924). In 1970, the first patent was published claiming that the hydrotalcite-like compounds are very good precursors for the hydrogenation catalysts (Brocker & Kainer, 1970). The first review paper dealing with the synthesis of LDHs was published by Reichle (1986). Two years later, Carrado et al., (1988) published a short review paper dealing with their synthesis, characterization and potential uses. However, a key and, now, classical extended review paper dealing with both as-prepared and calcined LDH materials in terms of synthesis, physicochemical characterization and applications were published several years later by Cavani, Trifirò and Vaccari (1991). At present, there still exist several controversies in terms of possible compositions, stoichiometry, ordering of metal cations within the layers, the stacking arrangement of the layers, the arrangement of anions and placement of water molecules in the interlayer galleries (Evans & Slade, 2006).

22.3 NANO-SCIENTIFIC DEVELOPMENT(S)

The basic layer structure of the LDHs is based on that of brucite-like material, $Mg(OH)_2$, where magnesium ions are octahedrally surrounded by hydroxide ions. When in these octahedral units some of the Mg^{2+} cations are isomorphously substituted with trivalent cations a positive charge appears. These edge-sharing octahedral units form infinite layers with the hydroxide ions in a close-packed arrangement. The layers stack on top of each other forming a three-dimensional structure. The positive charge from the layers is compensated by anions placed in the interlayer space together with crystallization water (Figure 22.1). The anions and water molecules can freely move in the interlayer space by breaking their bonds and forming new ones. The oxygen atoms of the water molecules and CO_3^{2-} groups are distributed around the axes of symmetry. The density of the anionic charge is equal to that of trivalent cations.

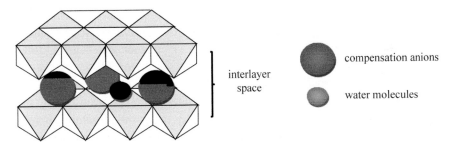

FIGURE 22.1 Schematic representation of LDH materials (adapted from Moustafa et al., 2015).

The brucite-like sheets can stack one on the other with two different symmetries: rhombohedral and hexagonal. The LDH specimens with rhombohedral symmetry have mainly been found in nature, while the hexagonal symmetry may be the high-temperature form of the rhombohedral one (Cavani et al., 1991). According to Lehn (1988), the LDH planes are "host species," and the interlayer anions and water molecules are "guest species."

Regarding the nature of the cations from the LDHs general formula $M_{1-x}^{2+}M_x^{3+}(OH)_2A_{x/n}^{n-}\cdot mH_2O$, only M^{2+} and M^{3+} ions with ionic radii close to that of Mg^{2+} can be accommodated in the holes of the close-packed arrangement of OH groups forming the brucite-like layers (Braterman et al., 2004; Cavani et al., 1991). Thus, M^{2+} are Mg^{2+}, Cu^{2+}, Ni^{2+}, Co^{2+}, Zn^{2+}, Fe^{2+} and Mn^{2+}, and M^{3+} are all the ions stable in air with atomic radii between 0.5 and 0.8 Å, such as Al^{3+}, Ga^{3+}, Fe^{3+}, Cr^{3+}, Co^{3+} etc. The peculiar behavior of Cu^{2+} that forms LDHs only in the presence of another bivalent cation can be attributed to its strong Jahn-Teller effect (Cavani et al., 1991). The optimum range for x is between 0.2 and 0.4. For values outside of this domain, the pure hydroxides or other compounds with different structures may be obtained. Basically, there is no limitation to the nature of the interlayered anions. The number, size, orientation, strength of the bonds between the anion and the hydroxyl groups determine the thickness of the interlayer region. LDHs with two different anions have been synthesized. In this case, there is an ordering of anions distribution in the interlayer region (Brindley and Kikkawa, 1979). Concerning the water molecules, these are located in the interlayer region in all the positions not occupied by anions.

Many combinations of divalent and trivalent cations can lead to LDHs but only one monovalent cation, Li^+, is able to form LDHs with $[LiAl_2(OH)_6]^+$ octahedral sheets because lithium has an ionic radius comparable to that of Mg^{2+}. With other monovalent cations, such as Na^+, K^+, NH_4^+, compounds of dawsonite-type will be formed (Serna et al., 1982). Recently, several papers reported the possibility of synthesizing LDHs containing M^{4+} ions including Zr^{4+} (Saber, 2007), Sn^{4+} (Velu et al., 1999), Ti^{4+} (Saber and Tagaya, 2003), and Si^{4+} (Saber, 2006).

LDHs can be prepared by several methods, such as co-precipitation (both at low and high supersaturation; involving separate nucleation and aging; urea hydrolysis), anionic exchange, rehydration by using the structural memory effect, hydrothermal, secondary intercalation (pre-pillaring method), and intercalation involving dissolution and re-coprecipitation procedures (Braterman et al., 2004; He et al., 2006). Other preparation methods less used are salt-oxide (or hydroxide), non-equilibrium aging, non-conventional aging, surface synthesis, templated synthesis, etc. (He et al., 2006). In all LDH obtaining methods the basic rule is to maintain the M^{2+}/M^{3+} molar ratio between 1.5 and 4 and $1/n < A^{n-}/M^{2+} < 1$ (Cavani et al., 1991). Preparation of hydrotalcite with another anion than CO_3^{2-} is difficult, because CO_2 is easily incorporated from the atmosphere.

They can be used as such or after thermal decomposition, in the form of mixed oxides. Mixed oxides are used due to their remarkable properties: high surface areas of about 200 m^2 g^{-1}, high basicity, but lower than that of MgO due to presence of trivalent element which usually shows a higher electronegativity, the possibility of forming a homogeneous mixture of oxides with very small crystal size, which by reducing leads to small and thermally stable metal crystallites, memory effect (Cavani et al., 1991).

One of the most interesting properties of the LDH materials is their *memory effect*. It consists in the ability of the materials obtained by the thermal decomposition of LDH precursors containing a volatile interlayered anion to reconstitute the original layered structure, either by adsorption of different anions or by exposure to air (Cavani et al., 1991; Tichit and

Coq, 2003). This behavior is closely linked to the heating temperature; at temperatures higher than 723-873 K, depending on the composition, a spinel phase appears and, consequently, the memory effect disappears (Tichit and Coq, 2003).

This class of materials can be well characterized by means of some physical and analytical techniques such as XRD, TG/DTA, IR, TEM, DSC, UV-VIS, NMR, etc. From all the techniques, the X-ray diffraction remains the main analytical procedure for their characterization.

22.4 NANOCHEMICAL APPLICATION(S)

These materials have found numerous applications in many areas such as catalysts, catalyst supports, adsorbents, in photochemistry, electrochemistry, industry, medicine, etc. (Cavani et al., 1991; Tichit and Coq, 2003; Li and Duan, 2006; Rives et al., 2014).

LDHs are important catalysts in organic reactions due to their basic character but also, for the transition metal-containing LDHs, due to their redox properties; they can be used in the as-synthesized LDH form or after different controlled thermal treatments (Braterman et al., 2004; Fan et al., 2014). In the mixed oxides, the weak base sites are associated with surface OH^- groups and the intermediate strength base sites are related to the oxygen bridging Mg^{2+} and Al^{3+}/M^{3+} cations, thus belonging both to Mg^{2+}–O^{2-} and Al^{3+}/M^{3+}–O^{2-} acid-base pairs. The higher electronegativity of Al^{3+} as compared to Mg^{2+} decreases the electron density and thus the nucleophilicity of the neighboring oxygen anions, and therefore favors the formation of medium strength basic sites in the LDH-derived mixed oxides (Bolognini et al., 2002). Strong Lewis basic sites can also exist in the LDH-derived mixed oxides due to the presence of low coordinated O^{2-} surface species. According to Sels et al., (2001), the LDHs can play as:

1. Redox catalysts in oxidation (aromatic compounds, alcohols, alkynes, etc.), reduction (nitro compounds, nitriles, alkynes and alkenes, phenol, hydrogen transfer reactions) and dehydrogenation reactions. The use of as-synthesized and calcined LDHs as redox catalysts in selective or total oxidation reactions was recently reviewed by Marcu et al. (2017).
2. Acid-base catalysts in addition on C = O bonds (aldol condensation and Claisen–Schmidt reactions, Knoevenagel reaction, Baeyer–Villiger oxidation), 1,4-addition on C = C and C≡C bonds (Michael

addition, Weitz–Scheffer epoxidation, cyanoethylation), addition on C≡N bonds (epoxidation, hydrolysis of nitriles), alkylation, acylation, reactions of epoxides (epoxide ring opening, polymerization of alkylene oxides), and decarboxylation.

3. Catalyst hosts in halide substitution reactions, oxidation with O_2 (intercalated and adsorbed organometallic complexes, polyoxometallate pillars), oxidation with single-oxygen atom donors (supported organometallic complexes, polyoxometallates, exchanged oxometallate anions); reduction (supported metal particles, supported organometallic complexes), polymerization, acid-catalyzed reactions, etc.

It is to be noted that due to their specific properties, LDHs are particularly good precursors for multifunctional catalysts possessing on their surface acid, base, and redox sites simultaneously, which are used in one-pot cascade reactions (Tichit et al., 2006).

As catalyst supports, they are ideal for many different active phases due to their textural properties. Their use as support begins in 1971 when LDHs containing different metal cations with carbonate as interlayer anions, calcined and partially or completely chlorinated, were reported to be effective as supports for Ziegler catalysts in the polymerization of olefins (Li and Duan, 2006). Nowadays, LDHs and their organic-derived products are supports for the immobilization of enzymes (Ren et al., 2002).

The LDHs are reliable materials as adsorbents and anion-exchangers due to the presence of anions in the interlayer region and to the thickness interlayer modification ability vs. dimension of guest anions (Figure 22.2). When the layer charge density is very high, the exchange reaction may become difficult. LDHs have greater affinities for multivalent anions compared to the monovalent anions (Miyata, 1983; Dong et al., 2014). These materials removed from solutions many organic or inorganic anionic species with very good results.

The first medical applications of LDHs materials, especially hydrotalcite, were as an antacid and anti-pepsin agents, these being an integral part even today in antacid medication. The driving force for this application is based on their anion exchange capacity. Recent studies have been focused on the intercalation and controlled release of pharmaceutically active compounds from LDH materials (Rojas et al., 2014). Different drugs were used as guest molecules such as diclofenac, naproxen, gemfibrozil, tolfenamic acid, etc. (Li and Duan, 2006). Notably, even DNA, genes, antibodies, vitamins, amino acids, proteins, etc. may be placed in the interlayer space (Rives et al., 2014; Kuthati et al., 2015).

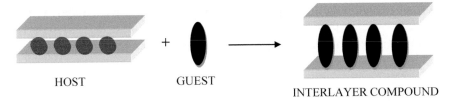

HOST   GUEST

INTERLAYER COMPOUND

FIGURE 22.2 The model of anionic exchange for LDHs (adapted from Hu et al., 2001).

A recent application of the LDHs materials is in electrochemistry due to their capability to exchange intercalated anions. The Ni-LDH can be used as an electrode for alkaline secondary cells (Jayashree and Vishnu Kamath, 2002). The LDH can be included even in nanocomposite polymer electrolytes (Liao and Ye, 2004).

LDHs compounds have been found useful as flame retardants in PVC and other polymers (Shanmuganathan and Ellison, 2014).

22.5 MULTI-/TRANS-DISCIPLINARY CONNECTION(S)

LDHs represent one of the most promising materials due to their low cost, relatively easy way of preparation and their compositional flexibility. From their use as catalysts and adsorbents, in industry or laboratories, to their use in medicine, the LDHs are everywhere. The LDHs present alongside memory effect even compression mouldability higher than any other inorganic material or microcrystalline cellulose, which is considered the representative material (Cavani et al., 1991). The LDHs are linked with culture because they find application as corrosion inhibitors in paints and coating compositions. They are also useful in the field of electronics, being used under sheet spacers for electrolytic capacitors. The study of LDHs and LDH-derived materials, from their synthesis to their applications passing through their physicochemical characterization, is highly multidisciplinary involving the cooperation of several scientific disciplines such as materials chemistry, physical chemistry, physics, engineering, etc.

22.6 OPEN ISSUES

Due to their basicity, memory effect, anionic exchange and other properties presented above, the use of LDHs materials is constantly increasing. The next challenge regarding the preparation of LDHs is for other cations with different valence than 2 and 3 to be involved, resulting in layered materials that benefit from the same properties listed above or showing new ones.

KEYWORDS

- **anionic clays**
- **hydrotalcite-like materials**
- **hydroxycarbonate layered solid**
- **LDHs**
- **solid base**

REFERENCES AND FURTHER READING

Allmann, R., (1968). The crystal structure of pyroaurite. *Acta Crystallographica Section B., 24,* 972–977.

Bolognini, M., Cavani, F., Scagliarini, D., Flego, C., Perego, C., & Saba, M., (2002). Heterogeneous basic catalysts as alternatives to homogeneous catalysts: Reactivity of Mg/Al mixed oxides in the alkylation of m-cresol with methanol. *Catalysis Today, 75,* 103–111.

Braterman, P. S., Xu, Z. P., & Yarberry, F., (2004). Layered double hydroxides (LDHs). In: Auerbach, S. M., Carrado, K. A., & Dutta, P. K., (eds.), *Handbook of Layered Materials* (pp. 373–474). Marcel Dekker: New York, USA.

Brindley, G. W., & Kikkawa, S., (1979). A crystal-chemical study of Mg. Al and Ni. N hydroxy-perchlorates and hydroxyl-carbonates. *American Mineralogist, 64,* 836–843.

Brocker, F. J., & Kainer, L., (1970). German Patent 2,024,282, to BASF AG, and UK Purenr 1,342,020 (1971), to BASF AG.

Carrado, K. A., Kostapapas, A., & Suib, S. L., (1988). Layered double hydroxides (LDHs). *Solid State Ionics, 26,* 77–86.

Cavani, F., Trifirò, F., & Vaccari, A., (1991). Hydrotalcite-type anionic clays: Preparation, properties, and applications. *Catalysis Today, 11,* 173–301.

Del Hoyo, C., (2007). Layered double hydroxides and human health: An overview. *Applied Clay Science, 36,* 103–121.

Dong, L., Ge, C., Qin, P., Chen, Y., & Xu, Q., (2014). Immobilization and catalytic properties of candida lipolytic lipase on the surface of organic intercalated and modified MgAl-LDHs. *Solid State Sciences, 31,* 8–15.

Evans, D. G., & Slade, R. C. T., (2006). Structural aspects of layered double hydroxides. In: Duan, X., & Evans, D. G., (eds.), *Layered Double Hydroxides* (Vol. 119, pp. 1–87). Springer-Verlag: Berlin, Germany.

Fan, G., Li, F., Evans, D. G., & Duan, X., (2014). Catalytic applications of layered double hydroxides: Recent advances and perspectives. *Chemical Society Reviews, 43,* 7040–7066.

Feitknecht, W., & Gerber, M., (1942). Zur kenntnis der doppelhydroxyde und basischen doppelsalze III. Über Magnesium-Aluminiumdoppelhydroxyd. *Helvetica Chimica Acta, 25*, 131–137.

He, J., Wei, M., Li, B., Kang, Y., Evans, D. G., & Duan, X., (2006). Preparation of layered double hydroxides. In: Duan, X., & Evans, D. G., (eds.), *Layered Double Hydroxides* (Vol. 119, pp. 89–119). Springer-Verlag: Berlin, Germany.

Hu, C., Li, D., Guo, Y., & Wang, E., (2001). Supermolecular layered double hydroxides. *Chinese Science Bulletin, 46*, 1061–1066.

Jayashree, R. S., & Vishnu, K. P., (2002). Layered double hydroxides of Ni with Cr and Mn as candidate electrode materials for alkaline secondary cells. *Journal of Power Sources, 107*, 120–124.

Kuthati, Y., Kankala, R. K., & Lee, C. H., (2015). Layered double hydroxide nanoparticles for biomedical applications: Current status and recent prospects. *Applied Clay Science, 112/113*, 100–116.

Lehn, J. M., (1988). Supramolecular chemistry - scope and perspectives molecules, supermolecules, and molecular devices (Nobel Lecture). *Angewandte Chemie International Edition in English, 27*, 89–112.

Li, F., & Duan, X., (2006). Applications of layered double hydroxides. In: Duan, X., & Evans, D. G., (eds.), *Layered Double Hydroxides* (Vol. 119, pp. 193–223). Springer-Verlag: Berlin, Germany.

Liao, C. S., & Ye, W. B., (2004). Structure and conductive properties of poly(ethylene oxide)/ layered double hydroxide nanocomposite polymer electrolytes. *Electrochimica Acta, 49*, 4993–4998.

Manasse, E., (1915). Idrotalcite e piroaurite. *Atti della Società Toscana di Scienze Naturali, Processi Verbali, 24*, 92–105.

Marcu, I. C., Urdă, A., Popescu, I., & Hulea, V., (2017). Layered double hydroxides-based materials as oxidation catalysts. In: Putz, M. V., & Mirica, M. C., (eds.), *Sustainable Nanosystems Development, Properties and Applications* (pp. 59–121). IGI Global: Hershey, PA, USA.

Miyata, S., (1983). Anion-exchange properties of hydrotalcite-like compounds. *Clays and Clay Minerals, 31*, 305–311.

Moustafa, E. A., Noah, A., Beshay, K., Sultan, L., Essam, M., & Nouh, O., (2015). Investigating the effect of various nanomaterials on the wettability of sandstone reservoir. *World Journal of Engineering and Technology, 3*, 116–126.

Reichle, W. T., (1986). Synthesis of anionic clay minerals (mixed metal hydroxides, hydrotalcite). *Solid States Ionics, 22*, 135–141.

Ren, L., He, J., Zhang, S., Evans, D. G., & Duan, X., (2002). Immobilization of penicillin G acylase in layered double hydroxides pillared by glutamate ions. *Journal of Molecular Catalysis B: Enzymatic, 18*, 3–11.

Rives, V., Del Arco, M., & Martín, C., (2014). Intercalation of drugs in layered double hydroxides and their controlled release: A review. *Applied Clay Science, 88/89*, 239–269.

Rojas, R., Jimenez-Kairuz, A. F., Manzo, R. H., & Giacomelli, C. E., (2014). Release kinetics from LDH-drug hybrids: Effect of layers stacking and drug solubility and polarity. *Colloids and Surfaces A: Physicochemical and Engineering Aspects, 463*, 37–43.

Saber, O., & Tagaya, H., (2003). New layered double hydroxide, Zn-Ti LDH: preparation and intercalation reactions. *Journal of Inclusion Phenomena and Macrocyclic Chemistry, 45*, 109–116.

Saber, O., (2006). Preparation and characterization of a new layered double hydroxide, Co-Zr-Si. *Journal of Colloid and Interface Science, 297*, 182–189.

Saber, O., (2007). Preparation and characterization of a new nano layered material, Co-Zr LDH. *Journal of Materials Science, 42*, 9905–9912.

Sels, B. F., De Vos, D. E., & Jacobs, P. A., (2001). Hydrotalcite-like anionic clays in catalytic organic reactions. *Catalysis Reviews, 43*, 443–488.

Serna, C. J., Rendon, J. L., & Iglesias, J. E., (1982). Crystal-chemical study of layered $[Al_2Li(OH)_6]^+X \cdot nH_2O$. *Clays and Clay Minerals, 30*, 180–184.

Shanmuganathan, K., & Ellison, C. J., (2014). Layered double hydroxides: An emerging class of flame retardants. In: Papaspyrides, C. D., & Kiliaris, P., (eds.), *Polymer Green Flame Retardants* (pp. 675–707). Elsevier: Amsterdam, The Netherlands.

Taylor, H. F. W., (1969). Segregation and cation-ordering in sjögrenite and pyroaurite. *Mineralogical Magazine, 37*, 338–342.

Tichit, D., & Coq, B., (2003). Catalysis by hydrotalcites and related materials. *CATTECH, 7*, 206–217.

Tichit, D., Gérardin, C., Durand, R., & Coq, B., (2006). Layered double hydroxides: Precursors for multifunctional catalysts. *Topics in Catalysis, 39*, 89–96.

Velu, S., Suzuki, K., Osaki, T., Ohashi, F., & Tomura, S., (1999). Synthesis of new Sn incorporated layered double hydroxides and their evolution to mixed oxides. *Materials Research Bulletin, 34*, 1707–1717.

Zelinsky, N., & Kommarewsky, W., (1924). Über die katalytischen Wirkungen des nickelierten Tonerde-Hydrats. *Berichte der Deutschen Chemischen Gesellschaft (A and B Series), 57*, 667–669.

CHAPTER 23

Living Active Centers (LACs)

A. M. KAPLAN and N. I. CHEKUNAEV

N.N. Semenov's Institute of Chemical Physics, Russian Academy of Sciences, Kosygin Street 4, Moscow 119991, Russian Federation, E-mail: amkaplan@mail.ru

23.1 DEFINITION OF LACS CONCEPT

The classic active center (A) (an ion, a radical, or an ion-radical) which is localized in "friable" zone of solids was named by authors as the *living active center* (LAC). It is known that the friable zones with the scale of nanosize exist near the structural defects (D) with the excess of a free volume (vacancies, dislocations, and microcracks in crystals and nanocavities in glasses). When approaching by a small distance (3–3.5 nm) the classical active center (A) and defect (D) the peculiar *complex* (C)–the *complex* (A...D) arises. Activation of the *chain solid-state reaction* (ChSR) occurs just in *friable zones* (FZs) of *nanoparticles of the complex* (C) (NPCs). Almost orientationally immobile and therefore chemically inert in dense zones of solids classic active centers (A) turn into LACs, when the centers (A) fall into the FZ of the NPC. The increasing of orientational mobility of reactants in NPC zones provides removal of the steric obstacles to the chemical interaction of LACs with neighboring particles of reagents. Activated in this manner, LACs provide realization of ChSRs.

23.2 HISTORICAL ORIGIN AND SCIENTIFIC DEVELOPMENT OF LACS CONCEPT

The introduction into the chemical science of a new LAC concept has been dictated by the need of the adequate description of the nontrivial (according to traditional notions) kinetic peculiarities of chain reactions in solids.

The LAC concept was proposed by Anatoly Kaplan in 1969. It should be noted the difference of sense of two close by the title of chemical concepts. The first one is the known concept in polymer chemistry "living polymerization" refers to the description of some process. The second concept (author's LAC concept) refers to a certain nanoparticle. The credibility to the author's concept was first confirmed by the results of experimental study of solid state polymerization (SP) of acrylonitrile in a regime of the samples heating with stops (Kaplan et al., 1969). Subsequently, the author's concept has been repeatedly confirmed in experimental studies of polymerization in crystals (e.g., Chachaty & Forchioni, 1972), in the glasses (e.g., Gerasimov Henry et al., 1973) and in studies of other ChSRs (e.g., Barkalov Igor et al., 1980).

In the development of new applications of the LAC concept for the analysis of peculiar nontrivial kinetic features of ChSRs, Anatoly Kaplan in 1989 has noted one additional important active agent in such processes. This is a mobile structural defect (MD), which is capable of transmuting orientationally motionless chemically inert active center A in the LAC (Kaplan, 1996).

Authors' further development of the LAC concept application for analyzing of ChSRs was compiling of the complete system of differential equations describing of the change of all active agents in such processes. To describe the ChSR kinetics, it is necessary to add to the traditional system of equations describing the chain reactions in the gas or in liquid phase two new differential equations. These are the equations describing variation with time of the concentration of the two most important active agents in the ChSR (LACs and MDs). Numerical solution to the complete system of equations with the help of modern computer technology has allowed analyzing the majority of nontrivial kinetic features of the SP already at the quantitative level (Kaplan & Chekunaev, 2012).

23.3   NANOCHEMICAL APPLICATION OF LACS CONCEPT

Because the LAC is the important nanoobject, determining the main features of the ChSR, the LAC concept was the basis for the authors' original kinetic model of the chain reactions development in solids

23.3.1 MAIN PROVISIONS OF THE KINETIC MODEL

The main provisions of the kinetic model of the chain reactions development in solids. Along with the LAC important object of the discussed model is a "friable" zone of NPC. In the discussed model as a structural defect involved in the formation of NPC was taken vacancy (Vac), because the Vac concentration is substantially higher than the concentrations of other structural defects in solids.

In accordance with the data (Rabinovich, 1968) diameter (d) of the "friable" "zone of NPC is equal $d = 2r \approx (5–6)\lambda$ (where $\lambda = 0.5–0.6$ nm is intermolecular distance). The rate constant for the interaction of LAC with neighboring reagent particles in the "friable" zone of solids is taken equal to the rate constant of the bimolecular chemical process: $K_x = K_0 \cdot \exp(-E_x/RT)$. Here, K_0 and E_x are, respectively, pre-exponential factor and the activation energy of a chemical act of interaction of indicated reagents. The development of ChSRs is performed in a large number of "friable" zones of NPCs in the systems studied. Classic active center, which left the "friable" zone of the NPC becomes orientationally immobile and chemically inert. If this takes place the chain process is stopped. However, the rapprochement with the active center of the mobile structural defect (MD) can revitalize such center. This will provide the continuation of the investigated process (Kaplan & Chekunaev, 2012).

The simplest non-trivial kinetic features of ChSRs will be considered in this article below on the example of the SP.

23.3.2 THE ROLE OF LAC IN GROWTH OF POLYMER CHAINS IN THE SP

Figure 23.1 shows a section of a sphere of radius $r = (2.5–3)\ \lambda$. The area inside the sphere occupies a "friable" zone of NPC. The growth of the individual polymer chain in the "friable" zone of the NPC carried by the mechanism of "chemical diffusion." According to this mechanism, the image of the polymer chain part formed in the NPC zone is a broken line in three-dimensional space. For clarity, Figure 23.1 shows only the monomer molecules, incorporated into the growing polymer chain, which propagate in two-dimensional space (in a plane).

The mean number N of new links of the polymer chain, which appeared in the NPC zone were determined using the Einstein-Smoluchowsky relation (E-SR) for diffusing particles $L^2 = \lambda^2 N$. Here $L = 2^{0.5}r = 2.5\lambda \cdot 2^{0.5}$ is the average and most probable length of a straight line between points A^*_{n+1} and A^*_{n+k}

(see Figure 23.1) on the surface of the sphere (with radius r) surrounding the NPC zone. By use of E-SR formula was calculated the number of links of the polymer chain which are formed in one NPC zone:

$$N = L^2/\lambda_p^2 = (L^2/\lambda_m^2) \cdot (\rho_p/\rho_m)^2 \approx (13\text{–}18) \cdot (1.2\text{–}1.3)^2 \approx (20\text{–}30).$$

Here are shown ρ_m and λ_m – monomer density and the distance between the monomer molecules in the solid state; ρ_p и λ_p – the same characteristics for the polymer.

The mechanism of "chemical diffusion" of active center A^*_{n+m} (m varies from 1 to $k + 1$) disposed at the end of the growing polymer chain in a zone of the NPC (a sphere at Figure 23.1), leads after some time to moving away of the active center A_{n+k+1} from the center of vacancy to a distance greater than r. The active center A_{n+k+1} becomes chemically inert ("congeals"), and chain process stops. However, the rapprochement of a mobile vacancy with this active center can reanimate it. The presence in the system of sufficient numbers of mobile vacancies causes the possibility of multiple returning of congealed polymer chain to its active growth.

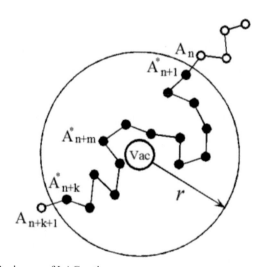

FIGURE 23.1   An image of LAC action.

Thus, use of the *LAC concept allowed predicts the previously unknown phenomenon of cyclic (with stops) growth of individual polymer chains in the solid state polymerization.*

In conclusion, we note that the successful application of the LAC conception for the explanation of nontrivial kinetic features of the solid state polymerization is noted in an extensive review of the radiation polymerization (Abram et al., 1973), in a monograph (Mark & Sergei, 1990) and in a textbook for universities (Evgeniy, 1978).

KEYWORDS

- classic active centers (ions, radicals, ion-radicals)
- congealing and reanimation of chains
- living active centers
- mobile and immobile nanodefects (vacancies, dislocations, microcracks)
- solid-phase chain reactions

REFERENCES

Abkin, A. D., Scheinker, A. P., & Gerasimov, G. N., (1973). Chapter 1, Radiation polymerization. In: Kargin, V. A., (ed.), *Radiation Chemistry of Polymers* (pp. 7–107). Publishing house "Science." Moscow.

Barkalov, I. M., Goldanskii, V. I., Kiryukhin, D. P., & Zanin, A. M., (1980). Kinetics and mechanism of the low- temperature chain hydrobromination of ethylene. *Chem. Phys. Lett., 73*(2), p. 273.

Brooke, M. A., & Pavlov, S. A., (1990). *Polymerization on the Surface of Solids* (p. 183). Publisher "Chemistry." Moscow. Soviet Union.

Chachaty, C., & Forchioni, A., (1972). NMR and ESR studies of the solid state polymerization of vinyl monomers. *J. Polymer Sci., Part A–1, 10,* p. 1905.

Denisov, E. T., (1978). *Kinetics of Homogeneous Chemical Reactions* (p. 367). Publisher "High School." Moscow. Soviet Union.

Gerasimov, G. N., Bespyatkina, T. A., et al., (1973). *Peculiarities of Growth and Termination of Chains in Polymerization in Glassy Matrices* (Vol. 209, p. 628). Dokl. Academy of Sciences of the USSR.

Kaplan, A. M., & Chekunaev, N. I., (2012). About the role played by mobile nanovoids in the stimulation of solid-state processes. *Russian Journal of Physical Chemistry B, 6*(3), pp. 407–415. © Pleiades Publishing, Ltd.

Kaplan, A. M., (1996). Thesis for the degree of doctor of chemical sciences. "Peculiarities of radiation and mechanical- stimulated processes in nonequilibrium condensed systems." *Inst. Chem. Phys. RAS, Moscow, 14.03.*

Kaplan, A. M., Kiryuhin, D. P., Barkalov, I. M., & Goldanskii, V. I., (1969). About the phenomenon of "congealing" and "reanimation" of polymer chains during post-polymerization of low-temperature polymorph modification of solid acrylonitrile. *Vysokomolek. Compound, 11, Brief Communications, 9*, p. 639.

Rabinovich, M. K., (1968). *Strength, Temperature, Time* (p. 160). Publisher "Nauka." Moscow. Soviet Union.

CHAPTER 24

Mathematical Chemistry

LOUIS V. QUINTAS and EDGAR G. DUCASSE

Mathematics Department, Pace University, New York, NY 10038, USA,
E-mail: lvquintas@gmail.com, educasse@pace.edu

24.1 DEFINITION

The application of any mathematical technique to the study of a chemistry topic.

24.2 ORIGINS AND HISTORICAL COMMENTS

The formal association of mathematics with chemistry has its origin in the 19th century with the work of the following chemists and mathematicians. The mathematician Arthur Cayley (1821–1895), with his work concerning the enumeration of the saturated hydrocarbon isomers, was the first to investigate this association (Cayley, 1857, 1874). Although not initially thought of as pioneers in this field, Archibald Scott Couper (1831–1892) and Friedrich August Kekulé (1829–1896), by representing a covalent bond with a line between a pair of atoms unwittingly gave birth, along with Cayley, to the application of graph theory to chemistry see p. 3 of Balaban (1976). In Helm (1894), Georg Helm made formal the association of mathematics and chemistry by the publication of a book on this topic. The linking of the words mathematics and chemistry became more frequent throughout the 20th century culminating in the establishment of a number of research journals on mathematical chemistry (MATCH, 1975; J. Math. Chem, 1987; Canadian J. Chemistry, 1951). Interest in this area is sufficiently high to sustain a number of additional journals (Int. J. Chem. Modeling, 2008; J. Molecular Modeling, 1995; J. Molecular Graphics and Modeling, 1998). This mathematical connection was further enhanced by the application to the chemistry of specialized areas of mathematics such as topology, computer science, probability/statistics, and graph theory.

The establishment of journals in mathematical chemistry was soon followed by conferences on this topic. One of the earliest conferences in Mathematics, Chemistry, and Computing was (MATH/CHEM/COMP, 1986) which originated in Dubrovnik, Croatia. Proceedings and books specializing in mathematical chemistry are plentiful, namely, (Balaban, 1976; Todeschini & Consonni, 2009; Bonchev & Rouvray, 2000; Trinajstić, 1992; Gutman & Polansky, 1986; King, 1983; Graovac et al., 1977; Kennedy & Quintas, 1988; King & Rouvray, 1987; J. Chem. Education, 1924). An organization with this specific interest, the International Academy of Mathematical Chemistry was founded to focus on and support the view that applying mathematical and computational methods to chemistry was and is a valuable area of research. Rather than a group of specialists in a specific field of science, it is a group of specialists from many fields joining together to apply their specialty to problems in chemistry. This acknowledges that this activity indeed constitutes a field called mathematical chemistry.

24.2.1   SPECIALIZED AREAS

(a)   Topology

By a topological application to chemistry, it is meant a topology application that uses methods that involve more than connectivity properties. The latter is better thought of as graph theory applications. Examples that illustrate this are found in stereochemistry which deals with a viewing of chemical structures in three or greater dimensional space. The authors (Walber, 1983; Dugundji, 1983; Mezey, 1983) in three representative papers found in the collection (King, 1983) one can find the following examples, respectively. A discussion of knotted cycles, consisting of the same atoms, and twisted ladders (2 x *n* meshes) also consisting of the same atoms are found in Walber (1983). The twisted ladders when twisted an even number of times correspond to cylinders and correspond to Möbius ladders when twisted an odd number of times. In both the knots and twisted ladders, their distinctions can be detected in 3-spaces in their embedding and relation to each other rather than in their connectivity. If similar structures or indeed other structures now consist of different atoms (corresponding to vertex labeled graphs) then, in addition to embedding questions, symmetry properties come into play involving automorphism groups of labeled graphs (Dugundji, 1983).

The third example, Mezey (1983) involving topology is that of work with open sets (fundamental in point-set topology) as it entities. This is consistent with the basic structures of quantum mechanics, where quantum mechanical particles, electrons, nuclei, and molecules are represented by wave packets and probability distributions. The latter can be described by open sets rather than by the classical concept of nuclear geometry of being represented by a single point in n-dimensional Euclidean space. The open set representation is analyzed in the setting of differentiable manifolds. Additional discussion of Möbius molecules in organic chemistry can be found in Day et al., (1983). Other embedding problems associate graphs with polymer molecules and take into consideration the notion of excluded volume, essentially, how much space does a molecule take in 3-space (Klein & Seitz, 1983). Additional examples of topological applications in chemistry can be found in King (1983) and King & Rouvray (1987).

(b) Computer Science

Initially, mathematical formulas were derived to count various things, such as isomers and other structures, and computers replaced adding machines to get these numbers, otherwise very tediously obtained. The drive to get computers faster, cheaper, and to do many other tasks than just number crunching has been productive, but limitations have led to deeper research in nanotechnology (Markoff, 2015). Obviously, many fields are involved in this research, including chemistry and consequently, mathematical chemistry (Jensen, 1980, 2002, 1985, 2006; Kennedy, 1983).

(c) Probability and Statistics

As research evolved to the molecular and then to the atomic level, probabilistic methods have played a significant role. Molecular models with bonds forming and breaking at random have led from chemical graphs to chemical random graphs to one might say, quantum chemical random graphs, the latter dealing with chemical structures below the molecular level. The evolution of some of these concepts is seen in Refs. (Gordon, 1962; Good, 1963; Gordon & Kennedy, 1973; Gordon & Ross-Murphy, 1975; Kennedy, 1981; Erdős & Rényi, 1960; Kennedy, 1983; Kennedy & Quintas, 1989; Balińska & Quintas, 2006).

(d)　Graph Theory

One of the earliest applications of mathematics to chemistry was that of graph theory. The literature on this is extensive (Balaban, 1976; Trinajstić, 1992; King, 1983; King and Rouvray, 1987; Randić, 1992). Just one of many applications of graph theory is that of topological indices (graph invariants and other mathematical invariants of chemical models). Credit for the first index reflecting the topological structure of a chemical graph is attributed to Wiener (1947). This index, called the Wiener number or the path number, is graph-theoretically the number of all distances between all pairs of distinct vertices of a graph (Gutman et al., 1990). Two of the best known topological indices are the Hosoya Index, also known as the Z index, introduced in Hosoya (1971, 2002), and the Randić Index, also known as the *connectivity index*, introduced in Randić (1975). Their definitions for a chemical graph G are most concisely given in graph theory language. Namely, the Hosoya Index is the number of matchings in G, where matching is a set of edges in G that do not share a vertex. Note the empty set of edges is counted as a matching. For the second index, for each edge $\{i, j\}$ in G, let $r\{i, j\}$ be the reciprocal of the square root of the product of the vertex degrees (number of edges incident to a vertex) of the vertices i and j. Then, the Randić Index is the sum of the $r\{i, j\}$ over all edges in G. These indices have been used to correlate chemical graphs with their associated physical properties. For example, strong correlations of these indices have been found for the boiling points and the branching of molecules. Similar correlations exist for more complex physical properties (Gutman et al., 1990). Many other topological indices have been defined and put into the context of applications in Rouvray (1983).

(e)　Biology

Nanochemical applications in biology are numerous with various disciplines involved. A cascade theoretical model for an antigen-antibody system has been explored in Refs. (Gross et al., 1988; Good, 1963; Kennedy, 1983). An elementary investigation concerning water can be found in (Quintas, 1983) and an in-depth study of liquids and solutions is found in Murrell & Boucher (1982). Graph invariants based on neighborhood symmetry has been used to analyze the inhibition of alcohols and the toxicity of barbiturates (Magnuson et al., 1983).

(f)   Pharmaceutical Applications

Correlating physical properties with the structure of pharmaceutical compounds has been of interest for some time. A commercial objective here is to find compounds with properties close to compounds protected by patents. If the properties of a discovered compound are close enough, these compounds are offered as alternatives to the protected compounds. This objective is closely intertwined with academic research. The works of Basak & Grunwald (1995) and Basak et al., (1996) are representative of this area.

(g)   Molecular Modeling

Almost all mathematical chemistry applications are done in the context of a molecular model. The simplest model is the graph model for a molecule. Another example is the nano-chemical approach where the modeling involves the electrons of the atoms in the graph model (Mezey, 1983). Additional examples can be found when the chemical structure is embedded in some topological or algebraic setting. If the surface of a sphere is subdivided into polygonal regions, the result is a polyhedron. In a chemical context, if the vertices of a polyhedron can be identified with the atoms of a molecule, one obtains a model called a polyhedral molecule (King, 1983). Algebraic models come up when one studies the symmetries of a molecular structure. The case of an asymmetric carbon, that is, a carbon atom that is bonded to four chemically distinct branches (ligands) is an extreme example. Permutation groups of labeled graphs, with the labeled parts corresponding to distinct chemical entities, make the labeled graph a model in an algebraic setting (Dugundji, 1983). For more details concerning molecular modeling see Leach (2001, 2008).

(h)   Molecular Transformations in the Laboratory

A fundamental part of chemistry research is the transformation of one compound into another compound. In this experimental science, there is much mathematics both in the engineering and in the analysis. Additionally, one finds mathematical chemistry applications in associated areas. For example, in the transforming of methane to higher order alkanes, it is of theoretical interest to view these transformations in models involving the entire saturated hydrocarbon network (DuCasse & Quintas, 2011; Cha et al., 2015).

The above are just some examples of topics that fall under the heading of mathematical chemistry, namely, the application of any mathematical technique to the study of a chemistry topic.

24.3.1   OPEN ISSUES

Since mathematical chemistry is an area of very active research, there are many open problems which can be located in each of the above-cited areas. No attempt will be made here to present such a list. It is of interest to note that enumeration problems, the earliest of mathematical chemistry questions, is still an active area of research. This is so because new and complex structures continue to be introduced. Thus the problem of how many of these structures exist is an ongoing challenge. For any chemically associated graph, there are a plethora of mathematical questions that can be asked in addition to enumeration questions (Cha et al., in Press).

KEYWORDS

- **mathematical chemistry**
- **molecular modeling**
- **pharmaceutical applications**
- **topology**
- **mathematical chemistry**

REFERENCES

Balaban, A. T., (1976). *Chemical Applications of Graph Theory* (pp. 1–389). Academic Press, New York.

Balaban, A. T., (2005). Reflections about Mathematical Chemistry. *Foundations of Chemistry, 7*, 289–306.

Balińska, K. T., & Quintas, L. V., (2006). *Random Graphs with Bounded Degree* (1–291). Publishing House of the Poznań University of Technology, Poznań.

Basak, S. C., & Grunwald, G. D., (1995). Predicting mutagenicity of chemicals using topological and quantum chemical parameters: A similarity-based study. *Chemosphere, 31*, 2529–2546.

Basak, S. C., Gute, B. D., & Drewes, L. R., (1996). Predicting blood-brain transport of drugs: A computational approach. *Pharmaceutical Research, 13*, 775–778.

Bonchev, D., & Rouvray, D. H., (2000). *Mathematical Chemistry Series.* Gordon and Breach, Amsterdam.

Canadian J. Chemistry. (1951). NRC Research Press, Established.

Cayley, A., (1857). On the theory of the analytical forms called trees. *Philos. Mag., 13*, 19–30, also in (1891), *Mathematical Papers* (Vol. 3, pp. 242–246), Cambridge.

Cayley, A., (1874). On the mathematical theory of isomers. *Philos. Mag., 67*, 444–446, also in (1895), *Mathematical Papers* (Vol. 9, pp. 202–204, 427–460). Cambridge.

Cha, S. H., DuCasse, E. G., Quintas, L. V., Kravette, K., & Mendoza, D. M., (2015). A graph model related to chemistry: Conversion of methane to higher alkanes. *International J. Chemical Modeling, 7*(1), 17–29.

Cha, S. H., DuCasse, E. G., Quintas, L. V., Kravette, K., & Shor, J. P., (In Press). Graphs whose vertices are forests with bounded degree: Open problems. *International J. Graph Theory and its Applications. Computational and Theoretical Chemistry*, Elsevier, Amsterdam, established (1985).

DuCasse, E. G., & Quintas, L. V., (2011). A graph model related to chemistry. *International J. Chemical Modeling, 3*(4), 449–454. Also appears in the book: *Advances in Chemical Modeling,* (2012). *3.* Nova Science Publications, Inc.

Dugundji, J., (1983). Qualitative stereochemistry. In: *Chemical Applications of Topology and Graph Theory, Studies in Physical and Theoretical Chemistry, 28,* 33–39. Elsevier, Amsterdam.

Erdős, P., & Rényi, A., (1960). On the evolution of random graphs. *Publ. Math. Inst. Hungarian Acad. Sci., 5*, 17–61.

Good, I. J., (1963). Cascade theory and the molecular weight averages of the sol fraction, *Proc. Roy. Soc. A., 272*, 54–59.

Gordon, M., & Kennedy, J. W., (1973). The graph-like state of matter: LCGI Scheme for the thermodynamics of alkanes and the theory of inductive inference. *J. C. S. Faraday II, 69*, 484–504.

Gordon, M., & Ross-Murphy, S. B., (1975). The structure and properties of molecular trees and networks. *Pure and Applied Chem., 43*, 1–26.

Gordon, M., (1962). Good's theory of cascade processes applied to the statistics of polymer distributions. *Proc. Roy. Soc. A., 268*, 240–259.

Graovac, A., Gutman, I., & Trinajstić, N., (1977). *Topological Approach to the Chemistry of Conjugated Molecules.* Lecture notes in chemistry No. 4, ISBN 978-3-540-08431, Springer-Verlag, Berlin.

Gross, R., Kennedy, J. W., Quintas, L. V., & Yarmush, M. L., (1988). Antigenesis: A cascade-theoretical analysis of the size distribution of antigen-antibody complexes. *Discrete Applied Math., 19*, 177–194.

Gutman, I., & Polansky, O. E., (1986). *Mathematical Concepts in Organic Chemistry.* ISBN 978-3-642-70984, Springer-Verlag, Berlin.

Gutman, I., Kennedy, J. W., & Quintas, L. V., (1990). Wiener numbers of random benzenoid chains. *Chemical Physics Letters, 173*(4), 403–408.

Helm, G., (1894). *The Principles of Mathematical Chemistry: The Energetics of Chemical Phenomena.* Translated by J. Livingston & R. Morgan, John Wiley & Sons, New York.

Hosoya, H., (1971). Topological index. A newly proposed quantity characterizing the topological nature of structural isomers of saturated hydrocarbons. *Bulletin Chemical Society of Japan, 44*(9), 2332.

Hosoya, H., (2002). The topological index Z before and after 1971. *Internet J. Molecular Design, 1*(9), 428–442.

IAMC, (2005). http://www.iamc-online.org/.

International J. Chemical Modeling, Nova Science Publishers, Hauppauge, NY, established (2008).

J. Chemical Education. American Chemical Society, established (1924).

J. Computational Chemistry. John Wiley & Sons, New York, established (1980).

J. Mathematical Chemistry. Springer, established (1987).

J. Molecular Graphics and Modeling. Elsevier, Amsterdam, established (1998).

J. Molecular Modeling. Springer, established (1995).

J. Theoretical and Computational Chemistry. World Scientific, established (2002).

Jensen, F., (2006). *Introduction to Computational Chemistry*. John Wiley & Sons, New York.

Kennedy, J. W., & Quintas, L. V., (1988). Applications of graphs in chemistry and physics. *Discrete Applied Mathematics, 19*, 1–416.

Kennedy, J. W., & Quintas, L. V., (1989). Probability models for random f-graphs, combinatorial mathematics (New York, 1985). *Ann. N.Y. Acad. Sci., 555*, 248–261.

Kennedy, J. W., (1981). Statistical mechanics and large random graphs. In: Hippe, Z., (ed.), *Data Processing in Chemistry* (pp. 113–132). Elsevier, Amsterdam.

Kennedy, J. W., (1983). The random-graph-like state of matter. *Computer Applications in Chemistry* (pp. 151–178). Elsevier, Amsterdam.

King, R. B., & Rouvray, D. H., (1987). *Graph Theory and Topology in Chemistry, Studies in Physical and Theoretical Chemistry* (Vol. 51, pp. 1–575). Elsevier, Amsterdam.

King, R. B., (1983). Chemical applications of topology and graph theory. *Studies in Physical and Theoretical Chemistry* (Vol. 28, pp. 1–494). Elsevier, Amsterdam.

King, R. B., (1983). The bonding topology of polyhedral molecules. *Chemical Applications of Topology and Graph Theory, Studies in Physical and Theoretical Chemistry* (Vol. 28, pp. 99–123). Elsevier, Amsterdam.

Klein, D. J., & Seitz, W. A., (1983). Graphs, polymer models, excluded volume, and chemical reality, *Chemical Applications of Topology and Graph Theory, Studies in Physical and Theoretical Chemistry* (Vol. 28, pp. 430–445). Elsevier, Amsterdam.

Leach, A. R., (2001). *Molecular Modeling: Principles and Applications* (2nd edn., pp. 1–773). Pearson Education EMA, Europe, the Middle East, and Africa.

Magnuson, V. R., Harris, D. K., & Basak, S. C., (1983). Topological indices based on neighborhood symmetry: Chemical and biological applications. *Chemical Applications of Topology and Graph Theory, Studies in Physical and Theoretical Chemistry* (Vol. 28, pp. 178–191). Elsevier, Amsterdam.

Markoff, J., (2015). Smaller, faster, cheaper, over. In: *New York Times*, B1.

MATCH, (1975). *Communications in Mathematical and Computer Chemistry*, Founded by Polansky, O. E. MAT/CHEM/COMP, http://mcc.irb.hr/.

Mezey, P. G., (1983). Reaction topology: Manifold theory of potential surfaces and quantum chemical synthesis design, *Chemical Applications of Topology and Graph Theory, Studies in Physical and Theoretical Chemistry* (Vol. 28, pp. 75–98). Elsevier, Amsterdam.

Murrell, J. N., & Boucher, E. A., (1982). *Properties of Liquids and Solutions*. John Wiley and Sons Limited, Chichester.

Quintas, L. V., (1983). A volume function for water based on a random lattice-subgraph model. *Chemical Applications of Topology and Graph Theory, Studies in Physical and Theoretical Chemistry* (Vol. 28, pp. 446–453). Elsevier, Amsterdam.

Randić, M., (1975). Characterization of molecular branching. *J. American Chemical Society, 97*(23), 6609–6615.

Randić, M., (1975). Characterization of molecular branching. *J. American Chemical Society, 97*(23), 6609–6615.

Randić, M., (2003). Chemical graph theory-Facts and fiction. *Indian J. Chemistry, 42A,* 1207–1218.

Restrepo, G., & Villaveces, J. L., (2012). Mathematical thinking in chemistry, *HYLE, 18,* 3–22.

Rouvray, D. H., (1983). Shall we have designs on topological indices? *Chemical Applications of Topology and Graph Theory, Studies in Physical and Theoretical Chemistry* (Vol. 28, pp. 159–177). Elsevier, Amsterdam.

Todeschini, R., & Consonni, V., (2009). *Molecular Descriptors for Chemoinformatics.* John Wiley & Sons, New York.

Trinajstić, N., & Gutman, I., (2002). Mathematical chemistry. *Croatica Chemica Acta, 75,* 329–356.

Trinajstić, N., (1992). *Chemical Graph Theory.* CRC Press, Boca Raton, Fl.

Walber, D. M., (1983). Stereochemical topology. *Chemical Applications of Topology and Graph Theory, Studies in Physical and Theoretical Chemistry* (Vol. 28, pp. 17–32). Elsevier, Amsterdam.

Wiener, H., (1947). Influence of interatomic forces on paraffin properties. *J. Chem. Physics, 15,* 766.

CHAPTER 25

Mechano(-)Synthesis

LORENTZ JÄNTSCHI[1] and SORANA D. BOLBOACĂ[2]

[1]Technical University of Cluj-Napoca, Romania,
Tel: +40-264-401775, E-mail: lorentz.jantschi@gmail.com

[2]Iuliu Hațieganu University of Medicine and Pharmacy Cluj-Napoca,
Romania

25.1 DEFINITION

Mechano(-)synthesis refers to a hypothetical chemical synthesis in which reaction outcomes are determined by the use of mechanical constraints to direct reactive molecules to specific molecular sites. In a Mech-synth process, reactive molecules would be attached to molecular mechanical systems, and their encounters would result from mechanical motions bringing them together in planned sequences, positions, and orientations. Mech-synth strongly favoring desired reactions by holding reactants together in optimal orientations for many molecular vibration cycles. In biology, the ribosome provides an example of a programmable Mech-synth device. There are presently no chemical syntheses which fully achieve this aim. Some atomic placement has been achieved with scanning tunneling microscopes.

25.2 HISTORICAL ORIGIN(S)

In conventional chemical synthesis or chemosynthesis, reactive molecules encounter one another through random thermal motion in a liquid or vapor, while Mech-synth aim is to position molecules in specific locations in a carefully chosen sequence. By holding and positioning molecules, controlling thus how the molecules react, building up complex structures with atomically precise control.

Mech-synth is inspired from biology. It is known that the ribosome is a complex molecular machine found within all living cells that serves as the site of biological protein synthesis (translation). Ribosomes link amino acids together in the order specified by messenger RNA (mRNA) molecules. From here it has been binged out the idea that is possible to synthesize chemical products by using only mechanical action. In 2013, a special issue of Chemical Society Reviews (Issue 18, Volume 42) was dedicated to the theme of mechanochemistry.

25.3 NANO-SCIENTIFIC DEVELOPMENT(S)

In the early times, Theophrastus of Eresus, an Aristotle's student wrote in about 315 B.C. a short booklet titled "On Stones" (the earliest surviving document related to chemistry) containing a reference on the reduction of cinnabar to mercury by grinding in a copper mortar with a copper pestle (Takacs, 2013).

Much later, in 1820, Faraday described the reduction of silver chloride by grinding with zinc, tin, iron, and copper in a mortar (Faraday, 1820), but the first systematic studies of Mech-synth reactions were carried out at the end of the 19th century (Spring, 1883; Lea, 1892) while the first documented application of mechanical stimulus to induce chemical reactions in organic systems is probably (Ling & Baker, 1893).

The beginning of the Mech-synth era can be considered when the technique of moving single atoms mechanically was firstly proposed (Drexler, 1986).

The principle of Mech-synth is relatively simple (see Figure 25.1), but the implementation of the principle is difficult, due to the scale (atomic scale).

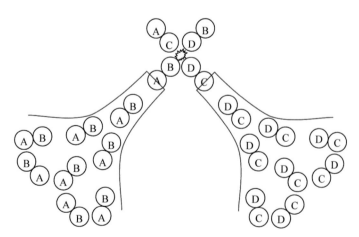

FIGURE 25.1 Mech-synth for driven substitution reaction.

It should be noted that the term mechanochemistry is sometimes confused with mechanosynthesis, which refers specifically to the machine-controlled construction of complex molecular products (Drexler, 1992). The principle of Mech-synth is relatively simple (see Figure 25.1), but the implementation of the principle is difficult, due to the scale (atomic scale). Since it is a typical machine-driven process, its use implies manipulation of manufacturing concepts, such as is exemplified in Table 25.1 (adapted from Table 1.1 from Drexler, 1991).

TABLE 25.1   Scale Differences Between Conventional, Miniaturized, Chemistry, Biochemistry, and Molecular Manufacturing.

Characteristic	Conventional	Miniaturized	Biochemistry	Chemistry	Mech-synth
Molecular precision	No	No	Yes	Yes	Yes
Positional control	Yes	Yes	Some	No	Yes
Feature scale	1 mm	1 μm	0.3 nm	0.3 nm	0.3 nm
Product scale	1 m	10 mm	10 nm	1 nm	1 μm
Defect rate	10^{-4}	10^{-7}	10^{-11}	10^{-2}	10^{-12}
Producing time	10^1 s	10^2 s	10^{-3} s	10^3 s	10^{-6} s
Output content	Materials & shapes		Monomer sequences	Atoms & bonds	

To work at a molecular level, the devices should be on the same scale. Therefore, it is a strong connection between Mech-synth and molecular machines concept. Most of the explorations to build Mech-synth have focused on using carbon, because it allows the building of structures of large size and finds its uses in medical and mechanical applications.

25.4   NANOCHEMICAL APPLICATION(S)

There is a growing body of studies for synthesizing diamond by mechanically removing/adding hydrogen atoms (Temelso et al., 2006) and depositing carbon atoms (De Federico & Jaime, 2006; Yin et al., 2007). Other applications include constructing of metal layers – sulfate intercalated Mg-Al layered double hydroxides (Mg-Al-SO$_4$-LDH) was successfully synthesized by the one-step mechanochemical activation method followed by subsequent water washing and aging in Fahami & Beall (2016), formation of highly incompressible and hardness materials – the formation of rhenium carbide

(Re$_2$C) from the elements was reported from mechanochemical treatment after 640 min of milling in Granados-Fitch et al. (2016), the formation of soft magnetic materials – soft magnetic Fe$_3$O$_4$/Ni$_3$Fe composite powder has been synthesized starting from Fe, Ni and Fe$_2$O$_3$ powders via mechanosynthesis and subsequent annealing in Marinca et al. (2015).

In regard with the diamond synthesis (Freitas & Merkle, 2008) proposes a minimal toolset (C/Ge/H DMS) for positional diamond mechanosynthesis consisting from three primary tool types and six auxiliary ones (see Figure 25.2, for further details see Freitas & Merkle, 2008) requiring system requiring only four simple feedstock molecules – CH$_4$ (methane) and C$_2$H$_2$ (acetylene) as the carbon sources, Ge$_2$H$_6$ (digermane) as the germanium source, and H$_2$ as a hydrogen source.

Primary tool types

Hydrogen abstraction, Hydrogen donation, Dimer placement

Auxiliary tool types

Adamantyl radical, Germano-adamantyl radical, Hydrogen transfer,

Methylene addition, Germylmethylene addition, Germylene addition

---

FIGURE 25.2   Minimal toolset for positional diamond Mech-synth.

25.5   MULTI-/TRANS-DISCIPLINARY CONNECTION(S)

To arrive at a Mech-synth one must use computational modeling, and here strict motion constraints reduce the number of configurations that must be examined, and because reactions that consume high-energy species can have large thermodynamic driving forces, making predicted behaviors relatively insensitive to errors in calculated model energies, leaving, therefore, a little room for computations at reasonable level of theory. Engineering of controlled processes passes the Mech-synth through the chemical engineering, and finally, the applications of the products include materials science where commonly these are tested.

25.6   OPEN ISSUES

The goal of one line of Mech-synth research focuses on overcoming the problems by calibration, and selection of appropriate synthesis reactions.

Some suggest include attempting to develop a specialized, very small machine tool that can build copies of it under the control of an external computer. In the literature, such a tool is called an assembler or molecular assembler. Once assemblers exist, geometric growth (directing copies to make copies) could reduce the cost of assemblers rapidly. Control by an external computer should then permit large groups of assemblers to construct large, useful projects to atomic precision. One such project would combine molecular-level conveyor belts with permanently mounted assemblers to produce a factory.

To manage the risk coming from various potential disasters arising from runaway replicators which could be built using mechanosynthesis, the UK Royal Society and UK Royal Academy of Engineering in 2003 commissioned a study to deal with these issues and larger social and ecological implications, led by mechanical engineering professor Ann Dowling. This was anticipated by some to take a strong position on these problems and potentials and suggest any development path to a general theory of so-called mechanosynthesis. It should be noted that current technical proposals for nanofactories do not include self-replicating nanorobots, and recent ethical guidelines would prohibit the development of unconstrained self-replication capabilities in nanomachines.

KEYWORDS

- **driven chemical reaction**
- **hypothetical chemical synthesis**
- **mechanochemistry**
- **molecular engineering**
- **molecular manufacturing**

REFERENCES AND FURTHER READING

De Federico, M., & Jaime, C., (2006). Free energy calculations (FEP and TI): Conformational preference of a cyclodextrinic [2]Catenane: A case study. *J. Comput. Theor. Nanosci., 3*(6), 874–879.

Drexler, E. K., (1986). *Engines of Creation: The Coming Era of Nanotechnology.* New York: Anchor Books, p. 320.

Drexler, E. K., (1991). Molecular machinery and manufacturing with applications to computation. *PhD Thesis in the Field of Molecular Nanotechnology* (PhD Advisor: Marvin L. MINSKY). Massachusetts Institute of Technology, p. 487, URL: https://dspace.mit.edu/handle/1721.1/27999?show=full.

Drexler, E. K., (1992). *Nanosystems: Molecular Machinery, Manufacturing and Computation.* Chichester: J. Wiley & Sons, p. 576.

Fahami, A., & Beall, G. W., (2016). Mechanosynthesis and characterization of Hydrotalcite-like Mg-Al-SO$_4$-LDH. *Mater. Lett., 165,* 192–195.

Faraday, M., (1820). Extracts, a method of preparing, by evaporation in vacuo, on the decomposition of chloride of silver by hydrogen and by zinc. *Q. J. Sci. Lit. Arts, 8,* 374–375.

Freitas, Jr., R. A., & Merkle, R. C., (2008). A minimal toolset for positional diamond mechanosynthesis. *J. Comput. Theor. Nanosci., 5,* 760–861.

Granados-Fitch, M. G., Juarez-Arellano, E. A., Quintana-Melgoza, J. M., & Avalos-Borja, M., (2016). Mechanosynthesis of rhenium carbide at ambient pressure and temperature. *Int. J. Refract. Metal. Hard Mater., 55,* 11–15.

Lea, C. M., (1892). Disruption of the silver haloid molecule by mechanical force. *Am. J. Sci., 43*(3), 527–531.

Ling, A. R., & Baker, J. L., (1893). Halogen derivatives of quinone. Part III. Derivatives of quinhydrone. *J. Chem. Soc. Trans., 63,* 1314–1327.

Marinca, T. F., Chicinaş, H. F., Neamţu, B. V., Isnard, O., & Chicinaş, I., (2015). Structural, thermal and magnetic characteristics of Fe$_3$O$_4$/Ni$_3$Fe composite powder obtained by mechanosynthesis-annealing route. *J. Alloy. Compound., 652,* 313–321.

Spring, W., (1883). Arsenide Formation by Pressure (In French). *Bull. Soc. Chim. Fr. 40,* 195–196. Formation of Some Sulfides by the Action of the Pressure. Considerations that Affect the Properties of Allotropic States of Phosphorus and Carbon (In French). *Bull. Soc. Chim. Fr., 40,* 641–647.

Takacs, L., (2013). The historical development of mechanochemistry. *Chem. Soc. Rev., 42,* 7649–7659.

Temelso, B., Sherrill, C. D., Merkle, R. C., & Freitas, R. A., (2006). High-level Ab initio studies of hydrogen abstraction from prototype hydrocarbon systems. *J. Phys. Chem. A., 110*(38), 11160–11173.

Yin, Z. X., Cui, J. Z., Liu, W., Shi, X. H., & Xu, J., (2007). Horizontal gene-substituted polymantane-based C2 Dimer placement tooltip motifs for diamond mechanosynthesis. *J. Comput. Theor. Nanosci., 4*(7), 1243–1248.

CHAPTER 26

Metal Nanoparticles: Silver

FRANCO CATALDO[1,2]

[1]*Actinium Chemical Research Institute, Via Casilina 1626A, 00133 Rome, Italy, Tel: +39 06-94368230, Fax: +39 06-94368230, E-mail: franco.cataldo@fastwebnet.it, cataldo.franco@fastwebnet.it*

[2]*Università degli Studi della Tuscia, Via S. Camillo de Lellis, Viterbo, Italy*

26.1 DEFINITION

Metal nanoparticles sols such as silver and gold nanoparticles in water and in other media are known since a long time as silver (Schaum et al., 1921; Wiegel et al., 1953) and gold (Zsigmondy et al., 1925; Svedberg et al., 1928) colloids. These colloidal solutions are characterized by their color which today we know as due to surface plasmon resonance (SPR) whose origin is attributed to the collective oscillation of the free conduction electrons induced by an interacting electromagnetic field as shown in Figure 26.1 (Link et al., 1999, 2003). Another feature of these solution regards the Tyndall effect which consists of the light scattering caused by the dispersed metal nanoparticles (Voyutsky, 1978). Of course many other metals and non-metals are able to form colloidal dispersions other than silver and gold, but currently, the largest attention by researchers is concentrated on Ag and Au nanoparticles.

As summarized by Mingos (2014), the metal nanoparticles can be classified into three main categories: clusters, nanodomains, and colloids. Clusters are defined as particles made by 30–150 atoms with diameters comprised between 0.3 and 1.7 nm. Nanodomains are made by 100 to 700,000 atoms with diameters comprised between 1.5 and 30 nm. Nanodomains can be distinguished into three sub-categories: nanoclusters, nanoparticles and nanocolloids considering, respectively, if the particles size are monodispersed, narrowly dispersed or widely dispersed (polydispersed). Finally, "colloidal" metal particles are characterized by 10^6 to 10^{11} atoms and diameters from 30 to >150 nm.

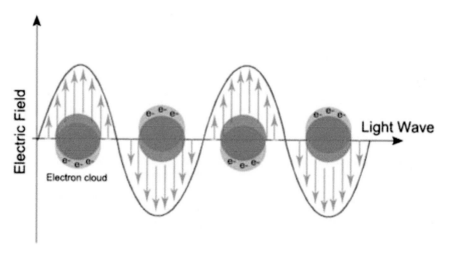

FIGURE 26.1   Mechanism of SPR electronic transition in metal nanoparticles.

26.2   HISTORICAL ORIGINS

Lively accounts of metal nanoparticles history can be found in many books (Johnson, 1949; Staudinger, 1950; Jirgensons et al., 1962; Voyutsky, 1978). Here we limit ourselves to only some selected milestone. In 1673 the German alchemist Andreas Cassium in Hamburg prepared the Purple of Cassius by reducing dilute solutions of $HAuCl_4$ with $SnCl_2$ obtaining a purple suspension of gold nanoparticles in water (Zsigmondy et al., 1925). In 1856, Michael Faraday was able to produce a colloidal gold hydrosol by reducing gold chloride with phosphorus (Johnson, 1949). The real breakthrough in the field of metal nanoparticles occurred with the development of colloidal chemistry worked out (among many others) by the famous chemists W. Ostwald (Nobel Laureate in 1909), R. Szigmondy (Nobel Laureate in 1925) who worked especially on colloidal gold and T. Svedberg (Nobel Laureate in 1926) for the discovery of the ultracentrifuge (Laylin, 1993). Another scientist to be remembered here is Gustav Mie who provided the theory for scattering and absorption by spherical particles explaining the change in color of the metal nanoparticles as a function of their dimensions (Voyutsky, 1978). The introduction of the concept of plasmon by Bohm and Pines (1953) and Shevchik (1974) as quantum plasma oscillations of the free electron gas density in metal nanoparticles led to a further insight in our understanding of the color of the nanoparticles and the mechanism of their electronic transitions. Henglein should be remembered here as the scientist who has paved the way to the developments in our ability to control the dimension, shape, and stability of the metal nanoparticles (Henglein, 1989).

26.3   NANO-SCIENTIFIC DEVELOPMENTS

Metal nanoparticles can be produced by a number of methods. The general chemical approach consists in reducing a dilute solution of the metal salt with an appropriate reducing agent (same approach as those used by Cassius and Faraday!). The advancements are related to our ability to control the particle size and shape, to control the stability of the hydrosols. Particularly popular is the contemporary approach to produce silver and gold nanoparticles using "green" reducing agents (Sharma et al., 2009; Tolaymat et al., 2010). For example, silver nanoparticles were produced using tannic acid as a reducing agent (Cataldo et al., 2013), but also the tannins of green and black tea are able to cause the formation of silver nanoparticles (Cataldo, 2014). The flavanols of *Hibiscus Sabdariffa* infusion (karkadé) are able to favor the formation of gold nanoparticles (Cataldo et al., 2016a). The most interesting thing of these green synthesis of silver or gold nanoparticles is the fact that the green, reducing agent is coming from renewable sources, and it is able not only to cause the reduction of the metal but also to act as capping agent, undergoing an adsorption on the metal particles surface and favoring the stabilization of the colloidal suspension. Using older terminology, it can be affirmed that the green, reducing agent acts also as a protective colloid (Cataldo, 2014; Cataldo et al., 2016a). Other methods employed to produce metal nanoparticles range from radiolysis in a solution of metal salts (Henglein et al., 1998; Henglein 1999; Cataldo et al., 2016a) which also permits to control the particles shape, to the submerged arc method between two metal electrodes. In the arc method, the two electrodes made with the metal desired to be dispersed are arced inside the liquid water or solution to obtain a metal lyosol. This method known also as electron sputtering, was developed by G. Bredig in 1898 and brought to perfection later by T. Svedberg (Jirgensons, 1962; Voyutsky, 1978). However, the lyosol prepared in this way suffers from a certain instability because of the lack of a stabilizing electrolyte or other agents which must be added purposely. Also in recent times, variants of the electron sputtering have been proposed involving water-submerged spark discharge (Tien et al., 2009). A less common method for the preparation of silver nanoparticles dispersion regards the use of high-intensity ultrasonic waves in order to fragment and disperse the metal particles (Darroudi et al., 2012). An approach which is not new, since was reported in the literature since a long time (Suslik, 1989). A complete summary on synthetic methods of metal nanoparticles can be found in review articles (Krutyakov et al., 2008; Olenin et al., 2011; Mingos, 2014).

26.4   NANOCHEMICAL APPLICATIONS

Metal nanoparticles have a number of current and potential applications due to their photocatalytic (Kamat, 2002) and unique catalytic properties (Johnson, 2003; Astruc, 2008; Xia et al., 2013). The Ag and Au nanoparticles can be quite easily surface functionalized using thiol end groups of alkyl chains or other functional groups (Jadzinsky et al., 2007) for biocatalytic and biosensors applications. In this context, it is useful to recall here that the SPR transition is highly sensitive to the environment surrounding the particle and can be used as a sensor of simple chemicals. For example, the interaction of ozone with spherical gold nanoparticles leads to a shift of the SPR toward longer wavelengths while the interaction of ozone with gold nanorods leads instead to a SPR band shift toward shorter wavelengths (Cataldo et al., 2016b). Consequently, by selecting the appropriate metal nanoparticle shape and knowing the entity of the SPR band shift as a function of the analyte concentration, it could be possible to determine the analyte concentration. A more sophisticated approach involves, of course, the surface functionalization of the nanoparticles to be used as sensing materials (Saha et al., 2012) or as carriers of medical drugs (Wang et al., 2012) or even medical drugs and radiation (Miller et al., 2013).

26.5   MULTI-/TRANS-DISCIPLINARY CONNECTIONS

Metal nanoparticles in general and in particular silver, copper and gold nanoparticles are known to have an antibacterial activity. This property is particularly pronounced in silver nanoparticles which can be considered the best colloidal metal as a bactericide (Lansdown, 2010). The bactericidal effect of silver is known since ancient times (Lansdown, 2010). The bactericidal effect of colloidal silver is firstly due to the production and release of Ag^+ ions which exert an oligodynamic effect on bacteria and protozoa. Even at high dilution, the Ag^+ ions bind to the –SH groups of enzymes inhibiting their activity and also causing the denaturation of proteins. Silver also reacts with the amino-, carboxyl-, phosphate-, and imidazole-groups and diminish the activities of lactate dehydrogenase and glutathione peroxidase

(Lansdown, 2010). Furthermore, silver nanoparticles may attach to the surface of membrane cell disturbing permeability and respiration function of the cell. Smaller Ag nanoparticles are more effective bactericidal that the larger nanoparticles since they can penetrate the bacterial cells and causing the disruption of the cell walls thereby causing the production of free radicals including ROS (reactive oxygen species) (Lansdown, 2010). The oligodynamic effect has been observed in living cells, algae, molds, spores, fungi, viruses, prokaryotic and eukaryotic microorganisms however silver nanoparticles are not very effective against viruses and especially against spores (Cataldo, 2014). The bactericidal properties of silver nanoparticles in both black and green tea infusions were confirmed on the Gram(–) bacterium *P. Aeruginosa.* Instead, the silver nanoparticles in both black and green tea infusions are not effective against the spores of *B. Subtilis* (Cataldo, 2014).

Gold nanoparticles are used in medicine, and they are known to be effective against arthritis but may found many other applications including diagnostics (Boisselier et al., 2009). Gold nanoparticles are used since long as contrasting agent in microscopy (Mingos, 2014).

KEYWORDS

- **developments**
- **gold**
- **history**
- **nanoparticles**
- **silver**
- **surface chemistry**

REFERENCES AND FURTHER READING

Astruc, D., (2008). *Nanoparticles and Catalysis*. Weinheim: Wiley-VCH.

Bohm, D., & Pines, D., (1953). A collective description of electron interactions: III. Coulomb interactions in a degenerate electron gas. *Physical Review, 92*, 609–625.

Boisselier, E., & Astruc, D., (2009). Gold nanoparticles in nanomedicine: Preparations, imaging, diagnostics, therapies, and toxicity. *Chemical Society Reviews, 38*, 1759–1782.

Cataldo, F., & Angelini, G., (2013). A green synthesis of colloidal silver nanoparticles and their reaction with ozone. *European Chemical Bulletin, 2*, 700–705.

Cataldo, F., (2014). Green synthesis of silver nanoparticles by the action of black or green tea infusions on silver ions. *European Chemical Bulletin, 3*, 280–289.

Cataldo, F., Ursini, O., & Angelini, G., (2016a). Synthesis of silver nanoparticles by radiolysis, photolysis and chemical reduction of $AgNO_3$ in Hibiscus sabdariffa infusion (karkadé). *Journal of Radioanalytical and Nuclear Chemistry, 307*, 447–455.

Cataldo, F., Ursini, O., & Angelini, G., (2016b). Colloidal gold nanoparticles: Interaction with ozone and analytical potential. *European Chemical Bulletin, 5*, 1–7.

Darroudi, M., Zak, A. K., Muhamad, M. R., Huang, N. M., & Hakimi, M., (2012). Green synthesis of colloidal silver nanoparticles by sonochemical method. *Materials Letters, 66*, 117–120.

Henglein, A., & Meisel, D., (1998). Radiolytic control of the size of colloidal gold nanoparticles. *Langmuir, 14*, 7392–7396.

Henglein, A., (1989). Small-particle research: Physicochemical properties of extremely small colloidal metal and semiconductor particles. *Chemical Reviews, 89*, 1861–1873.

Henglein, A., (1999). Radiolytic preparation of ultrafine colloidal gold particles in aqueous solution: Optical spectrum, controlled growth, and some chemical reactions. *Langmuir, 15*, 6738–6744.

Jadzinsky, P. D., Calero, G., Ackerson, C. J., Bushnell, D. A., & Kornberg, R. D., (2007). Structure of a thiol monolayer-protected gold nanoparticle at 1.1 Å resolution. *Science, 318*, 430–433.

Jirgenson, B., & Straumanns, M. E., (1962). *A Short Textbook of Colloid Chemistry* (pp. 1–7). Pergamon Press, Oxford.

Johnson, B. F., (2003). Nanoparticles in catalysis. *Topics in Catalysis, 24*, 147–159.

Johnson, P., (1949). *Colloid Science* (pp. 1–33). Clarendon Press, Oxford.

Kamat, P. V., (2002). Photophysical, photochemical and photocatalytic aspects of metal nanoparticles. *The Journal of Physical Chemistry B., 106*, 7729–7744.

Krutyakov, Y. A., Kudrinskiy, A. A., Olenin, A. Y., & Lisichkin, G. V., (2008). Synthesis and properties of silver nanoparticles: Advances and prospects. *Russian Chemical Reviews, 77*, 233–257.

Lansdown, A., (2010). *Silver in Healthcare: Its Antimicrobial Efficacy and Safety in Use.* Royal Society of Chemistry, Cambridge.

Laylin, K. J., (1993). *Nobel Laureates in Chemistry 1901–1992.* American Chemical Society, Washington, D.C.

Link, S., & El-Sayed, M. A., (1999). Spectral properties and relaxation dynamics of surface plasmon electronic oscillations in gold and silver nanodots and nanorods. *The Journal of Physical Chemistry B., 103*, 8410–8426.

Link, S., & El-Sayed, M. A., (2003). Optical properties and ultrafast dynamics of metallic nanocrystals. *Annual Review of Physical Chemistry, 54*, 331–366.

Miller, S. M., & Wang, A. Z., (2013). Nanomedicine in chemoradiation. *Therapeutic Delivery, 42*, 239–250.

Mingos, D. M. P., (2014). *Gold Clusters, Colloids and Nanoparticles I.* Springer. Dordrecht.

Neville, F., Pchelintsev, N. A., Broderick, M. J., Gibson, T., & Millner, P. A., (2009). Novel one-pot synthesis and characterization of bioactive thiol-silicate nanoparticles for biocatalytic and biosensor applications. *Nanotechnology, 20*, article N° 055612.

Olenin, A. Y., & Lisichkin, G. V., (2011). Metal nanoparticles in condensed media: Preparation and the bulk and surface structural dynamics. *Russian Chemical Reviews, 80*, 605–630.

Saha, K., Agasti, S. S., Kim, C., Li, X., & Rotello, V. M., (2012). Gold nanoparticles in chemical and biological sensing. *Chemical Reviews, 112*, 2739–2779.

Schaum, K., & Lang, H., (1921). Ueber die farbe von photochlorid und von kolloidem silber. I. *Colloid & Polymer Science, 28*, 243–249.

Sharma, V. K., Yngard, R. A., & Lin, Y., (2009). Silver nanoparticles: Green synthesis and their antimicrobial activities. *Advances in Colloid and Interface Science, 145*, 83–96.

Shevchik, N. J., (1974). Alternative derivation of the Bohm-Pines theory of electron-electron interactions. *Journal of Physics C: Solid State Physics, 7*, 3930–3936.

Staudinger, H., (1950). *Organische Kolloidchemie* (3rd edn.). Vieweg, F., & Sohn, Braunshweig, Germany, Chapters 1 and 2.

Suslik, K. S., (1989). *Ultrasound: It's Chemical, Physical and Biological Effects*, VCH Publishers, Weinheim.

Svedberg, T., & Tiselius, A., (1928). *Colloid Chemistry.* The Chemical Catalog Company, Inc. New York.

Tien, D. C., Tseng, K. H., Liao, C. Y., & Tsung, T. T., (2009). Identification and quantification of ionic silver from colloidal silver prepared by electric spark discharge system and its antimicrobial potency study. *Journal of Alloys and Compounds, 473*, 298–302.

Tolaymat, T. M., El Badawy, A. M., Genaidy, A., Scheckel, K. G., Luxton, T. P., & Suidan, M., (2010). An evidence-based environmental perspective of manufactured silver nanoparticle in syntheses and applications: A systematic review and critical appraisal of peer-reviewed scientific papers. *Science of the Total Environment, 408*, 999–1006.

Voyutsky, S., (1978). *Colloid Chemistry*. Mir Publishers, Moscow, Chapter 8.

Wang, A. Z., Langer, R., & Farokhzad, O. C., (2012). Nanoparticle delivery of cancer drugs. *Annual Review of Medicine, 63*, 185–198.

Wiegel, E., (1953). Über die Farben des kolloiden silbers und die miesche theorie. *Zeitschrift für Physik, 136*, 642–653.

Xia, Y., Yang, H., & Campbell, C. T., (2013). Nanoparticles for catalysis. *Accounts of Chemical Research, 46*, 1671–1672.

Zsigmondy, R., & Thiessen, P. A., (1925). *Das Kolloide Gold*. Akademische Verlagsgesell-schaft: Leipzig, Germany.

CHAPTER 27

Mobile Nanodefects

A. M. KAPLAN[1] and N. I. CHEKUNAEV[2]

[1]*N.N. Semenov's Institute of Chemical Physics, Russian Academy of Sciences, Russian Federation, E-mail: amkaplan@mail.ru*

[2]*Kosygin Street 4, Moscow 119991, Russian Federation*

27.1 DEFINITION

This chapter proposes the conception of mobile nanodefects as a universal catalyst. This chapter considers the different ways of realization by the mobile nanodefects (MNDs) of the catalytic activity in various physical and chemical processes in solids. It has also been shown that MNDs with an excess of free volume demonstrate the catalytic activity when they approach orientationally immobile chemically inert classic active centers in solids. Only the participation of mobile electron traps (MElTr) of nanosize in the process of electrons transfer in the irradiated semiconductor allowed us to explain the nontrivial effect – the independence of the velocity of movements along such semiconductor of electrons having different binding energies with electronic traps which captured these electrons. With the help of the catalytic activity of MND, the nontrivial features of the industrial important energy-saving high-temperature process of polymers shear grinding (HTSGr) were explained.

27.2 HISTORICAL ORIGIN AND SCIENTIFIC DEVELOPMENT

In 1989, Kaplan proposed MND as one type of catalyst that promotes the creation of new active centers of the polymerization in solids (Kaplan, 1996).

In 1995, the authors explained the non-trivial effect of the independence of the electron transfer rate for electrons that having different binding energies with IENTr. This effect has manifested another type of catalysis of

MND (MENTr). The MENTr accelerates electron transfer between IENTr and EAc (Kaplan and Chekunaev, 2012).

The third type of catalysis by MND was proposed by Anatoly Kaplan (Kaplan, 1996) for an explanation of non-trivial features of grinding of polymers; when such polymers expose to intense shear stresses. Such grinding was seen as the result of joint action on the polymers simultaneously two connected chemical stage of MND (mobile vacancies and cracks of supercritical size).

27.3   NANOCHEMICAL APPLICATION

The manifestation of the catalytic activity of MND, caused by the excess free volume of such defects, has been demonstrated in the authors' article "Unbranched Chain Reactions in Solids and in Viscous Media: Nontraditional Methods and Analysis" (Chapter 42 in Vol. 2 of "New Frontiers in Nanochemistry).

One fundamental problem of chemical physics is the study of the mechanism of solid-phase redox reactions with electron transfer on the abnormally large distance. A promising approach to the investigation of such reactions is the study of electrons (e_{tr}), which are trapped by IETr in solid dielectrics after the cessation of their irradiation.

Interesting in this area of research is the experimental study presented in Buxton & Kemsley (1975) of release of e_{tr} from INTr in the irradiated glassy matrix of 9.5M aqueous LiCl solution $(LiCl)_g$ at $T = 138K = T_g - 10K$ at different time moments (t_0 and t_1, when the difference $t_1 - t_0 = 150$ ns) after the cessation of the matrix; where T_g is the glass transition temperature of this matrix. The kinetics of the e_{tr} release from INTr with different binding energy (EB) of e_{tr} with INTr was characterized by parameter "fractional loss" (FL)$_i = V[E_i(\lambda_i)]$ $= 1-n[E_i(\lambda_i),t_1]/n[E_i(\lambda_i),t_0]$ [see the experimental curve (point ∘ in Figure 27.1), where EB $= E_i(\lambda_i)$ in the interval (1.25 ÷ 2.6 eV)].

The release of e_{tr} from INTr can be carried out in five ways. However, at low temperature (138K) and EB = 1.25 ÷ 2.6 eV, it is impossible for the transition of e_{tr} in the zone of electronic conduction. The authors showed that the second possible path of transfer of electrons, captured by the primary traps (IEITrs), to electronic acceptors (ElAcs) due to mutual diffusion (IEITrs) to (ElAcs) at low temperature T = 138K is also absent. The remaining three ways of e_{tr} release from INTr with the use of the tunnel effect are: (1) the transfer of e_{tr} from INTr to IAc by one act of tunneling on large distances (≥1.5 nm) (e.g., Khairutdinov & Zamaraev, 1989); (2) the transfer of e_{tr} from INTr to IAc by many successive acts of tunneling through the system of immobile electron traps between INTr and EAc (Hamill & Funabashi, 1977); and (3) release of the e_{tr} by the authors' model presented in Kaplan (1996).

For the above three marked models, the three calculated curves are shown in Figure 27.1. All the three marked models were selected in the calculations with physically reasonable best fitting parameters, allowing to obtain the best approximation of such curves to the experiment. It is seen that only curve 3, built by the authors model, is in an excellent agreement with the experimental curve.

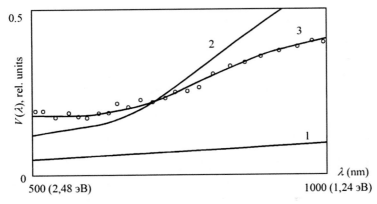

FIGURE 27.1 Experimental data ($\circ$) and calculated dependencies (1, 2, 3) of the parameter $V(E_i(\lambda_i))$ as a function of the wavelength of $(LiCl)_g$ absorption spectrum. 1 – direct e_{tr} tunneling from initial electron traps to EAc; 2 – multiple e_{tr} tunneling on a system of immobile traps; 3 – according to the authors model.

The mentioned agreement of curve 3 with the experimental curve is obliged to the inclusion by the authors in the studied process of a small amount of mobile electron traps with concentration [METr] $\approx 10^{-2} \cdot$[ETr]. This ratio of concentrations exists in a glassy matrix $(LiCl)_g$ at temperatures not much different from the temperature of its glass transition T_g. This is due to a wide range of nanodefects mobility in micro-inhomogeneous glasses (e.g., Emanuel & Buchachenko, 1987).

According to the authors' model, the studied system takes place in the tunneling transfer of e_{tr} from IETr to mobile trap (METr). Then METr transfers an electron from IETr to EAc by diffusion mechanism and passes this electron to EAc if EMETr < EAc. The calculated curve 3 obtained at the fitting parameter: [METr]$\cdot D_{METr} = 2 \times 10^{12}$ cm^{-1}s^{-1}, where D_{METr} is diffusion coefficient of METr. For the validity of the numerical value of this fitting parameter, see in Kaplan (1996).

In the Semenov's Institute of Chemical Physics of the RAS under the guidance of Prof. V. G. Nikol'skii was designed to drive the practical implementation of a valuable method of high-temperature shear grinding (HTSGr) of polymers (Wolfson & Nikol'skii, 1994).

The most important nontrivial features of these studies are: (1) HTSGr method is not effective for the grinding of homogeneous materials, but works great when grinding heterogeneous composites, including semicrystalline polymers; (2) the nontraditional temperature dependence of the polymers grinding rate with maximum at elevated temperatures $[T_m - (10°C \div 20°C)]$ was observed when using HTSGr. T_m is the melting point of the individual polymers or temperature of the beginning of chemical decomposition of the cross-linked polymer composites; (3) HTSGr of polymers is the most energy-saving method compared to known polymers grinding methods.

The known models of the grinding of a stressed polymer cannot explain these HTSGr features (Narisava, 1987). Therefore, the authors developed an original authors model (Kaplan & Chekunaev, 2010) to describe the grinding of the polymers by HTSGr method in the field of intensive tensile or shear stresses in the air (Kaplan & Chekunaev, 2012). In this, the earlier model was proposed to consider grinding of polymers in the air by HTSGr method as a result of the branched mechanochemical chain process in the studied system. This process consists of successively physical and chemical elementary stages. The physical stage of the process is a rapid spread in the studied system of cracks of a supercritical size, which leads to the formation of chemically active radicals agents (Regel and Slutsker, 1974). The chemical interaction of radicals with each other or with chemically active impurities in the polymer (e.g., with oxygen) causes the formation of vacancies. The flow of the mobile vacancies (MV) to the existing polymers immobile precritical cracks (T_{PCr}) may turn such cracks in the mobile supercritical cracks (T_{SCr}) under certain conditions (Pines, 1959). The small transverse dimension of supercritical cracks (less than 30 nm) in polymers allows us to consider such cracks as the nano-objects (Regel & Slutsker, 1974).

A more detailed analysis of the kinetics of the chain process of destruction of the mechanically loaded inhomogeneous amorphous-crystalline polymer was conducted in Kaplan and Chekunaev (2012). This analysis allowed us to estimate the values of optimal temperatures and stresses acting on the polymer, which is necessary for providing the most energy-saving grinding of the polymers due to avalanche mode of cracking of the polymers. The avalanche mode of polymers cracking is described by the equation (1) under the condition $f > \varphi$.

$$dn/dt = (f - \varphi)\, n - k_t n^2 \tag{1}$$

Here, n is the concentration of stopped supercritical cracks (T_a); f is the rate constant of the initiation of new cracks T_a near the border of stopped cracks T_a; φ is the deactivation rate constant of T_a; and k_t is the constant rate of the chain termination.

Note important factors distinguish the authors' model from the well-known models of fracture of mechanically stressed materials.

In 1999, the authors noticed that the grinding of strained polymers could be carried out by exposure to the polymers of cracks of supercritical size (TSCr). However, it has been shown that supercritical cracks cannot multiply inhomogeneous materials and therefore, cannot grind the materials. Unlike homogeneous materials in heterogeneous ones, supercritical cracks can stop and become by active centers (T_a). One such supercritical crack is able to initiate several new, similar cracks. This possibility allows, under the condition $(f > \varphi$ in formula 1), to initiate the avalanche-like multiplication process of supercritical cracks (T_a) in heterosystems. A large number of supercritical cracks formed in heterogeneous polymers when exposed to them intense shear stress leads to intensive grinding of such polymers. Thus, the first nontrivial feature of the method (HTGr) noted on p. 310 was explained (Kaplan & Chekunaev, 2012).

In 1999, the authors established a necessary condition for the multiplication of supercritical cracks in strained heterogeneous polymers—the participation in such a process of mobile nanodefects (vacancies). The diffusion rate of vacancies is $W \sim \exp(-E_v/k_B T)$ (here E_v is the activation energy of vacancy diffusion, k_B is the Boltzmann constant, T is the temperature). Therefore, the rate of multiplication of cracks T_a in strained heterogeneous polymers increases with increasing temperature to their melting temperature range (ΔT). At these temperatures, the heterogeneous polymer is homogenized. And in a homogeneous polymer, as shown above, supercritical cracks do not multiply. As a result, the temperature dependence of the cracks T_a concentration should be of an extreme nature with a maximum temperature near the melting temperature range (ΔT) of the polymer under study. The authors first performed computer calculations of the temperature dependence of the concentration of T_a cracks which are present in a mechanically loaded low-density polyethylene (LDPE). The results of such calculations are shown in Figure 27.2. This figure shows an increase of T_{SCr} concentration in mechanically-loaded LDPE to a certain temperature (85–90°C) above which this concentration is sharply reduced. Since the rate of the chain grinding of polymers is proportional to the concentration of T_{SCr}, the calculated curves in

Figure 27.2 explain the second of the previously unexplained features of the HTSG method, i.e., the polymers grinding rate with a maximum at elevated temperatures (85–95°C).

Finally, in the article by Kaplan & Chekunaev (2010), the first explanation of the most important feature of HTSGr method was given, i.e., the most energy-saving method for grinding polymers. Above it was shown that grinding of strained heterogeneous polymers using the HTSGr method is carried out by supercritical cracks. The high speed ($V = 0.38Ss = 800$ m/s, Ss – sound speed) of the effect of such cracks on the studied polymers causes the effect of a significant decrease in the relative elongation of samples of such polymers (in 30–40 times) in the HTSGr apparatus compared to other methods. The authors showed that the noted effect provides about the same (in 30–40 times) energy saving during the grinding of polymers using the HTSGr method compared to other known methods.

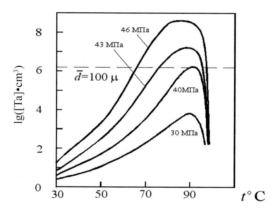

FIGURE 27.2   The calculated temperature dependences of the concentration of the "active" supercritical cracks formed in the LDPE samples at different mechanical stresses σ_0.

KEYWORDS

- **active crack of supercritical size**
- **mobile and immobile electron traps**
- **mobile nanodefects with the excess of free volume**
- **universal catalysts**

REFERENCES

Buxton, G. V., & Kemsley, K. J., (1975). *Chem. Soc. Farad. Trans. I., 717*. p. 568.

Emanuel, N. M., & Buchachenko, A. L., (1987). *Chemical Physics of Polymer Degradation and Stabilization (New Concepts in Polymer Science),* (p. 336). Publisher: VSP International Science Publishers.

Hamill, W. H., & Funabashi, K., (1977). *Phys. Rev., 16*, p. 5523.

Kaplan, A. M., & Chekunaev, N. I., (2010). "Theoretical foundations of the grinding of heterogeneous materials." *Theoretical Foundations of Chemical Engineering, 4*(3), pp. 339–347.

Kaplan, A. M., & Chekunaev, N. I., (2012). About the role played by mobile nanovoids in the stimulation of solid-state processes. *Russian Journal of Physical Chemistry B, 6*(3), pp. 407–415. © Pleiades Publishing, Ltd.

Kaplan, A. M., (1996). Thesis for the degree of doctor of chemical sciences. "Peculiarities of radiation and mechanical- stimulated processes in nonequilibrium condensed systems." *Inst. Chem. Phys. RAS, Moscow, 14.03.*

Khairutdinov, R. F., Zamaraev, K. I., & Zhadanov, V. P., (1989). Comprehensive chemical kinetics. *Electron Tunneling in Chemistry,* (1st edn), *30*, p. 358. Elsevier Science. eBook ISBN: 9780080868240.

Narisava, I., (1987). *Strength of Polymeric Materials* (Ohmsha, Tokyo, 1982, Khimiya, Moscow).

Pines, B. J., (1959). *Russian Journal of Solid State Physics, 1*(2), p. 265.

Wolfson, S. A., & Nikolskii, V. G., (1994). *Polymer Science, Ser. B, 36*(6), pp. 861–874.

CHAPTER 28

Nanoporous Carbon

LORENTZ JÄNTSCHI[1] and SORANA D. BOLBOACĂ[2]

[1]*Technical University of Cluj-Napoca, Romania*

[2]*Iuliu Hațieganu University of Medicine and Pharmacy Cluj-Napoca, Romania, E-mail: sbolboaca@gmail.com*

28.1 DEFINITION

Nanoporous carbon is a form of carbon processed to have pores producing a high ratio between the available surface area and the volume. The size of the pores depends on the method used for preparation and are classified (Sing et al., 1985) in micropores (<2 nm), mesopores (2–50 nm), and macropores (>50 nm). The term nanopore embraces the above three categories of pores, with an upper limit of 100 nm (Thommes et al., 2015).

28.2 HISTORICAL ORIGIN(S)

Going back to its ancestors (activated charcoal and activated carbon), porous carbons are the oldest adsorbents known, the use of it being described in Egypt in 1550 BC (Cooney, 1980). In Europe, the historical origin of activated carbon uses dates before 1900 (Wrench, 1931), its primary use being in the purification of drug and pharmaceutical products, while the first industrial production of activated carbons started in 1913 in the US (Baker et al., 1992). Systematic studies of nanoporous carbons (nPorC) began shortly after (Debye & Scherrer, 1917).

28.3   NANO-SCIENTIFIC DEVELOPMENT(S)

Today, nPorC can be prepared from organic natural (wood, nutshells, peat, lignite, coal, and petroleum coke) and synthetic (polyacrylonitrile, polyvinylidene chloride, polyvinyl chloride, and polyfurfuryl alcohol) and inorganic (metallic carbides) precursors while the pre-treatment of the natural precursors generally involves a reaction with a chemical reagent. The activation generally involves a thermal decomposition in the inert or oxidative atmosphere, while for metallic carbides the removal of reaction wastes is conducted with a hydrogenation. These processes may be combined in different successions (see Figure 28.1) for archiving a specific purpose, which is generally driven by the size and the distribution of the pores.

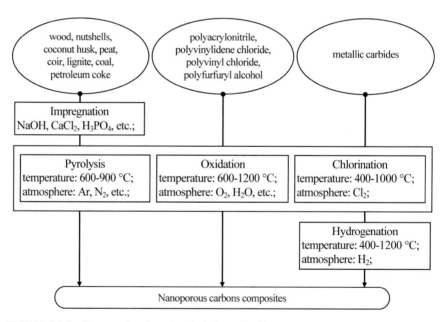

FIGURE 28.1   From carbon-based materials to nPorC.

The macrostructure of the nPorC can be templated (for instance with a mesoporous silica template) when the outcome (nPorC) is obtained after a supplementary step (of washing with a solvent) for removing the template (for an example, see Jong et al., 2002). Auxiliary steps, such as short (15–90 minutes) treatment with NH_3 at 400–700°C temperature may be involved having as purpose the increase of the specific activated surface (as suggested in Gogotsi et al., 2006).

When one desires controlled pore size and distribution, then the using of and/or starting from templates are the alternatives, when also the tunable parameters include a proper selection of the oxidative gas. Thus, following the procedure proposed in Barsoum & Gogotsi (2004), as heavier the halogen is used (F_2, Cl_2, Br_2, I_2), the larger resulted pores are obtained.

Carbon aerogels, belonging to nPorC, with a very low density (ranging in 0.035–0.1 g/cm^3) can be obtained from organic aerogels (for different routes of organic aerogels preparation please, see Pekala, 1989; Pekala et al., 1995; Biesmans et al., 1998; Li & Guo, 2000; Zhang et al., 2003) by following a recipe similar with the one given in Figure 28.1, where the thermal decomposition is conducted with N_2 at 900°C in Czakkel et al. (2005), and in CO_2 at 950°C in Kabbour et al. (2006). It has porosity over 50%, with pore diameter under 100 nm and surface areas ranging in 400–1000 m^2/g.

The thermal treatment plays an essential role in the producing of the nPorC. Table 28.1 adapted from Mohun (1959) provides terminology of the outcome.

TABLE 28.1 Thermal Treatment of Materials to Obtain nPorC

Temperature range	Outcome
600–700°C	'chars,' condensed molecular
700–1000°C	'cokes,' substantial amounts of carbon in disorganized phase
1000–2000°C	'baked' carbons
2000–2500°C	partially graphitized
>2500°C	polycrystalline graphite

28.4 NANOCHEMICAL APPLICATION(S)

The development of nPorC materials is driven by their uses and vice versa. Their applications include:

1. Filtering purposes, such as separation, purification, and barriers.
    - In Gilbert et al. (1982), nPorC for gas chromatography and high-performance liquid chromatography is produced by impregnating a suitable silica gel or porous glass with a phenol-formaldehyde resin mixture. After polymerization within the pores of the template material, the polymer is converted to glassy carbon by heating in an inert atmosphere to about 1000°C. The silica template is then removed by alkali to and

finally the material is fired in an inert atmosphere at a high temperature in the range 2000–2800°C to anneal the surface, remove micropores, and depending upon the temperature, produce some degree of graphitization. In Corbin et al., (2001), nPorC with additives including TiO_2, SiO_2, and ethylene glycol synthesized by pyrolysis of one or more layers of polymer containing materials, wherein at least one of the pyrolyzed layers is a polymer and additive mixture, on a porous substrate so as to produce a thin mixed matrix film with pores for the separation of small molecules. Ribeiro et al., (2008) provided adsorption equilibrium data of CO_2, CH_4, and N_2 at 299 K to 423 K temperature and 0 kPa to 700 kPa pressure.

2. For storage, such as electricity storage and matter storage.
   - *Energy storage.* Oh et al., (2000) described a method of producing nPorC materials with pore sizes from 2 nm to 20 nm. Rufford et al., (2009) described a method of reacting waste coffee grounds (or sugar cane bagasse) with an activating agent in an environment including at least one inert gas. Malentin et al., (2012) includes additional doping the surface of the pores with nitrogen atoms ($>10\% N_2$).
   - *Fluids storage.* Dimeo et al., (2008) describes the producing of nPorC composite having an average pore diameter of less than 10 nm, and when nPorC incorporates boron provides a hydrogen storage medium otherwise the use is as a chlorine storage medium (when according to the claims, the actual capacity of the porous carbon exceeds the capacity of the high-pressure liquid containment volume by approximately 30%). Also, according to the same claims, in addition to chlorine, the approach may be applied to other industrial gases. Carruthers et al., (2011) is described a nPorC prepared for fluid storage, dispensing and desulfurization (claiming a fill density) >400 g/L, measured in $gAsH_3$ at 25°C and 650 mmHg per litter nPorC, or >30% pores sizes in 0.3–0.72 nm and >20% micropores <2 nm or been formed by pyrolysis and optional activation, and having density of from 0.80 g/cm^3 to 2.0 g/cm^3.
   - *Solid-like storage.* Ting et al., (2015) reported confinement of molecular hydrogen in nPorC with characteristics commensurate with solid H_2 at temperatures up to 67 K above the liquid-vapor critical temperature of bulk H_2 occurring at 0.02 MPa pressure.

3. For transformations, such as an actuator.
    -   In Biener et al., (2010), an actuator is designed to include an nPorC aerogel exhibiting charge-induced reversible strain, when nPorC is used as an electrode immersed in an electrolyte. When a small voltage relative to a reference electrode is applied to the nPorC, an actuation appears.
4. For catalyze, such as the one prospected in Yu et al. (2010), reviewed in Su et al. (2013), and prospected in Qi & Su (2014).
5. For transport, such as condensed hydrocarbons (Falk et al., 2015).

Uses of nPorC of continuing importance include catalysis (Sharma, 1991), gas masks (Bakajin et al., 2011), and water treatment (Torad et al., 2014), capture of volatile organic compounds (Dziura et al., 2014) as well as newer includes fuel cell electrodes (Ji et al., 2010), environmental remediation (Chartier et al., 2014), rechargeable batteries (Zhao et al., 2015), H_2 (Jae et al., 2012), and CH_4 (Choi et al., 2016) storage, and controlled drug delivery (Saha et al., 2015).

28.5 MULTI-/TRANS-DISCIPLINARY CONNECTION(S)

Along with its use as reinforcements for composites, where the content of NanoC is below 5%, and where the integrated performances are based on the matrix materials (Ajayan & Tour 2007).

Assembling NanoC into macroscopic superstructures is a strategy to devise multifunctional materials (for further details see Shimoda et al., 2002; Zhou et al., 2002; Lee et al., 2011; and Wu et al., 2012).

On the one hand, zeolites are another group of porous structures, and on the other hand, are used as templates to provide nPorC. Four zeolites with large cages are given in Table 28.2.

It is difficult to evaluate the specific surface area of nPorC (see Stoeckli, 1990). A series of measurement and estimation strategies were designed, including:

1. *Particle size distribution from diffraction.* Laser diffraction spectroscopy (monochromatic light from a He-Ne laser can be used) uses the diffraction patterns (Fraunhofer diffraction equation, see Stewart & Korff, 1928) of the laser beam to measure geometrical dimensions of particles (see Figure 28.2). Particles of a given size diffract light through a given angle increasing with decreasing size (see McCave et al., 1986).

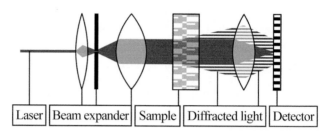

| Laser | Beam expander | Sample | Diffracted light | Detector |

FIGURE 28.2   Particle size distribution from diffraction analysis.

TABLE 28.2   Zeolites with Large Pores

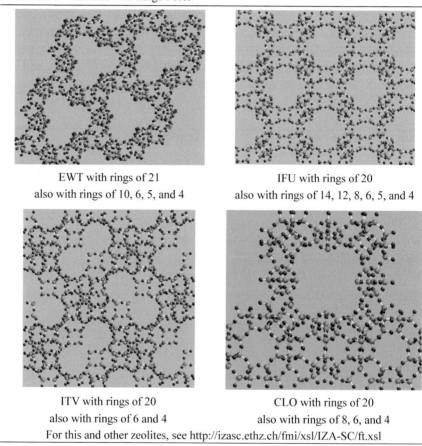

EWT with rings of 21
also with rings of 10, 6, 5, and 4

IFU with rings of 20
also with rings of 14, 12, 8, 6, 5, and 4

ITV with rings of 20
also with rings of 6 and 4

CLO with rings of 20
also with rings of 8, 6, and 4

For this and other zeolites, see http://izasc.ethz.ch/fmi/xsl/IZA-SC/ft.xsl

2.   *Specific surface area estimated from experimental measurements.*
The specific surface area can be estimated from experimental
measurements using different levels of theory (see Table 28.3).

TABLE 28.3   Monolayer and Multilayer Site Binding Models

Model	Langmuir	Brunauer–Emmett–Teller
Labels	p_A: partial pressure of adsorbate A; $[S]_0$, $[S]$, $[AS]$: concentrations of all (initial), free, and occupied sites; k_1, k_{-1}, K_{eq}: rate and equilibrium constants;	$[A_iS]$ concentration of the sites with i layers of A; p, p_0: equilibrium and saturation pressures of A; v, v_m: absorbed A total and monolayer volumes; E_1, E_L: heats for 1st layer and liquefaction;
Derive	$A_{(g)} + S_{(s)} \rightleftharpoons AS_{(s)}$ $k_1 p_A[S] = v_1 = v_{-1} = k_{-1}[AS]$ $\rightarrow K_{eq} p_A[S] = [AS]$ $[S]_0 = [S] + [AS] \rightarrow [AS]/[S]_0$ $= K_{eq} p_A/(1 + K_{eq} p_A)$ Rearranging, $p_A/[AS] = [S]_0^{-1} p_A + [S]_0^{-1} K_{eq}^{-1}$	$A_{(g)} + A_{i-1}S_{(s)} \rightleftharpoons A_iS_{(s)}$ $k_i p_A[A_{i-1}S] = v_{i-1} = v_{-i} = k_{-i}[A_iS]$ $k_1 = \exp(-E_1/RT)$, $k_i = \exp(-E_L/RT)$ $c = \exp(E_1/RT - E_L/RT)$ $(p_0/p-1)^{-1}v^{-1} = (v_m^{-1} - v_m^{-1}c^{-1})p/p_0 + v_m^{-1}c^{-1}$
Plots	$y = p_A/[AS]$ $\hat{y} = ax+b$ $x = p_A$ $[S]_0 = a^{-1}$ $\rightarrow K_{eq} = a/b$	$y = v^{-1}(x^{-1}-1)^{-1}$ $\hat{y} = ax+b$ $x = p/p_0$ $v_m = (a+b)^{-1}$ $\rightarrow c = 1 + a/b$
SSA	$A_{Langmuir} = [S]_0 \cdot N_A \cdot s/m(S)$	$A_{total} = v_m \cdot N_A \cdot s/V_0(A)$; $A_{BET} = A_{total}/m(S)$

Specific surface area (A_{Lng} or A_{BET}): accessible area of solid surface per unit mass of material

s = cross-section (occupied surface area) of one molecule of adsorbate;

m(S) = mass of adsorbent; $N_A = 6.023 \times 10^{23}$ mol^{-1}; $V_0(A)$: molar volume of A;

Extending Langmuir theory (Langmuir, 1918) for monolayer absorption to multilayer absorption, the Brunauer-Emmett-Teller theory (or BET, see Brunauer et al., 1938) assumes (1) gas molecules physically adsorb on a solid in layers infinitely; (2) there is no interaction between each adsorption layer; and (3) the Langmuir theory can be applied to each layer (see Table 28.3). A modified BET technique has been adopted as ISO-9277:2010 (ISO 2010) and as ASTM-D6556-14 (ASTM, 2014).

3. *Qualitative surface analysis with microscopy.* Scanning electron microscopy (see Figure 28.3) is involved for porosity visualization (for instance refer Liu et al., 2016).

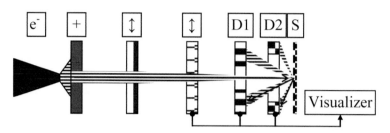

FIGURE 28.3   Porosity visualization with scanning electron microscopy.

4. *Profiling from X-ray scattering (see Figure 28.4).* Small angle X-ray scattering brings information about phenomena impacting large distance range such as pore filling, and wide angle X-ray scattering is used to study short (interatomic) distances (Trognko et al., 2015).

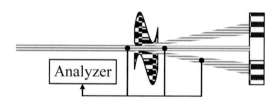

FIGURE 28.4   X-ray scattering for profiling.

5. *Mercury's porosimetry and liquid nitrogen physisorption.* Within a vacuum, mercury is intruded at high pressure into the sample (by using a porosimeter, see Figure 28.5). The pore size is determined based on the external pressure needed to force the liquid into the sample (Abell et al., 1999). A similar experimental device is used for liquid N_2 absorption at $-196°C$ (Oschatz, 2014). Derived parameters include total pore surface area, pore size distribution, and total pore volume.

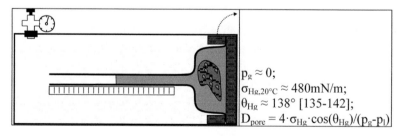

$p_g \approx 0;$
$\sigma_{Hg,20°C} \approx 480mN/m;$
$\theta_{Hg} \approx 138° [135-142];$
$D_{pore} = 4 \cdot \sigma_{Hg} \cdot \cos(\theta_{Hg})/(p_g-p_l)$

FIGURE 28.5   Porosimetry from mercury absorption.

6. *Measurement of the heat of adsorption.* Adsorption heat can be obtained from heating/cooling profile (see Figure 28.6) with an absorbed reference gas (such as *n*-butane, see Wollmann et al., 2012) into the pores. The specific surface area can be retrieved after calibration with reference materials (Wollmann et al., 2011).

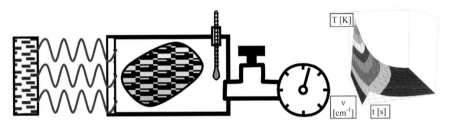

FIGURE 28.6   Porosimetry from thermal measurements.

28.6 OPEN ISSUES

A series of models were designed to estimate the absorption space, specific surface area as well as the pore sizes distribution. For further details, see Emmett (1948), Dubinin (1960), and Biggs & Buts (2006).

28.7 NANOCARBONS

So-called nanocarbons refer graphitic materials with at least one dimension below 100 nm: fullerenes, nanodiamonds (diamonds with a size below 1 µm), nano-onions (onion-like carbon), CNTs (carbon nanotubes, single-walled–SWCNT and multi-walled–MWCNT), nanofoams, graphene, nanofibers, nanoribbons, etc. (see Figure 28.7).

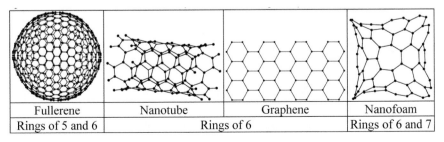

Fullerene	Nanotube	Graphene	Nanofoam
Rings of 5 and 6	Rings of 6		Rings of 6 and 7

FIGURE 28.7   Nanocarbons.

KEYWORDS

- **activated carbon**
- **carbon aerogels**
- **nanocarbon**
- **porous carbon**

REFERENCES AND FURTHER READING

Abell, A. B., Willis, K. L., & Lange, D. A., (1999). Mercury intrusion porosimetry and image analysis of cement-based materials. *Journal of Colloid and Interface Science, 211*, 39–44.

Ajayan, P. M., & Tour, J. M., (2007). Nanotube composites. *Nature, 447*, 1066–1068.

ASTM, (2014). *Standard Test Method for Carbon Black-Total and External Surface Area by Nitrogen Adsorption.* ASTM D6556–14. doi: 10.1520/D6556–14.

Bakajin, O., Holt, J., Noy, A., & Park, H. G., (2006). *Membranes for Nanometer-Scale Mass Fast Transport.* Patent US8038887 B2 from Oct. 18, 2011.

Baker, F. S., Miller, C. E., & Repik, E. D., (1992). *Kirk-Othmer Encyclopedia of Chemical Technology, 4.* New York: J. Wiley.

Barsoum, M. M., & Gogotsi, Y., (2004). *Nanoporous Carbide-Derived Carbon with Tunable Pore Size.* Patent WO2005007566 A3 from Jun. 30, 2005.

Biener, J., Baumann, T. F., Shao, L., & Weissmueller, J., (2010). *Nanoporous Carbon Actuator and Methods of Use Thereof.* Patent US2010230298 A1 from Sept. 16, 2010.

Biesmans, G., Mertens, A., Duffours, L., Woignier, T., & Phalippou, J., (1998). Polyurethane-based organic aerogels and their transformation into carbon aerogels. *Journal of Non-Crystalline Solids, 225*, 64–68.

Biggs, M. J., & Buts, A., (2006). Virtual porous carbons: What they are and what they can be used for. *Molecular Simulation, 32*(7), 579–593.

Brunauer, S., Emmett, P. H., & Teller, E., (1938). Adsorption of gases in multimolecular layers. *Journal of the American Chemical Society, 60*(2), 309–319.

Carruthers, J. D., Dimeo, F. Jr., & Bobita, B., (2011). *Nanoporous Articles and Methods of Making Same.* Patent US2011220518 A1 from Sept. 15, 2011.

Chartier, Y., Emmanuel, J., Pieper, U., Prüss, A., Rushbrook, V. P., Stringer, R., Townend, W., Wilburn, S., & Zghondi, R., (2014). *Safe Management of Wastes from Health-Care Activities.* Geneva: World Health Organization.

Choi, P. S., Jeong, J. M., Choi, Y. K., Kim, M. S., Shin, G. J., & Park, S. J., (2016). A review: Methane capture by nanoporous carbon materials for automobiles. *Carbon Letters, 17*(1), 18–28.

Cooney, O., (1980). *Activated Charcoal: Antidotal and Other Medical Uses.* New York: M. Dekker, p. 160.

Corbin, D. R., Foley, H. C., & Shiflett, M. B., (2001). *Mixed Matrix Nanoporous Carbon Membranes.* Patent EP1292380 A1 from Mar. 19, 2003.

Czakkel, O., Marthi, K., Geissler, E., & Lászlo, K., (2005). Influence of drying on the morphology of resorcinol-formaldehyde-based carbon gels. *Microporous and Mesoporous Materials, 86*, 124–133.

Debye, P., & Scherrer, P., (1917). X-ray interference produced by irregularly oriented particles. III. Constitution of graphite and amorphous carbon. *Physikalische Zeitschrift, 18*, 291–301.

Dimeo, F. Jr., Carruthers, J. D., Wodjenski, M. J., McManus, J. V., & Marzullo, J., (2007). *Nanoporous Carbon Materials, and Systems and Methods Utilizing Same.* Patent WO2007136887 A2 from Jul. 24, 2008.

Dubinin, M. M., (1960). The potential theory of adsorption of gases and vapors for adsorbents with energetically nonuniform surfaces. *Chemical Reviews, 60*(2), 235–241.

Dziura, A., Marszewski, M., Choma, J., Souza, L. K. C., De Osuchowski, Ł., & Jaroniec, M., (2014). Saran-derived carbons for CO_2 and benzene sorption at ambient conditions. *Ind. Eng. Chem. Res., 53*(40), 15383–15388.

Emmett, P. H., (1948). Adsorption and pore-size measurements on charcoals and whetlerites. *Chemical Reviews, 43*(1), 69–148.

Falk, K., Coasne, B., Pellenq, R., Ulm, F. J., & Bocquet, L., (2015). Subcontinuum mass transport of condensed hydrocarbons in nanoporous media. *Nature Communications, 6*, 6949(7p).

Gilbert, M. T., Knox, J. H., & Kaur, B., (1982). Porous glassy carbon, a new columns packing material for gas chromatography and high-performance liquid chromatography. *Chromatographia, 16*, 138–146.

Gogotsi, Y., Yushin, G., Hoffman, E. N., & Barsoum, M. M., (2006). *Process for Producing Nanoporous Carbide-Derived Carbon with Large Specific Surface Area.* Patent WO2007062095 A1 from May 31, 2007.

ISO, (2010). *Determination of the Specific Surface Area of Solids by Gas Adsorption–BET Method.* ISO 9277, 2010 (E). Geneva: International Organization for Standardization.

Ji, X., Lee, K. T., Holden, R., Zhang, L., Zhang, J., Botton, A. G., Couillard, M., & Nazar, L. F., (2010). Nanocrystalline intermetallics on mesoporous carbon for direct formic acid fuel cell anodes. *Nature Chemistry, 2*, 286–293.

Jong, S. Y., Jin, G. L., & Seok, C., (2002). *Method for Preparing Nanoporous Carbons with Enhanced Mechanical Strength and the Nanoporous Carbons Prepared by the Method.* Patent US7326396 B2 from Feb. 5, 2008.

Kabbour, H., Baumann, T. F., Satcher, J. H. Jr., Saulnier, A., & Ahn, C. C., (2006). Toward new candidates for hydrogen storage: High-surface-area carbon aerogels. *Chemistry of Materials, 18*(26), 6085–6087.

Langmuir, I., (1918). The adsorption of gases on plane surfaces of glass, mica, and platinum. *Journal of the American Chemical Society, 40*(9), 1361–1403.

Lee, S. H., Lee, D. H., Lee, W. J., & Kim, S. O., (2011). Tailored assembly of carbon nanotubes and graphene. *Adv. Funct. Mater., 21*, 1338–1354.

Li, W. C., & Guo, S., (2000). Preparation of low-density carbon aerogels from a cresol/formaldehyde mixture. *Carbon, 38*(10), 1520–1523.

Liu, X., Wang, C., Wu, Q., & Wang, Z., (2016). Magnetic porous carbon-based solid-phase extraction of carbamates prior to HPLC analysis. *Microchimica Acta, 183*(1), 415–421.

Maletin, Y., Stryzhakova, N., Zelinskyi, S., Gromadskyi, D., & Tychyna, S., (2012). *Method for Selecting Nanoporous Carbon Material for Polarizable Electrode, Method for Manufacturing Such Polarizable Electrodes and Method for Manufacturing Electrochemical Double Layer Capacitor.* Patent US2013139951 A1 from July 6, 2013.

McCave, I. N., Bryant, R. J., Cook, H. F., & Coughanowr, C. A., (1986). Evaluation of a laser-diffraction-size analyzer for use with natural sediments: Research method paper. *Journal of Sedimentary Petrology, 56*(4), 561–564.

Mohun, W. A., (1959). Mineral *Active Carbon and Process for Producing Same.* Patent US3066099 A from Nov. 27, 1962.

Oh, S. M., Hyeon, T. H., & Han, S. J., (2000). *Method for Preparing Nanoporous Carbon Materials and Electric Double-Layer Capacitors Using Them.* Patent US6515845 B1 from Feb. 4, 2003.

Oschatz, M., (2014). *New Routes Towards Nanoporous Carbon Materials for Electrochemical Energy Storage and Gas Adsorption.* PhD Thesis (Ph.D. Advisor: Kaskel, S.), Technical University of Dresda.

Pekala, R. W., (1989). Organic aerogels from the polycondensation of resorcinol with formaldehyde. *Journal of Materials Science, 24,* 3221–3227.

Pekala, R. W., Alviso, C. T., Lu, X., Gross, J., & Fricke, J., (1995). New organic aerogels based upon a phenolic-furfural reaction. *Journal of Non-Crystalline Solids, 188*(1/2), 34–40.

Qi, W., & Su, D., (2014). Metal-free carbon catalysts for oxidative dehydrogenation reactions. *ACS Catalysis, 4,* 3212–3218.

Ribeiro, R. P., Sauer, T. P., Lopes, F. V., Moreira, R. F., Grande, C. A., & Rodrigues, A. E., (2008). Adsorption of CO_2, CH_4, and N_2 in activated carbon honeycomb monolith. *Journal of Chemical & Engineering Data, 53*(10), 2311–2317.

Rufford, T. E., Jurcakova, D., Zhu, Z., & Lu, G. Q., (2009). *Nanoporous Carbon Electrodes and Supercapacitors Formed There From.* Patent WO2010020007 A1 from Feb. 25, 2010.

Saha, D., Moken, T., Chen, J., Hensley, D. K., Delaney, K., Hunt, M. A., et al., (2015). Micro-/mesoporous carbons for controlled release of antipyrine and indomethacin. *RSC Advances, 5,* 23699–23707.

Sharma, B. K., (1991). *Industrial Chemistry (Including Chemical Engineering).* Meerut: Goel.

Shimoda, H., Oh, S. J., Geng, H. Z., Walker, R. J., Zhang, X. B., McNeil, L. E., & Zhou, O., (2002). Self-assembly of carbon nanotubes. *Adv. Mater., 14,* 899–901.

Sing, K. S. W., Everett, D. H., Haul, R. A. W., Moscou, L., Pieroti, R. A., Rouquerol, J., & Siemieniewska, T., (1985). Reporting physisorption data for gas/solid systems with special reference to the determination of surface area and porosity (IUPAC Recommendations 1984). *Pure and Applied Chemistry, 57*(4), 603–619.

Stewart, J. Q., & Korff, S. A., (1928). The refractive index of sodium vapor and the width of the D lines in absorption. *Physical Review, 32,* 676–680.

Stoeckli, H. F., (1990). Microporous carbons and their characterization: The present state of the art. *Carbon, 28*(1), 1–6.

Su, D. S., Perathoner, S., & Centi, G., (2013). Nanocarbons for the development of advanced catalysts. *Chemical Reviews, 113,* 5782–5816.

Thommes, M., Kaneko, K., Neimark, A. V., Olivier, J. P., Rodriguez-Reinoso, F., Rouquerol, J., & Sing, K. S. W., (2015). Physisorption of gases, with special reference to the evaluation of surface area and pore size distribution (IUPAC Technical Report). *Pure and Applied Chemistry, 87*(9/10), 1051–1069.

Ting, V. P., Ramirez-Cuesta, A. J., Bimbo, N., Sharpe, J. E., Noguera-Diaz, A., Presser, V., Rudic, S., & Mays, T. J., (2015). Direct evidence for solid-like hydrogen in a nanoporous carbon hydrogen storage material at supercritical temperatures. *ACS Nano, 9*(8), 8249–8254.

Torad, N. L., Hu, M., Ishihara, S., Sukegawa, H., Belik, A. A., Imura, M., Ariga, K., Sakka, Y., & Yamauchi, Y., (2014). Direct synthesis of MOF-derived nanoporous carbon with magnetic Co nanoparticles toward efficient water treatment. *Small, 10,* 2096–2107.

Trognko, L., Lecante, P., Ratel-Ramond, N., Rozier, P., Daffos, B., Taberna, P. L., & Simon, P., (2015). TiC-carbide derived carbon electrolyte adsorption study byways of X-ray scattering analysis. *Materials for Renewable and Sustainable Energy, 4*, 17(6p).

Wollmann, P., Leistner, M., Grählert, W., Stoeck, U., Grünker, R., Gedrich, K., et al., (2011). High-throughput screening: Speeding up porous materials discovery. *Chemical Communications, 47*(18), 86–94.

Wollmann, P., Leistner, M., Grählert, W., Throl, O., Dreisbach, F., & Kaskel, S., (2012). Infrasorb: Optical detection of the heat of adsorption for high throughput adsorption screening of porous solids. *Microporous and Mesoporous Materials, 149*(1), 86–94.

Wrench, J., (1931). Origin, properties, and uses of activated carbon. *Oil & Fat Industries, 8*(12), 441–453.

Wu, D., Zhang, F., Liang, H., & Feng, X., (2012). Nanocomposites and macroscopic materials: Assembly of chemically modified graphene sheets. *Chem. Soc. Rev., 41*, 6160–6177.

Yang, S. J., Jung, H., Kim, T., & Park, C. R., (2012). Recent advances in hydrogen storage technologies based on nanoporous carbon materials. *Progress in Natural Science: Materials International, 22*(6), 631–638.

Yu, D., Nagelli, E., Du, F., & Dai, L., (2010). Metal-free carbon nanomaterials become more active than metal catalysts and last longer. *The Journal of Physical Chemistry Letters, 1*, 2165–2173.

Zhang, R., Li, W., Liang, X., Wu, G., Lü, Y., Zhan, L., Lu, C., & Ling, L., (2003). Effect of hydrophobic group in polymer matrix on porosity of organic and carbon aerogels from sol-gel polymerization of phenolic resole and methylolated melamine. *Micropor. Mesopor. Mater., 62*(1/2), 17–27.

Zhao, Q., Lu, Y., Zhu, Z., Tao, Z., & Chen, J., (2015). Rechargeable lithium-iodine batteries with iodine/nanoporous carbon cathode. *Nano Letters, 15*(9), 5982–5987.

Zhou, O., Shimoda, H., Gao, B., Oh, S., Fleming, L., & Yue, G., (2002). Materials science of carbon nanotubes: Fabrication, integration, and properties of macroscopic structures of carbon nanotubes. *Acc. Chem. Res., 35*, 1045–1053.

CHAPTER 29

Nanospheres–Nanocubes–Nanowires

CRISTINA MOŞOARCĂ,[1] RADU BĂNICĂ,[1] and PETRICA LINUL[1,2]

[1]National Institute of Research and Development for Electrochemistry and Condensed Matter, Timisoara, 300569, Romania, Tel. +40752197759, E-mail: mosoarca.c@gmail.com, radu.banica@yahoo.com

[2]Politechnica University Timisoara, 300006, Romania

29.1 DEFINITION

Currently, metal nanostructures are emerging from shape-controlled synthesis, the predefined morphology being the aim. Due to the practical importance of metal nanostructures, we will try to approach the metal nanoparticles topic. We will mostly discuss the Ag and Au nanostructures because of their resistance to corrosion. Metal nanoparticles are used in applications involving electronics, photonics, photography, catalysis, information storage, biological labeling, optoelectronics, and imaging.

Generally, NSs-NCs-NWs, are defined as being reduced size objects that behave as a single element with respect to their transport and properties. NSs-NCs-NWs are particles with size ranging from 1 to 10 μm. NSs and NCs are belonging to the class of ultrafine and fine particles ranging between 1 nm and 1 μm, while the NWs can exceed the 10 μm limit of course class particles. Nowadays metal nanoparticles shape and size control still represents a great challenge, especially due to the possibility of controlling the optical properties of materials. Noble metals nanoparticles (Ag, Au, Pt) have the capacity of interacting with UV-vis light through a localized surface plasmon resonance (LSR) which is excited therefore being sensitive to the nanoparticles sizes and shapes.

29.2 HISTORICAL ORIGINS AND NANO-SCIENTIFIC DEVELOPMENTS

Nanoparticles have a long history that reaches the 9th century when artisans were decorating the surface of pots in Mesopotamia. The glittering at the surface of the pots was made of a metallic film containing silver, gold and copper nanoparticles dispersed evenly in the matrix of the ceramic coating (Reiss et al., 2010). In his paper, Experimental Relations of Gold (and Other Metals) to Light, Michael Faraday gave the first scientific description of nanometer-scale metals and their optical properties (Michael Faraday, 1857).

Nanoparticle research is a developing scientific area with many applications in various fields like biomedicine, electronics, and optics (Taylor et al., 2013).

The shape controlled nanostructures synthesis is mostly performed by the polyol process (Xia et al., 2009) in ethylene glycol (EG) accompanied by varying different parameters, like molar ratio between the metallic ions and the stabilizers, concentration of the precursors, addition of helper agents and temperature of the reaction. The reaction conditions are greatly influencing the shapes of the final structures. The reactant injection rate and the reaction atmosphere are very important for the architecture of the resulting products (Wiley et al., 2004), this is due to the higher reactivity exhibited in the presence of oxygen which leads to preferentially etching of the twinned particles. Contrariwise, without oxygen, there is no oxidation etching of the twinned particles resulting the generation of NWs. Ionic species and stabilizers are also influencing the shape and size of the nanoparticles. In the case of silver NWs, polyvinyl pyrrolidone (PVP) is the most common stabilizer used for their synthesis (Sun et al., 2002). If the amount of PVP added is low or absent, the product will consist mainly from NSs. The molecular weight of PVP is influencing the result; by adding a lower molecular weight PVP, the obtained products will most probably have irregular shapes. The addition of Cl^-, Br^-, Fe^{3+} support the NCs, NWs and bipyramids formation. To obtain a high yield of nanostructures with predefined shapes, without a large amount of by-products, the reactant conditions must be carefully supervised and controlled (Jin et al., 2003, 2012) (Figure 29.1).

Silver NCs were synthesized by Sun and Xia (2002) by reducing the silver nitrate with ethylene glycol in the presence of PVP (polyvinylpyrrolidone). According to the results of this study, the product morphology is due, in a great extent, to the reaction conditions, among which, the temperature, $PVP-AgNO_3$ molar ratio, and $AgNO_3$ concentration, has a very important contribution. When the temperature was reduced to 120°C or increased to

190°C from 161°C, the obtained shapes of the nanostructures were irregular. Im et al. (2005) synthesized silver NCs in the presence of HCl and PVP at 140°C using ethylene glycol as reducing agent for $AgNO_3$.

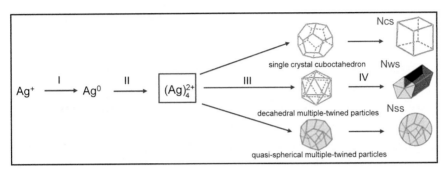

FIGURE 29.1 Silver ions reduction by EG (I); silver cluster formation (II); seeds nucleation (III), the formation of NCs, nanorods or NWs and NSs (IV), redrawn and adapted from Wiley et al., (2005.

In Figure 29.2, Nss-Ncs are presented in high-resolution microscopy along with their sizes reported to the structural model. Xu et al., (2006) reported a study upon the connection between the shape of metal nanostructures and their catalytic activity in different organic and inorganic environments.

Poinerm et al. (2013) used leaf extracts from *Eucalyptus macrocarpa* to synthesize, at room temperature, silver nanocubes. The leaf extracts represented both the stabilizing and the reducing agent. They obtained both spherical and cubic silver nanoparticles, as showed in their TEM analysis. The silver NSs had sizes between 10 and 100 nm, while the silver NCs were smaller, ranging from 10 to 50 nm, the predominant shapes where NCs with larger sizes ranging from 50 nm to 1 μm (Figure 29.3).

AL-Thabaiti surveyed the effect of polymers and surfactants on silver nanoparticles. They utilized polyvinyl alcohol (PVA) and cetyltrimeth-ylammonium bromide (CTAB) as stabilizers and thiosulfate as reducing agent. The concentration of thiosulfate in solution is an important factor in controlling the synthesis of nanomaterials with different optical properties. The presence and absence of stabilizers is a determining factor regarding the shape of the obtained structures, as shown in Figure 29.3, leading to spherical, hexagonal plates or rod-like shapes (AL-Thabaiti et al., 2013).

Liang synthesized uniform spherical NSs with a diameter of 54 nm using PVP as capping agent and PEG as both solvent and reducing agent, at 260°C (Liang et al., 2010). Qin obtained NSs in aqueous bath at 30°C using citrate as stabilizer and ascorbic acid as reducing agent. They observed that the size of the NSs decreased from 73 nm to 31 nm with the increase of the pH from 6 to 11.5. The researchers found a correlation between temperature and shape of the synthesized nanoparticles, by keeping the samples at 100°C for 2 hours they obtained more spherical structures (Qin et al., 2010).

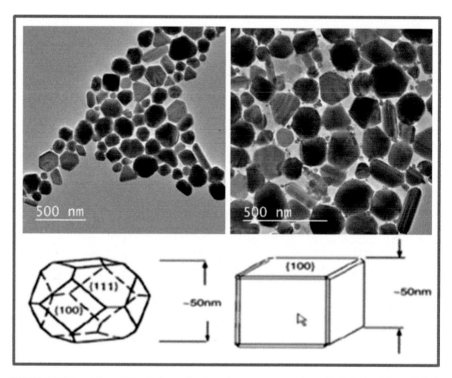

FIGURE 29.2　Silver NSs, NCs and other shapes supported on Cu-TEM grids and their structural models, adapted from Xu et al., 2006.

Small size NSs, less than 20 nm, where produced by using glucose for the silver nitrate reduction and gelatin as a capping agent. The colloidal solution, containing NSs with sizes ranging from 5 to 24 nm, remained stable at room temperature for more than 3 months (Aguilare et al., 2011).

Silver NWs with an average length of around 4 µm and a diameter of 50 nm and have been synthesized by the polyol method. The synthesized NWs were deposited on a flexible and transparent support of polyethylene terephthalate (PET). The conductivity and adhesion of the silver NWs were improved by interposing a thin film of polymethyl methacrylate (PMMA) between the layer of NWs and the PET substrate (Banica et al., 2016).

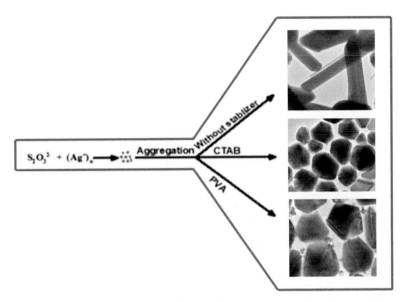

FIGURE 29.3 Formation and the stabilizers influence upon the silver nanoparticles shape (rods, spherical and hexagonal), adapted from AL-Thabaiti et al., 2013.

Using a silane-mediated approach, Yu et al., synthesized ultrathin scale (1–2 nm) gold NWs. The NWs contain stacking fault defects that evolved at sustained electron exposure leading to breaking and necking. Their work was very important for the ultrathin metal NWs structure understanding at atomic level, which could have major effects on the possible applications (Yu et al., 2016).

29.4 MULTI-/TRANS-DISCIPLINARY CONNECTION(S)

Nowadays the scientific focus is turned towards various polymeric and metal nanoparticles, as small size chemically modified and with controlled geometry nanodiscs, nanorods, NSs, NCs, and NWs are attracting more interest, especially in healthcare. The general aim is to produce biocompatible compounds which are combining the nontoxic biomaterials, like chitosan, albumin, carbon nanotubes, lipopolysaccharides and polysaccharides with metal and noble metal ions (Au, Ag, Pt) in order to further explore the medical applications that can derive from these procedures (Rani, 2015).

The coupling of distinct metals physicochemical properties made bimetallic nanoparticles an important area of interest for numerous applications. Kim et al. (2015) synthesized ordered Au-Cu NWs and utilized them in a nanoelectrocatalytic glucose detection application.

CuI was used as a precursor, as shown in Figure 29.4, and Au NWs were the preferred substrate for the ordered intermetallic Au-Cu NWs synthesis. CuI can generate in gas-phase the following reaction products: Cu_3I_3, $2Cu$, I_2, Cu_2, from which Cu_3I_3 and $2Cu$ can provide Cu atoms by collision-induced dissociation or by direct collision with the Au NWs. The resulting Cu atoms will form Au-Cu NWs by diffusion into the Au NWs.

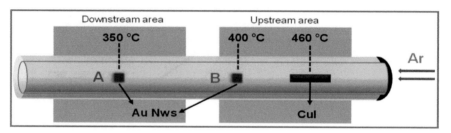

FIGURE 29.4   Experimental setup schematics for the ordered Au-Cu NWs synthesis. The setup contains a horizontal quartz tube (yellow) furnace with two heating areas. Au_3Cu and $AuCu_3$ NWs are the synthesis results of the reaction between CuI (precursor) and Au NWs (substrate) placed at positions A and B, adapted from Kim et al. (2015).

The green preparation of Ncs/nanodiscs by using jasmine oil as biocatalyst was shown to be non-toxic, eco-friendly, non-allergic, cheap and safe. This novel method can be utilized for preparing bovine serum albumin nanostructures that may be used for pharmaceutical, clinical and therapeutic applications employing optimizations and loading of other metals and drugs, antibiotics, enzymes, and spliced genes. This may be a path for becoming drug and gene delivery carriers (Rani, 2015).

Jebali et al. (2013) evaluated *in vitro* the antifungal activity of silver and gold nanostructures, NCs, NSs, and NWs, on three different Candida species (*C. albicans, C. glabrata,* and *C. tropicalis*). They concluded that silver and gold NCs had higher antifungal activity against the tested Candida species than NSs and NWs. A few isolates were resistant to gold and silver NSs and NWs, none of the isolates tolerated silver and gold NCs.

29.5   HISTORICAL COMMENT

Although there is no specific period of time at which the beginning of *NSs-NCs-NWs* study, as a unified topic, can be identified, the book of Michael Faraday (Faraday, 1857) is the first attempt to treat from a scientific point of view such a subject.

29.6 OPEN ISSUES

Tang and Tsuji (2010) extended the microwave polyol method by adding an air-assisted technique in which bubbling air and nitrogen are used for preparing the Ag nanostructures. Figure 29.5 (A, B) depicts TEM images of Ag NWs with sizes ranging in between 5 and 30 µm in length and 40 nm diameter obtained under bubbling air. Furthermore, they have obtained mixtures of NCs, NWs, and bipyramids using bubbling N_2 as shown in Figure 29.5 (C, D). The yield of NWs prepared without bubbling gas was midway (75%) between that obtained under bubbling N_2 (20%) and air, which rendered a very high output, reaching up to 90%.

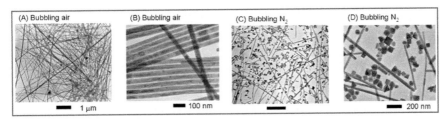

FIGURE 29.5 TEM images of Ag nanoparticles obtained from $AgNO_3$ (46.5mM)/PVP (264 mM) /KCl (0.3 mM) /EG mixtures at 198°C by bubbling air (A,B) and N_2 (C,D) (Tang and Tsuji, 2010). Reprinted with permission. © 2010 Tang X, Tsuji M. Published [short citation] under CC BY-NC-SA 3.0 license. Available from: http://dx.doi.org/10.5772/39491.

AgCl NCs were gown on the sidewalls of Ag NWs (Khaligh and Gold-thorpe, 2015). The researchers utilized a simple low-temperature annealing process (100–150°C) and a vacuumed environment. The NCs size and density are affected by several factors, like the annealing temperature, the time length in which the annealing process is conducted and the diameter of the NWs. The hybrid nanostructure, as the authors are calling it, may have various applications in photovoltaics, photocatalysis, and Raman (Khaligh and Goldthorpe, 2015).

Recently tellurium based nanomaterials have been accepted as being biocompatible effective antibacterial agents. Researchers are trying to integrate nanoparticles based on this p-type semiconductor into carbon fiber fabrics. They developed Te and Te-Au NWs, with controlled length by adjusting the concentrations of the TeO_2 precursor and of the N_2H_4 reducing agent (Figure 29.6).

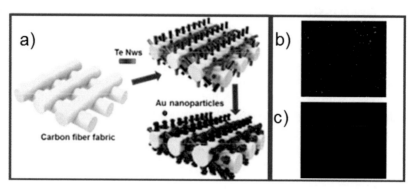

FIGURE 29.6   Schematics of (a) carbon fiber fabric with Te-NWs and Te-Au nanoparticles; fluorescence images of (b) live (green), and (c) dead (red) bacteria (Reprinted from Cho et al., 2016. Copyright © 2016 by the authors; licensee MDPI, Basel, Switzerland.).

Since Au nanoparticles have a selective growth on the Te NWs, the Te-Au Nws-based carbon fiber fabrics where easily obtained. The latter showed increased inhibition potential towards *Escherichia coli* and *Staphylococcus aureus* when compared to Te Nws-based fabrics. After a short time treatment, 30 minutes, of the carbon fiber fabrics with Te-Au NWs only 10% of bacteria were still alive, thus opening a very promising research area field of antimicrobial clothing products (Cho et al., 2016).

In order to enhance the nanocatalyst capacity for nitrophenol reduction reaction, the authors described a new class of Au nanoalloys. The 4-nitrophenol reduction to aminophenol was spectroscopically studied in aqueous solution on 9.1 nm AuCuPt core-shell NSs, 8.3 nm AuCu NSs and 7.8 nm Au NSs (Mehmood et al., 2016) (Figure 29.7).

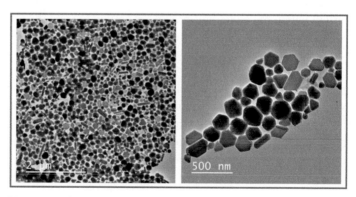

FIGURE 29.7   TEM images of Ag NSs clusters.

The rate constants, entropies and activation energies of the reaction catalyzed at room temperature by the core-shell NSs of AuCu@Pt has values that differ from the reaction catalyzed by the pure metal NSs. The results show that the catalysis performed by nanoparticles is occurring at the highest efficiency on the NSs surface.

The nitrophenol reduction was probably controlled by the Au surface properties and by the Au alloy nanoparticles. The difference in catalytic activity of the Au, AuCu and AuCuPt NSs is possibly due to their morphological differences, where the triangular shape indicates a monometallic Pt shell with an alloy core which are responsible by the synergistic enhancement of the reaction (Mehmood et al., 2016).

The production of nanoscale gaps between metallic surfaces by the assembly of plasmonic nanoparticles is an inexpensive method of generating hotspots when illuminated (intense electric field strength in small volumes). Rabin and Lee (2012) are presenting high-throughput assembly methods of silver NCs into small clusters at predefined locations onto a certain substrate. As shown in Figure 29.8, the silver NCs are grouped in linear assemblies of dimers and clusters of polymers with face-to-face configurations.

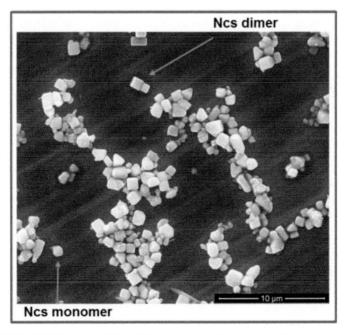

FIGURE 29.8 SEM image of silver NCs monomers, dimers, and polymers

Therefore, metallic nanostructures, with numerous applications in various fields like textile industry, electronics, spectroscopy, healthcare, green chemistry and so on, will continuously be a topic of interest for science worldwide.

KEYWORDS

- **nanogeometry**
- **nanoparticle**
- **nanoshapes**
- **plasmonic**

REFERENCES

Aguilare, M. A. M., Martin, E. S. M., Arroyo, L. O., Portillo, G. C., & Espindola, E. S. J., (2011). Synthesis and characterization of silver nanoparticles: Effect on phytopathogen Colletotrichum gloesporioides. *Nanopart. Res., 13*, 2525–2532.

AL-Thabaiti, S. A., Malik, M. A., Al-Youbi, A. A. O., Khan, Z., & Hussain, J. I., (2013). Effects of surfactant and polymer on the morphology of advanced nanomaterials in aqueous solution. *Int. J. Electrochem. Sci., 8*, 204–218.

Banica, R., Ursu, D., Svera, P., Sarvas, C., Rus, S. F., Novaconi, S., Kellenberger, A., Racu, A. V., Nyari, T., & Vaszilcsin, N., (2016). Electrical properties optimization of silver nanowires supported on polyethylene terephthalate. *Particulate Science and Technology: An International Journal, 34*, 217–222.

Chou, T. M., Ke, Y. Y., Tsao, Y. H., Li, Y. C., & Lin, Z. H., (2016). Fabrication of Te and Te-Au nanowires-based carbon fiber fabrics for antibacterial applications. *Int. J. Environ. Res. Public Health, 13*, 1–9.

Faraday, M., (1857). Experimental relations of gold (and other metals) to light. *Philosophical Transactions of the Royal Society of London, 147*, 145–181.

Im, S. H., Lee, Y. T., Wiley, B., & Xia, Y., (2005). Large-scale synthesis of silver nanocubes: The role of HCl in promoting cube perfection and monodispersity. *Angew. Chem. Int. Ed., 44*, 2154–2157.

Jebali, A., Hajjar, F. H. E., Pourdanesh, F., Hekmatimoghaddam, S., Kazemi, B., Masoudi, A., Daliri, K., & Sedighi, N., (2013). Silver and gold nanostructures: Antifungal property of different shapes of these nanostructures on *Candida* species. *Medical Mycology, 52*, 65–72.

Jin, R. C., Cao, Y. C., Hao, E. C., Metraux, G. S., Schatz, G. C., & Mirkin, C. A., (2003). Controlling anisotropic nanoparticle growth through plasmon excitation. *Nature, 425*, 487–490.

Khaligh, H. H., & Goldthorpe, I. A., (2015). A simple method of growing silver chloride nanocubes on silver nanowires. *Nanotechnology, 26*, 381002.

Kim, S. I., Eom, G., Kang, M., Kang, T., Lee, H., Hwang, A., Yang, H., & Kim, B., (2015). Composition-selective fabrication of ordered intermetallic Au–Cu nanowires and their application to nanosize electrochemical glucose detection. *Nanotechnology, 26*, 245702.

Lee, S. Y., Jeon, H. C., & Yang, S. M., (2012). Unconventional methods for fabricating nanostructures toward high-fidelity sensors. *J. Mater. Chem.*, *22*, 5900–5913.

Liang, H., Wang, W., Huang, Y., Zhang, S., Wei, H., & Xu, H., (2010). Controlled synthesis of uniform silver nanospheres. *J. Phys. Chem. C.*, *114*, 7427–7431.

Mehmood, S., Janjua, N. K., Saira, F., & Fenniri, H., (2016). AuCu@Pt nanoalloys for catalytic application in reduction of 4-nitrophenol. *Journal of Spectroscopy*, 1–8.

Poinern, G. E. J., Chapman, P., Shah, M., & Fawcett, D., (2013). Green biosynthesis of silver nanocubes using the leaf extracts from Eucalyptus macrocarpa. *Nano Bull.*, *2*, 1–7.

Qin, Y., Ji, X., Jing, J., Liu, H., Wu, H., & Yang, W., (2010). Size control over spherical silver nanoparticles by ascorbic acid reduction. *Colloids Surf. A.*, *372*, 172–176.

Rabin, O., & Lee, S. Y., (2012). SERS substrates by the assembly of silver nanocubes: High-throughput and enhancement reliability considerations. *J. of Nanotech.*, 1–12.

Rani, K., (2015). Green Preparation of Jasmine Oil Mediated Bovine Serum Albumin. *Sch. Acad. J. Pharm.*, 5, 9–11.

Reiss, G., & Hutten, A., (2010). *Magnetic Nanoparticles Handbook of Nanophysics: Nanoparticles and Quantum Dots*. CRC Press, Taylor & Francis, United States of America.

Sun, Y., & Xia, Y., (2002). Large-scale synthesis of uniform silver nanowires through a soft, self-seeding, polyol process. *Adv. Mater.*, *14*, 833–837.

Sun, Y., & Xia, Y., (2002). Shape-controlled synthesis of gold and silver nanoparticles. *Science, 298*, 2176–2179.

Tang, X., & Tsuji, M., (2010). Syntheses of silver nanowires in liquid phase. Nanowires Science and Technology, 25-42, Published [short citation] under CC BY-NC-SA 3.0 license.

Taylor, R., Coulombe, S., Otanicar, T., Phelan, P., Gunawan, A., Lv, W., Rosengarten, G., Prasher, R., & Tyagi, H., (2013). Small particles, big impacts: A review of the diverse applications of nanofluids. *Journal of Applied Physics, 113*, 011301.

Wiley, B., Herricks, T., Sun, Y. G., & Xia, Y. N., (2004). Polyol synthesis of silver nanoparticles: Use of chloride and oxygen to promote the formation of single-crystal, truncated cubes, and tetrahedrons. *Nano Lett.*, *4*, 1733–1739.

Wiley, B., Sun, Y., Mayers, B., & Xia, Y., (2005). Shape-controlled synthesis of metal nanostructures: The case of silver. *Chem. Eur. J.*, *11*, 454–463.

Xia, Y., Xiong, Y., Lim, B., & Skrabalak, S. E., (2009). Shape-controlled synthesis of metal nanocrystals: Simple chemistry meets complex physics? *Angew. Chem. Int. Ed. Engl.*, *48*, 60–103.

Xu, R., Wang, D., Zhang, J., & Li, Y., (2006). Shape-dependent catalytic activity of silver nanoparticles for the oxidation of styrene. *Chem. Asian J.*, *1*, 888–893.

Yu, Y., Cui, F., Sun, J., & Yang, P., (2016). Atomic structure of ultrathin gold nanowires. *Nano Lett.*, *16*, 3078–3084.

CHAPTER 30

Polyynes

FRANCO CATALDO[1,2]

[1]*Actinium Chemical Research Institute, Via Casilina 1626A, 00133 Rome, Italy*

[2]*Università degli Studi della Tuscia, Via S. Camillo de Lellis, Viterbo, Italy, Tel: +39 06-94368230, Fax: +39 06-94368230, E-mail: franco.cataldo@fastwebnet.it, cataldo.franco@fastwebnet.it*

30.1 DEFINITION

Polyynes or oligoynes are sp-hybridized carbon chains characterized by a sequence of single and triple bonds, as reported in Figure 30.1 (Cataldo, 2005). Very long polyyne chains, with almost infinite length, were defined as "carbyne," but such very long chains were never isolated in pure form. A theoretical distinction was also made between α-carbyne made by an acetylenic chain of infinite length and β-carbyne made by an infinite carbon chain of cumulated double bonds (Heimann et al., 1997). However, thermodynamic calculations show that the cumulenic structure is by far unfavored with respect to the acetylenic structure (Cataldo, 1997). In general, polyynes are produced with a limited chain length which at most can reach 30 carbon atoms or so (Szafert and Gladysz, 2003; Chalifoux and Tykwinski, 2009), and the chain must be end-capped by end groups. Bulkier (in sterical sense) end-groups ensure higher length and stability to the acetylenic carbon chains. In the absence of end groups, the polyyne chains undergo a very rapid crosslinking reaction (Casari et al., 2004).

Also, cyclic polyynes are known as well (Diederich et al., 2005; Stang et al., 2008) and cyclic polyynes are thought to be the precursors of fullerenes as demonstrated by Rubin et al. (1998). Polyyne chains segments are naturally occurring in a large variety of plants and fungi extracts (Bohlmann et al., 1973).

H-(C≡C-)$_n$-H   Polyynes

H-(C≡C)$_n$-C≡N   Monocyanopolyynes

N≡C-(C≡C)$_n$-C≡N   Dicyanopolyynes

R-(C≡C-)$_n$-R'   Generic polyynes with

                        generic end groups

FIGURE 30.1   Chemical structures of stable polyynes.

30.2   HISTORICAL ORIGINS

As already reported in the definition section, polyyne chain segments are found in several plants and fungi extracts (Bohlmann et al., 1973). However, one of the first synthesis and isolation of hydrogen-terminated polyynes was achieved by a relatively complex step synthesis approach in 1972 (Eastmond et al., 1972; Johnson and Walton, 1972). Much later, while studying the acid hydrolysis of oxidized dicopper acetylide and diacetylide Cataldo (1998a and 1998b) obtained a mixture of polyynes in a hydrocarbon solvent and recognized the mixture as formed by polyynes. In addition, the other product of acid hydrolysis of copper acetylide and diacetylide was a carbonaceous matter containing carbynoid structures, i.e., a mixture of sp, sp^2, and sp^3 hybridized carbon as recognized by solid-state ^{13}C-NMR spectroscopy (Cataldo et al., 1999) and by Raman spectroscopy (Cataldo, 2008). Tsuji et al., (2002) reported the formation of hydrogen capped polyynes by laser ablation of graphite targets submerged in a solvent. The same polyynes mixture was then produced by Cataldo (2003) in the easiest way by striking the electric carbon arc immersed in various solvents. When the carbon arc was struck submerged in hydrocarbon solvents and alcohols only hydrogen capped polyynes were produced (Cataldo, 2004a), but when the arc was struck in acetonitrile, then monocyanopolyynes were obtained (Cataldo, 2004b). Finally, running the arc in liquid nitrogen yielded dicyanopolyynes (Cataldo, 2004c). The explanation of these results is based on the spectral evidence that at the carbon arc temperature (say >4000 K) the elemental carbon is vaporized as C$_2$ diradical which is then able to polymerize into chains and to catch atomic hydrogen from the atomized hydrocarbon solvent molecules. When the solvent is acetonitrile, then CN radicals are formed as well and are attached to the acetylenic chains forming monocyanopolyynes. In liquid nitrogen, the N_2 molecule is cracked into atomic nitrogen end capping the carbon chains and yielding dicyanopolyynes (Cataldo, 2005). The kinetics of polyynes formation can be followed quite easily with spectral methods, and polyynes concentration in solvent as high as 10^{-2} to 10^{-3} M can be achieved (Cataldo, 2007). At higher concentration, the polyynes in solution become unstable and decompose or crosslink. Dicyanopolyynes are the most labile compounds and can be stored for just one day in the air at room temperature. Monocyanopolyynes have a shelf life of a couple of days or so while hydrogen capped polyynes in dilute solutions can be stored for weeks.

There are now available also alternative routes to the carbon arc in the polyynes synthesis in one shot using oxidative coupling reactions (Cataldo, 2005b).

30.3   NANO-SCIENTIFIC DEVELOPMENTS

Nonlinear optical properties were found on polyynes (Slepkov et al., 2004) and this property can be enhanced by attaching different end groups one electron donor and the other electron acceptor in a push-pull arrangement (Jaquemin et al., 1997) leading also to hyper-polarizability. Polyyne chains are also known as "molecular wires," and unusual electrical conductance was measured on these systems (Crljen et al., 2007). Polyynes were found as a very promising class of molecular wires for integration into electronic circuitry (Wang et al., 2009). Polyynes were also successfully introduced inside the case of a single wall carbon nanotube (Nishide et al., 2006).

30.4   NANOCHEMICAL APPLICATIONS

Polyynes are important key models to test and applying the surface enhanced Raman spectroscopy (Lucotti et al., 2006; Tabata et al., 2006). Diphenylpolyynes were synthesized (Cataldo et al., 2010a, 2010b) as well as dinaphthylpolyynes (Cinquanta et al., 2011) and employed to investigate the charge transfer with metal nanoparticles. The occurrence of a charge transfer between carbon wires and metal nanoparticles (both in liquids and supported on surfaces) was evidenced by Raman and surface-enhanced Raman scattering as a softening of the vibrational stretching modes (Milani et al., 2011). This is interpreted, with the support of density functional theory calculations of the Raman modes, as a modification of the bond length alternation of carbon atoms in the wire. As a consequence of the charge transfer, carbon wires rearrange their structure toward a more equalized geometry which corresponds to a tendency toward a cumulenic structure (i.e., all double bonds). These observations open potential perspectives for developing carbon-based atomic devices with tunable electronic properties (Fazzi et al., 2013).

Polyynes in general and certain enediynes, in particular, are of great interest in medicinal chemistry because through the Bergmann cyclization reaction can impair the DNA, and hence they have potential as an active principle in cancer therapy (Basak et al., 2008) and in other fields like anti-biotics (Maretina et al., 2006).

30.5 MULTI-/TRANS-DISCIPLINARY CONNECTIONS

Polyynes are thought to be the precursors of fullerenes not only in the carbon arc under helium or in laser ablation experiments (Heat et al., 1987) but also in space (Kroto et al., 1987, 1993). Indeed polyynes and cyanopolyynes were detected by radioastronomy in the circumstellar shell of late-type carbon-rich stars (Kroto et al., 1987; Helling et al., 1996; Kroto et al.1993). The production of polyynes and cyanopolyynes with the submerged electric arc seems real to achieve the same conditions occurring in the circum-stellar environment of carbon-rich stars since the relative abundances of the polyynes obtained with the arc are coincident with those measured by radioastronomy (Cataldo, 2004d, 2006b, 2006c).

KEYWORDS

- chemistry
- developments
- history
- oligoynes
- polyynes
- structure

REFERENCES AND FURTHER READING

Basak, A., Roy, S. K., Roy, B., & Basak, A., (2008). Synthesis of highly strained enediynes and dienediynes. *Current Topics in Medicinal Chemistry, 8*, 487–504.

Bohlmann, F., Burkhardt, T., & Zdero, C., (1973). *Naturally Occurring Acetylenes*. London: Academic Press.

Casari, C. S., Bassi, A. L., Ravagnan, L., Siviero, F., Lenardi, C., Piseri, P., & Milani, P., (2004). Chemical and thermal stability of carbyne-like structures in cluster-assembled carbon films. *Physical Review B., 69*, 075422.

Cataldo, F., & Capitani, D., (1999). Preparation and characterization of carbonaceous matter rich in diamond-like carbon and carbyne moieties. *Materials Chemistry and Physics, 59*, 225–231.

Cataldo, F., (1997). On the enthalpy of formation of the most known carbon allotropes. *Fullerenes, Nanotubes, and Carbon Nanostructures, 5*, 1615–1620.

Cataldo, F., (1998a). Structural relationships between dicopper diacetylide (Cu-C≡ CC≡ C-Cu) and dicopper acetylide (Cu-C≡ C-Cu). *European Journal of Solid State and Inorganic Chemistry, 35*, 281–291.

Cataldo, F., (1998b). From dicopper diacetylide (Cu-C≡CC≡C-Cu) to carbyne. *European Journal of Solid State and Inorganic Chemistry, 35*, 293–304.

Cataldo, F., (2003). Simple generation and detection of polyynes in an arc discharge between graphite electrodes submerged in various solvents. *Carbon, 41*, 2671–2674.

Cataldo, F., (2004a). Synthesis of polyynes in a submerged electric arc in organic solvents. *Carbon, 42*, 129–142.

Cataldo, F., (2004b). Polyynes and cyanopolyynes synthesis from the submerged electric arc: About the role played by the electrodes and solvents in polyynes formation. *Tetrahedron, 60*, 4265–4274.

Cataldo, F., (2004c). Polyynes: A new class of carbon allotropes. About the formation of dicyanopolyynes from an electric arc between graphite electrodes in liquid nitrogen. *Polyhedron, 23*, 1889–1896.

Cataldo, F., (2004d). Cyanopolyynes: Carbon chains formation in a carbon arc mimicking the formation of carbon chains in the circumstellar medium. *International Journal of Astrobiology, 3*, 237–246.

Cataldo, F., (2005a). *Polyynes: Synthesis, Properties, and Applications*. CRC Press, Boca Raton, FL.

Cataldo, F., (2005b). A method for synthesizing polyynes in solution. *Carbon, 43*, 2792–2800.

Cataldo, F., (2006a). On the action of ozone on polyynes and monocyanopolyynes: Selective ozonolysis of polyynes. *Fullerenes, Nanotubes, and Carbon Nanostructures, 14*, 1–8.

Cataldo, F., (2006b). Polyynes and cyanopolyynes: Their synthesis with the carbon arc gives the same abundances occurring in carbon-rich stars. *Origins of Life and Evolution of Biospheres, 36*, 467–475.

Cataldo, F., (2006c). Monocyanopolyynes from a carbon arc in ammonia: About the relative abundance of polyenes series formed in a carbon arc and those detected in the circumstellar shells of AGB stars. *International Journal of Astrobiology, 5*, 37–43.

Cataldo, F., (2008). The role of Raman spectroscopy in the research on sp-hybridized carbon chains: Carbynoid structures polyynes and metal polymides. *Journal of Raman Spectroscopy, 39*, 169–176.

Cataldo, F., Ravagnan, L., Cinquanta, E., Castelli, I. E., Manini, N., Onida, G., & Milani, P., (2010b). Synthesis, characterization, and modeling of naphthyl-terminated sp carbon chains: Dinaphthylpolyynes. *The Journal of Physical Chemistry B., 114*, 14834–14841.

Cataldo, F., Strazzulla, G., & Iglesias-Groth, S., (2008a). UV photolysis of polyynes at λ = 254 nm and at λ> 222 nm. *International Journal of Astrobiology, 7,* 107–116.

Cataldo, F., Ursini, O., & Angelini, G., (2007). Kinetics of polyynes formation with the submerged carbon arc. *Journal of Electroanalytical Chemistry, 602,* 82–90.

Cataldo, F., Ursini, O., Angelini, G., & Strazzulla, G., (2008b). Polyynes decomposition with γ radiation. *Fullerenes, Nanotubes and Carbon Nanostructures, 16,* 272–281.

Cataldo, F., Ursini, O., Angelini, G., Tommasini, M., & Casari, C., (2010a). Simple synthesis of α, ω-diarylpolyynes part 1, diphenylpolyynes. *Journal of Macromolecular Science, Part A: Pure and Applied Chemistry, 47,* 739–746.

Chalifoux, W. A., & Tykwinski, R. R., (2009). Synthesis of extended polyynes: Toward carbyne. *Comptes Rendus Chimie, 12,* 341–358.

Cinquanta, E., Ravagnan, L., Castelli, I. E., Cataldo, F., Manini, N., Onida, G., & Milani, P., (2011). Vibrational characterization of dinaphthylpolyynes: A model system for the study of end-capped sp carbon chains. *The Journal of Chemical Physics, 135,* 194501.

Crljen, Ž., & Baranović, G., (2007). Unusual conductance of polyyne-based molecular wires. *Physical Review Letters, 98,* 116801.

Diederich, F., Stang, P. J., & Tykwinski, R. R., (2005). *Acetylene Chemistry: Chemistry, Biology, and Material Science.* Wiley-VCH, Weinheim.

Eastmond, R., Johnson, T. R., & Walton, D. R. M., (1972). Silylation as a protective method for terminal alkynes in oxidative couplings: A general synthesis of the parent polyynes H (C≡C) nH (n = 4–10, 12). *Tetrahedron, 28,* 4601–4616.

Fazzi, D., Scotognella, F., Milani, A., Cataldo, F., Brida, D., Manzoni, C., Cinquanta, E., & Lüer, L., (2013). Ultrafast spectroscopy of linear carbon chains: The case of dinaphthylpolyynes. *Physical Chemistry Chemical Physics, 15,* 9384–9391.

Heath, J. R., Zhang, Q., O'Brien, S. C., Curl, R. F., Kroto, H. W., & Smalley, R. E., (1987). The formation of long carbon chain molecules during laser vaporization of graphite. *Journal of the American Chemical Society, 109,* 359–363.

Heimann, R. B., Evsvukov, S. E., & Koga, Y., (1997). Carbon allotropes: A suggested classification scheme based on valence orbital hybridization. *Carbon, 35,* 1654–1658.

Helling, C., Jorgensen, U. G., Plez, B., & Johnson, H. R., (1996). Formation of PAHs, polyynes, and other macromolecules in the photosphere of carbon stars. *Astronomy and Astrophysics, 315,* 194–203.

Heymann, D., (2005). Thermolysis of the polyyne C_8H_2 in hexane and methanol: Experimental and theoretical study. *Carbon, 43,* 2235–2242.

Jacquemin, D., Champagne, B., & André, J. M., (1997). Electron correlation effects upon the static (hyper) polarizabilities of push-pull conjugated polyenes and polyynes. *International Journal of Quantum Chemistry, 65,* 679–688.

Johnson, T. R., & Walton, D. R. M., (1972). Silylation as a protective method in acetylene chemistry: Polyyne chain extensions using the reagents, $Et_3Si(C≡C)_mH$ (m = 1, 2, 4) in mixed oxidative couplings. *Tetrahedron, 28,* 5221–5236.

Kroto, H. W., Heath, J. R., Obrien, S. C., Curl, R. F., & Smalley, R. E., (1987). Long carbon chain molecules in circumstellar shells. *The Astrophysical Journal, 314,* 352–355.

Kroto, H. W., Walton, D. R. M., Jones, D. E. H., & Haddon, R. C., (1993). Polyynes and the formation of fullerenes. *Philosophical Transactions of the Royal Society of London A: Mathematical, Physical and Engineering Sciences, 343,* 103–112.

Lucotti, A., Tommasini, M., Del Zoppo, M., Castiglioni, C., Zerbi, G., Cataldo, F., & Bottani, C. E., (2006). Raman and SERS investigation of isolated sp carbon chains. *Chemical Physics Letters, 417,* 78–82.

Maretina, I. A., & Trofimov, B. A., (2006). Enediyne antibiotics and their models: New potential of acetylene chemistry. *Russian Chemical Reviews, 75*, 825–845.

Milani, A., Lucotti, A., Russo, V., Tommasini, M., Cataldo, F., Li Bassi, A., & Casari, C. S., (2011). Charge transfer and vibrational structure of sp-hybridized carbon atomic wires probed by surface-enhanced Raman spectroscopy. *The Journal of Physical Chemistry C., 115*, 12836–12843.

Nishide, D., Dohi, H., Wakabayashi, T., Nishibori, E., Aoyagi, S., Ishida, M., & Shinohara, H., (2006). Single-wall carbon nanotubes are engaging linear chain $C_{10}H_2$ polyyne molecules inside. *Chemical Physics Letters, 428*, 356–360.

Rogers, D. W., Matsunaga, N., McLafferty, F. J., Zavitsas, A. A., & Liebman, J. F., (2004). On the lack of conjugation stabilization in polyynes (polyacetylenes). *The Journal of Organic Chemistry, 69*, 7143–7147.

Rubin, Y., Parker, T. C., Pastor, S. J., Jalisatgi, S., Boulle, C., & Wilkins, C. L., (1998). Acetylenic cyclophanes as fullerene precursors: Formation of $C_{60}H_6$ and C_{60} by laser desorption mass spectrometry of $C_{60}H_6(CO)_{12}$. *Angewandte Chemie International Edition, 37*, 1226–1229.

Slepkov, A. D., Hegmann, F. A., Eisler, S., Elliott, E., & Tykwinski, R. R., (2004). The surprising nonlinear optical properties of conjugated polyyne oligomers. *The Journal of Chemical Physics, 120*, 6807–6810.

Stang, P. J., & Diederich, F., (2008). *Modern Acetylene Chemistry*. Wiley-VCH, Weinheim.

Szafert, S., & Gladysz, J. A., (2003). Carbon in one dimension: Structural analysis of the higher conjugated polyynes. *Chemical Reviews, 103*, 4175–4206.

Tabata, H., Fujii, M., Hayashi, S., Doi, T., & Wakabayashi, T., (2006). Raman and surface-enhanced Raman scattering of a series of size-separated polyynes. *Carbon, 44*, 3168–3176.

Tsuji, M., Tsuji, T., Kuboyama, S., Yoon, S. H., Korai, Y., Tsujimoto, T., & Mochida, I., (2002). Formation of hydrogen-capped polyynes by laser ablation of graphite particles suspended in solution. *Chemical Physics Letters, 355*, 101–108.

Wang, C., Batsanov, A. S., Bryce, M. R., Martin, S., Nichols, R. J., Higgins, S. J., & Lambert, C. J., (2009). Oligoyne single molecule wires. *Journal of the American Chemical Society, 131*, 15647–15654.

CHAPTER 31

Powder Electrodeposition

MARIUS C. MIRICA,[1] MARINA A. TUDORAN,[1,2] and MIHAI V. PUTZ[1,2]

[1]*Laboratory of Renewable Energies-Photovoltaics, R&D National Institute for Electrochemistry and Condensed Matter, Dr. A. Paunescu Podeanu Str. No. 144, Timisoara, RO–300569, Romania*

[2]*Laboratory of Structural and Computational Physical Chemistry for Nanosciences and QSAR, Biology-Chemistry Department, West University of Timisoara, Pestalozzi Street No. 44, Timisoara, RO–300115, Romania*
Tel: +40-256-592638, Fax: +40-256-592620,
E-mail: mv_putz@yahoo.com, mihai.putz@e-uvt.ro

31.1 DEFINITION

The electrodepositions, also known as electroplating method, is based on the electrochemical reduction reaction on cathodes and represents an electrochemical deposition of metallic ions from an electrolyte solution which contains minimum one type of metal. By choosing proper values for the parameters that influence the quality of the metallic deposit, it is possible to obtain a metallic deposit that is porous, brittle, and non-adherent. This deposit will detach from the electrode, and the final result of electrolysis will be a suspended metallic powder. In short, powder electrodepositing can be seen as an electrochemical process which has as result powder deposits on the conductive materials surface.

31.2 HISTORICAL ORIGIN(S)

The electrodeposition method was performed for the first time by Luigi V. Brugnatelli in 1805 (Pasa & Munford, 2005) using a Volta pile, soon after its discovering by Allesandro Volta in 1799 (Zangari, 2015; Hunt, 1973), the first deposited material being gold from a solution whit dissolved gold to a metal object surface (Pasa & Munford, 2005). After 40 years, potassium cyanide was discovered as an appropriate electrolyte for silver and gold electroplating by Jon Wright, on the mid-1800 being developed many other solutions for depositing metals and alloys (zinc, brass, tin, nickel) to enhance corrosion resistance, increase their hardness and to make an improvement in their appearance (Pasa & Munford, 2005; Zangari, 2015).

In 1931, Volmer and Erdey-Grutz proposed a method of metals electro-crystallization on a smooth surface-based on the occurrence model of a new phase from a supersaturated phase (Mirica et al., 2006). Starting with 1940, electrodeposition was used also in electronic industry, first by the gold electrodeposition for electronic compounds (Pasa & Munford, 2005), and along with the information revolution, IBM (International Business Machines) further developed the electrodeposition of magnetic materials (Romankiw, 1997) in order to extend the microfabrication method capabilities (Krongelb et al., 1998; Zangari, 2015).

In the last years, there were many developments in the electrodeposition process: description of cathodic and anodic reaction, developing of direct current (DC) power supplies, developing of acid electrolytes bath safer than ones based on cyanide which were poisonous, new deposition process models, which involves nucleation and growth on the working electrode and kinetics of charge transfer, etc. (Pasa & Munford, 2005). The constant needs for improvement in the performance of energy conversion devices and microelectronics (Zangari & Fukunaka, 2014) leads the electrodeposition to its third revolution (Zangari, 2015), this process could be now used in manufacturing thermoelectric materials and nanostructures (Xiao et al., 2008), thin films magnetic sensors (Heremans, 1993), nanostructures for energy conversion and photovoltaics (Yan et al., 2009; Atwater & Polman, 2010; Lombardi et al., 2007; Vukmirovic et al., 2011).

31.3   NANO-SCIENTIFIC DEVELOPMENT(S)

Electrodeposition process occurs in an electrochemical cell (Figure 31.1) consists of two conductive or semiconducting electrodes, the working electrode, and the counter electrode, immersed in an electrolyte. The working electrode is also known as a cathode and is represented by the object where the electrodeposition occurs, both being immersed in an electrolyte. The counter electrode is also known as anode and represents that part necessary in order to the electrical circuit to be completed. The electrolyte consists of an aqueous solution with negative and positive ions in which the metal salts are dissolving (Pasa & Munford, 2005).

The electric current passing the electrochemical cell determines the electrodeposits formation and appears due to the charged spin motion, i.e., migration and diffusion, towards polarized electrodes surface in the presence of an external voltage. The electrochemical reactions are different from the typical chemical reactions by the fact that the exchange of charge occurs between the electrode and chemical species (Pasa & Munford, 2005).

Three different electrodeposition methods used in, for example, the Ni-Fe coating, such as DC, PC, and PRC (Torabinejad et al., 2017). The *DC (direct current) electrodeposition* is usually applied on metals and alloys and is characterized by the fact that electrical current is uninterrupted and continuously to the system. In case of *PC (pulse current) electrodeposition*, on the cathodic surface is formed a negatively charged layer as the current reach zero in a periodic manner. The layer thickness is increased until a specific value, remaining constant after that, and hinders the ions diffusion. The ions diffuse toward the cathode surface as a result of this layer discharging in off-time (Sajjadnejad et al., 2014). On the other hand, the oxidation reaction in the *PRC (pulse reverse current) electrodeposition* occur on the cathode surface, a change in polarity determining the current necessary to ions reduction to reach zero. Usually, this method is utilized in reducing Fe anomalous deposition (Torabinejad et al., 2017).

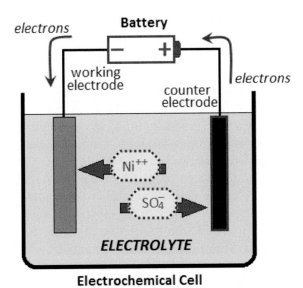

FIGURE 31.1 Schematic representation of an electrochemical cell. Redrawn and adapted after Pasa and Munford (2005).

The metals cathodic deposition represents a simultaneous process of both discharge and crystallization, being also known as the electro-crystallization due to the fact that crystallization process occurs in the presence of electric charge (Mirica et al., 2006). The mechanism occurs in two successively phases: hydrate ion deposition in the crystalline network and crystallization (tridimensional crystal growth). In other words, in the electrodeposition process, the cation is first reduced on the substrate surface in the presence of a current or tension with adatom formation and its migration on the surface towards a favorable energetic point. The atoms which are next deposited will clutter next to the first one with new phase nucleus formation (Mirica et al., 2006). There can be one or more nuclei which can grow in parallel or perpendicular on the surface. The first layers formation determines the deposit structure and adhesion, being easier to deposit on the same substrate than on different substrates (Mirica et al., 2006; Vaszilcsin, 1995; Rubinstein, 1995; Firoiu, 1983).

The metallic electrode surface can contain a lot of defects, e.g., steps (extremely important in crystal growth), corners, vacancies in steps, and voids. While interacting with a flat area of the surface, the adion will keep the maximum number of water molecules of hydrate ion when interacting with a flat zone of the surface, progressively decreasing, the minimum number of retained water molecules being occurred on fixating in a void (Mirica et al., 2006). The charge transfer may occur in a similar manner, i.e., by fixing on a flat area of the electrode, due to the fact that this type of area has the highest frequency and the hydrate ion assume to exhibit a minimum deformation on flat areas corresponding to a minimum energy variation (Paunovic, 1998). By diffusion on the electrode surface, the adion, which do not possess the ze_0 complete charge of the ion from solution, can reach in the stair areas, corner and finally end up in a void, gradually losing the hydration water molecules and coordinating instead of the metal atoms, the process being finalized with its integration in network (Mirica et al., 2006).

If in solution is introduced a crystal of the same substance or of a substance with a crystalline network similar to the separation crystals, the degree of supersaturation necessary to the crystallization process decrease semnificativly and is related with the consuming work for a crystal germ formation in solution (Mirica et al., 2006). For a cubic crystal, one can define formation work A_g of a tridimensional germ as following:

$$A_g = \frac{4\sigma^3}{3} \cdot \frac{V^2}{\left(G_g \cdot G_\infty\right)^2} \tag{1}$$

with V as the crystal molecular volume, G_g as the germ-free enthalpy, G_∞ as the free enthalpy of a crystal with infinite dimension, σ as the crystal surface tension and ω as a geometric factor depending on the crystal geometry. The A_g decrease significantly if the new phase occurs on a separation surface already existent, as in the formation of mono- or bi-dimensional germ (Mirica et al., 2006).

The atoms different positions responsible for the germ ulterior growth are not equivalent from the energetic point of view on a different point of the crystal (Rubinstein, 1995). Studies conducted on the growth of already existed crystals sown that the cations discharges do not occur on the whole facet of the growing crystal, but only on its specific active regions, i.e., the most active region due to the atoms from cations discharge being attracted in high number by the bigger number of atoms from the crystalline network (I, II, and III), the energy necessary for deposition in this being minimum. After completing the most active zone, atoms start to fix on the less active area leading to new two active areas formation (Mirica et al., 2006). This growth pattern continues until the whole crystal facet is completely covered by metallic atoms. From that moment atoms can be placed in the crystalline network in corners of the complementary facet, where the atoms from network exert an even higher attraction and the gradual coverage of the crystal facet is resuming, the deposition process being made in successive layers (Koryta et al., 1993).

Several adatoms assembled together constitute the nucleus, and if the overpressure is too small, the probability of nucleus formation is also small, and the process may continue by growing the preexistent germs (Mirica et al., 2006). On sufficiently high values of the overpressure, the current density and bi-dimensional germs number are increasing as in equation:

$$\log|i| = -A\frac{1}{|\eta_{k,2}|} + B \tag{2}$$

with A, B constants, i the current density and $_{k,2}$ the overpressure necessary bi-dimensional germs formation.

A similar expression was proposed by Erdy-Grutz and Volmer in order to correlate the overpressure necessary tri-dimensional germs formation with the current density:

$$\log|i| = -A\frac{1}{\eta_{k,3}^2} + B \tag{3}$$

with A, B constants, i the current density and $_{k,3}$ the overpressure necessary tri-dimensional germs formation (Mirica et al., 2006).

The nucleation process, i.e., the nuclei formation on the electrode surface can be described using the equation:

$$N = N_0 \left(1 - e^{-At}\right) \tag{4}$$

with A the nucleation constant and N_0 the nucleation circle number.

There are two limit cases of the relation (3). First, instantaneous nucleation occurs when $N = N_0$ and $At \gg 1$ when an overpressure is applied and shows when a certain number of nuclei are formed at the beginning of a process which is not modified during its progress or when the nuclei occupy all the active centers at the beginning of electrolysis. The second case is also called the progressive/continuous nucleation and respects the conditions: $N = AN_0t$ and $At \ll 1$ (Mirica et al., 2006).

Being an electrode process, the reaction speed of electrodeposition is determined by the tension applied to the electrode, with effect also on deposit structure, i.e., a small overpressure leads to an increasing amount of time in which a deposit with a perfect crystalline structure can be obtained (Oniciu & Constantinescu, 1989; Firoiu, 1983; Hine, 1985). The grains size is determined by the report between the formation speed of crystal-lization germs and the growth speed of the already formed crystals. Also, a larger number of germs formed in the time unit lead to a lower size of the crystalline granules in the metal deposition. The probability of germs formation increases with increasing the cathodic polarizability. When the crystal evolution reaches the electrode limits, the growth does not stop, due to the fact that the real deposit area of crystal support present defects which can initiate new steps, e.g., shear displacements which progress by rotating the edge (helicoidally movement) (Mirica et al., 2006).

During the electrocrystalization, the crystal faces characterized by different crystallographic indices present different growth speeds. The different deposition speed will lead to a particular phenomenon in which the faces characterized by small growing speeds tend to develop while the faces characterized by high speeds tends to disappear. The structure is also defined by the texture, i.e., the crystal orientation in the cathodic deposits (Mirica et al., 2006). Some times during electrodeposition, some spectacular and complicated structures occur such as dendrites and acicular crystals with importance in practical applications. In metals electrodeposition, there are several types of growth: texture, dendrites, spiral, blocks, layers, pyramid, cubic layers, edge, and acicular crystals. If polycrystalline structure growth after a systematic orientation, it may occur a macroscopic pattern of the model or texture on the cathode surface (Mirica et al., 2006). The electrode-position morphology can be affected by different factors: solution composition (temperature and pH including), operation temperature, electrolyte flowing speed, cathode overpressure and potential, current density and the substrate (Antropov, 2001; Oniciu & Constantinescu, 1989; Paunovic, 1998; Golumbioschi, 1995).

The amount of deposited material through the electrodeposition process can be measured by measuring the current from the electrochemical cell, starting from the fact that this process implies the electrons transfer to an electrode. Considering that there is no other reaction occurring in parallel, the *working electrode reaction* in the aqueous electrolyte can be seen as simple metal (M) reduction:

$$M^{n+} + ne^- \rightarrow M^0 \tag{5}$$

in other words, M^{n+} gains n electrons and is reduced to M^0 (Pasa & Munford, 2005).

The charge Q (in coulombs) can be used to calculate the quantity of electrodeposited material and is obtained by multiplying N times the charge of n electrons as in formula:

$$Q = Nne \tag{6}$$

with $e = -1.6 \times 10^{-19}$ C the electron charge (Pasa & Munford, 2005).

The deposit thickness *h* (m) of the electrode surface with a given area *A* (m^2) can be determined using the formula:

$$h = \frac{MQ}{ndAF} \tag{7}$$

with *M* the atomic weight, *d* the density and $F = 96485.34$ C the Faraday's constant (Pasa & Munford, 2005).

There are three types of mass transport mechanisms which occur in an electrochemical cell: diffusion, migration, and convection. Diffusion usually occurs close to the electrode surface and is given a variation in species concentration with position (concentration gradients). The migration process is characteristic to the electrolyte bulk and is determined by the ionic motion, which appears as a result of the applied electric field caused by the potential drop in the cell. In convection, the natural density gradients (i.e., caused by fluctuations in temperature and concentration) determine the fluid to flow in a controlled way; by mechanically stirring the electrolyte one can also create convection (Pasa & Munford, 2005).

31.4 NANOCHEMICAL APPLICATION(S)

The electrodeposition process can be described at a macroscopic scale using the kinetics thermodynamic laws which are applied in electrochemistry (Zangari, 2015). This way, for a given reaction $A^{z+} + ze \rightarrow A_{crystal}$, the electrochemical equilibrium calculations can be applied to calculate the free metal ions A^{z+} concentration and the redox potential E_{eq} given by the Nernst equation as follow:

$$E_{eq}(A) = E^0_{A^{z+}} + (RT/zF) \cdot \ln a_{A^{z+}} \qquad (8)$$

with T as the absolute temperature, F = 96845 A·s/mol as the Faraday constant, R = 8.314 J/molK as the gas constant, z as the exchanged electrons number, $a_{A^{z+}}$ as the ion A^{z+} activity and $E^0_{A^{z+}}$ as the standard redox potential for A (Bard & Faulkner, 2000a; Zangari, 2015).

If the potential applied to the substrate electrode E_{appl} has a more negative value than $E_{eq}(A)$, the metal electrodeposition can occur, with an overpotential (the film formation driving force) given by the relation:

$$\eta = E_{appl} - E_{eq}(A) \qquad (9)$$

If the applied voltage has a more positive value than the $E_{eq}(A)$ the metal can be dissolved in a solution or can form an oxide (Zangari, 2015).

The ionic species and the electrode are involved in an electron transfer which leads to metal ion reduction, the relation between the overpotential, and the current density is explained using the Butler-Volmer (Bard & Faulkner, 2000a) as following:

$$j = j_0 \left[\exp(-\alpha f \eta) - \exp((1-\alpha) f \eta) \right] \qquad (10)$$

whit α the transfer coefficient, η the overpotential, j_0 the exchanged current density, j the current density, and $f = F/RT$ (Zangari, 2015).

Different approximations of the Butler-Volmer relation can be applied, in restricting conditions, to high and small overpotentials. A high overpotential, i.e., $|\eta| > 120$ mV, is also known as the Tafel approximation and can be expressed as (Zangari, 2015):

$$j = j_0 \left[\exp(-\alpha f \eta) \right] \qquad (11)$$

The latter two equations (10 and 11) can be applied only for monovalent free ions reduction (such as Ag^+) involving a one-electron transfer, for others metal ion reduction being necessary a multi-step mechanism and several chemical and charge transfer steps (Zangari, 2015).

For the metal reduction at the electrode and for a given current density j, the growth rate of a pure metal film has the formula:

$$\frac{dm}{dt} = \frac{Mj}{zF} \tag{12}$$

with z the oxidation state and M the atomic weight of the element, dm/dt the mass deposited per unit area and time (Zangari, 2015).

From the atomic point of view, electrodeposition can be described as a multi-step process involving charge transfer and ionic species transportation towards the electrode, ions transformation into absorbed atoms and incorporation of atoms into the growing crystal (Zangari, 2015).

There are several factors which influence the metal cathodic deposition, which can be interpreted using different methods and analytical techniques of mass and surface which studies the physic, magnetic and electric properties of metals and alloys (Mirica et al., 2006).

In case of simple salts solutions electrolysis, the cathodic deposits nature and the electrode polarization degree are determined mostly by the *type of the deposited metal*. There are three classes of metals which can be deposited in aqueous solution. The first class, *normal metals*, consist of metals deposited without overpressure or on overpressure not higher than several millivolts, on usual density current, e.g., silver, thallium, lead, cadmium and tin. On industrial current densities, these metals present high exchange currents and form rough rubbish with coarse grains with dimension in the range of several tenths of a micron (Mirica et al., 2006). The *intermediate class*, e.g., zinc, copper, and bismuth, is characterized by smaller exchange currents than the previous group, an overpressure which can reach about tents of volts and forms thin deposits with grains the average size of 10µ (Mirica et al., 2006). The third class, *inherent metals*, e.g., iron present the highest overpressure (several tents of volts), small values of exchange currents and are deposited on the cathode as dense systems with fine-grains (Mirica et al., 2006).

Electrolyte composition appears to be also considered. Studies in this area determined that cathodic deposition of metals form simple salt solutions is affected by the nature of salt anion, the influence on overpressure and deposit nature being increased in case of metals which do not imply a higher polarizability (Mirica et al., 2006). For anions, the tendency of forming with rough rubbish with coarse grains increase and the overpressure decrease in the following order:

$$PO_4^{3-} > NO_3^- > SO_4^{2-} > ClO_4^- > NH_2SO_3^- > Cl^- > Br^- > I^- \tag{13}$$

An increase in the metal overpressure is also determined by the indifferent cations (which are usually hydrogen ions), and also by the presence of active superficial cations of the tetrasubstituted ammonium type (Antropov, 2001).

In electrodeposition, it is necessary to add several *substances with superficially active properties,* e.g., molecular or ionic substances, due to the fact that the process is sensitive to solution impurities. There are *leveling agents* for surface smoothing (Heitz & Kreysa, 1986; Pasquale et al., 2005), *gloss agents* for improving the luster (McFadden et al., 2003; Tzanavaras et al., 2005; Haatja et al., 2003), and can modify several other properties such as hardness, porosity (Herbert et al., 2005; Hasegawa et al., 2005; Wheeler et al., 2004) and the capacity of hydrogen absorption (Mirica et al., 2006).

Due to high polarizability, smooth structures can also be obtained without using superficially active substances using *complexation agents* (Antropov, 2001). The cathodic process appears to undergo significant modifications when the added organic/inorganic substances form complexes with the deposited metal by changing the hydrate ions in complexions, shifting the metal equilibrium potential towards more negative values by reducing its free ions concentration (Mirica et al., 2006). By properly choosing the complexation agents one can modify the equilibrium potential for different metallic ions from solution leading to their co-deposition (Krastev et al., 2003) resulting in alloys (Senna et al., 2003) or their maximum possible separation. The presence of complexes in solution influence also the overpressure size and the cathodic deposits nature, i.e., by moving from simple complex electrolytes determine an increase in overpressure, a decrees in deposited grains size and the cease of dendrites formation and growth (Antropov, 2001; Firoiu, 1983; Koryta et al., 1993; Yang et al., 2005).

On the other hand, by increasing the *current density* one will obtain deposits with smooth grains, but on a much higher value, the deposits will present dendritic aspect due to a considerable decrease of the metallic ions concentration from the cathode vicinity. When the current density value is in the zone of diffusion limit current, the deposits will have sponges aspects and will be partially separated from cathode during the electrolysis (Mirica et al., 2006). Another factor which needs to be considered is the variation of current density between different parts of electrodes, which affects the mass transport and the cations deposition accessibility (Krastev et al., 2003).

The *temperature and stirring* (Mirica et al., 2006) can also influence the process, the structure becoming rough with increasing the temperature, but one can also obtain smooth deposits in the presence of stirring when proper current density is selected (Koryta et al., 1993; Haatja et al., 2003).

The crystals size and orientation are affected by the increase of *deposition thickness*, i.e., in the beginning of electrolysis there are formed small crystals but some of them stop growing during the process (Mirica et al., 2006), the growing of metal crystals increases with increasing the layer thickness (Koryta et al., 1993; Yang et al., 2005).

The *hydrogen resulted on cathode* is also an influencing factor, due to the fact that adheres on its surface and prevents the metal deposition (Mirica et al., 2006). When the hydrogen bubble is presented on the entire electrolysis process, the obtained deposit will be porous with decreased corrosion resistance (Gabe, 1997), when the bubble is separated after a while the metallic layer will present "pitts." The deposit fragility is also determined by the presence of hydrogen in dissolved form in metal (Koryta et al., 1993).

Electrodeposition implies that the deposition only occurs in the presence of an electrical connection to the external circuit, which allows an area selective deposition. This characteristic is very useful to fill high-aspect ratio templates and when materials are needed to be grown on previously determined patterns, i.e., for growing nanowires with nanometric diameter. An appropriate template is a nanoporous membrane obtained by anodic oxidation of aluminum (Pasa & Munford, 2005) resulting in alumina layers with parallel nanopores which can be further filled through electrodeposition (see Figure 31.2).

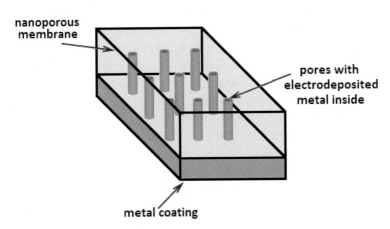

FIGURE 31.2 Schematic representation of an alumina membrane with nanopores. Redrawn and adapted after Pasa and Munford (2005).

Another template used in nanowires electrodeposition is the surface of a single crystal. The silicone surface, for example, presents large terraces with parallel steps which can acts like deposition sites for Au electrodeposition at low deposition rates, the nanometric size wires being formed along their edges (Pasa & Munford, 2005).

Electrodeposition became a very used method to produce nano-objects due to the fact that allows the controlling of the amount of atoms which are deposited, e.g., deposition of atoms clusters or layers with few to hundreds of nanometers thickness. The multilayered structure is a very common example of electrodeposited nanostructure and is obtained by applying alternatively two potentials over the two salts from the electrolyte. When one of the repeating layers is a magnetic material, the multilayer structure is called a magnetic multilayer (Pasa & Munford, 2005). On the other hand, while the classical electrochemical method for obtaining a porous metallic deposit poorly adherent which can be further mechanically removed from the cathode and grounded, by cathode vibration one can obtain the metallic powder directly in suspension, this way being decreased the working time and the production cost (Mirica et al., 2006).

31.5 MULTI-/TRANS-DISCIPLINARY CONNECTION(S)

The electrodeposition method can be used to synthesize nanocrystalline nickel, and many researchers conducted studies in this area. The nanocrystalline nickel structure, properties, and synthesis were studied by Erb and his co-workers (Erb, 1995; El-Sherik & Erb, 1995), they obtaining nickel with a grain size of about 6–100 nm using *pulse electrodeposition* and by adding saccharin in the Watts-type bath. Nanocrystalline nickel with a grain of about 15-80 m was obtained by Lei and his team (Zhu et al., 2002; Lei et al., 2002) through high-frequency pulse current using an electrolyte which contains sulphamate and saccharin (Zhu et al., 2008).

In their work, Zhu and his co-workers (2008) developed a new mechanical electrodeposition technology in order to obtain bright deposits of nickel with a smooth surface without being necessary to use additives. Their system presents a cathode unit, an electrolyte bath, a heater, a temperature controller, a stirrer, and a power supply (Zhu et al., 2008). The space between the anode and the cathode mandrel they insert ceramic and glass beads (hard isolated and insoluble particles) which move around and remove the impurities and the hydrogen bubbles when the cathode mandrel rotates smoothing this way the deposition surface. By disturbing the electrocrystallization and the charge transfer reaction the grain size of the deposits can also be reduced (Zhu et al., 2008). The authors determined that the deposits quality can be improved by inhibiting the nodules and pinholes from the deposit surface using this technology, the final "product" consisting in a compact nickel coating which presents a bright and smooth surface with grain size less than 100 nm and higher microhardness–400 HV (Zhu et al., 2008).

Raney nickel alloys (e.g., NiZn or NiAl) represents one of the advantageous electrocatalysts which is used for fuel cells (Larminie & Dicks, 2003) or in hydrogen gas releasing from water (Lohrberg & Kohl, 1984), due to the fact that is inexpensive and presents a high surface area on which noble metals can be sputtered or deposited. In their work, Cooper and Botte (2006) describe the procedure of creating high surface area electrode by electro-depositing Raney nickel powder to a titanium grid. The electrodeposition process occurs in a nickel plating bath which continues suspended nickel particles, the nickel from a nickel anode is suspended in the bath being electro-plated on a titanium grid-cathode (Cooper & Botte, 2006). The nickel ions reduction form the anode and plating bath produce a deposit in which Raney nickel particles are incorporated when the current is crossing the solution. This way, there are two materials produces by electrodeposition in the final deposit, Raney nickel particles, and pure nickel (Cooper & Botte, 2006).

The *thin metal/alloy film depositing technology* plays an important role in manufacturing the microprocessors, computers, and microelectronic devices in general (Dini & Snyder, 2010; Dukovic, 1994). Studies in this area conducted at IBM had as results the implementation of copper electroplating technology, a process called damascene copper electroplating, to fabricate the chip interconnected structures (Andricacos, 1998a, 1998b, Ye et al., 1992). Recent developments in microelectronics propose the electrodeposited bonding bump as high-density interconnection microconnectors.

Laser-enhanced electroplating method can be successfully applied to microcircuits, in repair and engineering design scheme and also for high-speed and maskless selective plating (Hsiao & Wan, 1991, von Gutfeld et al., 1982; Puippe et al., 1981; Bindra et al., 1987, 1988). Electroforming with acid nickel or copper is used in advanced design rockets and other applications to fabricate the outer shells from the regeneratively cooled thrust chambers (Dini & Snyder, 2010).

The *Electrochemical metal reduction* can also be applied to metal extraction from their ores, also called electrometallurgy, and for molds reproduction so that the objects can be formed in their final shape directly, the so-called electroforming, electrocrystalization as the obtained metallic deposits are crystalline (Gamburg & Zangari, 2011a). The electrocrystallization process has several applications: in producing high purity metals, Cu deposition is used in microelectronic interconnects, gold and gold alloy deposition produce electrical contacts for electronic circuits, magnetic recording heads are made from Ni-Fe alloys, etc. In the last years, the electrodeposition process is more and more studied and connected with energy technology and information, sensors, microelectronics, and microsystems, the commercial synthesized materials having different orders of magnitude, e.g., copper interconnects with 20–100 nm width or tin-coated strips of hundreds of meters (Gamburg & Zangari, 2011a).

31.6   OPEN ISSUES

During the electroplating operations, there are several problems which may occur having as result a deposit with inferior quality, such as: when the current density is to low, the anodes are not properly located and when different article on a rake is screening each other the surface will not be completely covered by the deposit or the deposit will be non-uniform; poor quality deposits are obtained also when harmful impurities are present in the rinsing bath or when there are insufficient intermediate articles rinsing; the presence of particle in suspension determined by the anodic sludge is usually announced by its dark color or by a coarse surface appearing; when the current density is too high because the agitation is not sufficient or the cathodic area is incorrect, at the edges/protruting parts of the article may appears high roughen, dendrites and "burning"; during the mechanical process (or, some times, spontaneous) the phenomena of poor adhesion of the deposits may appear determining crack formation and peeling off of the film, due to insufficient cleaning or inadequate preliminary surface treatment (Gamburg & Zangari, 2011b).

The electroless deposition represents an alternative solution to the typical electrodeposition method and is known in the literature as a deposition of different types of coating without external current source, from aqueous solution. While metal electrodeposition is a cathodic reaction as implying the metal ions reduction, the metals electroless deposition requires the presence of reducing agents and can be realized by displacement deposition or by autocatalytic deposition. Same, the oxides electrodepositing is an anodic process which results from metal oxidation, while the electroless deposition of compounds is realized in the presence of oxidizing agents (Djokić & Cavallotti, 2010). In other words, the electroless deposition can be seen as a chemical process based on the oxidoreduction reactions.

In the *displacement deposition*, also called galvanic or immersed plating, a more noble metals (M_2) form a solution of metallic ions ($M_2^{z_2^+}$) in which a less noble metal (M_1) is immersed having as a result of the metal deposition on the metal surface as powder or a continuous film (Djokić & Cavallotti, 2010), as in the following reactions:

a) $\quad M_2^{z_2^+} + z_2 e^- \rightarrow M_2^o$ \hfill (14)

b) $\quad M_1^o \rightarrow M_1^{z_1^+} + z_1 e^-$ \hfill (15)

c) $\quad M_2^{z_2^+} + \dfrac{z_2}{z_1} M_1^o \rightarrow M_2^o + \dfrac{z_2}{z_1} M_1^{z_1^+}$ \hfill (16)

The galvanic displacement deposition take place only on the less noble metal surface, in this case, the less noble metal (M_1) can be considered the reducing agent while the more noble metal (M_2) will be seen as the oxidizing agent. This type of deposition is usually used on the systems: Au/Ni, Au/Ag, Ag/Zn, Pt/Co, Pt/Fe, Cu/Fe, etc. (Djokić & Cavallotti, 2010).

On the other hand, in the autocatalytic deposition, the galvanic displacement deposition only occurs at the less noble metal surface (the reducing agent). This type of reaction can occur on both surface object and in bulk solution, the reaction being catalyzed by the metal deposition, resulting in powder with different sizes and shapes. The following reaction properly describes the metal M electroless deposition with the reducing agent R^{n-} (Djokić & Cavallotti, 2010):

$$M^{z+} + R^{n-} \rightarrow M + R^{n-z}$$ \hfill (17)

In other words, in the previous reaction, the reducing agent ions R^{n-} are oxidized with R^{n-z} formation and the metallic ions M^{z+} are reduced being obtained the metal M.

The electroless deposition can be applied almost to electrodeposited every type of metals/alloys from aqueous solution in proper condition and using appropriate reducing agents. This method has the advantage of not requiring an external current, which, along with the fact that the deposits products are in most of the time "nano"-crystalline, make the electroless deposition an important area of the modern technology (Djokić & Cavallotti, 2010).

ACKNOWLEDGMENT

This contribution is part of the Nucleus-Programme under the project "Deca-Nano-Graphenic Semiconductor: From Photoactive Structure to The Integrated Quantum Information Transport," PN-18-36-02-01/2018, funded by the Romanian National Authority for Scientific Research and Innovation (ANCSI).

KEYWORDS

- **aqueous electrolyte**
- **counter electrode**
- **electrochemical cell**
- **working electrode**

REFERENCES AND FURTHER READING

Andricacos, P. C., (1998a). Electroplated copper wiring on IC chips. *The Electrochemical Society Interface, 7*, 23.

Andricacos, P. C., (1998b). Copper on-chip interconnections, a breakthrough in electrodeposition to make better chips. *Interface (Electrochemical Society), 8*, 32–39.

Antropov, L. I., (2001). *Theoretical Electrochemistry* (pp. 354, 475–487). University Press of the Pacific Honolulu, Hawaii.

Atwater, H. A., & Polman, A., (2010). Plasmonics for improved photovoltaic devices. *Nature Materials, 9*, 205–213.

Bard, A. J., & Faulkner, L. R., (2000a). Potentials and thermodynamics of cells. In: *Electrochemical Methods: Fundamentals and Applications* (2nd edn., pp. 44–86). Wiley: New York, NY, USA.

Bard, A. J., & Faulkner, L. R., (2000b). Kinetics of electrode reactions. In: *Electrochemical Methods: Fundamentals and Applications* (2nd edn., pp. 87–136). Wiley: New York, NY, USA.

Bindra, P., Arbach, G. V., & Slimming, U., (1987). On the mechanism of laser enhanced plating of copper. *Journal of the Electrochemical Society, 134*(11), 2893–2900.

Bindra, P., Light, D., Arbach, G. V., & Stimming, U., (1988). In: Romankiw, L. T., & Osaka, T., (eds.), *Proc. Electrochemical Technology in Electronics, Electrochemical Society, Pennington, NJ, 88–23*, p. 341.

Cooper, M., & Botte, G. G., (2006). Optimization of the electrodeposition of Raney nickel on titanium Substrate. *Journal of Materials Science, 41*, 5608–5612.

Dini, J. W., & Snyder, D. D., (2010). Electrodeposition of copper. In: Schlesinger, M., & Paunovic, M., (eds.), *Modern Electroplating, Fifth Edition* (pp. 33–77). John Wiley & Sons, Inc, Chapter 2.

Djokić, S. S., & Cavallotti, P. L., (2010). Electroless deposition: Theory and applications. In: Djokić, S. S., (ed.), *Electrodeposition: Theory and Practice. Modern Aspects of Electrochemistry* (Vol. 48, pp. 251–289). Springer Science + Business Media, LLC.

Dukovic, J. O., (1994). Current distribution and shape change in electrodeposition of thin films for microelectronic fabrication. In: Gerischer, H., & Tobias, C. W., (eds.), *Advances in Electrochemical Science and Engineering* (Vol. 3, p. 117). VCH, Germany.

El-Sherik, A. M., & Erb, U., (1995). Synthesis of bulk nanocrystalline nickel by pulsed electrodeposition. *Journal of Materials Science, 30*, 5743–5749.

Erb, U., (1995). Electrodeposited nanocrystals: Synthesis, properties and industrial applications. *Nanostructured Materials, 6*(5/8), 533–538.

Firoiu, C., (1983). *Tehnologia Proceselor Electronice* (pp. 63–67). Editura Didactica si Pedagogica Bucuresti.

Gabe, D. R., (1997). The role of hydrogen in metal electrodeposition processes. *Journal of Applied Electrochemistry, 27*(8), 908–915.

Gamburg, Y. D., & Zangari, G., (2011a). Introduction to electrodeposition: Basic terms and fundamental concepts. In: *Theory and Practice of Metal Electrodeposition* (pp. 1–25). Springer Science + Business Media, LLC, Chapter 1.

Gamburg, Y. D., & Zangari, G., (2011b). Technologies for the electrodeposition of metals and alloys: Electrolytes and processes. In: *Theory and Practice of Metal Electrodeposition* (pp. 265–316). Springer Science + Business Media, LLC, Chapter 13.

Golumbioschi, F., (1995). *Tehnologia Proceselor Electrochimice–Curs* (pp. 139–143). Universitatea Tehnica din Timisoara.

Haataja, M., Srolovitz, D. J., & Bocarsly, A. B., (2003). Morphological stability during electrodeposition: II. Additive effects. *Journal of the Electrochemical Society, 150*(10), C708–C716.

Hasegawa, M., Negishi, Y., Nakanishi, T., & Osaka, T., (2005). Effects of additives on copper electrodeposition in submicrometer trenches. *Journal of the Electrochemical Society, 152*(4), C221–C228.

Hebert, K. R., Adhikari, S., & Houser, J. E., (2005). Chemical mechanism of suppression of copper electrodeposition by poly(ethylene glycol). *Journal of the Electrochemical Society, 152*(5), C324–C329.

Heitz, E., & Kreysa, G., (1986). *Principles of Electrochemical Engineering* (pp. 25–35). VCH-Verlagsgesellschaftmb H, Weinheim.

Heremans, J., (1993). Solid state magnetic field sensors and applications. *Journal of Physics D: Applied Physics, 26*, 1149–1168.

Hine, F., (1985). *Electrode Process and Electrochemical Engineering* (pp. 231–233). Plenum Press, New York.

Hsiao, M. C., & Wan, C. C., (1991). The investigations of laser-enhanced copper plating on a good heat conducting copper foil. *Journal of the Electrochemical Society, 138*(8), 2273–2278.

Hunt, L. B., (1973). The early history of gold plating. *Gold Bulletin, 6*, 16–27.

Koryta, J., Dvorak, J., & Kavan, L., (1993). *Principles of Electrochemistry* (pp. 368–377). John Wiley & Sons Ltd., New York.

Krastev, I., Valkova, T., & Zielonka, A., (2003). Effect of electrolysis conditions on the deposition of silver–bismuth alloys. *Journal of Applied Electrochemistry, 33*(12), 1199–1204.

Krongelb, S., Romankiw, L. T., & Tornello, J. A., (1998). Electrochemical process for advanced package fabrication. *IBM Journal of Research and Development, 42*, 575–586.

Larminie, J., & Dicks, A., (2003). *Fuel Cell Systems Explained* (p. 135). Wiley, West Sussex.

Lei, W. N., Zhu, D., & Qu, N. S., (2002). Synthesis of nanocrystalline nickel in pulse deposition. *Transactions of the Institute of Metal Finishing, 80*(6), 168–171.

Lohrberg, K., & Kohl, P., (1984). Preparation and use of Raney-Ni activated cathodes for large scale hydrogen production. *Electrochimica Acta, 29*(11), 1557–1561.

Lombardi, I., Marchionna, S., Zangari, G., & Pizzini, S., (2007). Effect of Pt particle size and distribution on photoelectrochemical hydrogen evolution by p-Si photocathodes. *Langmuir, 23*, 12413–12420.

McFadden, G. B., Coriell, S. R., Moffat, T. P., Josell, D., Wheeler, D., Schwarzacher, W., & Mallett, J., (2003). A mechanism for brightening: Linear stability analysis of the curvature-enhanced coverage model. *Journal of the Electrochemical Society, 150*(9), C591–C599.

Mirica, N., Deagos, A., Mirica, M. C., Iorga, M., & Macarie, C., (2006). *Metode Electrochimice de Recuperare a Ionilor Metalici din Solutii* (Eng.: *Electrochemical Methods of Metallic Ions Recovery from Solutions*) (pp. 67–81). Editura Mirton (Mirton Publisher House), Timisoara, Romania.

Oniciu, L., & Constantinescu, E., (1989). *Electrochimie si Coroziune* (Eng.: *Electrochemistry and Corrosion*) (pp. 116–119, 212–228). Didactic and Pedagogical Publishing House, Bucharest, Romania.

Pasa, A. A., & Munford, M. L., (2005). Electrodeposition, In: Lee. S., (ed.), *Encyclopedia of Chemical Processing* (pp. 821–832). CRC Press Taylor & Francis.

Pasquale, M. A., Barkey, D. P., & Arvia, A. J., (2005). Influence of additives on the growth velocity and morphology of branching copper electrodeposit. *Journal of the Electrochemical Society, 152*(3), C149–C157.

Paunovic, M., (1998). *Fundamentals of Electrochemical Deposition* (pp. 1–3, 39–52, 64–67, 73–75, 90–93). John Wiley & Sons Inc., New York.

Puippe, J. C., Acosta, R. E., & Von Gutfeld, R. J., (1981). Investigation of laser-enhanced electroplating mechanisms. *Journal of the Electrochemical Society, 128*(12), 2539–2545.

Romankiw, L. T., (1997). A path: From electroplating through lithographic masks in electronics to LIGA in MEMS. *Electrochimica Acta, 42*, 2985–3005.

Rubinstein, I., (1995). *Physical Electrochemistry: Principles, Methods and Applications* (pp. 20–25). Marcel Dekker Inc., New York.

Sajjadnejad, M., Mozafari, A., Omidvar, H., & Javanbakht, M., (2014). Preparation and corrosion resistance of pulse electrodeposited Zn and ZneSiC nanocomposite coatings. *Applied Surface Science, 300*, 1–7.

Senna, L. F., Díaz, S. L., & Sathler, L., (2003). Electrodeposition of copper-zinc alloys in pyrophosphate-based electrolytes. *Journal of Applied Electrochemistry, 33*(12), 1155–1161.

Torabinejad, V., Aliofkhazraei, M., Assareh, S., Allahyarzadeh, M. H., & Sabour, R. A., (2017). Electrodeposition of Ni-Fe alloys, composites, and nanocoatings–A review. *Journal of Alloys and Compounds, 691*, 841–859.

Tzanavaras, A., Young, G., & Gleixner, S., (2005). The grain size and microstructure of jet-electroplated damascene copper films. *Journal of the Electrochemical Society, 152*(2), C101–C107,

Vaszilcsin, N., (1995). *Electrochimie–Curs* (pp. 64–67, 111, 120–125). Universitatea Politehnica din Timisoara.

Von Gutfeld, R. J., Acosta, R. E., & Romankiw, L. T., (1982). Laser-enhanced plating and etching: Mechanisms and applications. *IBM Journal of Research and Development, 26*(2), 136–144.

Vukmirovic, M. B., Bliznakov, S. T., Sasaki, K., Wang, J. X., & Adzic, R. R., (2011). Electrodeposition of metals in catalyst synthesis: The case of platinum monolayer electrocatalysts. *The Electrochemical Society Interface, 20*, 33–40.

Wheeler, D., Moffat, T. P., McFadden, G. B., Coriell, S., & Josell, D., (2004). Influence of a catalytic surfactant on roughness evolution during film growth. *Journal of the Electrochemical Society, 151*(8), C538–C544,

Xiao, F., Hangarter, C., Yoo, B., Rheem, Y., Lee, K. H., & Myung, N. V., (2008). Recent progress in electrodeposition of thermoelectric thin films and nanostructures. *Electrochimica Acta, 53*, 8103–8117.

Yan, R., Gargas, D., & Yang, P., (2009). Nanowire photonics. *Nature Photonics, 3*, 569–576.

Yang, L., Luan, B., Cheong, W. J., & Shoesmith, D., (2005). Sono-immersion deposition on magnesium alloy. *Journal of the Electrochemical Society, 152*(3), C131–C136.

Ye, X., DeBonte, M., Celis, J. P., & Roos, J. R., (1992). Role of overpotential on texture, morphology and ductility of electrodeposited copper foils for printed circuit board applications. *Journal of the Electrochemical Society, 139*(6), 1592–1600.

Zangari, G., & Fukunaka, Y., (2014). Electrochemical processing and materials tailoring for advanced energy technology. *Journal of the Electrochemical Society, 161*, Y5–Y7.

Zangari, G., (2015). Electrodeposition of alloys and compounds in the era of microelectronics and energy conversion technology. *Coatings, 5*, 195–218.

Zhu, D., Lei, W. N., Qu, N. S., & Xu, H. Y., (2002). Nanocrystalline electroforming process. *CIRP Annals-Manufacturing Technology, 51*(1), 173–176.

Zhu, Z. W., Zhu, D., & Qu, N. S., (2008). Mechanical electrodeposition of bright nanocrystalline nickel. *Science China Technological Sciences, 51*(7), 911–920.

CHAPTER 32

Q-Wiener Index

ASMA HAMZEH and ALI IRANMANESH

*Department of Mathematics, Faculty of Mathematical Sciences,
Tarbiat Modares University, P.O. Box: 14115-137, Tehran, Iran,
Tel: +982182883447; Fax: +982182883493;
E-mail: iranmanesh@modares.ac.ir*

32.1 DEFINITION

Suppose G is a simple graph; as usual, the distance between the vertices u and v of G is denoted by $d_G(u,v)$ ($d(u,v)$ for short). It is defined as the length of a minimum path connecting them. We let $d_G(v)$ be the degree of a vertex v in G, and the eccentricity, denoted by $\varepsilon(v)$, is defined as the maximum distance from vertex v to any other vertex. The diameter of a graph G, denoted by d_G, is the maximum eccentricity over all vertices in a graph G. Let $d(G,k)$ be the number of pairs of vertices of a graph G that are at distance k. Note that $d(G,0)$ and $d(G,1)$ represent the number of vertices and edges, respectively. The Wiener index is the sum of distances between all pairs of vertices of a connected graph. Let q be a positive real number, $q \neq 1$. Three different variants of the q-Wiener index were defined as

$$W_1(G,q) = \sum_{\{u,v\} \subseteq V(G)} [d(u,v)]_q,$$

$$W_2(G,q) = \sum_{\{u,v\} \subseteq V(G)} [d(u,v)]_q \, q^{d_G - d(u,v)},$$

$$W_3(G,q) = \sum_{\{u,v\} \subseteq V(G)} [d(u,v)]_q \, q^{d(u,v)}.$$

where

$$[k]_q = \frac{1-q^k}{1-q} = 1 + q + q^2 + \cdots + q^{k-1}.$$

Obviously, $\lim_{q \to 1}[k]_q = k$, and therefore:

$$\lim_{q\to1}W_1(G,q) = \lim_{q\to1}W_2(G,q) = \lim_{q\to1}W_3(G,q) = W(G).$$

The three q-Wiener indices are mutually related as:

$$W_2(G,q) = q^{d_G-1}W_1(G,\frac{1}{q}),\qquad(1)$$

$$W_3(G,q) = (1+q)W_1(G,q^2) - W_1(G,q).\qquad(2)$$

In addition, three q-Wiener indices can be written as:

$$W_1(G,q) = \sum_{k\geq1}[k]_q d(G,k),$$

$$W_2(G,q) = \sum_{k\geq1}[k]_q q^{d_G-k} d(G,k),$$

$$W_3(G,q) = \sum_{k\geq1}[k]_q q^k d(G,k).$$

32.2   HISTORICAL ORIGIN(S)

The earliest q-analog studied in detail is the basic hypergeometric series, which was introduced in the 19th century (Gasper et al., 2004).

A q-analog is, roughly speaking, a theorem or identity in the variable q that gives back a known result in the limit, as $q \to 1$ (from inside the complex unit circle in most situations).

q-Analogs find applications in a number of areas, including the study of fractals and multifractal measures, and expressions for the entropy of chaotic dynamical systems. q-Analogs also appear in the study of quantum groups and in q-deformed superalgebras (Andrews, 1986; Roman, 1985). q-Analogs of the Wiener index, motivated by the theory of hypergeometric series. Three different variants of the q-Wiener index were considered in Zhang et al., (2012).

32.3   NANO-SCIENTIFIC DEVELOPMENT(S)

Let v and u be two vertices of the graph G and let their distance be d. The shortest path between v and u can be viewed as a sequence d mutually incident edges, $e_1, e_2, \ldots, e_d$, such that v is an end-vertex of e_1, u and end-vertex of e_d. So, we can go from v to u in d steps, along the edges $e_1, e_2, \ldots, e_d$. Suppose that the contribution of the first step is unity, of the second step is q of the third step q^2, of the i-th step is q^{i-1}. The contribution obtained by moving along the entire shortest path would then be $1 + q + q^2 + \ldots + q^{d-1}$. This observation may serve for an interpretation of the invariants W_1, and after an obvious modification, also of W_2 and W_3. If the parameter q, is chosen to be positive and less than unity, then the q-analogs of the Wiener index would provide models for measuring interactions between individual atoms in a molecule which are known to decrease with their distance.

32.4 NANOCHEMICAL APPLICATION(S)

Mathematical chemistry is a branch of theoretical chemistry for discussion and prediction of the molecular structure using mathematical methods without necessarily referring to quantum mechanics. Chemical graph theory is a branch of mathematical chemistry which applies graph theory to mathematical modeling of chemical phenomena. This theory had an important effect on the development of the chemical sciences.

A topological index is a numeric quantity from the structural graph of a molecule. Usage of topological indices in chemistry began in 1947 when chemist Harold Wiener developed the most widely known topological descriptor, the Wiener index, and used it to determine physical properties of types of alkanes known as paraffin.

Diudea and his co-authors were the first scientists considered topological indices of nanostructures into account. In some research paper, he and his team computed the Wiener index of armchair and nanotubes. In this section, we bring some results of q-Wiener indices, for chemical graphs and nanostructures.

The q-Wiener index $W_1(G, q)$ is polynomial in q, and

$$W_1(G,q) = \sum_{k=0}^{d_G-1} \sum_{j=k+1}^{d_G} d(G,j)q^k. \tag{3}$$

This equation shows us that the coefficients of q^k in $W_1(G, q)$, is exactly the numbers of edges of G that have been weighted with q^k. In Table 32.1 are given the coefficients of the polynomial $W_1(G, q)$, for some alkanes, according to Eq. (3).

From Table 32.1 we see that 2, 2, 3-trimethyloctane and 2, 2, 4-trimethyloctane have equal Wiener indices $W(G)$, but different $W_1(G, q)$. The same is true for 2, 3, 3-trimethyloctane and 2, 3, 4-trimethyloctane. This hints toward possible advantages of the q-Wiener indices over the ordinary Wiener index. In Hamzeh et al., (2014), authors computed the q-Wiener indices of several classes of chemical graphs that a number of these results is given in following:

TABLE 32.1   The Coefficients q^k ($0 \leq k \leq 6$), Pertaining to q^k in Eq. (3)

Alkane	a_0	a_1	a_2	a_3	a_4	a_5	a_6	W(G)
2-methyloctane	36	28	20	14	9	5	2	114
3-methyloctane	36	28	20	13	8	4	1	110
4-methyloctane	36	28	20	13	7	3	1	108
2,2-dimethyloctane	45	36	25	18	12	7	3	146
2,3-dimethyloctane	45	36	26	17	11	6	2	143
2,4-dimethyloctane	45	36	26	18	10	5	2	142
3,3-dimethyloctane	45	36	25	16	10	5	1	138
3,4-dimethyloctane	45	36	26	16	9	4	1	137
4,4-dimethyloctane	45	36	25	16	8	3	1	134
2,2,3-trimethyloctane	55	45	32	21	14	8	3	178
2,2,4-trimethyloctane	55	45	32	23	13	7	3	178
2,3,3-trimethyloctane	55	45	32	20	13	7	2	174
2,3,4-trimethyloctane	55	45	33	21	12	6	2	174
3,3,4-trimethyloctane	55	45	32	19	11	5	1	168
3,4,4-trimethyloctane	55	45	32	19	10	4	1	166
2,2,4,4-tetramethyloctane	66	55	39	28	14	7	3	212
2,3,4,5-tetramethyloctane	66	55	41	26	14	6	2	210

Let G be a graph rooted at vertex w and let n be a positive integer. The bridge graph $B_d(G,w)$ is the graph obtained by taking d copies of G and by connecting the vertex w of the i-th copy of G to the vertex w of the $i + 1$-th copy of G by an edge for $i = 1,2,\ldots,d-1$, as shown in Figure 32.1.

Let G be a graph rooted at vertex w, then

$$W_1(B_d(G,w),q) = dW_1(G,q)$$

$$+ \frac{1}{2}\sum_{k=1}^{d-1}\left(k(k+1)q^{d-k-1}\right)\left[Q_G(w) + \frac{1}{3}Q_G(w)(|V(G)|-1)\right]$$

$$+ \frac{1}{2}\sum_{k=1}^{d-1}\left(k(k+1)q^{d-k-1}\right)\left[\frac{1}{3}Q_G^2(w) + \frac{1}{3}(|V(G)|-1)^2 + |V(G)|\right]$$

$$+ \sum_{k=1}^{d-1}(kq^{d-k})\left[D_G(w,q) + \frac{2}{3}Q_G(w)D_G(w,q) + \frac{1}{3}(|V(G)|-1)D_G(w,q)\right]$$

$$+D_G(w,q)\frac{d(d-1)}{2}\left[\frac{1}{3}Q_G(w)+1+\frac{2}{3}(|V(G)|-1)\right].$$

That for vertex $x \in V(G)$, $D_G(x,q) = \sum_{x \neq u \in V(G)}[d(u,x)]_q$ and $Q_G(x) = \sum_{x \neq u \in V(G)}q^{d(u,x)}$.

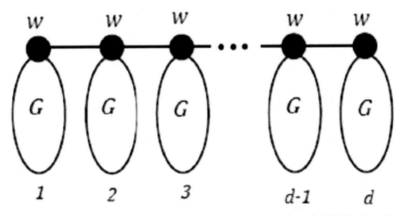

FIGURE 32.1 The bridge graph $B_d(G,w)$. Reprinted from Hamzeh et al., (2014). Open Access.

Now, consider the square comb lattice $Cq(N)$ with open ends where $N = n^2$ is the number of vertices of this graph (see Figure 32.2). This graph can be represented as the cluster $P_n\{P_n\}$, where the root of P_n is on its vertex of degree 1.

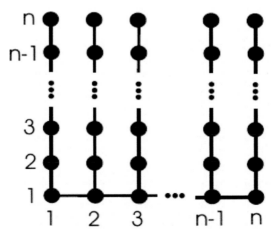

FIGURE 32.2 The square comb lattice $Cq(N)$. Reprinted from Hamzeh et al., (2014). Open Access.

The q-Wiener index of the square comb lattice $Cq(N)$ with $N = n^2$ vertices is given by:

$$W_1(Cq(N),q) = \frac{1}{6}\sum_{k=1}^{n-1}k(k+1)q^{n-k-1}\left[n^2 + 4n + 1 + (\sum_{k=1}^{n-1}q^k)^2 + (n+2)\sum_{k=1}^{n-1}q^k\right]$$

$$+\frac{1}{3}\sum_{k=1}^{n-1}kq^{n-k}\sum_{k=1}^{n-1}[k]_q\left[2\sum_{k=1}^{n-1}q^k + n + 2\right]$$

$$+\frac{n(n-1)}{6}\sum_{k=1}^{n-1}[k]_q\left[2n + 1 + \sum_{k=1}^{n-1}q^k\right].$$

A caterpillar or caterpillar tree is a tree in which all the vertices are within distance 1 of a central path. If we delete all pendent vertices of a caterpillar tree, we reach to a path. So caterpillars are thorn graphs whose parent graph is a path, see Figure 32.3.

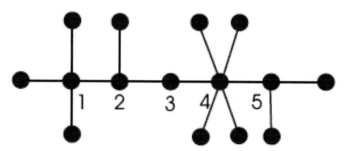

FIGURE 32.3   The caterpillar tree. Reprinted from Hamzeh et al., (2014). Open Access.

The q-Wiener index of caterpillar tree $P_n^*(p_1,\cdots,p_n)$ is given by:

$$W_1(P_n^*(p_1,p_2,\cdots,p_n),q) = \sum_{i=1}^{n}\left[p_i + 1 + p_i \quad q\right] + W_1(P_n,q)$$

$$+ \sum_{\substack{i,j=1 \\ i<j}}^{n} p_i p_j [d_{P_n}(w_i,w_j)+2]_q + \sum_{i=1}^{n} p_i[D_{P_n}(w_i,q)+Q_{P_n}(w_i)].$$

Caterpillar trees are used in chemical graph theory to represent the structure of benzenoid hydrocarbon molecules. For example, for positive integer $p \le 3$, the caterpillar tree $P_n^*(p,2,2,\cdots,2,p)$ is the molecular graph of certain hydrocarbon. Specially, $P_2^*(3,3)$, $P_3^*(3,2,3)$ and $P_2^*(3,2,2,3)$ are the molecular graphs of Ethane, Propane and Butane.

The q-Wiener index of Ethane, Propane and Butane, are given by:

$$W_1(P_2^*(3,3),q) = 9q^2 + 21q + 28,$$
$$W_1(P_3^*(3,2,3),q) = 9q^3 + 27q^2 + 45q + 55,$$
$$W_1(P_4^*(3,2,2,3),q) = 9q^4 + 21q^3 + 44q^2 + 64q + 91.$$

32.5 MULTI-/TRANS-DISCIPLINARY CONNECTION(S)

q-Analogs find applications in a number of areas, including the study of fractals and multi-fractal measures, and expressions for the entropy of chaotic dynamic systems. q-Analogs also appear in the study of quantum groups and in q-deformed superalgebras.

KEYWORDS

- **distance**
- **Wiener index**
- **q-Wiener index**

REFERENCES AND FURTHER READING

Andrews, G. E., (1986). *q-Series, Their Development and Application in Analysis, Number Theory, Combinatorics, Physics, and Computer Algebra* (3rd edn.). Publisher: Am. Math. Soc., Providence.

Gasper, G., & Rahman, M., (2004). *Basic Hypergeometric Series* (3rd edn.). Publisher: Cambridge Univ. Press, Cambridge.

Hamzeh, A., Iranmanesh, A., Reti, T., & Gutman, I., (2014). Chemical graphs constructed of composite graphs and their q-Wiener index. *MATCH Commun. Math. Comput. Chem., 72*, 807–833.

Liu, J., Gutman, I., Mu, Z., & Zhang, Y., (2012). q-Wiener index of some compound trees. *Appl. Math. Comput., 218*, 9528–9535.

Roman, S., (1985). More on the umbral calculus, with emphasis on the q-umbral calculus. *J. Math. Anal. Appl., 107*, 222–254.

Zhang, Y. S., Gutman, I., Liu, J. G., & Mu, Z. C., (2012). q-Analog of Wiener index. *MATCH Commun. Math. Comput. Chem., 67*, 347–356.

CHAPTER 33

Randic Index and Its Variants: Applications

MONIKA SINGH,[1] HARISH DUREJA,[2] and A. K. MADAN[3]

[1]Anand College of Pharmacy, Keetham, Agra–282007, India

[2]Faculty of Pharmaceutical Sciences, M.D. University, Rohtak–124001, India

[3]Faculty of Pharmaceutical Sciences, Pt. B.D. Sharma University of Health Sciences, Rohtak–124001, India, Tel.: +91-9896346211, E-mail: madan_ak@yahoo.com

ABSTRACT

Randic indices (connectivity indices) have been widely utilized by the scientific community for a long time. Randic index and its variants have proved to be useful tools for correlating chemical structures with vast but diverse information/data leading the to development of models for prediction of vital physicochemical, biological or biopharmaceutical properties of organic molecules. The combinatorial design approach appears to be a useful platform for the design of medicinal compounds. The concept of molecular connectivity prompted the construction of a variety of novel molecular descriptors (MDs). The aim of the present work was to review the applications of Randic index and its variants. Randic index and its variants offer a vast potential in isomer discrimination, similarity/dissimilarity, drug design, (quantitative) structure-activity/toxicity/property relationships, lead optimization, and combinatorial library design.

33.1 INTRODUCTION

In recent times, people are much concerned regarding the impact of chemicals on human health as well as on the environment. Therefore, chemicals should be used only if they have been tested to be safe. As a consequence, legislation (REACH, i.e., Registration, Evaluation, Authorization, and Restriction of Chemicals) has been introduced in Europe to ensure the safety of chemicals that are used in a significant quantity (Gasteiger, 2016). A substitute to the 'real' world of synthesis and screening of compounds/chemicals in the laboratory is an *in silico* 'virtual' world of data, hypothesis, analysis, and design that reside inside a computer. Mathematical models that distinguish between active and inactive compounds can be developed by using the (quantitative) structure-activity/toxicity property/pharmacokinetic [(Q)SAR/(Q)STR/(Q)SPR/(Q)SPkR] relationship studies. It provides an opportunity to identify the novel subsets of active compounds. This *in silico* approach manages to save the time and expenses of either synthesizing or screening the inactive compounds/chemicals (Estrada and Pena, 2000).

A methodical study of topological indices (TIs) among other molecular descriptors, is one of the most remarkable approach in various other fields of science and technology with its diverse applications (Lokesha et al., 2013). The Randić index is the one of the popular, most often applied, and widely studied among all other TIs (Gutman, 2013). Numerous research papers (Randic, 2008; Randic, 2001) and books (Gutman & Furtula, 2008) are devoted to this structural descriptor. Randic introduced a bond-additive topological index for characterizing molecular branching termed as branching index. This topological index was denoted by the symbol χ. In the literature, this symbol was known by different names such as the Randic index and most often, the *molecular connectivity index* (Randic, 1975; Todeschini & Consonni, 2000).

Molecular connectivity is mainly concerned with the information inherent in a chemical structural diagram, i.e., presence of atoms and connection between the atoms by means of covalent bonds. These descriptors have the advantage of being adaptable to extremely rapid algorithms on small computers (Boyd, 2001). Molecular connectivity descriptors have the advantage of being easily calculated by using extremely rapid algorithms (Ivanciuc, 2013). The χ values decrease within a set of alkane isomers with corresponding increase in branching (Diudea et al., 1999). This index was renamed as the *connectivity index* and generalized to *connectivity indices* of various orders, starting with the *zero order connectivity ($^0\chi$) index* (Kier & Hall, 1976; Bonchev, 2000) as given in Eq. (1).

$$^0\chi = \sum_{edges\ ij}^{n} 1/\sqrt{v_i} \qquad (1)$$

Kier and Hall further gave detailed account for *molecular connectivity index* as per equation (2) and generalized connectivity Index (equation 3) (Basak et al., 1988; Kier & Hall, 2001; Janezic et al., 2006).

$$\chi = \sum_{edges\ ij}^{n} 1/\sqrt{v_i v_j} \tag{2}$$

$$\chi = \sum_{edges\ ij}^{n} 1/\sqrt{v_i v_j \ldots\ldots v_h} \tag{3}$$

Bollobas and Erdös obtained *general Randić index* R_α by generalizing this index and by replacing $1/2$ with any real number α, where d_i and d_j are degrees of ith vertex and degree of jth vertex, respectively (Bollobas & Erdos, 1998; Andova et al., 2011) (see Eq. (4)).

$$R_\alpha(G) = \sum_{ij \in E(G)}^{n} \left(d_i d_j\right)^\alpha \tag{4}$$

After modification of Randic index, other *bonding connectivity indices* and *valence connectivity indices* were developed (Kier & Hall, 1976; Basak et al., 1988). Kier & Hall also introduced variable connectivity index in 1977 (Kier & Hall, 1977). The *molecular connectivity index* is the most widely used graph invariant which has been found to have a high degree of accuracy in prediction of activity. Goel & Madan (1995) proposed a topochemical modification of *molecular connectivity index* termed as *atomic molecular connectivity index* (χ^A) to take care of heteroatoms. Afterwards, it was renamed as *molecular connectivity topochemical index* (Dureja & Madan, 2005) (see Eq. (5)).

$$\chi^c = \sum_{ij}^{n} \left(v_i^c v_j^c\right)^{-1/2} \tag{5}$$

Ernesto Estrada conceived a new topological index and named it *atom-bond connectivity index* which was conveniently abbreviated by *ABC* (Estrada et al., 1998).

$$ABC = \sum_{ij} \sqrt{\frac{v_i + v_j - 2}{v_i v_j}} \tag{6}$$

Kier and Hall (1983) also examined biological SAR applications of *path connectivity index*. It is defined as:

$$CI = \sum_{ij} \frac{1}{\sqrt{v_i + v_j}} \tag{7}$$

The encoding of the chemical structure in a non-empirical way is an important feature of molecular connectivity (Kier & Hall, 2001). Mathematical properties of the molecular connectivity indices have also been studied (Pogliani, 2002; Araujo & Rada, 2000; Xueliang, 2008; Andova et al., 2013). Singh and Madan proposed *refined general Randic indices* (both topostructural and topochemical indices).

33.2   REFINED GENERAL RANDIC INDICES

The *refined general Randic indices* can be defined as the summation of the quotients of the inverse of the product of the degree of each vertex on every edge in the hydrogen-suppressed molecular graph as per the following equation:

$$R_N = \sum_{ij}^{n} k_m \left(v_i v_j \right)^{-N} \tag{8}$$

where v_i and v_j are degrees of ith vertex and j^{th} vertex, respectively, and N is equal to 2, 3, 4 for refined *general Randic index–1, 2, 3*, respectively (denoted by R_2, R_3, R_4). A constant k_m was incorporated, which ensures low index values without compromising with the discriminating power. The value of k_m is 1, 10, 100 for refined *general Randic index-1, 2, 3*, respectively and m is 1, 2, 3 for refined *general Randic index-1, 2, 3*, respectively (Singh et al., 2013).

The topochemical versions of R_N can be similarly represented by the following equation:

$$R_N^c = \sum_{ij}^{n} K_m \left(v_i^c v_j^c \right)^{-N} \tag{9}$$

where v_i^c and v_j^c are chemical degree of ith vertex and chemical degree of jth vertex, respectively, and N is equal to 2, 3, 4 for refined *general Randic topochemical index-1, 2, 3*, respectively (denoted by R_2^c, R_3^c, R_4^c). The value of k_m is 1, 10, 100 for refined *general Randic topochemical index-1, 2, 3*, respectively and m is 1, 2, 3 for refined *general Randic topochemical index-1, 2, 3*, respectively (Singh et al., 2013).

The ratio of highest to lowest value for all possible structures of the same number of vertices is called *discriminating* power. Indices R_3 and R_4 belong to fourth-generation TIs, whereas R_4^c is a fifth-generation TI (Singh et al., 2013). The measure of an index to differentiate between the relative positions of atoms in a molecule is known as *degeneracy*. All *refined general Randic topochemical indices* exhibit negligible degeneracy. Moreover, these indices are not intercorrelated with χ^c (Singh et al., 2013).

33.3 APPLICATIONS OF REFINED GENERAL RANDIC INDICES

Applications of Randic index and its variants have been illustrated/exemplified in Table 33.1. There is no doubt that the Randić index and its modifications are the most studied as well as most often applied in (Q)SAR/(Q)STR/(Q)SPR/(Q)SPkR studies.

TABLE 33.1 Illustrative Applications of Randic Index and Its Variants

Sr. No.	Index name	Biological activity	References
1	Randic index/ molecular connectivity index (MCI)	Boiling points of alkanes; Anti-HIV activity; QSTR estimation of chemicals; Topological models for prediction of adductability of branched aliphatic compounds in urea; QSPR and QSAR applications	Randic, 1975; Bajaj et al., 2005; Lather & Madan, 2005; Basaka et al., 2003; Thakral and Madan, 2006; Kier & Hall, 1976; Kier & Hall, 1986
2	Zero-order connectivity index	Design of polymers with optimal levels of macroscopic properties	Camarda and Maranas, 1999
3	Variable connectivity index	Multiple linear regression analysis (MLRA) for SAR studies	Randić, 2001
4	Path connectivity index	Prediction of antihistaminic activity	Duarta et al., 2001
5	General Randić index R_a	Ordering of alkane isomers; QSPR and QSAR applications	Gutman et al., 2002; Kier & Hall, 1976; Kier & Hall, 1986
6	Bonding connectivity indices	Characterization of Molecular Structure	Basak and Gute, 1997
7	Valence connectivity indices	Agonistic activity of human A_3 adenosine receptor ligands; Characterization of Molecular Structure	Sharma et al., 2009; Basak and Gute, 1997
8	Molecular connectivity topochemical index (Dureja & Madan, 2005)	Prediction of anti-inflammatory activity; Prediction of HIV integrase Inhibitory activity; Prediction of cyclin-dependent kinase 2 inhibitory activity; Prediction of h5-HT_{2A} Receptor Antagonistic Activity	Goel & Madan, 1995; Gupta and Madan, 2012; Dureja & Madan, 2005; Dureja & Madan, 2006

TABLE 33.1   *(Continued)*

Sr. No.	Index name	Biological activity	References
		Prediction of telomerase inhibitory activity	Dureja & Madan, 2007; Dureja & Madan, 2008
		Prediction of antitumor activity	
9	*Refined general Randić indices*	CB_2 cannabinoid receptor agonist activity	Singh et al., 2013; Singh, 2014
		Prediction of hiCE and hCE1 inhibitory activities	

Note: The definitions of most of these indices are provided in Kier & Hall (1976), Kier & Hall (1986), and Todeschini & Consonni (2000).

33.4   CONCLUSIONS

Randic indices and its variants are simple topological indices which can be easily calculated by means of mathematical matrices. These indices have been extensively used in (quantitative) structure-activity/toxicity property/pharmacokinetic relationship [(Q)SAR/(Q)STR/(Q)SPR/(Q)SPkR] studies and assisted in the designing of new drugs and chemical products. The simplicity, ease of calculation and sensitivity towards branching render Randic index and its variants to be promising tools for isomer discrimination, similarity/dissimilarity, drug design, quantitative structure-activity/structure-property relationships, lead optimization, and combinatorial library design. Randic index and its variants will continue to enjoy wide use in (Q)SAR/(Q)STR/(Q)SPR/(Q)SPkR) studies so as to assist drug design.

KEYWORDS

- **molecular descriptors**
- **Randic index**
- **refined general Randic indices**
- **topological descriptors**

REFERENCES

Andova, V., Knor, M., Potocnik, P., & Skrekovski, R., (2011). On a variation of Randic index. *IMFM. Preprint Series, 49*, 1156–1168.

Andova, V., Knor, M., PotoCnik, P., & Skrekovski, R., (2013). On a variation of the Randic index. *Australas. J. Combin., 56*, 61–75.

Araujo, O., & Rada, J., (2000). Randic index and lexicographic order. *J. Math. Chem., 27*(3), 201–212.

Bajaj, S., Sambi, S. S., & Madan, A. K., (2005). Topological models for prediction of anti-HIV activity of acylthiocarbamates, *Bioorg. Med. Chem., 13*(9), 3263–3268.

Basak, S. C., & Gute, B. D., (1997). Characterization of molecular structure using topological indices. *SAR & QSAR Environ. Res., 7*, 1–21.

Basak, S. C., Magnuson, V. R., Niemi, G. J., & Regal, R. R., (1988). Determining structural similarity of chemicals using graph-theoretic indices. *Dis. App. Math., 19*, 17–44.

Basaka, S. C., Gutea, B. D., Millsa, D., & Hawkins, D. M., (2003). Quantitative molecular similarity methods in the property/toxicity estimation of chemicals: A comparison of arbitrary versus tailored similarity spaces. *J. Mol. Struct. (Theochem), 622*, 127–145.

Bollobas, B., & Erdös, P., (1998). Graphs of external weights. *Ars. Combinatoria, 50*, 225–233.

Bonchev, D., (2000). Overall connectivity's/topological complexities: A new powerful tool for QSPR/QSAR. *J. Chem. Inf. Comput. Sci., 40*, 934–941.

Boyd, D. B., (2001). Introduction and foreword to the special issue commemorating the 25th anniversary of molecular connectivity as a structure description system. *J. Mol. Graph. Model., 20*, 1–3.

Camarda, K. V., & Maranas, C. D., (1999). Optimization in polymer design using connectivity indices. *Ind. Eng. Chem. Res., 38*(5), 1884–1892.

Diudea, M. V., Gutman, I., & Jantschi, L., (1999). *Molecular Topology, Topological Indices* (p. 112). NOVA Science Publications.

Duarta, M. J., Garcia-Domenechb, R., Anton-Fosa, G. M., & Galvez, J., (2001). Optimization of a mathematical topological pattern for the prediction of antihistaminic activity. *J. Comp. Aided Mol. Des., 15*, 561–572.

Dureja, H., & Madan, A. K., (2005). Topochemical models for prediction of cyclin-dependent kinase 2 inhibitory activity of indole–2-ones. *J. Mol. Model., 11*, 525–531.

Dureja, H., & Madan, A. K., (2006). Models for the prediction of h5-HT$_{2A}$ receptor antagonistic activity of arylindoles: Computational approach using topochemical descriptors. *J. Mol. Graph. Mod., 25*, 373–379.

Dureja, H., & Madan, A. K., (2007). Topochemical models for prediction of telomerase inhibitory activity of flavonoids. *Chem. Biol. Drug Des., 70*, 47–52.

Dureja, H., & Madan, A. K., (2008). Prediction of antitumor activity of N-(phenylsulfonyl) benzamides: computational approach using topochemical descriptors. *Leonardo J. Sci., 7*, 214–231.

Estrada, E., & Pena, A., (2000). *In silico* studies for the rational discovery of anticonvulsant compounds. *Bioorganic & Medicinal Chemistry, 8*, 2755–2770.

Estrada, E., Torres, L., Rodríguez, L., & Gutman, I., (1998). An atom-bond connectivity index: Modeling the enthalpy of formation of alkanes. *Indian J. Chem., 37A*, 849.

Gasteiger, J., (2016). Chemoinformatics: Achievements and challenges. *A Personal View, Molecules, 21*, 151–166.

Goel, A., & Madan, A. K., (1995). Structure-activity study on anti-inflammatory pyrazole carboxylic acid hydrazide analogs using molecular connectivity indices. *J. Chem. Inf. Comput. Sci., 35*, 510–514.

Goel, A., & Madan, A. K., (1995). Structure-activity study on anti-inflammatory pyrazole carboxylic acid hydrazide analogs using molecular connectivity indices. *J. Chem. Inf. Comput. Sci., 35*, 510–514.

Gupta, M., & Madan, A. K., (2012). Diverse models for the prediction of HIV integrase inhibitory activity of substituted quinolone carboxylic acids. *Arch. Pharm. Chem. Life Sci., 345*(12), 989–1000.

Gutman, I., & Furtula, B., (2008). *Recent Results in the Theory of Randić Index.* Univ. Kragujevac, Kragujevac.

Gutman, I., (2013). Degree-based topological indices. *Croat. Chem. Acta, 86*(4), 351–361.

Gutman, I., Araujo, O., & Rada, J., (2000). An identity for Randić connectivity index and its application. *Acta Chim. Hung. –Models Chem., 137*, 653–658.

Gutman, I., Vidović, D., & Nedić, A., (2002). Ordering of alkane isomers by means of connectivity indices. *J. Serb. Chem. Soc., 67*(2), 87–97.

Ivanciuc, O., (2013). Chemical graphs, molecular matrices and topological indices in chemoinformatics and quantitative structure-activity relationships. *Curr. Comput. Aided Drug Des., 9*(2), 153–163.

Kier, L. B., & Hall, L. H., (1976). *Molecular Connectivity in Chemistry and Drug Research.* Academic Press: New York.

Kier, L. B., & Hall, L. H., (1977). *Eur. J. Med. Chem., 12*, 307.

Kier, L. B., & Hall, L. H., (1983). General definition of valence delta-values for molecular connectivity. *J. Pharm. Sci., 72*, 1170.

Kier, L. B., & Hall, L. H., (1986). *Molecular Connectivity in Structure-Activity Analysis.* Wiley: New York.

Kier, L. B., & Hall, L. H., (2001). Molecular connectivity: Intermolecular accessibility and encounter simulation. *J. Mol. Graph. Model., 20*, 76–83.

Lather, V., & Madan, A. K., (2005). Topological models for the prediction of anti-HIV activity of dihydro (alkylthio) (naphthylmethyl) oxopyrimidines. *Bioorg. Med. Chem., 13*(5), 1599–1604.

Lokesha, V., Shetty, B. S., Ranjini, P. S., Cangul, I. N., & Cevik, A. S., (2013). New bounds for randic and GA indices. *J. Inequal. Appl., 180*, 2–7.

Pogliani, L., (2002). Higher-level descriptors in molecular connectivity. *Croat. Chem. Acta, 75*(2), 409–432.

Randic, M., (1975). Characterization of molecular branching. *J. Am. Chem. Soc., 97*, 6609–6615.

Randic, M., (2001). The connectivity index 25 years after. *J. Mol. Graf. Model., 20*, 19–35.

Randic, M., (2008). On history of the Randic index and emerging hostility toward chemical graph theory. *MATCH -Comm. Math. Comput. Chem., 59*, 5–124.

Sharma, S., Sharma, B. K., Sharma, S. K., Singh, P., & Prabhakar, Y. S., (2009). Topological descriptors in modeling the agonistic activity of human A_3 adenosine receptor ligands: The derivatives of 2-chloro-N^6-substituted–4'-thioadenosine–5'-uronamide. *Eur. J. Med. Chem., 44*(4), 1377–1382.

Singh, M., (2014). *Development of New Generation Molecular Descriptors for Accelerating in Silicodrug Discovery Process.* PhD Thesis, M. D. U. Rohtak, Haryana, India.

Singh, M., Gupta, S., Lather, V., & Madan, A. K., (2013). Development of models for CB$_2$ cannabinoid receptor agonist activity using refined general Randic indices. *Int. J. Chemical Modeling, 5*(4), 419–455.

Thakral, S., & Madan, A. K., (2006). Topological models for prediction of adductability of branched aliphatic compounds in urea. *J. Incl. Phenom. Macro., 56*, 405–412.

Todeschini, R., & Consonni, V., (2000). *Handbook of Molecular Descriptors*. Wiley-VCH, Weinheim.

Xueliang, L., (2008). A new survey on the randic index. *MATCH– Comm. Math. Comput. Chem., 59*(1), 127–156.

CHAPTER 34

Silicon Carbide Nanowires

MARCO NEGRI,[1,2,3] GIOVANNI ATTOLINI,[1] PAOLA LAGONEGRO,[1,2]
and MATTEO BOSI[1]

[1]*IMEM-CNR Institute, Parco Area delle Scienze 37A 43124 PARMA, Italy*

[2]*Department of Mathematical, Physical and Computer Sciences,
Parma University, Viale delle Scienze 7/A, I-43124 Parma, Italy*

[3]*Axcelis Technologies, 108 Cherry Hill Drive, Beverly, MA, USA,
E-mail: negri.m1@gmail.com*

34.1 DEFINITION

Silicon carbide (or carborundum) is a covalent compound of silicon and carbon, and its chemical formula is SiC. More than 250 crystalline forms (or polytypes) of SiC are known, and each one of them has a unique structure that can be viewed as a different stacking sequence of crystalline layers. Each polytype shows different physical properties such as the band gap, mobility, dielectric constant, etc.

A structure with two dimensions less than 100 nm and no constraints on the third dimension is referred to as a nanowire. Semiconductor nanowires of different chemical species have been synthesized and, among them, Silicon Carbide nanowires are studied for the interesting properties of the material coupled with the unique features of quasi-one-dimensional nanostructures.

34.2 HISTORICAL ORIGINS

In 1907, Henry J. Round, an assistant of Guglielmo Marconi, reported a 'bright glow' from diodes made of carborundum (Round, 1907), the first reported observation of electroluminescence from a semiconductor. Since then, the bulk synthesis of silicon carbide has been gaining increasing attention not only for the mechanical uses of the material, but also for the electronic properties. The efficiency of SiC electronic devices was low because of the poor crystalline quality and, in addition to that, no synthesis method to obtain SiC nanostructures were known.

In 1964, R. S. Wagner and W. C. Ellis theorized the vapor-liquid-solid (VLS) growth mechanism, and they proposed it as an explanation for silicon whiskers growth from the gas phase (Wagner & Ellis, 1964). In the same article, they mentioned "crystals of SiC […] may have grown by the VLS mechanism" as a byproduct due to the impurities present in the reactor. Other groups tried to use the same mechanism expressly for silicon carbide, and in the following years, they reported the synthesis of SiC whiskers (Ryan et al., 1967; Shaffer, 1966). Since then, VLS growth mechanism has been the main route to fabricate SiC microwhiskers. Only in recent years other mechanisms, some of them catalyst-free, where reported.

Moving to the nanometer scale, the term "nanowire" gradually replaces "whiskers" in the literature.

After the work of Iijima (1991) focused an unprecedented interest in the carbon nanostructures, the conversion of carbon nanotubes into SiC nanowires was demonstrated (Dai et al., 1995; Zhou & Seraphin, 1994) through reactions between silicon precursors such as SiO or SiI_2 and carbon nanotubes.

For many decades, SiC whiskers or nanowires have been studied for the incorporation in composite materials (metal matrix especially) in order to enhance the strength, toughness, and stiffness thanks to their superior mechanical properties (Christman, Needleman, & Suresh, 1989; McDanels, 1985; Nardone & Prewo, 1986).

In recent times, many other possible applications (treated in the following) have been studied and with the miniaturization of electronic devices, a brand new field for the nanostructures electronics arose.

34.3 NANO-SCIENTIFIC DEVELOPMENTS

As said, the first method developed to synthesize SiC NWs was from the vapor phase, taking advantage of the so-called VLS growth mechanism. Nowadays growth from the vapor phase is still the most widespread and well-established method, but also other techniques are gaining attention in the quest to find the highest yield, reliability, morphological and structural control and the possibility to tune the electrical properties.

34.3.1 *GROWTH FROM VAPOR PHASE*

Epitaxy is a well-established technique used to grow films over a substrate (see Chapter 43), and it is commonly used also for silicon carbide deposition. In order to obtain the growth of a quasi-one-dimensional structure instead of a two-dimensional one, it is necessary to have preferential sites on the substrate surface acting as nucleation points for the vapor deposition.

The phase transitions and the states of matter involved in the growth give the name to the mechanism of the nanowires synthesis, so, for SiC nanowires from the vapor phase, the to-date proposed growth mechanisms are Vapor-Liquid-Solid (VLS), Vapor-Solid-Solid (VSS), and Vapor-Solid (VS).

Only some growth mechanism as VLS and VSS need a catalyst acting as nucleation point, and it is possible to find catalyst particles at the end of the growth. If the growth occurs from the top it is called "float growth," and the particles remain on the top of the wires. In case of a "root growth," it is possible to find the catalyst particles on the bottom of the wires at the end of the growth (Kolasinski, 2006).

In the *VLS mechanism* (Givargizov, 1975; Wacaser et al., 2009; Wagner & Ellis, 1964), the growth takes advantage of nanoclusters in the liquid phase at the process temperature acting as catalysts. In this case, the synthesis must be carried out at a temperature above the melting point of the catalyst (usually the eutectic point of the system if the material is an alloy).

The catalyst can be dispersed in the form of nanoparticles but, most frequently, the so-called "dewetting" process is exploited. Owing to the surface tension, a thin film deposited on a substrate is a metastable phase and tends to agglomerate as soon as the temperature is raised over the melting point of the film material or alloy formed by the film and the substrate. This results in the formation of islands all over the substrate surface that can act as a nucleation sites for the VLS growth.

As shown in Figure 34.1, the gas phase precursor is incorporated in the liquid droplets and forms a solution. When supersaturation is reached, precipitation occurs, and the growth takes place.

It is well-known that the melting temperature of small particles can be lower than the corresponding bulk material (Buffat & Borel, 1976), but in some cases the low temperature used and the size of the nanoparticles exclude the possibility for these particles to melt (Attolini et al., 2014; Tong Goh & Abdul Rahman, 2014). This proves that sometimes also a solid particle can act as a nucleation point. If the nucleation particle remains in solid phase because the process temperature is lower than the melting temperature, the mechanism is called vapor-solid-solid (VSS). A schematic view of the process is reported in Figure 34.2.

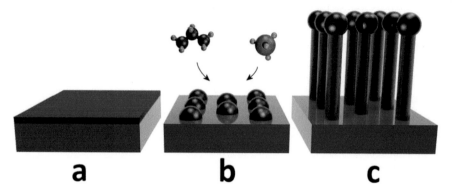

FIGURE 34.1   Schematic view of the dewetting process followed by a VLS growth mechanism for the synthesis of SiC nanowires. (a) A substrate covered with a thin film of the catalyst material is used. (b) Temperature is raised, and the dewetting occurs: the catalyst melts and forms small droplets on the surface. The precursors (in this case silane and propane are depicted) are incorporated into the droplets and growth occurs. (d) The final result is a nanowires array with catalyst particles on top in case of a "float growth" process.

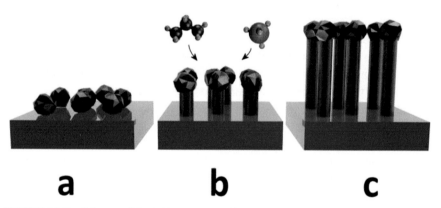

FIGURE 34.2   Scheme of the VSS process. Solid particles deposited on the substrate prior to the growth (a) act as nucleation sites for the gaseous precursors (b) and cause the growth of the nanowires (c).

Nevertheless, the presence of a catalyst (usually a metal) on the nanowires can be deleterious for electronic applications, and this is the reason why catalyst-free growth mechanisms are studied. The *VS* growth mechanism allows to obtain a one-dimensional growth without the aid of a catalyst. Owing to the high temperature of an epitaxial process, incoming adatoms have high energy on the substrate surface and high mobility as a consequence. The motion of these adatoms stops when a point with lower surface energy is reached, in this way defects or impurities can act as nucleation points for a uniaxial growth. The typical example, also depicted in Figure 34.3, is the growth starting from a screw dislocation (Burton, Cabrera, & Frank, 1951), resulting in a helical SiC nanowire (Bechelany et al., 2007; Zhang, Wang, & Wang, 2002).

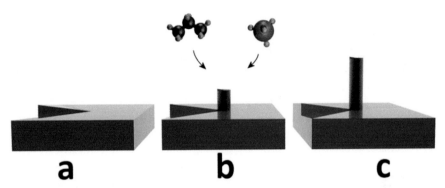

FIGURE 34.3 (a) Graphical representation of a screw dislocation on a crystalline substrate. (b) Sometimes this can act as nucleation site for the uniaxial growth from the vapor phase. (c) The nanowire itself acts as the propagation of the dislocation.

Another synthesis technique to obtain SiC nanowires from the vapor phase is the conversion of carbon nanostructures (such as nanotubes) to SiC nanowires using gaseous silicon precursors like silane (Mo et al., 2004) or using powders like SiO (Dai et al., 1995; Zhou & Seraphin, 1994) or Si (Wallis, Patyk, & Zerda, 2008) at high temperature. The problem is to control the morphology of the species obtained. The specular counterpart of this technique consists of the SiC formation by the carburization of silicon nanostructures (Ollivier et al., 2013).

34.3.2 OTHER GROWTH MECHANISMS AND TECHNIQUES

In the Solid-Liquid-Solid (SLS) mechanism the precursor is provided from the (solid) substrate (see Figure 34.4). In the case of SiC, the silicon substrate is the source of SiC. A metal-rich catalyst remains at the base of the nanowires and it is the vehicle for the reaction between the silicon of the substrate and the carbon provided from a carbon-rich catalyst like Ni-C (Xing et al., 2001) or Fe-C (T. Yang et al., 2003) or from the vapor phase (Ryu, Tak, & Yong, 2005). In the latter case, the mechanism can be defined as a hybrid of SLS and VLS. The SLS has always been observed as a root growth because at least one of the precursors must be continuously provided from the substrate. This is the reason because the term "root growth" is sometimes erroneously used as a synonym of SLS, but some VLS mechanism with root growth were also observed (Joshi et al., 2008; Kolasinski, 2006).

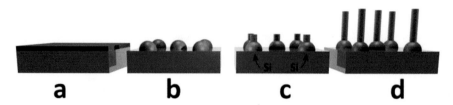

FIGURE 34.4   General scheme of a dewetting process followed by an SLS mechanism. (a) A thin film deposited on the substrate prior to the growth melts and forms droplets (b) on the surface. Both the precursors or only one (e.g., Silicon in the case depicted above) are provided from the substrate (c) and promote the growth of the nanowires from the bottom. The final morphology (d) is showing the root-growth mechanism with the catalyst particles at the bottom of the wires.

Other techniques are gaining attention for their relative simplicity: the polymers conversion into SiC nanowires allows to obtain big quantities of nanowires. Some polymers used as a precursor contain both carbon and silicon and can origin the SiC nanowires. The uniaxial growth can be obtained thanks to the presence of a catalyst like Fe (Feng, Ma, & Yang, 2012; Li et al., 2009). The starting material can also be prepared via the sol-gel process and treated with a carbothermal reduction to obtain the SiC growth (Liang et al., 2000). Frequently in these cases, the mechanism is still VLS, but the precursors are not provided in the gaseous phase with a CVD apparatus.

The so-called top-down approaches for the fabrication of SiC nanowires have proven to be valid alternatives to more traditional techniques. The microfabrication methods are well established for silicon, and they allowed to obtain short gate length MOSFET using Si NWs (Ernst et al., 2011), but silicon carbide has a strong resistance to chemical etching, and plasma etching is not straightforward as well. Nevertheless, inductive coupled SF_6/ O_2 plasma etching proved to be suitable to obtain nanopillars with aspect ratios of about 1:7 (Choi et al., 2012). The employment of top-down techniques has several advantages like the possibility to have well-ordered structures with uniform diameter and length. Further development of the Bosch/STS process (Laermer & Schilp, 1996) might obtain nanostructures with higher aspect ratios.

Although the aforementioned methods are the most used, other techniques to obtain nanowires such as arc-discharge (Li et al., 2002), laser ablation (Shi et al., 2000) and direct chemical reaction at low-temperature and high-pressure (Lu et al., 1999) have been developed. To enter in the details of these techniques is beyond the aim of this work.

While all of the mentioned synthesis methods are used to obtain SiC nanowires, many of them allow to obtain more complex nanostructures based on SiC nanowires, such as core-shell structures obtained with self-assembled growth mechanism (Negri et al., 2015) or by multiple steps procedure such as chemical vapor deposition, carburization, etc. (Zhou et al., 2000).

34.4   NANOCHEMICAL APPLICATIONS

Semiconductor nanowires are widely studied as catalyst supports, photocatalysts, and electrocatalysts for the high surface to volume ratio. The quest to find new supports for catalytic active phases is gaining increasing attention for the possibility to pave the way to new industrial processes or improve the yield of the existing ones. The advantages of using silicon carbide instead of traditional supports like alumina are linked to the chemical inertness of the material, both because the support has to withstand reactions in sometimes harsh chemical environment to allow the recovery of the active phase at the end of catalyst lifetime and because any chemical interaction between the support and catalytic active phase may lead to a decrease in the catalyst performance (Burtin et al., 1987; Zwinkels et al., 1993). Moreover, the good thermal SiC conductivity can help in dispersing the heat formed during the reaction. One of the drawbacks of the well-established alumina support is its thermal insulation leading to the formation of hot spots that may hinder the catalytic process. Even if these problems regarding alumina can be partially overcome using carbon-based supports which are suitable for certain processes, the weak resistance to oxidation of the latter pushes towards the investigation of more inert materials.

The use of more conventional SiC structures as catalyst support is nowadays well developed, and there are companies (SICAT http://www.sicatcatalyst.com) already using them for industrial processes such as coal/gas-to-liquid, Fischer-Tropsch, methane conversion, selective oxidation, selective hydrogenation, biomass conversion, wastewater purification, and photocatalysis. Using nanowires might lead to higher yield owing to the superior surface-to-volume ratio.

SiC nanostructures proved to be an excellent support for Ni catalyst in low-temperature selective oxidation of H_2S into elemental sulfur (Nhut, 2002; Pham-Huu et al., 2001), Pd nanoparticles for methanol combustion (Guo et al., 2009). As a support of noble catalyst as Pt and Pd, SiC NWs are also studied for fuel cells applications (Niu & Wang, 2009; Tong et al., 2011).

Many research groups have been studying SiC NWs as an *active catalyst*. Zhou et al. (2006) found that SiC nanowires accelerate the photodegradation of acetaldehyde under UV irradiation. SiC NWs proved also to be a valid photocatalytic agent for water splitting under UV or visible irradiation for hydrogen production (Hao et al., 2012; Liu et al., 2012).

34.5 MULTI-/TRANS-DISCIPLINARY CONNECTIONS

As other nanostructures do, nanowires exhibit outstanding properties not possessed by their bulk counterparts. The effect of the reduced size along two dimensions on the mechanical properties was measured on nanowires of different materials (Cuenot et al., 2004). Basing on the excellent mechanical properties of bulk silicon carbide one can infer that its nanostructured counterpart might show remarkable mechanical properties: nanomechanical studies observed indeed a super-plasticity of the NWs (Zhang et al., 2007), but even before this discovery one of the first applications of SiC NWs was as reinforcement for *composite materials*.

SiC nanowires embedded in a polymer matrix (Meng et al., 2010; Nhuapeng et al., 2008) sensibly enhance tensile strength and bending strength of the composite. Ceramic/SiC composite was studied (Jirabor-vornpongsa et al., 2013) with a measurable enhancement of properties like thermal conductivity, chemical, and thermal stability. Even SiC/SiC composite was fabricated (Yang et al., 2004) and superior mechanical properties were found.

The excellent biocompatibility of silicon carbide (Saddow, 2012) encouraged many research groups to focus on *biomedical applications*. SiC nanocrystals (quasi-one-dimensional) already demonstrated to be excellent labels for bioimaging (Fan et al., 2008) and the *in-vitro* studies of the cytocompatibility of core-shell SiC-SiO_2 nanowires (Cacchioli et al., 2014) encouraged to use SiC nanowires in medical nanosystems. Rossi et al. studied a system based on SiC/SiO_2 core-shell nanowires conjugated with an organic photosensitizer, a tetracarboxyphenyl porphyrin derivative, to be used as an oxidizing agent for photodynamic therapy in cancer treatments (Rossi et al., 2015). SiC NWs seem to be promising also as biosensors: the well-known silicon functionalization chemistry allowed Fradetal et al. to functionalize SiC nanowires with desoxyribonucleic acid (L. Fradetal et al., 2014) and to use them as FET sensor for DNA detection (Louis

Fradetal et al., 2013). Williams et al. demonstrated the possibility to functionalize silicon carbide nanowires with streptavidin protein thus creating a NW-based sensing platform for the multiplexed electrical detection of bioanalytes (Williams et al., 2012).

The use of nanowires as *field emission transistors* (FET) is another promising application to obtain small electronic semiconductor devices suitable to develop sensors (Rogdakis et al., 2007). From the first attempts (Seong et al., 2004; Zhou et al., 2006) many improvements have been achieved to obtain better performances, but still many issues have to be overcome, like obtaining ohmic or rectifying behavior of metal contacts deposited on SiC depending on the need is not trivial but it is essential to fabricate NW FET (Jang et al., 2008). Another problem is the control of n or p-type doping of the channel (Li et al., 2014). The unintentional doping level measured to-date in the SiC NWs is still too high to use ohmic contacts as source and drain, because it is impossible to fully deplete the nanowire in the off state, so Schottky barrier contacts are still needed to have satisfactory device performances (Rogdakis et al., 2011).

Other interesting applications of SiC NWs are the use as *microwave absorbers* to avoid electromagnetic interference (Chiu, Yu, & Li, 2010), as electrodes for *micro-supercapacitors* (Alper et al., 2013) to be used as energy source for micrometric devices or as *anode material* for lithium ion batteries showing increased capacity and lifetime thanks to the ability to retain the integrity over multiple charge-discharge cycles (Yang et al., 2013).

The use of nanowires for *nanoelectromechanical* devices (Feng et al., 2010) must be mentioned, and the field emission properties of SiC nanowires (Wong et al., 1999) make SiC NWs very promising for electron-emitting device applications (displays, etc.).

34.5   OPEN ISSUES

As demonstrated, the research progress on SiC nanowires is strictly connected to achievements in many other fields such as fundamental physics or biotechnology. Thanks to the interesting silicon carbide properties coupled with those of one-dimensional nanostructures many promising applications may emerge in the future.

Nevertheless, even if preliminary experiments at laboratory level are very promising and fruitful, scaling the technologies up to industrial production keeping the unique, peculiar properties is still a challenge. The assessment of a single nanowire property is a hint on the material properties, but collective properties may be also sensibly different and unpredictable.

In addition to that, conflicting studies on the toxicity of one-dimensional nanostructured materials and, in particular on SiC nanowires (Cacchioli et al., 2014; Morimoto et al., 2003; Ogami et al., 2001) leaves still room for deepening the knowledge about possible problems or opportunities linked to the development of this field.

KEYWORDS

- **bottom-up**
- **Epitaxy**
- **Quantum Wires**
- **semiconductor nanostructures**
- **top-down**

REFERENCES AND FURTHER READING

Alper, J. P., Kim, M. S., Vincent, M., Hsia, B., Radmilovic, V., Carraro, C., & Maboudian, R., (2013). Silicon carbide nanowires as highly robust electrodes for micro-supercapacitors. *Journal of Power Sources, 230*, 298–302. doi: 10.1016/j.jpowsour.2012.12.085.

Attolini, G., Rossi, F., Negri, M., Dhanabalan, S. C., Bosi, M., Boschi, F., & Salviati, G., (2014). Growth of SiC NWs by vapor phase technique using Fe as catalyst. *Materials Letters, 124*, 169–172. doi: 10.1016/j.matlet.2014.03.061.

Bechelany, M., Brioude, A., Stadelmann, P., Ferro, G., Cornu, D., & Miele, P., (2007). Very long SiC-based coaxial nanocables with tunable chemical composition. *Advanced Functional Materials, 17*(16), 3251–3257. doi: 10.1002/adfm.200700110.

Buffat, P., & Borel, J. P., (1976). Size effect on the melting temperature of gold particles. *Physical Review A., 13*(6), 2287–2298. doi: 10.1103/PhysRevA.13.2287.

Burtin, P., Brunelle, J. P., Pijolat, M., & Soustelle, M., (1987). Influence of surface area and additives on the thermal stability of transition alumina catalyst supports. II: Kinetic model and interpretation. *Applied Catalysis, 34*, 239–254. doi: 10.1016/S0166–9834(00)82459-8.

Burton, W. K., Cabrera, N., & Frank, F. C., (1951). The growth of crystals and the equilibrium structure of their surfaces. *Philosophical Transactions of the Royal Society A: Mathematical, Physical and Engineering Sciences, 243*(866), 299–358. doi: 10.1098/rsta.1951.0006.

Cacchioli, A., Ravanetti, F., Alinovi, R., Pinelli, S., Rossi, F., Negri, M., & Salviati, G., (2014). Cytocompatibility and cellular internalization mechanisms of SiC/SiO2 nanowires. *Nano Letters*, doi: 10.1021/nl501255m.

Chiu, S. C., Yu, H. C., & Li, Y. Y., (2010). High electromagnetic wave absorption performance of silicon carbide nanowires in the gigahertz range. *The Journal of Physical Chemistry C.*, *114*(4), 1947–1952. doi: 10.1021/jp905127t.

Choi, J. H., Latu-Romain, L., Bano, E., Dhalluin, F., Chevolleau, T., & Baron, T., (2012). Fabrication of SiC nanopillars by inductively coupled SF 6 /O 2 plasma etching. *Journal of Physics D: Applied Physics*, *45*(23), 235204. doi: 10.1088/0022-3727/45/23/235204.

Christman, T., Needleman, A., & Suresh, S., (1989). An experimental and numerical study of deformation in metal-ceramic composites. *Acta Metallurgica*, *37*(11), 3029–3050. doi: 10.1016/0001-6160(89)90339-8.

Cuenot, S., Frétigny, C., Demoustier-Champagne, S., & Nysten, B., (2004). Surface tension effect on the mechanical properties of nanomaterials measured by atomic force microscopy. *Physical Review B.*, *69*(16), 165410. doi: 10.1103/PhysRevB.69.165410.

Dai, H., Wong, E. W., Lu, Y. Z., Fan, S., & Lieber, C. M., (1995). Synthesis and characterization of carbide nanorods. *Nature*, *375*(6534), 769–772. doi: 10.1038/375769a0.

Ernst, T., Barraud, S., Tachi, K., Vizioz, C., Magis, T., Brianceau, P., & Cassé, M., (2011). Ultra-dense silicon nanowires: A technology, transport, and interfaces challenges insight (invited). *Microelectronic Engineering*, *88*(7), 1198–1202. doi: 10.1016/j.mee.2011.03.102.

Fan, J., Li, H., Jiang, J., So, L. K. Y., Lam, Y. W., & Chu, P. K., (2008). 3C-SiC nanocrystals as fluorescent biological labels. *Small (Weinheim an Der Bergstrasse, Germany)*, *4*(8), 1058–1062. doi: 10.1002/smll.200800080.

Feng, W., Ma, J., & Yang, W., (2012). Precise control on the growth of SiC nanowires. *Cryst. Eng. Comm.*, *14*(4), 1210–1212. doi: 10.1039/C2CE06569J.

Feng, X. L., Matheny, M. H., Zorman, C. A., Mehregany, M., & Roukes, M. L., (2010). Low voltage nanoelectromechanical switches based on silicon carbide nanowires. *Nano Letters*, *10*(8), 2891–2896. doi: 10.1021/nl1009734.

Fradetal, L., Stambouli, V., Bano, E., Pelissier, B., Choi, J. H., Ollivier, M., & Pignot-Paintrand, I., (2014). Bio-functionalization of silicon carbide nanostructures for SiC nanowire-based sensors realization. *Journal of Nanoscience and Nanotechnology*, *14*(5), 3391–3397. doi: 10.1166/jnn.2014.8223.

Fradetal, L., Stambouli, V., Bano, E., Pelissier, B., Wierzbowska, K., Choi, J. H., & Latu-Romain, L., (2013). First experimental functionalization results of SiC nanopillars for biosensing applications. *Materials Science Forum*, *740–742*, 821–824.

Givargizov, E. I., (1975). Fundamental aspects of VLS growth. *Journal of Crystal Growth*, *31*, 20–30. doi: 10.1016/0022-0248(75)90105-0.

Guo, X. N., Shang, R. J., Wang, D. H., Jin, G. Q., Guo, X. Y., & Tu, K. N., (2009). Avoiding loss of catalytic activity of Pd nanoparticles partially embedded in nanoditches in SiC nanowires. *Nanoscale Research Letters*, *5*(2), 332–337. doi: 10.1007/s11671-009-9484-6.

Hao, J. Y., Wang, Y. Y., Tong, X. L., Jin, G. Q., & Guo, X. Y., (2012). Photocatalytic hydrogen production over modified SiC nanowires under visible light irradiation. *International Journal of Hydrogen Energy*, *37*(20), 15038–15044. doi: 10.1016/j.ijhydene.2012.08.021.

Iijima, S., (1991). Helical microtubules of graphitic carbon. *Nature*, *354*(6348), 56–58. doi: 10.1038/354056a0.

Jang, C. O., Kim, T. H., Lee, S. Y., Kim, D. J., & Lee, S. K., (2008). Low-resistance ohmic contacts to SiC nanowires and their applications to field-effect transistors. *Nanotechnology*, *19*(34), 345203. doi: 10.1088/0957-4484/19/34/345203.

Jiraborvornpongsa, N., Imai, M., Yoshida, K., & Yano, T., (2013). Effects of trace amount of nanometric SiC additives with wire or particle shapes on the mechanical and thermal properties of alumina matrix composites. *Journal of Materials Science, 48*(20), 7022–7027. doi: 10.1007/s10853-013-7511-6.

Joshi, R. K., Yoshimura, M., Tanaka, K., Ueda, K., Kumar, A., & Ramgir, N., (2008). Synthesis of vertically aligned Pd 2 Si nanowires in microwave plasma enhanced chemical vapor deposition system. *The Journal of Physical Chemistry C., 112*(36), 13901–13904. doi: 10.1021/jp8050752.

Kolasinski, K. W., (2006). Catalytic growth of nanowires: Vapor-liquid-solid, vapor-solid–solid, solution–liquid–solid and solid-liquid-solid growth. *Current Opinion in Solid State and Materials Science, 10*(3), 182–191.

Laermer, F., & Schilp, A., (1996). *Method of Anisotropically Etching Silicon*, USA. US patent number US5501893A.

Li, G., Li, X., Chen, Z., Wang, J., Wang, H., & Che, R., (2009). Large areas of centimeters-long SiC nanowires synthesized by pyrolysis of a polymer precursor by a CVD route. *The Journal of Physical Chemistry C., 113*(41), 17655–17660. doi: 10.1021/jp904277f.

Li, S., Wang, N., Zhao, H., & Du, L., (2014). Synthesis and electrical properties of p-type 3C–SiC nanowires. *Materials Letters, 126*, 217–219. doi: 10.1016/j.matlet.2014.04.072.

Li, Y., Xie, S., Zhou, W., Ci, L., & Bando, Y., (2002). Cone-shaped hexagonal 6H–SiC nanorods. *Chemical Physics Letters, 356*, 325–330.

Liang, C., Meng, G., Zhang, L., Wu, Y., & Cui, Z., (2000). Large-scale synthesis of β-SiC nanowires by using mesoporous silica embedded with Fe nanoparticles. *Chemical Physics Letters, 329*(3/4), 323–328. doi: 10.1016/S0009–2614(00)01023-X.

Liu, H., She, G., Mu, L., & Shi, W., (2012). Porous SiC nanowire arrays as stable photocatalyst for water splitting under UV irradiation. *Materials Research Bulletin, 47*(3), 917–920. doi: 10.1016/j.materresbull.2011.12.046.

Lu, Q., Hu, J., Tang, K., Qian, Y., Zhou, G., Liu, X., & Zhu, J., (1999). Growth of SiC nanorods at low temperature. *Applied Physics Letters, 75*(4), 507. doi: 10.1063/1.124431.

McDanels, D. L., (1985). Analysis of stress-strain, fracture, and ductility behavior of aluminum matrix composites containing discontinuous silicon carbide reinforcement. *Metallurgical Transactions A., 16*(6), 1105–1115. doi: 10.1007/BF02811679.

Meng, S., Jin, G. Q., Wang, Y. Y., & Guo, X. Y., (2010). Tailoring and application of SiC nanowires in composites. *Materials Science and Engineering: A., 527*(21/22), 5761–5765. doi: 10.1016/j.msea.2010.05.045.

Mo, Y. H., Shajahan, M. D., Lee, Y. S., Hahn, Y. B., & Nahm, K. S., (2004). Structural transformation of carbon nanotubes to silicon carbide nanorods or microcrystals by the reaction with different silicon sources in RF-induced CVD reactor. *Synthetic Metals, 140*(2/3), 309–315. doi: 10.1016/S0379–6779(03)00381-3.

Morimoto, Y., Ding, L., Oyabu, T., Hirohashi, M., Kim, H., Ogami, A., & Tanaka, I., (2003). Expression of Clara cell secretory protein in the lungs of rats exposed to silicon carbide whisker *in vivo. Toxicology Letters, 145*(3), 273–279. doi: 10.1016/S0378–4274(03)00308-4.

Nardone, V. C., & Prewo, K. M., (1986). On the strength of discontinuous silicon carbide reinforced aluminum composites. *Scripta Metallurgica, 20*(1), 43–48. doi: 10.1016/0036-9748 (86)90210-3.

Negri, M., Dhanabalan, S. C., Attolini, G., Lagonegro, P., Campanini, M., Bosi, M., & Salviati, G., (2015). Tuning the radial structure of core-shell silicon carbide nanowires. *Cryst. Eng. Comm., 17*(6), 1258–1263. doi: 10.1039/C4CE01381F.

Nhuapeng, W., Thamjaree, W., Kumfu, S., Singjai, P., & Tunkasiri, T., (2008). Fabrication and mechanical properties of silicon carbide nanowires/epoxy resin composites. *Current Applied Physics, 8*(3/4), 295–299. doi: 10.1016/j.cap.2007.10.074.

Nhut, J., (2002). Synthesis and catalytic uses of carbon and silicon carbide nanostructures. *Catalysis Today, 76*(1), 11–32. doi: 10.1016/S0920–5861(02)00206-7.

Niu, J. J., & Wang, J. N., (2009). Synthesis of macroscopic SiC nanowires at the gram level and their electrochemical activity with Pt loadings. *Acta Materialia, 57*(10), 3084–3090. doi: 10.1016/j.actamat.2009.03.014.

Ogami, A., Morimoto, Y., Yamato, H., Oyabu, T., Akiyama, I., & Tanaka, I., (2001). Short term effect of silicon carbide whisker to the rat lung. *Industrial Health, 39*(2), 175–182. doi: 10.2486/indhealth.39.175.

Ollivier, M., Latu-Romain, L., Martin, M., David, S., Mantoux, A., Bano, E., & Baron, T., (2013). Si-SiC core-shell nanowires. *Journal of Crystal Growth, 363*, 158–163.

Pham-Huu, C., Keller, N., Ehret, G., & Ledoux, M. J., (2001). The first preparation of silicon carbide nanotubes by shape memory synthesis and their catalytic potential. *Journal of Catalysis, 200*(2), 400–410. doi: 10.1006/jcat.2001.3216.

Rogdakis, K., Bano, E., Montes, L., Bechelany, M., Cornu, D., & Zekentes, K., (2011). Rectifying source and drain contacts for effective carrier transport modulation of extremely doped SiC nanowire FETs. *IEEE Transactions on Nanotechnology, 10*(5), 980–984. doi: 10.1109/TNANO.2010.2091147.

Rogdakis, K., Bescond, M., Bano, E., & Zekentes, K., (2007). Theoretical comparison of 3C-SiC and Si nanowire FETs in ballistic and diffusive regimes. *Nanotechnology, 18*(47), 475715. doi: 10.1088/0957–4484/18/47/475715.

Rossi, F., Bedogni, E., Bigi, F., Rimoldi, T., Cristofolini, L., Pinelli, S., & Salviati, G., (2015). Porphyrin conjugated SiC/SiOx nanowires for X-ray-excited photodynamic therapy. *Scientific Reports, 5*, 7606. doi: 10.1038/srep07606.

Round, H. J., (1907). A note on carborundum. *Electrical World, 49*, 309.

Ryan, C. E., Berman, I., Marshall, R. C., Considine, D. P., & Hawley, J. J., (1967). Vapor-liquid-solid and melt growth of silicon carbide. *Journal of Crystal Growth, 1*(5), 255–262.

Ryu, Y., Tak, Y., & Yong, K., (2005). Direct growth of core-shell SiC-SiO(2) nanowires and field emission characteristics. *Nanotechnology, 16*(7), 370–374. doi: 10.1088/0957–4484/16/7/009.

Saddow, S. E., (2012). *Silicon Carbide Biotechnology: A Biocompatible Semiconductor for Advanced Biomedical Devices and Applications.* Waltham, MA: Elsevier, ISBN: 9780128030059.

Seong, H. K., Choi, H. J., Lee, S. K., Lee, J. I., & Choi, D. J., (2004). Optical and electrical transport properties in silicon carbide nanowires. *Applied Physics Letters, 85*(7), 1256. doi: 10.1063/1.1781749.

Shaffer, P. T. B., (1966). SiC whiskers. *Ceramic Age, 82*, 46.

Shi, W., Zheng, Y., Peng, H., Wang, N., Lee, C. S., & Lee, S. T., (2000). Laser ablation synthesis and optical characterization of silicon carbide nanowires. *Journal of the American Ceramic Society, 83*(12), 3228–3230. doi: 10.1111/j.1151–2916.2000.tb01714.x.

Tong, G. B., & Abdul, R. S., (2014). Synthesis of nickel-catalyzed Si/SiC core-shell nanowires by HWCVD. *Journal of Crystal Growth, 407*, 25–30. doi: 10.1016/j.jcrysgro.2014.09.004.

Tong, X., Dong, L., Jin, G., Wang, Y., & Guo, X. Y., (2011). Electrocatalytic performance of Pd nanoparticles supported on SiC nanowires for methanol oxidation in alkaline media. *Fuel Cells, 11*(6), 907–910. doi: 10.1002/fuce.201100017.

Wacaser, B. A., Dick, K. A., Johansson, J., Borgström, M. T., Deppert, K., & Samuelson, L., (2009). Preferential interface nucleation: An expansion of the VLS growth mechanism for nanowires. *Advanced Materials, 21*(2), 153–165. doi: 10.1002/adma.200800440.

Wagner, R. S., & Ellis, W. C., (1964). Vapor-liquid-solid mechanism of single crystal growth. *Applied Physics Letters, 4*(5), 89. doi: 10.1063/1.1753975.

Wallis, K. L., Patyk, J. K., & Zerda, T. W., (2008). Reaction kinetics of nanostructured silicon carbide. *Journal of Physics: Condensed Matter, 20*(32), 325216. doi: 10.1088/0953-8984/20/32/325216.

Williams, E. H., Schreifels, J. A., Rao, M. V., Davydov, A. V., Oleshko, V. P., Lin, N. J., & Koshka, Y., (2012). Selective streptavidin bioconjugation on silicon and silicon carbide nanowires for biosensor applications. *Journal of Materials Research, 28*(01), 68–77. doi: 10.1557/jmr.2012.283.

Wong, K. W., Zhou, X. T., Au, F. C. K., Lai, H. L., Lee, C. S., & Lee, S. T., (1999). Field-emission characteristics of SiC nanowires prepared by chemical vapor deposition. *Applied Physics Letters, 75*(19), 2918. doi: 10.1063/1.125189.

Xing, Y., Hang, Q., Yan, H., Pan, H., Xu, J., Yu, D., & Feng, S., (2001). Solid-liquid–solid (SLS) growth of coaxial nanocables: Silicon carbide sheathed with silicon oxide. *Chemical Physics Letters, 345*(1–2), 29–32. doi: 10.1016/S0009-2614(01)00768-0.

Yang, T., Chen, C., Chatterjee, A., Li, H., Lo, J., Wu, C., & Chen, L., (2003). Controlled growth of silicon carbide nanorods by rapid thermal process and their field emission properties. *Chemical Physics Letters, 379*(1/2), 155–161. doi: 10.1016/j.cplett.2003.08.001.

Yang, W., Araki, H., Kohyama, A., Thaveethavorn, S., Suzuki, H., & Noda, T., (2004). Process and mechanical properties of in situ silicon carbide-nanowire-reinforced chemical vapor infiltrated silicon carbide/silicon carbide composite. *Journal of the American Ceramic Society, 87*(9), 1720–1725. doi: 10.1111/j.1551-2916.2004.01720.x.

Yang, Y., Ren, J. G., Wang, X., Chui, Y. S., Wu, Q. H., Chen, X., & Zhang, W., (2013). Graphene encapsulated, and SiC reinforced silicon nanowires as an anode material for lithium-ion batteries. *Nanoscale, 5*(18), 8689–8694. doi: 10.1039/c3nr02788k.

Zhang, H. F., Wang, C. M., & Wang, L. S., (2002). Helical crystalline SiC/SiO 2 core-shell nanowires. *Nano Letters, 2*(9), 941–944. doi: 10.1021/nl025667t.

Zhang, Y., Han, X., Zheng, K., Zhang, Z., Zhang, X., Fu, J., & Wang, Z. L., (2007). Direct observation of super-plasticity of beta-SiC nanowires at low temperature. *Advanced Functional Materials, 17*(17), 3435–3440. doi: 10.1002/adfm.200700162.

Zhou, D., & Seraphin, S., (1994). Production of silicon carbide whiskers from carbon nanoclusters. *Chemical Physics Letters, 222*(3), 233–238.

Zhou, W. M., Fang, F., Hou, Z. Y., Yan, L. J., & Zhang, Y. F., (2006). Field-effect transistor based on/spl beta/-SiC nanowire. *IEEE Electron Device Letters, 27*(6), 463–465. doi: 10.1109/LED.2006.874219.

Zhou, W., Yan, L., Wang, Y., & Zhang, Y., (2006). SiC nanowires: A photocatalytic nanomaterial. *Applied Physics Letters, 89*(1), 013105. doi: 10.1063/1.2219139.

Zhou, X., Zhang, R., Peng, H., Shang, N., Wang, N., Bello, I., & Lee, S., (2000). Highly efficient and stable photoluminescence from silicon nanowires coated with SiC. *Chemical Physics Letters, 332*(3–4), 215–218. doi: 10.1016/S0009-2614(00)01145-3.

Zwinkels, M. F. M., Järås, S. G., Menon, P. G., & Griffin, T. A., (1993). Catalytic materials for high-temperature combustion. *Catalysis Reviews, 35*(3), 319–358. doi: 10.1080/01614949308013910.

CHAPTER 35

Specific Adjacency in Bonding

MIHAI V. PUTZ[1,2] and MARINA A. TUDORAN[2]

[1]*Laboratory of Structural and Computational Physical Chemistry for Nanosciences and QSAR, Biology-Chemistry Department, West University of Timisoara, Pestalozzi Street No. 44, Timisoara, RO-300115, Romania, Tel: +40-256-592638, Fax: +40-256-592620, E-mails: mv_putz@yahoo.com or mihai.putz@e-uvt.ro*

[2]*Laboratory of Renewable Energies-Photovoltaics, R&D National Institute for Electrochemistry and Condensed Matter, Dr. A. Paunescu Podeanu Str. No. 144, Timisoara, RO-300569, Romania*

35.1 DEFINITION

The Specific Adjacency in Bonding (SAIB) is a generic name given to express a specific type of bonding in organic molecules, such as polycyclic aromatic hydrocarbons, and express the carbon and hydrogen atoms arrangements towards the simple or the double bond between two carbon atoms from the benzenic ring.

35.2 HISTORICAL ORIGIN(S)

In modern science, the QSAR (Quantitative Structure-Activity Relationship) method has become more and more used in theoretical studies, the reactivity indices being of a good use due to the fact that can describe the chemical reactivity with observable quantities such as the electrons number or the total valence energy. The first index, *electronegativity* (χ), is the chemical property which describes the atom tendency to attract electrons to him. When the equilibrium charged system is considered, the electronegativity can be seen as an instantaneous variation of total energy, with the formula (Putz, 2012b, 2013; Putz & Dudaș, 2013):

$$\chi = -\frac{\varepsilon_{HOMO} - \varepsilon_{LUMO}}{2} \tag{1}$$

where ε_{HOMO} represents the HOMO (Highest Occupied Molecular Orbital) energy and ε_{LUMO} the LUMO (Lowest Unoccupied Molecular Orbital) energy.

Another index, *chemical hardness* (η) represents the instantaneous electronegativity charge of atoms in a molecule and is usually used to measure the stability of a molecule, following the formula (Putz, 2012b, 2013; Putz & Dudaş, 2013):

$$\eta = \frac{\varepsilon_{LUMO} - \varepsilon_{HOMO}}{2} \tag{2}$$

In chemical reactivity analysis, due to the conceptual differences between the energetic level characterizing the electronegativity and the energetic interval characterizing the chemical hardness (Putz, 2012b, 2013), these two forces can be considered as bases for a chemical orthogonal space $\{\chi, \eta |\ \chi \perp \eta\}$ (Figure 35.1).

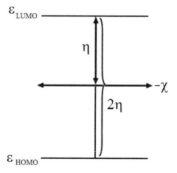

FIGURE 35.1   The chemical orthogonal space described by electronegativity and chemical hardness. Redrawn and adapted after Putz et al., (2013), Putz (2012b) and Putz (2013).

However, the full occupied electronic state can be reached only with the aid of other indices, the chemical power, and the electrophilicity. In this context, *chemical power* (π) is considered as the dynamic charge of atoms in a molecule, between molecular fragments or between adducts in a chemical bond. In other words, due to the charge transfer, the chemical power realizes the minimization of HOMO levels in bonding without changing the spin, and has the formula (Putz & Dudaş, 2013):

$$\pi = \frac{\chi}{2\eta} \tag{3}$$

The *electrophilicity* (ω) measures, in a system, the energy consumed for the manifestation of chemical power in a chemical orthogonal space, was developed in order to characterize the electrophilic/nucleophilic charge transfer by accepting or donating electrons, and has the formula (Putz & Dudaş, 2013):

$$\omega = \frac{\chi^2}{2\eta} = \pi \cdot \chi \tag{4}$$

At the molecular level, electronegativity (5) and chemical hardness (6) can be also determined using an iterative formula (Bratsch, 1985) based on the principle of chemical potential equalization applied to the atoms equalization in the molecule, with the following working formulas (Tudoran & Putz, 2015):

$$\chi_{\text{Molecule}} = \frac{n_{Atoms}}{\sum\limits_{A} \dfrac{n_A}{\chi_A}} \tag{5}$$

$$\eta_{\text{Molecule}} = \frac{n_{Atoms}}{\sum\limits_{A} \dfrac{n_A}{\eta_A}} \tag{6}$$

However, when applied to polycyclic aromatic hydrocarbons, the values obtained for electronegativity and chemical hardness for carbon atom, simple and double carbon bonds are the same, the only difference seems to appear between the hetero-bond of carbon and hydrogen (see Table 35.1). Even the graph theory presents a sort of limitation, such as the adjacency matrix returns information only about the presence or absence of bond between atoms, while the distance matrix counts the atoms vicinity (Tudoran & Putz, 2015).

TABLE 35.1 Electronegativity and Chemical Hardness for Carbon and Hydrogen Atoms and the Bonds Between Them Obtained by Applying the Equations (5) and (6) (Tudoran & Putz, 2015)

χ					η				
C	C–C	C = C	H	C–H	C	C–C	C = C	H	C–H
6.24	6.24	6.24	7.18	6.68	4.99	4.99	4.99	6.45	5.63

Starting from these limitations, Tudoran & Putz (2015) developed a new method of molecular characterization, the specific adjacency in bonding, in which the adjacency matrix, from topology, is enriched with reactivity indices for each type of specific bond manifested in aromatic compounds.

35.3 NANO-SCIENTIFIC DEVELOPMENT(S)

The Specific Adjacency in Bonding method is based on the idea that the simple and double bonds can be found in six different molecular valence medium. The simple carbon bond can be in one of the three forms (Figure 35.2): in adjacency with two hydrogen atoms (sHH), with two carbon atoms (sCC) and in hetero-adjacency with a carbon atom and a hydrogen atom (sHC) (Tudoran & Putz, 2015).

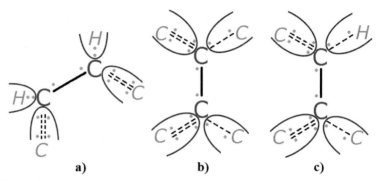

FIGURE 35.2 Simple carbon bond in three different valence mediums: (a) sHH, (b) sCC, and (c) sHC. Redrawn and adapted after Tudoran & Putz (2015).

The same rule is also applied to the double carbon-carbon bond (Figure 35.3), being in adjacency with two hydrogen atoms (dHH), with two carbon atoms (dCC) or in hetero-adjacency with a carbon atom and a hydrogen atom (dHC).

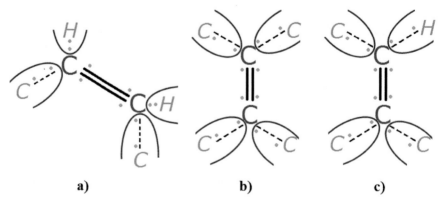

FIGURE 35.3 Simple carbon bond in three different valence mediums: (a) sHH, (b) sCC, and (c) sHC. Redrawn and adapted after Tudoran & Putz (2015).

On the other hand, the atoms in the molecule equation can be reinterpreted as the type of bonds in the molecule as follows (Tudoran & Putz, 2015):

a)
$$\chi_{\text{adjacency-in-bonding}} = \frac{n_{\text{adjacency-in-bonding}}}{\sum_{\text{type-of-bond}} \frac{n_{\text{type-of-bond}}}{\chi_{\text{type-of-bond}}}} \qquad (7)$$

b)
$$\eta_{\text{adjacency-in-bonding}} = \frac{n_{\text{adjacency-in-bonding}}}{\sum_{\text{type-of-bond}} \frac{n_{\text{type-of-bond}}}{\eta_{\text{type-of-bond}}}} \qquad (8)$$

with $n_{\text{adjacency-in-bonding}}$, the number of bonds of the specific type, $n_{\text{type-of-bond}}$, the number of bonds, $\chi_{\text{type-of-bond}}$, and $\eta_{\text{type-of-bond}}$ the values obtained from the literature.

Applied to each type of bond, Eqs. (7) and (8) can be used to obtain the specific calculation formula for electronegativity and chemical hardness (Tudoran & Putz, 2015):

- **Simple carbon-carbon bond:**

a)
$$\begin{cases} \chi_{\text{sHH}} = \dfrac{7}{\dfrac{1}{\chi_{C-C}} + \dfrac{2}{\chi_{C-H}} + \dfrac{2}{\chi_{C=C}}} \\ \eta_{\text{sHH}} = \dfrac{7}{\dfrac{1}{\eta_{C-C}} + \dfrac{2}{\eta_{C-H}} + \dfrac{2}{\eta_{C=C}}} \end{cases} \qquad (9)$$

b)
$$\begin{cases} \chi_{\text{sCC}} = \dfrac{7}{\dfrac{3}{\chi_{C-C}} + \dfrac{2}{\chi_{C=C}}} \\ \eta_{\text{sCC}} = \dfrac{7}{\dfrac{3}{\eta_{C-C}} + \dfrac{2}{\eta_{C=C}}} \end{cases} \qquad (10)$$

c)
$$\begin{cases} \chi_{\text{sHC}} = \dfrac{7}{\dfrac{2}{\chi_{C-C}} + \dfrac{1}{\chi_{C-H}} + \dfrac{2}{\chi_{C=C}}} \\ \eta_{\text{sHC}} = \dfrac{7}{\dfrac{2}{\eta_{C-C}} + \dfrac{1}{\eta_{C-H}} + \dfrac{2}{\eta_{C=C}}} \end{cases} \qquad (11)$$

- **Double carbon-carbon bond**:

a)
$$\begin{cases} \chi_{dHH} = \dfrac{6}{\dfrac{2}{\chi_{C-C}} + \dfrac{2}{\chi_{C-H}} + \dfrac{1}{\chi_{C=C}}} \\[4ex] \eta_{dHH} = \dfrac{6}{\dfrac{2}{\eta_{C-C}} + \dfrac{2}{\eta_{C-H}} + \dfrac{1}{\eta_{C=C}}} \end{cases} \tag{12}$$

b)
$$\begin{cases} \chi_{dCC} = \dfrac{6}{\dfrac{4}{\chi_{C-C}} + \dfrac{1}{\chi_{C=C}}} \\[4ex] \eta_{dCC} = \dfrac{6}{\dfrac{4}{\eta_{C-C}} + \dfrac{1}{\eta_{C=C}}} \end{cases} \tag{13}$$

c)
$$\begin{cases} \chi_{dHC} = \dfrac{6}{\dfrac{3}{\chi_{C-C}} + \dfrac{1}{\chi_{C-H}} + \dfrac{1}{\chi_{C=C}}} \\[4ex] \eta_{dHC} = \dfrac{6}{\dfrac{3}{\eta_{C-C}} + \dfrac{1}{\eta_{C-H}} + \dfrac{1}{\eta_{C=C}}} \end{cases} \tag{14}$$

Using the values obtained for electronegativity and chemical hardness, one can also determine the chemical power and the electronegativity values for the six types of bonds, which can be further utilized in studies (see Table 35.2) (Tudoran & Putz, 2015).

This way, the simple and double carbon-carbon bond was characterized in a novel manner, which considers not only the type of bond, but also the number of electrons involved in bonding (Tudoran & Putz, 2015).

35.4 NANOCHEMICAL APPLICATION(S)

In their work, Putz and his co-workers considered a series of 16 PAHs molecules with linear (benzene, naphthalene, anthracene, tetracene, penta-cene) and angular (phenanthrene, benzo [c] phenanthrene, dibenzo [c,g] phenanthrene, benzo [a] anthracene, dibenzo [a,j] anthracene, naphtho [a] benzo [j] anthracene, naphtho [o] 5-helicene) and cluster (coronene, pyrene, perylene, benzo [g,h,i] perylene) conformations (Arey & Atkinson, 2003; Di-Toro et al., 2000; Abdel-Shafy & Mansour, 2016) as working set.

TABLE 35.2 Electronegativity, Chemical Hardness Chemical Power and Electrophilicity Values Obtained by Applying the SAIB Method of the Six Types of Bonds (Tudoran & Putz, 2015)

		SIMPLE C-C BOND		
	χ	η	π	ω
sHH	8.9724	7.31879	0.61297	5.49981
sCC	8.736	6.986	0.625251	5.46219
sHC	8.85262	7.14852	0.619192	5.48147
		DOUBLE C = C BOND		
	χ	η	π	ω
dHH	7.69063	6.27325	0.61297	4.71412
dCC	7.488	5.988	0.625251	4.68188
dHC	7.58796	6.12731	0.619192	4.69841

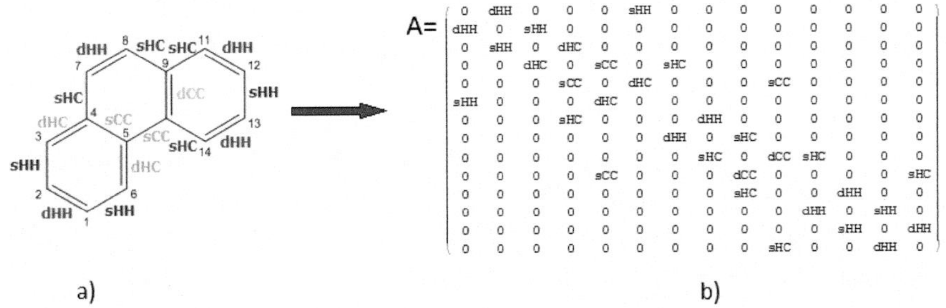

FIGURE 35.4 Phenanthrene: (a) colored graph; (b) colored adjacency matrix. Redrawn and adapted after Tudoran & Putz (2015).

For each molecule was constructed the correspondent graph and the adjacency matrix. In the next step, the adjacency matrix was further "colored" by replacing the unity values with the electronegativity values for each type of bond (see Figure 35.4 for phenanthrene case). The overall electronegativity for each PAH molecule was determined from the colored matrix spectra, i.e., $\{eigen_{min}A(\chi), \ldots, eigen_{max}A(\chi)\}_{PAH}$ after the formula (Tudoran & Putz, 2015):

$$\chi_{Molecule/PAH}^{eigen} = \frac{eigen_{max}A(\chi) - eigen_{min}A(\chi)}{2} \quad (15)$$

The same principle was applied for calculating the overall chemical hardness, chemical power, and electrophilicity values. This way, the SAIB method combines the min-max formulation from the chemical reactivity theory with the eigenvalues of the topological matrix from the chemical graph theory (Tudoran & Putz, 2015).

In a previous study, Putz and his co-workers (2013) considered the lipophilicity (LogP) a parameter for measuring the PAHs propensity to bind to the receptor by penetrating the cellular wall and calculate its values for each PAH molecule from the working set using AM1 semi-empirical computational methods[1]. Same values were used in SAIB study (Tudoran & Putz, 2015) in correlation with the obtained reactivity indices as a measure of the ecotoxicity influence on *in vivo* and *in vitro* organisms. The efficiency of the SAIB method was confirmed by the obtained results, in terms of Pearson correlation coefficients, the topo-reactivity calculation determining an overall higher correlation than the case when the reactivity indices were obtained using the quantum mechanic calculation of HOLO-LUMO energies applied to the same set of PAHs (Putz et al., 2013).

In the final step, the study was completed by producing the reactivity indices hierarchy of causes towards the correlated lipophilicity searching across the endpoints the least path, i.e., the Euclidian length for all the correlation schemes obtained at the previous step (Putz & Lacrama, 2010; Putz 2012a).

In literature, the chemical reactivity "dogma," i.e., the natural order of action (see equation SAIB-16) of the reactivity indices assumes that the electronegativity is the first in action, being associated with the chemical potential equalization of atoms in the molecule. Next, the chemical resistance is adjusting to charge exchange through chemical hardness followed by chemical power, which realizes the charge transfer. At the end of the process, the electrophilicity acts as the energy consumed by the system to reach the equilibrium state (Tudoran & Putz, 2015).

$$\chi \to \eta \to \pi \to \omega \qquad (16)$$

The least path obtained using the SAIB method can be associated with the most probable way in which the mechanistic chain of chemical reactivity is produced and follow the order:

$$\omega \to \pi \to \chi \to \eta \qquad (17)$$

In the topo-reactivity analysis, the process starts with the energy charge exchange first, starting with the electrophilicity, then occurs the fragments custom equalization in molecule and polarizability deformation, the electronegativity being the final "force" in action. As a conclusion, one can

---

[1]Program Package, HyperChem 7.01; Hypercube, Inc.: Gainesville, FL, USA, 2002.

state that the topological information (adjacency, in this case) determine at the beginning, an increase in the energy and charge follow over the entire structure, which is then distributed different sites with respecting the principle of electronegativity equalization and chemical hardness maximum of atoms in molecules (Putz, 2012a; Parr et al., 1978; Chattaraj et al., 1991; Putz, 2011).

35.5 MULTI-/TRANS-DISCIPLINARY CONNECTION(S)

The aromaticity character of benzenoid systems can be studied if their conformation is considered, i.e., their Kekulé, Clar, and Fries structures. A Kekulé structure represents the structural formula of benzene in which single bonds alternate with double bonds with the property that every carbon atom is tetravalent and incident with exactly one double bond. Fries structure is defined as a Kekulé structure maximum number of double bonds where most of the double bonds are common to two rings. A Clar structure one presents the maximum number of benzenoid rings with inscribed circles and, at the same time, there are no circles in adjacent rings (Balaban, 2004).

Putz considers another approach in describing the aromatic character by considering as compact finite difference the approximation of electronegativity and chemical hardness, i.e., using the molecular frontier orbitals, the bond forming stage being described by the superposition of atoms in molecule based on the electronegativity equalization principle, while the post-bonding molecular stability being based on the maximum hardness principle, using to indices: *absolute aromaticity* (A_a) as the difference between the atomic information superposition in bonding (Π_{AIM}) and the information of the molecular orbital (Π_{MO}) (Putz, 2010a) and *compactness aromaticity* (A_c) (Putz, 2010b) as the report between the pre-bonding stage information from (atoms-in-molecules Π_{AIM}) and the post-bonding stage information (molecular orbitals Π_{MO}) (Putz & Tudoran, 2016).

In a recent work, Putz and Tudoran (2016) redefine the absolute and compactness aromaticity using the SAIB method, by replacing with Atom-in-Bonding (AIB) the pre-bonding stage Atoms-in-Molecule (AIM) and with the HOMO-LUMO value (HL) the post-bonding stage the Molecular (MOL) manifestation of a given property (Π) (Putz 2010a, 2010b). In this context, for the new absolute aromaticity is obtained the formula:

$$A_{abs} = \left[\Pi_{AIM} \to \Pi_{AIB} \right] - \left[\Pi_{Mol} \to \Pi_{Homo-Lumo(H-L)} \right] = \Pi_{AIB} - \Pi_{HL} \qquad (18)$$

and, in the same manner, for new compactness aromaticity is obtained the formula:

$$A_{comp} = \frac{\Pi_{AIM} \rightarrow \Pi_{(AIB)}}{\Pi_{Mol} \rightarrow \Pi_{Homo-Lumo(H-L)}} = \frac{\Pi_{AIB}}{\Pi_{HL}} \tag{19}$$

When the electronegativity index is taken as reference, the equations (18) and (19) are rewritten as:

a) $\left(A_\chi \right)_{abs} = \chi_{AIB} - \chi_{HL}$ (20)

b) $\left(A_\chi \right)_{comp} = \dfrac{\chi_{AIB}}{\chi_{HL}}$ (21)

The absolute aromaticity can be equalized with compactness aromaticity, resulting in the corrected absolute aromaticity index, if the following correction is applied:

$$AC_\chi = \log \frac{S}{A_{abs}} \tag{22}$$

where $S = 27.21$ represents the electron Volts conversion constant and A_{abs} represents the absolute aromaticity.

For each PAH molecule from the working set (used in previous work, see Putz et al. (2013) and Tudoran & Putz (2015)) all the Kekulé structures were generated (Figure 35.5), and the topo-reactivity indices were determined using the SAIB method.

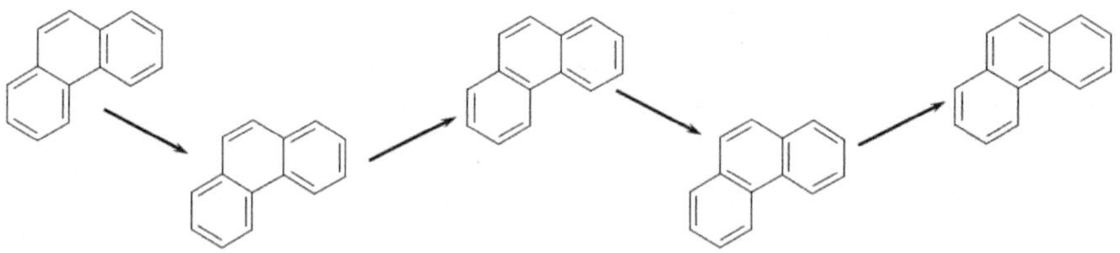

FIGURE 35.5 All Kekulé structures of phenanthrene. Redrawn and adapted after Putz and Tudoran (2016).

The electronegativity was further utilized to calculate the absolute aromaticity, the compactness aromaticity and the corrected absolute aromaticity using Eqs. (20)–(22). The maximum value of corrected absolute can be associated with the highest aromatic character, meaning that a distinction between isomers can be made using the topo-reactivity method.

By selecting the highest aromatic isomer for all of the 16 molecules and using the obtained values in correlation with lipophilicity values (LogP) collected from literature, the results show that the interaction mechanism takes a totally different path, chemical power being the first in action while chemical hardness is closing the circuit with. In other words, one can say that the specific conformation adopted by molecules in interactions is not necessarily the most aromatic one (Tudoran & Putz, 2015).

35.6 OPEN ISSUES

Polycyclic aromatic hydrocarbons represent a class of organic compounds composed only from carbon and hydrogen atoms, very spread in nature as pollutants with carcinogenic and mutagenic effects (Grimmer, 1983; Ramdahl & Bjorseth, 1985; Dias, 1987; Siegmann & Siegmann, 1998) and in the interstellar medium as carriers of the UIR (unidentified infrared) bands and possibly being responsible for the diffuse absorption infrared bands of absorption (Putz et al., 2013; Allamandola et al., 1985; Leger & Puget, 1984; Salama & Allamandola, 1993; Cossart-Magos & Leach, 1990). Studies in this area were leading to the theory according which PAHs can be considered the main structures to produce graphite particles following the same principle as their occurrence in space through fullerene synthesis reduced by size of several PAHs-based nanocompounds (Braga et al., 1991; Iglesias-Groth, 2004; Joblin et al., 1992).

The SAIB method presented by Tudoran & Putz (2015) was, is a topo-reactivity method based on replacing the 1 and 0 values (existence and respectively non-existence of a bond between atoms) in the adjacency matrix whit the reactivity indices obtained from each type of bond and was, until now, successfully applied only to polycyclic aromatic hydrocarbons (finite systems) with the maximum of 30 carbon atoms. As a result of studying 1D polymers at the beginning and later multidimensional systems (Cataldo et al., 2010), the extension of topological methods (TM) based on distance matrix invariants to infinite systems appears as a good possibility due to their behavior of the polynomial type. It appears that TM simulations are able to isolate the regions stabilized by topology connections at the lattice specific critical size (Putz et al., 2015). In this context, the SAIB method can be further improved if the adjacency matrix can be replaced with the distance matrix, i.e., if one can determine a recurrence formula which can determine the type of bond only by knowing the distance between the neighbors for a reference vertex/atom from the lattice.

KEYWORDS

- **adjacency matrix**
- **chemical graph theory**
- **polycyclic aromatic hydrocarbons**
- **reactivity indices**

REFERENCES AND FURTHER READING

Abdel-Shafy, H. I., & Mansour, M. S. M., (2016). A review on polycyclic aromatic hydrocarbons: Source, environmental impact, effect on human health and remediation. *Egyptian Journal of Petroleum, 25*, 107–123.

Allamandola, L. J., Tielens, A. G. G. M., & Barker, J. R., (1985). Polycyclic aromatic hydrocarbons and the unidentified infrared emission bands: auto exhaust along the Milky Way! *Astrophysical Journal Letters, 290*, L25–L28.

Arey, J., & Atkinson, R., (2003). Photochemical reactions of PAH in the atmosphere. In: Douben, P. E. T., (ed.), *PAHs: An Ecotoxicological Perspective* (pp. 47–63). John Wiley and Sons Ltd, New York.

Balaban, A. T., (2004). Clar formulas: How to draw and how not to draw formulas of polycyclic aromatic hydrocarbons. *Polycyclic Aromatic Compounds, 24*, 83–89.

Braga, M., Larsson, S., Rosen, A., & Volosov, A., (1991). Electronic transition in C60 - On the origin of the strong interstellar absorption at 217 NM. *Astronomy and Astrophysics, 245*, 232–238.

Bratsch, S. G., (1985). A group electronegativity method with Pauling units. *Journal of Chemical Education, 62*, 101–103.

Cataldo, F., Ori, O., & Iglesias-Groth, S., (2010). Topological lattice descriptors of graphene sheets with fullerene-like nanostructures. *Molecular Simulation, 36*, 341–353.

Chattaraj, P. K., Lee, H., & Parr, R. G., (1991). Principle of maximum hardness. *Journal of the American Chemical Society, 113*, 1854–1855.

Cossart-Magos, C., & Leach, S., (1990). Polycyclic aromatic hydrocarbons as carriers of the diffuse interstellar bands - Rotational band contour tests. *Astronomy and Astrophysics, 233*, 559–569.

Dias, J. R., (1987). *Handbook of Polycyclic Hydrocarbons*. Elsevier, Amsterdam.

Di-Toro, D. M., McGrath, J. A., & Hansen, D. J., (2000). Technical basis for narcotic chemicals and polycyclic aromatic hydrocarbon criteria. I. Water and tissue. *Environmental Toxicology and Chemistry, 19*, 1951–1970.

Grimmer, G., (1983). *Environmental Carcinogens: Polycyclic Aromatic Hydrocarbons: Chemistry, Occurrence, Biochemistry, Carcinogenity*. CRC Press: Boca Raton-Florida.

Iglesias-Groth, S., (2004). Fullerenes and buckyonions in the interstellar medium. *Astrophysical Journal 608*, L37–L40.

Joblin, C., Leger, A., & Martin, P., (1992). Contribution of polycyclic aromatic hydrocarbon molecules to the interstellar extinction curve. *Astrophysical Journal, 393*, L79–L82.

Leger, A., & Puget, J. L., (1984). Identification of the 'unidentified' IR emission features of interstellar dust? *Astronomy and Astrophysics, 137*, L5–L8.

Parr, R. G., Donnelly, R. A., Levy M., & Palke, W. E., (1978). Electronegativity: The density functional viewpoint. *The Journal of Chemical Physics, 68*, 3801–3808.

Putz, M. V., & Dudaş, N. A., (2013). Variational principles for mechanistic quantitative structure–activity relationship (QSAR) studies: Application on uracil derivatives' anti-HIV action. *Structural Chemistry, 24*, 1873–1893.

Putz, M. V., & Lacrămă, A. M., (2007). Introducing spectral structure-activity relationship (S-SAR) analysis. Application to ecotoxicology. *International Journal of Molecular Sciences, 8*(5), 363–395.

Putz, M. V., & Tudoran, M. A., (2016). Carbon-based specific adjacency-in-bonding (SAIB) isomerism driving aromaticity. *Fullerenes, Nanotubes and Carbon Nanostructures, 24*(12), 733–748.

Putz, M. V., (2010a). On absolute aromaticity within electronegativity and chemical hardness reactivity pictures. *MATCH, 64*, 391–418.

Putz, M. V., (2010b). Compactness aromaticity of atoms in molecules. *International Journal of Molecular Sciences, 11*, 1269–1310.

Putz, M. V., (2011). Electronegativity and chemical hardness: Different patterns in quantum chemistry. *Current Physical Chemistry, 1*(2), 111–139.

Putz, M. V., (2012a). *QSAR & SPECTRAL-SAR in Computational Ecotoxicology* (p. 242). Apple Academics, Ontario, Canada.

Putz, M. V., (2012b). Chemical orthogonal spaces. In: *Mathematical Chemistry Monographs* (Vol. 14, p. 240). University of Kragujevac, Serbia.

Putz, M. V., (2013). Chemical orthogonal spaces (COSs): From structure to reactivity to biological activity. *International Journal of Chemical Modeling, 5*(1), 1–34.

Putz, M. V., Tudoran, M. A., & Ori, O., (2015). Topological organic chemistry: From distance matrix to Timisoara excentricity. *Current Organic Chemistry, 19*, 249–273.

Putz, M. V., Tudoran, M. A., & Putz, A. M., (2013). Structure properties and chemical-bio/ ecological of PAH interactions: From synthesis to cosmic spectral lines, nanochemistry, and lipophilicity-driven reactivity. *Current Organic Chemistry, 17*, 2845–2871.

Ramdahl, T., & Bjorseth, J., (1985). *Handbook of Polycyclic Aromatic Hydrocarbons* (2nd edn.). Marcel Dekker, New York.

Salama, F., & Allamandola, L. J., (1993). Neutral and ionized polycyclic aromatic hydrocarbons, diffuse interstellar bands and the ultraviolet extinction curve. *Journal of the Chemical Society, Faraday Transactions, 89*, 2277–2284.

Siegmann, K., & Siegmann, H. C., (1998). Molecular precursor of soot and quantification of the associated health risk. In: Moran-Lopez, (ed.), *Current Problems in Condensed Matter.* Plenum Press, New York.

Tudoran, M. A., & Putz, M. V., (2015). Molecular graph theory: From adjacency information to colored topology by chemical reactivity. *Current Organic Chemistry, 19*, 359–386.

CHAPTER 36

Stone-Wales Rotations

OTTORINO ORI[1-3]

[1]*Actinium Chemical Research, 00133 Rome, Italy*

[2]*Laboratory of Computational and Structural Physical Chemistry for Nanosciences and QSAR, West University of Timişoara, 300115, Timisoara, Romania*

[3]*Laboratory of Renewable Energies-Photovoltaics, National Institute for R&D in Electrochemistry and Condensed Matter INCEMC-Timisoara, 300569 Timisoara, Romania, E-mail: Ottorino.Ori@gmail.com*

36.1 DEFINITION

Stone-Wales rotations represent class of topological defects and topological modifications—the *topological defects* may actually detect in the honeycomb structure of the graphene layers and, from a formal point of view, a large class of *topological modifications* applicable to three-connected networks such as sp^2-carbon nanoclusters, fullerenes, carbon nanotubes, graphene, and others. Some of these *SW* modifications, including certain sequences of *SW* rotations, have been widely investigated but, in general, topological and physical properties of the large majority of *SW* flips still remain to be studied. Even their symbolism needs some work in order to get to a universal definition.

36.2 HISTORICAL ORIGINS

More than three decades ago, in their epochal paper on "Theoretical studies of icosahedral C_{60} and some related structures" (Stone & Wales, 1986), the two scientists have introduced a basal transformation of the buckminsterfullerene surface that, *by rotating one single bond*, causes two hexagons and two pentagons exchanging their positions and "hence the pentagons can migrate around the surface of the fullerene shell" via subsequent *SW* rotations (Kumeda & Wales, 2003), evidencing the ability of C_n fullerenes to rearrange into different isomers having the same number of hexagonal faces (n/2-10), just by varying the position of the 12 pentagons, in a sort of spherical version of the cube of Rubik. Downhill/uphill barriers for the *SW* flip connecting $C_{60}-I_h$ buckminsterfullerene and the next-most-stable isomer with C_{2v} symmetry stay in the 1.55/6.91 eV range, respectively (Austin et al., 1995). The authors reported in their *SW* "map" of the C_{60} isomer space that $C_{60}-I_h$ is connected to 1709 of the possible 1812 C_{60} fullerene isomers through a maximum of thirty SW transformations, connecting in such a way a 94% of all C_{60} valid cages by proper sequences of *SW* moves. In the same period (Babic et al., 1995) greatly expanded the ability of *SW* rotations in *mapping* the isomeric space of fullerenes by introducing *linear generalized SW* rotations (*gSW*) able to connect *every* C_{60} fullerene to the buckminsterfullerene stable cage. Similar results have been derived in Ōsawa et al. (1998). In the recent years, new studies on the subject have been devoted to *linear* and *radial gSW* rearrangements of fullerene surfaces (Ori et al., 2014) and graphenic layers (Ori et al., 2011a, 2011b; Ori & Putz 2014). Original methods for generating fullerenes *at random* also testify the usefulness of *SW* rotations to generate the isomer space of a given C_n fullerene (Plestenjak et al., 1996; Kirby 2000). Although the knowledge of the *SW rotations* increases year after year, future investigations and simulations are needed to fully understand all the ways these topological transformations have to modify 3-connected networks.

Till now we have spoken about *SW* rotations in *spherical* carbon nanosystems. The first thing to notice on *planar or cylindrical* 3-connected networks is that the presence of non-hexagonal rings of carbon atoms it is considered the key element for explaining the diverse morphologies of graphene-based materials. In fact, not only pentagons (e.g., the key constituent of the hollow fullerene molecules) but also squares, heptagons, octagons and even higher carbon polygons represent important topological defects in graphene and carbon nanotubes. A relevant role is played by the simplest of these defects, called the Stone-Thrower-Wales (*STW*) defect, that is originated by a *SW* operation interchanging four adjacent hexagons with two pentagon-heptagon 5|7 pairs – see next section. Since the pentagon and heptagon induce respectively positive and negative curvature in the honeycomb pristine network, the pentagon-heptagon pair 5|7 remain planar. This has been experimentally shown (Thrower, 1969) in the pioneering study on dislocation defects produced by graphite irradiation. The formation energy of the *STW* defect has been reported in the range of 2.29 eV and largely depending from many factors, like the catalyzing action of interstitials defects or ad-atoms present in the hexagonal networks - see (Ori et al., 2011b; Heggie et al., 2016) and related references.

36.3 NANO-SCIENTIFIC DEVELOPMENTS

The "original" *SW* rearrangement swaps the position of two pentagons and two hexagons on the fullerene surface. This mechanism rotates the *pyracylene unit* shown in the graph *G* in Figure 36.1a by cutting-and-reconstructing the two labeled bonds in order to reach, as a final effect, the 90° rotation of the central bond without changing the bond structure surrounding the pyracylene unit. Figure 36.1b visualizes the general *SW* local rearrangement that transforms a group of four proximal polygons with p, q, r, s three-connected atoms in four $p - 1$, $q + 1$, $r - 1$, $s + 1$ new rings by adopting the representation given by Kirby (2000). With some arbitrariness, we selected for that general rotation the $SW_{q|r}$ notation. This operator changes the *internal* connectivity of four carbon rings *without changing* the surrounding lattice. $SW_{q|r}$ transformations are *reversible* and *isomeric* by preserving, in fact, the total number of the carbon atoms belonging to the rotating faces $v = p + q + r + s - 8$ and the total number of carbon bonds $e = v + 3$. The celebrated *STW* rotation, typically occurring in graphene, correspond to the $SW_{6|6}$ operator shown in Figure 36.1c.

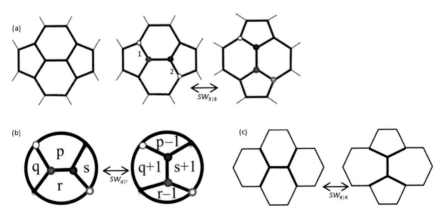

FIGURE 36.1 Examples of relevant *SW* rotations. (a) The celebrated $SW_{5|6}$ transformation interchanges two hexagons and two pentagons in the graph G by cutting the two labeled bonds in order to rotate, after the creation of two new bonds, the central bond between blue and red atoms; (b) the same *SW* mechanism applied to generic rings, the action of the general local $SW_{q|r}$ operator is described; (c) another peculiar case, the $SW_{6|6}$ operator that, starting from 4 hexagons, generates the *STW* defect, the pentagon-heptagon double pair 5|7|7|5. All *SW* rotations are fully isomeric and reversible.

Moving into the dual space G^*, in which the nodes correspond to the carbon atoms rings (Ori et al., 2011a, 2011b), *SW* transformations are more easy to visualize and to automatize by computer programs. Figures 36.2 and 36.3 show well the advantages of operating in G^*. In the dual space the generic $SW_{q|r}$ flip corresponds to the reversible rotation of the bond at the center of the quartet of faces, making the study of *SW* defects generation (Figure 36.2a) and propagation (Figure 36.2b) a straightforward task. Interested readers may consult recent collections (Ori & Putz, 2014) and (Ori et al., 2014) on various sorts of *SW* transformations applicable to the polygonal tessellation of sp²-carbon nanosystems, like fullerenes, nanotubes, graphenic layers and tori, even in the one-surface Möbius and Klein bottle forms.

In particular, going back to Figure 36.2, we like to polarize the attention on the peculiar topological defect, the *SW* wave (*SWw*), one may *topologically* generate in the hexagonal regions present on the surfaces of fullerenes, nanotubes or graphene (Ori et al., 2011a, 2011b), see Figure 36.2. The initial $SW_{6|6}$ rotation (Figure 36.2a) transforms four 6-rings (shaded polygons) in two pentagons (red) and two heptagons (green) by rotating the blue-arrowed bond. The action of that $SW_{6|6}$ operator is illustrated by Figure 36.2 in both representations, direct G and dual G^*, of the hexagonal lattices. After the first $SW_{6|6}$ rotation it is evident that the 5|7 pairs may migrate "pushed" by another *SW* move, namely the $SW_{6|7}$ flip, applied to the blue-arrowed edge that in G coincides with the pentagon-hexagon chemical bond, whereas in the dual graph G^* corresponds to the edge between the 7-connected vertex and the hexagon of the 6|6 pair, as the central column of Figure 36.2 graphically explains. In Figure 36.2, always a blue(red)-arrowed edge indicate the edge to be rotated in order to pass to the next-right(previous-left) configurations. In particular the $SW_{6|7}$ flip swaps the position of the 5|7 pair with the pair 6|6 of connected hexagons, driving the linear diffusion of the 5|7 pair in the graphenic mesh. Figure 36.2b shows, in fact, the result of the swapping mechanism involving 5|7 and 6|6 couples. The repeated action of the $SW_{6|7}$ operator on the blue-arrowed bonds of Figure 36.2b propagates the topological *SWw* in the hexagonal lattices. The Figure 36.2b represents the vertical diffusion of *SWw* dislocation dipole after $\eta = 1$ rearrangements driven by the $SW_{6|7}$ operators, producing characteristic pattern spacing the two 5|7 pairs that represent a good theoretical model describing the boundary between graphenic fragments. Obviously, being all *SW* moves *fully reversible and isomeric*, the *SWw* mechanism is also able to *annihilate* the dislocation dipole by reconstructing backwards a perfect hexagonal mesh. The ability of the *SW* operators to modify 3-connected lattices is evidenced once more in Figure 36.3 where an original mechanism leading to the generation of the

5|8|5 defect in graphenic systems is shown. This linear twofold-symmetrical defect, consisting in an octagon sandwiched by two pentagonal rings, normally explained by invoking the divacancy mechanism, may also arise from proper sequences of *SW isomeric* rotations in the graphenic layer preserving both, the number of carbon atoms and bonds of the pristine nano-structure. The complete treatment of these *Rubik-cube*-like topics may be found in the original article (Ori & Putz, 2014).

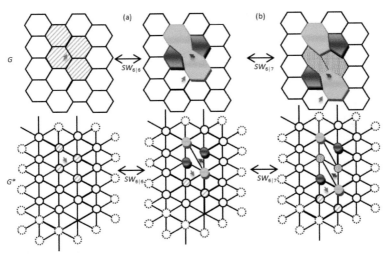

FIGURE 36.2 The figure shows the actions of the *SW* transformations in the direct *G* and dual spaces *G**. The blue(red) arrows point to the bonds that have to be rotated to generate the configuration on the right(left). (a) Details on how the $SW_{6|6}$ works to form the double pair 5|7, starting from 4 hexagons. (b) The application of $SW_{6|7}$ one operator ($\eta = 1$) moves the 5|7 pair exchanging its position with one 6|6 pair (gray shaded). The *SW* rotation of the blue-arrowed bonds will further enlarge the dislocation dipole size η.

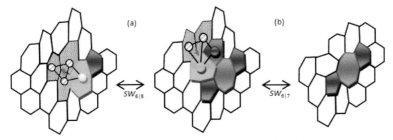

FIGURE 36.3 The figure shows the generation of the 5|8|5 defects by using a proper combination of *SW* transformations. The blue arrows point to the bonds to be rotated to produce the configuration on the right. (a) The $SW_{6|7}$ operator has the role of transforming the green heptagon in the octagon. (b) The $SW_{6|7}$ rotation plays then its usual role to move the 5|7 pair away leaving the 5|8|5 triplet surrounded by only hexagons.

We conclude this parade of *endless* classes of *SW* rotations by presenting an interesting family of *SW* transformations that rotates a *rhombic* region included between two pivotal hexagon/pentagon pairs. Figure 36.4 shows some examples of these *rhombic* generalized *SW* transformations $SW^R(\eta)$ depending on their size η, the integer number of internal hexagons bridging the two pivotal hexagons (Ori et al., 2014). One may find these *rhombic arrangements* of faces on the surfaces of the fullerenes and other carbon nanosystems. Clearly, the $\eta = 0$ case corresponds to the basal operator $SW_{5|6}$ given in Figure 36.1a.

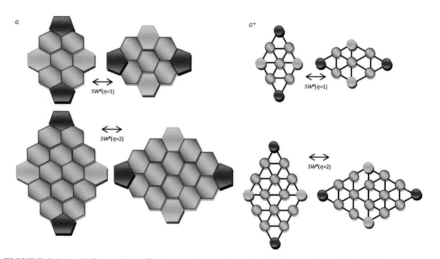

FIGURE 36.4 Effects of $SW^R(\eta)$ transformations in the direct G and dual G^* spaces for sizes $\eta = 1$ and $\eta = 2$. Red, blue and gray elements represent, respectively, pivotal pentagons, pivotal hexagons and internal hexagons. For this class of *rhombic* rearrangements, the size η equals the number of 6-rings bridging pivotal hexagons.

The elegance of the SW transformations represented in this chapter somehow reflects the *inherent and profound beauty* of the topological methods applied to chemistry. See Schwerdtfeger et al. (2015) on fullerenes topology for an exhaustive overview of this matter.

36.4 MULTI-/TRANS-DISCIPLINARY CONNECTIONS

As we have shown here, *SW* rotations are *effective graphical tools* for generating valid isomers of carbon nanostructures. Their relevance we think is far to be fully understood, being the topic still largely unexplored.

In other scientific sectors, polygonal tessellations of 3-connected systems are quite "popular" in Nature, where biological tissues especially do provide multiple examples of tessellations of cells whose topology is modifiable by *SW* rearrangements. An interesting applications to neuronal tissues is described in Tozzi et al., (2017) in which an original method to model the signal activation/propagation/latency in neural networks based on $SW_{6|6}$ and $SW_{6|7}$ is proposed; really a promising sector in which the use of SW will certainly offer a cascade of correlations between bio-systems, topology and functionality.

36.5 OPEN ISSUES

As reported already for the Wiener index (see the item of the present dictionary), topological defects play a specific role in *enhancing* the 'bondonic' effects governed by long-range interactions (Putz & Ori, 2012, 2014). These bosonic quasi-particles promise to unveil a full range of new physical phenomena in carbon nanosystem.

KEYWORDS

- **generalized Stone-Wales rotations**
- **Stone-Wales rotations**
- **SW wave**

REFERENCES AND FURTHER READING

Austin, S. J., Fowler, P. W., Manolopoulos, D. E., & Zerbetto, F., (1995). The stone-wales map for C_{60}. *Chem. Phys. Lett., 235*(1/2), 146–151.

Heggie, M. I., Haffenden, G. L., Latham, C. D., & Trevethan, T., (2016). The stone–wales transformation: From fullerenes to graphite, from radiation damage to heat capacity. *Phil. Trans. R. Soc. A., 374*, 20150317. http://dx.doi.org/10.1098/rsta.2015.0317.

Kirby, E. C., (2000). On the partially random generation of fullerenes. *Croatica Chemica Acta, 73*(4), 983–991.

Kumeda, Y., & Wales, D. J., (2003). Ab initio study of rearrangements between C60 fullerenes. *Chemical Physics Letters, 374*, 125–131.

Ori, O., & Putz, M. V., (2014). Isomeric formation of 5|8|5 defects in graphenic systems. *Fullerenes, Nanotubes and Carbon Nanostructures, 22*(10), 887–900.

Ori, O., Cataldo, F., & Graovac, A., (2011a). On topological modeling of 5|7 structural defects drifting in graphene. *Carbon Bonding and Structures* (pp. 43–55). Springer Netherlands.

Ori, O., Cataldo, F., & Putz, M. V., (2011b). Topological anisotropy of stone-wales waves in graphenic fragments. *Int. J. Mol. Sci.*, *12*(11), 7934–7949.

Ori, O., Putz, M. V., Gutman, I., & Schwerdtfeger, P., (2014). Generalized Stone-Wales transformations for fullerene graphs derived from Berge's switching theorem. *Ante Graovac—Life and Works, Mathematical Chemistry Monographs*, *16*, 259–272.

Ōsawa, E., Ueno, H., Yoshida, M., Slanina, Z., Zhao, X., Nishiyama, M., & Saito, H., (1998) Combined topological and energy analysis of the annealing process in fullerene formation. Stone-Wales interconversion pathways among IPR isomers of higher fullerenes. *Journal of the Chemical Society, Perkin Transactions, 2*(4), 943–950.

Plestenjak, B., Pisanski, T., & Graovac, A., (1996). Generating fullerenes at random. *Journal of Chemical Information and Computer Sciences, 36*(4), 825–828.

Putz, M. V., & Ori, O., (2012). Bondonic characterization of extended nanosystems: Application to graphene's nanoribbons. *Chemical Physics Letters*, *548*, 95–100.

Putz, M. V., & Ori, O., (2014). Bondonic effects in group-IV honeycomb nanoribbons with Stone-Wales topological defects. *Molecules, 19*(4), 4157–4188.

Schwerdtfeger, P., Wirz, L. N., & Avery, J., (2015). The topology of fullerenes. *WIREs Comput. Mol. Sci.*, *5*, 96–145. doi: 10.1002/wcms.1207.

Stone, A. J., & Wales, D. J., (1986). Theoretical studies of icosahedral C60 and some related species. *Chem. Phys. Lett.*, *128*, 501–503.

Thrower, P. A., (1969). The study of defects in graphite by transmission electron microscopy. *Chemistry and Physics of Carbon, 5*, 217–320.

Tozzi, A., Peters, J. F., & Ori, O., (2017). Fullerenic-topological tools for honeycomb nanomechanics. Toward a fullerenic approach to brain functions. *Fullerenes, Nanotubes and Carbon Nanostructures*, *25*(4), 282–288.

CHAPTER 37

Tetrahedral Dual Coordinate System: The Static and Dynamic Algebraic Model

JAMES SAWYER and MARLA WAGNER

Six Dimension Design, Buffalo, N.Y., USA

37.1 DEFINITION

The tetrahedral dual coordinate system (TDCS) is found within the atomic structure of square and triangular grid planes of crystallography as (100) and (111) Miller lattice system, and in organic stereochemistry compounds with tetrahedral and ligand bonding. There is a need for the normalization of the geometry of atoms that integrates the triangular and square planes into a dual coordinate system based on polyhedrons. Extra algebraic dimensions are applied to the edges and axis of symmetry of the tetrahedral geometry found in chemistry which includes three, four, five, and six-fold symmetry. The TDCS derived by applying four triangular algebraic planes to the tetrahedron of organic chemistry with six edges, represented algebraically as (r, s, t, u, v, w). These edges become six number lines with six directions in atomic space as six dimensions (6D). The static cuboctahedral coordinate model is the intersection of four (111) triangular/hexagonal planes on the grids (r, s, t), (s, u, v), (r, v, w), and (t, u, w). The square plane edges are now represented algebraically in terms of (r, s, t, u, v, w) with three (x, y, z) axes of symmetry orthogonal to square planes or octahedral vertices. The dynamic geometric motion of a dual coordinate system is facilitated by the rotation of the eight (111) triangular planes on the four axes of the vector equilibrium model. The resultant motion produces the complex nanocrystal, piezoelectric, superconductive and quasicrystal polyhedrons such as icosahedron, and pentagonal dodecahedron. A 6D tetrahedron is viewed on a three-directional, three-dimensional triangular (3DT) plane (see Figure 37.1).

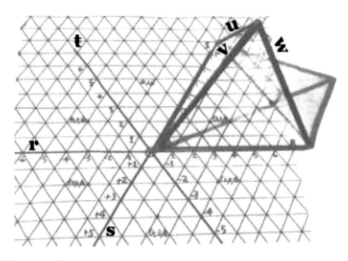

FIGURE 37.1   Tetrahedron on a 6D triangular lattice.

37.2   HISTORICAL ORIGINS: EUCLID'S PROBLEMS WITH EQUILATERAL TRIANGLES TO EINSTEIN'S DUAL-COORDINATES

The need for a triangular-based coordinate system was first illustrated by the mathematician Charles L. Dodgson (1885). Albert Einstein in General Relativity (GR) developed a dual-like coordinate system representing the dynamic properties of atomic mass in outer space (Albert Einstein, 1914–1917).

A multi-dimensional, mathematical, geometric system is needed to describe the triangulated properties of atomic particles and crystals in quantum space. The 1800's curriculum of mathematics of Euclid noted the disadvantages of the orthogonal 90-degree coordinate system to describe the properties of triangular polygons in space. The study of x-ray crystallography by scientists presents an opportunity to understand the polyhedral geometry of atomic nature. Albert Einstein postulated a multiple coordinate system from his observations of Special Relativity (SR) and applied it to General Relativity theory (GR), composed of "Koordinate Systeme (K) Galilean and Koordinate Systeme (K') Geodesic." Einstein developed a Cartesian-like calculus in GR and applied a set of field equations to form a dual-like variation of triangular geodesic space (Fuller, 1975), which was applied to the atomic space of gravity fields. R. B. Fuller defined the polyhedral geometry of the tetrahedron, octahedron, cuboctahedron, icosahedron, and the vector equilibrium polyhedrons without Cartesian algebra. Linus Pauling recognized the four axes of symmetry of tetrahedral carbon bonding in organic chemistry and recognized the close-packed hexagonal theory of nucleons (Pauling, 1960, 1966). Fritsche and Benfield further developed the Mean Coordinate numbers from Linus Pauling's Magic Numbers theory [MN] and documented the atomic nanoclusters structural geometry of tetrahedron, octahedron, cuboctahedron and icosahedron (Fritsche and Benfield, 1993; Pauling, 1901–1994).

37.3 APPLICATION OF A DUAL COORDINATE SYSTEM IN X-RAY CRYSTALLOGRAPHY

The DCS applies algebra to the triangular planes of Geometric Frustration (GF) with the Six-Dimensional String Theory [SDS]. The GF that Jean-François Sadoc and Rémy Mosseri opened in their scientific research revealed electro-magnetic properties of atomic triangular space in naturally occurring spin ice materials (Sadoc & Mosseri, 2006). GF triangular lattice and tetrahedral crystal symmetry introduced the need for a postulate system coordinated with 60-degree angles to represent the electromagnetic properties of atomic structure. GF attempted to define the five-fold electromagnetic properties of atomic particles with a triangular coordinate grid system. The additional research of platinum nanocrystals revealed patterns of triangular grid planes coordinated into polyhedrons such as tetrahedron, octahedral, cuboctahedron. The nanocrystals also included the square cubic geometry of the Cartesian (x, y, z) philosophy of space. Nature has thus revealed a fundamental dual polyhedral geometry of square and triangular coordinated space. The recent writing of Jean-François Sadoc and Rémy Mosseri on GF opened scientific research into the electromagnetic properties of atomic triangular space and in naturally occurring spin ice materials (Sadoc & Mosseri, 2006). The additional research of platinum nanocrystals by Lawrence Berkley National Laboratory reveals triangular coordinated patterns of octahedral, cuboctahedral, and cubic geometry involved in the process of catalysis (Lawrence Berkley National Laboratory July 22, 2015). Nature has revealed a fundamental geometric property of polyhedral dimensions related to square and triangular coordinated space. At the Hauptman laboratory there is a realization for a combined x-ray diffraction and neutron diffraction analysis system to study the complex protein geometry of cancer cells (Herbert A. Hauptman, 1917–2011). The futuristic combination of x-ray diffraction and neutron diffraction analysis system should be used to study the complex protein geometry of cancer cells, protein cells and their apoptosis mechanism.

37.4　SIX DIMENSIONAL STRING THEORY (SST)

TDSC represents the true geometry of String Theory. Six dimensions (6D) as a mathematical system has a phenomenological relationship with other fields such as mathematics, physics, biology, engineering, and architecture. The phenomena of extra dimensions in space as presented in six-dimensional string theory by F. David Peat (1989) is applied to a DCS by replacing the cube of the Cartesian method with the cuboctahedron with 3, 4, and 6 axes of symmetry, representing a new philosophy of triangular and square space.

String Theory represents the existence of extra dimensions, such as six and ten dimensions, without a primary coordinate system to mathematically describe the quantum properties of physics. The DCS includes the dimensions of atomic space with three and six dimensions. There are also four axes of rotational symmetry which explain the unique properties of the quasicrystals which have fifteen extra dimensions in the icosahedron and pentagonal dodecahedron. Examples of 6D engineering and architecture are observed in the tetrahedron and octahedron truss system in the Paris Louvre and the Leonardo Da Vinci Airport Rome.

37.5　NORMALIZE THE DIMENSIONAL ANALYSIS OF CRYSTALLOGRAPHY

The application of dual six-dimensional triangular coordinate system with three orthogonal axes of symmetry is discussed below.

Presently, the Miller index [MI] orthogonal (h k l) lattice planes are divided into seven polyhedral formulas of crystallography; however the coordination numbers see two separate interatomic types of distances, and bonding based on triangular/hexagonal planes and square planes. The x-ray crystallography and neutron crystallography represent two distinct types of diffraction techniques used for the analysis of complex organic proteins. Presently, the two separate systems square (h, k, l) and triangular as (r, s, t, u, v, w) need to be unified into a common coordinate system. Our present computer hardware and software coordinate only square orthogonal systems. The extra dimensions of equilateral triangular grids are not represented in [MI] space.

There is a need for a TDCS with extra dimensions to describe the recent theories such as six and ten-dimensional string theory. We see the existence of an extra-dimensional dual coordinate system which includes

Three Dimensions Cartesian (3D) with triangular 6D. There is a theoretical coordinated dual chemical equilibrium balance between the bonding electro-negativity forces looping out from the spinning electrons in 6D and internal looping via ligand bonding into nucleons in 3D. This synchronizes with the dual cuboctahedron [111] triangular cuboctahedron planer bonding and the secondary [100] internal square bonding found in the nano-imaging technique of platinum cuboctahedron nanocrystals. Protein binding in organic tetrahedral compounds also exists.

• *Six-dimension theory includes the merging of four equilateral triangular grids with three orthogonal (x, y, z) axes of symmetry, geometrically modeled on the cuboctahedron.*

It is based on a six-dimensional triangular coordinate system composed of four hexagonal planes as 3DT planes, which merge into cuboctahedron (DCS) (see Figure 37.2; Sawyer & Wagner, 2014). The intersection of four equilateral grids as a static lattice structure produces dihedral angles of 70.53° for the enclosed tetrahedrons and 109.47° for enclosed octahedrons (see Figures 37.3 and 37.4). This system can be used to normalize the Miller Indices system of polyhedrons found in crystallography.

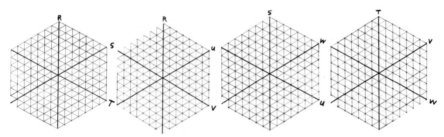

FIGURE 37.2 Four triangular 3DT planes (r, s, t), (r, u, v), (s, w, u) and (t, v, w).

• *The tetrahedron as six dimensional (6D) needs to be normalized as a unit volume of one by National Institute of Standards (NIST)*

This six-dimensional (6D) coordinate system defines algebraic unit one as a tetrahedron. When the volume of a tetrahedron is one unit, the cube is three units, the octahedron is four units, the rhombic dodecahedron is six units, and the cuboctahedron is 20 units. This dramatically simplifies the calculus of atomic geometry and unifies mathematics with organic chemistry and crystallography. The tetrahedron is the foundation of organic chemistry. NIST needs to normalize the term "dimension" in regards to the polyhedral geometry of quantum nanoparticles in atomic space. A demonstration figure of a rare tetrahedral diamond is placed on a tetrahedral-octahedral truss grid (see Figure 37.5). There are many classifications for the exterior geometry of atomic crystals; however, the application of Euler's formula to these geometries reveals three, four and six-fold symmetry.

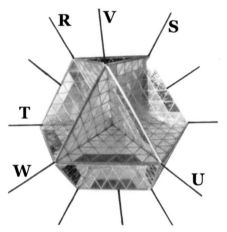

FIGURE 37.3 Six-dimensional cuboctahedron with four intersecting triangular lattices (r, s, t, u, v, w). The six edges are colored: r = red, s = sulfur (yellow), t = tangerine, u = ultramarine, y = yellow, w = white.

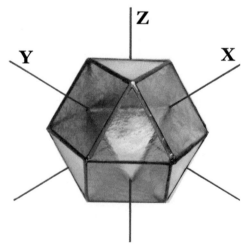

FIGURE 37.4 Six-dimensional cuboctahedron with inner octahedron (x, y, z) based on three axes of symmetry.

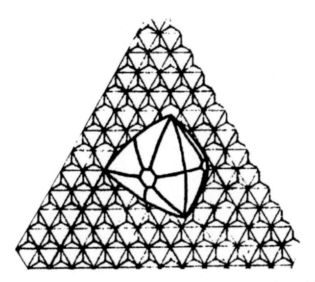

FIGURE 37.5 Tetrahedral diamond crystal on the tetrahedron-octahedron grid (this is rare, as diamonds are primarily in octahedral symmetry).

The Vector Equilibrium Dual Coordinate System (VEDCS) mechanism presents the dynamic properties found in the transition element compounds with electromagnetic and geometric properties with 3, 4, and 6 axes of symmetry. The new five-fold symmetry of quasicrystals can be visualized in real atomic space with the dual six-dimensional coordinate system. $Al_{63}Cu_{24}Fe_{13}$ icosahedrite crystal is an example of a quasicrystal due to its five-fold symmetry (Bindi et al., 2011). At high temperatures with rapid cooling, each of the cuboctahedron's six square (a) faces collapse along the (x, y, z) axes into two equilateral triangles, forming a 15-directional icosahedron. This is in congruence with Buckminster Fuller's Vector Equilibrium VE icosahedron. Fuller developed a VE triangular coordinated system to represent eight triangular planes in rotation on four axes of symmetry (without the algebra needed to determine the positions of atoms in space) which is represented as a DCS.

- *Dynamic geometric mechanism for the formation of a quasicrystal utilizing the [VEDCS] four axes of rotation of dual coordinate system as vector equilibrium (VE)*

Each set of parallel triangles rotates 30° on an axes of symmetry and each edge maintains parallel status with the opposite triangle edge on the same axis of symmetry. Four axes of symmetry are maintained for four sets of triangular planes, relative to phase angle and parallel plane properties. Six axes of symmetry are maintained through cuboctahedron→icosahedron VE transformation; however, diagonal distances between square faces decrease (see Figure 37.6).

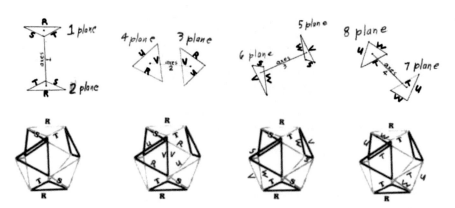

FIGURE 37.6 Rotation of parallel sets of equilateral triangles in vector equilibrium (VE).

The cuboctahedron has four sets of parallel (r, s, t, u, v, w) edges for a total of 24 edges. Under transformation, six sets of parallel (r, s, t, u, v, w) edges rotate 30° clockwise, and six sets of parallel (r^1, s^1, t^1, u^1, v^1, w^1) edges rotate 30° counterclockwise. The clockwise and counterclockwise sets of edges are no longer parallel to each other, resulting in twelve sets of parallel edges and the formation of six new edges in VE (Fuller, 1975). We contend that two of the new edges are parallel to the x-axis, two of the new edges are parallel to y-axis and two of the new edges are parallel to z-axis (see Figure 37.7).

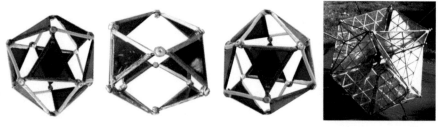

FIGURE 37.7 The cuboctahedron is in the neutral position between positive icosahedron (30° clockwise rotation) and negative icosahedron (30° counterclockwise rotation); dynamic cuboctahedron model.

We contend that the interaction of three axes of symmetry with six axes of symmetry is the mechanism for producing five-fold symmetry. We contend that there is six-dimensional, four-fold rotational symmetry in quasicrystals. Two-fold symmetry is defined by viewing the icosahedron on edge; we define the icosahedron edges as having 15 two-fold symmetries. Three-fold symmetry is defined by viewing icosahedral triangular faces; we define the icosahedral face as having ten three-fold symmetries. Five-fold symmetry is defined by the rotation of a cuboctahedron on four axes of symmetry, which produces six new edges that are normal to the three-dimensional coordinate system (x, y, z).

There exist two mechanisms of rotation simultaneously acting on positive and negative icosahedrons. The clockwise or counterclockwise rotation is relative to the observer. The 30 edges of a positive icosahedron are 90° out of phase from the 30 edges of a negative icosahedron. Therefore, the sum of edges is 60 (24 + 6 + 24 + 6 = 60).

- *The formation of a pentagonal dodecahedron utilizing (VEDCS) whereby four axes of symmetry rotation of eight isosceles tetrahedrons*

The quasicrystal is formed under extreme conditions of rapid cooling and includes the transition elements nickel, manganese, iron, titanium, cobalt, and copper. Quasicrystal alloys are found between the 2-3, 2-3-4, 3-4, 3-5, and 4-5 atomic shells of chemistry. The twelve outer spinning electrons are represented with six-dimensional mathematical symmetry. The six inner quantum loop gravity fields are activated by octahedral ligand bonding. The second image in, Figure 37.8 represents a VE motion into icosahedral symmetry, producing a pentagonal dodecahedron from eight isosceles tetrahedrons. The competing forces in quasicrystals produce more elaborate geometric forms such as the pentagonal dodecahedron.

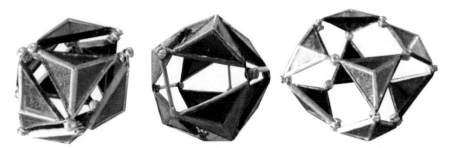

FIGURE 37.8 Rotational mechanism that leads to the pentagonal dodecahedron for quasicrystals.

- *Duals rhombic dodecahedron and cuboctahedron are geometrically related to (6D) cuboctahedron model*

Polyhedron duals have an interchangeable relationship between faces and vertices or vertices and edges. Six dimension (6D) theory views this polyhedral relationship as three, four and six axes of symmetry. The rhombic dodecahedron and cuboctahedron are duals between the vertices of the cuboctahedron and the twelve diagonals of the rhombic dodecahedron (see Figure 37.9). Garnet crystals are an example in nature that forms in the geometry of a rhombic dodecahedron.

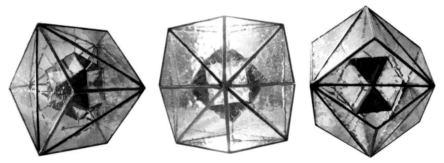

FIGURE 37.9   Three views of the rhombic dodecahedron with an inner cuboctahedron.

- *Duals icosahedron and pentagonal dodecahedron with 30 orthogonal edge symmetry are geometrically related to the (VEDCS)*

The icosahedral edges are orthogonal to the pentagonal dodecahedral edges on (x, y, z) axes of symmetry. Every (x, y, z) edge of the icosahedron has a complimentary (x, y, z) edge for the pentagonal dodecahedron at a 90° angle rotation, if analyzed only on axes of symmetry (see Figure 37.10).

30 edges of icosahedron + 30 edges of pentagonal dodecahedron = 60 edges, where each icosahedral edge has a complementary orthogonal edge for a pentagonal dodecahedron (see Figure 37.11). The intersection of an icosahedron with 30 edges and a dodecahedron with 30 edges produces a polyhedron similar and parallel to 60 of the edges of a twin isosceles pyritohedron of crystallography (see Figure 37.12). Quasicrystals of equivalent diameter approximate pentagonal dodecahedrons.

Pyritohedron and twin pyritohedron (see Figure 37.12), and icosahedron (see Figure 37.10) have common symmetry patterns based on sets of 15 parallel edges derived for the dynamic motion of VEDCS. The pyritohedron has 30 primary edges, and the twin pyritohedron has 60 primary edges (30 of which are parallel to the edges of the pyritohedron). The pentagonal dodecahedron has 15 parallel edges, for a total of 30 edges. The icosahedron also has 15 parallel edges for a total of 30 edges. The 60 edges of the twin pyritohedron are parallel to the combination of icosahedron and dodecahedron edges, or edges of a positive icosahedron and a negative icosahedron.

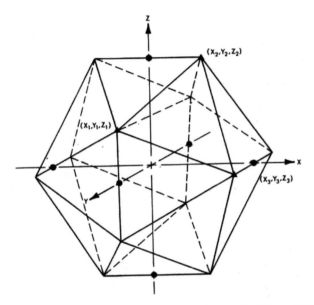

FIGURE 37.10 Icosahedron (x, y, z) axes (Reprinted from *J. D. Clinton 1971*).

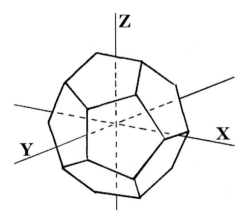

FIGURE 37.11 Pentagonal dodecahedron with equilateral pentagons and (x, y, z) axes.

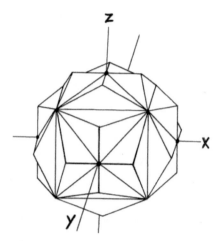

FIGURE 37.12   Twin pyritohedron.

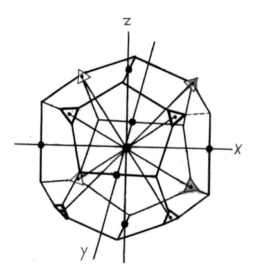

FIGURE 37.13   Pyritohedron with pentagons of two edge lengths, and eight (111) 3DT planes on four axes of rotational symmetry (Cotton, 1990).

- *The barium atomic model, including geometric properties of truncated rhombic dodecahedron, cuboctahedron and VE*

The barium atom is based on the VE icosahedron-cuboctahedron transformation of isosceles tetrahedrons in the superconductive $y_1ba_2cu_3o_7$ compound (see Figure 37.14). The Fermi model of Barium as a truncated rhombic dodecahedron has similar geometric properties of the cuboctahedron (see Figure 37.15).

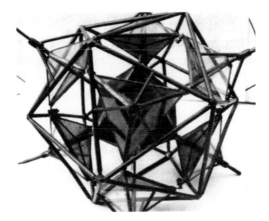

FIGURE 37.14   Barium model (James Sawyer, 1992).

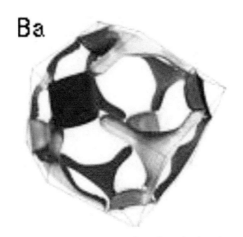

FIGURE 37.15   Fermi model of Barium (Fermi Surface database).

• *The primary issue of a dual coordinate system based on 6D triangular geometry with three orthogonal (3D) axis of symmetry is the present philosophical acceptance of the Cartesian (x, y, z) three dimensional coordinate system (3D)*

Presently our entire internet, computer, iPhones, and GPS mapping system are programmed with square pixels. We need to be able to visualize atomic space with triangular pixels. We should, however, be able to see triangular atomic space by applying additional 3, 4, and 6 axes of symmetry to x-ray crystallography.

- *The VEDCS is the mechanism for the five-fold symmetry of a TDCS*

There is a fundamental dynamic equilibrium balance in piezoelectric compounds, nano-catalytic compounds, and superconductive compounds. The dynamic VEDCS with 6D polyhedrons and three axes of symmetry 3D forces geometrically describes the electromagnetic properties of these dynamic compounds. The electromagnetic compounds all include the transition family of elements and triangular polyhedral geometry; therefore, they are acting dynamically in a DCS. The (P) application of intuition to the Cartesian algebra (3D) of the cube as a mathematical object leads to an additional Euclidean mathematical analysis of the tetrahedron in space as 6D algebra. X-ray crystallography of proteins as crystals and catalysis of platinum nanoparticles need a fundamental DCS to define and position the 3, 4, and 6-dimensional properties of these complex atomic polyhedrons in atomic space.

KEYWORDS

- **coordinates**
- **dimensions of space**
- **polyhedrons**
- **triangular and square lattice**
- **x-ray diffraction crystallography**

REFERENCES AND FURTHER READING

Bindi, L., Steinhardt, P. J., Yao, N., & Lu, P. J. I., (2011). *Al63Cu24Fe13, the First Natural Quasicrystal- American Mineralogist*, www.degruyter.com.

Buckminster, R., (1975/1979). *Fuller Synergetics.*

Clinton, J. D., (1971). *Advanced Structural Geometry Studies.* NASA Contractor CR–1735.

Cotton, F. A., (1990). *Chemical Applications of Group Theory* (3rd edn., p. 63). New York: Wiley.

David, F. P., (1989). *Super Strings and the Theory of Everything.*

Dodgson, C. L., (1885). *Euclid and the Modern Rivals*, pp. 169–172, 194.

Fritsche, H. G., & Benfield, R. E., (1993). Exact analytical formulae for mean coordination numbers in clusters. *Z. Phys. D26*, 15–17.

Herbert, H., (1917–2011). *Hauptman Woodward Medical Research.* http://www.hwi.buffalo. edu/about_hwi/nobel_laureate/hauptman_work.html.

Knox, A. J., Klein, M. J., & Schulmann, R. The collected papers of Albert Einstein, writings 1914–1917, DOC. 30. *The Foundation of the General Theory of Relativity,* pp. 146–200.

Lawrence Berkley National Laboratory, (2015).

Morrison and Boyd, (1992). *Organic Chemistry.*

Pauling, L., (1960). *The Nature of the Chemical Bond.* Cornell University Press.

Pauling, L., (1991–1994). *Selected Scientific Papers* (Vol. 1). By Linus Pauling, Barclay Kamb.

Sadoc, J. F., & Mosseri, R., (2006). *Geometric Frustration.*

Sawyer, J. G., & Wagner, M. J., (2013). *De Revolution VII: Encyclopedia of Six Dimensions.* Self-published, Buffalo, New York.

Sawyer, J., (1992). Icosahedral valence bond order in superconductive crystals. *World Conference on Superconductivity,* pp. 883–897.

University of Florida Data Base. http://www.phys.ufl.edu/fermisurface/.

Yarris, L., (2015). *A Most Singular Nano-Imaging Technique Berkeley Lab's Single Provides Images of Individual Nanoparticles in Solution.*

CHAPTER 38

Topological Efficiency Index

OTTORINO ORI[1-3]

[1]*Actinium Chemical Research, 00133 Rome, Italy*

[2]*Laboratory of Computational and Structural Physical Chemistry for Nanosciences and QSAR, West University of Timişoara, 300115, Timişoara, Romania*

[3]*Laboratory of Renewable Energies-Photovoltaics, National Institute for R&D in Electrochemistry and Condensed Matter INCEMC-Timisoara, 300569, Timişoara, Romania, E-mail: Ottorino.Ori@gmail.com*

38.1   DEFINITION

Topological efficiency index (TEI) is a graph invariant introduced in 2009 as a measure of the overall topological stability (TS) of small fullerenes; since then it has been applied to investigate stable configurations of large fullerenes and other organic compounds with n carbon atoms including infinite graphenic layers. Its formulation derives by a straight concept: every molecular graph G contains one or more vertices v characterized by their minimal contribution $\underline{w}$ to the Wiener index W. Then, under the basal hypothesis of considering those nodes v particularly stable (then *efficient*) the choice to compare the average w_{av} Wiener index to $\underline{w}$ appears a logic and extremely fast way to rank the topological stability of G in respect to "similar" diagrams such, for example, the isomers of a given molecule or the graphs of defective honeycomb planes. TEI – symbolized with ρ – corresponds to the ratio $\rho = w_{av}/\underline{w}$ with $w_{av} = W/n$ and it provides a quantitative measure of the topological efficiency of the molecular structure under the assumption that the minimal nodes are the most efficient vertices of G. By definition $\rho \geq 1$. The full understanding of the present concept also depends on the knowledge of the item WIE (Wiener Index) of the present dictionary.

38.2   HISTORICAL ORIGINS

Since the late 80's the stability of C_n fullerene molecules is a complex puzzle for theoreticians that attack this target by using symmetry and structural considerations as well as sophisticated ab-initio quantum mechanical simulations. In their work on C_{28} isomers (Ori et al., 2009) authors introduced a new molecular invariants ρ to describe a peculiar topological feature of a molecular graph initially baptized as *topological efficiency*. Nowadays in many contexts ρ is also called *topological roundness*. The name itself suggests the idea that this topological invariant is capable to measure the (average) "symmetry" of the whole graph. In fact, ρ assumes the minimum value $\rho = 1$ (maximum efficiency) for the buckminsterfullerene $C_{60}(I_h)$ a spherical molecule with icosahedral symmetry made by 60 symmetry-equivalent carbons; that is, by far the most stable fullerene. ρ is a rational number that expresses the topological symmetry embedded in the molecular graph. During these years index ρ has been successfully applied to determine the stability of other fullerenes and various kinds of graphenic planes, originating yet the interest of various theoreticians, Iranian scientists especially, attracted by its formal and practical properties. Like the Wiener index (WIE), the topological efficiency index evidences the intimate relationship between topology and molecular stability.

38.3   NANO-SCIENTIFIC DEVELOPMENTS

Considering an organic molecule having a chemical graph G (e.g., the diagram formed by the sole n carbon atoms), the original definition of ρ made by Ori et al., (2009) is

$$\rho(G,n) = \frac{W(G,n)}{n\underline{w}} \tag{1}$$

The Wiener index W and $\underline{w} = \min\{w_i\}$ (i.e., the minimal contribution to it) appearing in Eq. (1) are given by Eqs. (1) and (8) of the Chapter 44 of the present volume, respectively. By considering the average contribution to the Wiener index

$$w_{av} = \frac{W(G,n)}{n} \tag{2}$$

an equivalent formula for computing ρ is derived

$$\rho(G,n) = \frac{w_{av}}{\underline{w}} \tag{3}$$

Diagrams of carbon nanostructures with topological efficiency $\rho = 1$ are given in Figure 38.1. Some of these structures are formed by one set of symmetry-atoms, others show instead chemical graphs G formed by two or more independent sites. Infinite tori with square or honeycomb tessellations will also exhibit, in any dimension D, the signature $\rho = 1$ of a perfect topological roundness. C_{40} and C_{240} fullerenes, both with three independent sets of nodes, still show a $\rho = 1$ cage. To understand the mechanism that makes the structure maximally efficient $\rho = 1$ even in the presence of multiple independent sites we consider the example of the dual graph G^* of $C_{60}(I_h)$ made by 12 pentagons and 20 hexagons with two sets of topologically independent vertices. Partition in topologically independent sets is driven by the Wiener Weights $\{b_{ik}\}$ of the graph vertices i. In such a way, the 12 pentagons have $\{b_{ik}\}_P = \{5,10,10,5,1\}$ whereas the 20 hexagons belong to a different class characterized by $\{b_{ik}\}_H = \{6,9,9,6,1\}$. What causes $\rho = 1$ also in the presence of two topologically independent sites is the fact that the contributions w_P and w_H to the Wiener index of the dual C_{60} $W(G^*) = 1280$, have the same value $w_P = w_H = w = 40$. Thus, all sites are minimal vertices also in this case. Computation of these invariants is easily done by applying Eq. (8) to WW $\{b_{ik}\}_P$ and $\{b_{ik}\}_H$. In Figure 38.2, both $C_{60}(I_h)$ graphs are represented.

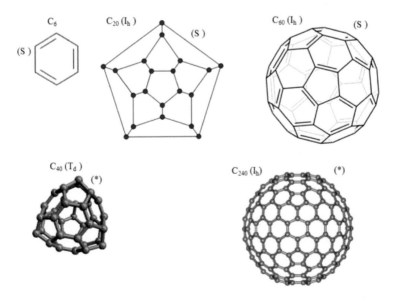

FIGURE 38.1 Structures with unitary topological efficiency. The list includes chemical graphs formed by symmetry-equivalent atoms; the two starred fullerenes are formed by sets of atoms with different symmetries.

The definition of ρ given in Eq. (3) naturally leads to a new invariant the *extreme topological efficiency* index ρ^E

$$\rho^E(G,n) = \frac{\underline{\underline{w}}}{\underline{w}} \tag{4}$$

where quantity $\underline{\underline{w}} = \max\{w_i\}$ corresponds to the contribution to W coming from the maximal graph nodes. Index ρ^E has been introduced in 2010 by Damir Vukicevic, Split University, Croatia in a private communication at the time of the joint studies on fullerene C_{66} (Vukicevic et al., 2011).

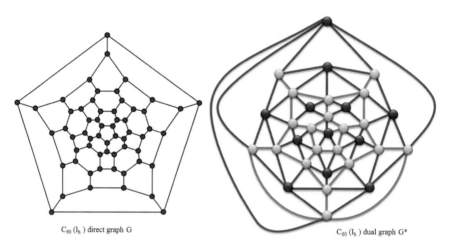

$C_{60} (I_h)$ direct graph G $C_{60} (I_h)$ dual graph G*

FIGURE 38.2   Graphs of C_{60} icosahedral fullerene are represented in the direct G and dual G^* space. Both graphs have $\rho = 1$. In G all atoms share the string $\{b_{ik}\} = \{3,6,8,10,10,10,8,3,1\}$ whereas in G^* red (green) nodes show Wiener Weights $\{b_k\}_P = \{5,10,10,5,1\}$ ($\{b_k\}_H = \{6,9,9,6,1\}$).

Topological descriptor ρ^E obeys to the same constraint $\rho^E \geq 1$ holding for ρ, looking like a sort ratio between two axes of an ellipsoid. The structural information contained in $\underline{w}$ and $\underline{\underline{w}}$ is very rich since the related $\{b_{ik}\}$ sets store the population of k-neighbors in all k-coordination shell of minimal and maximal nodes. Due to its definition Eq. (4) of Chapter 44 of the present volume) invariant ρ^E provides an improved evaluation for the topological molecular roundness, and sporadically it is also named *topological roundness index*. Cases with $\rho = 1$ also show $\rho^E = 1$.

We know (see Eq. (3) of Chapter 44 of the present volume) that for infinite systems index W follows a polynomial curve of powers of $n^{1/D}$ of grade $2D+1$, therefore w_i contributions have the following asymptotic rule.

$$w_i \cong \frac{W(G,n)}{n} \cong n^{(D+1)/D} \tag{5}$$

where a polynomial powers of $n^{1/D}$ is grade D+1. Both ρ and ρ^E tend to constant values in the limit of large n. The compression factor f_W given in Eq. (6) of Chapter 44 of the present volume has been introduced in Cataldo et al. (2011) to discriminate the asymptotic behavior of periodic infinite lattices and it appears to be connected to ρ by the following conjectured relation

$$f_W^{-1} = Lim_{n\to\infty}\rho \tag{6}$$

The numerical conjectured constraint for compression factor upper limit $f_W \leq 0.75$ implies, *for infinite periodic open lattices*, the lower limit for the asymptotic topological efficiency ρ_∞.

$$\rho_\infty \geq 4/3 \tag{7}$$

We have already mentioned the $\rho_\infty = 1$ limit that holds for *closed* square and honeycomb D-tori, whereas $\rho_\infty = 4/3$ characterizes infinite open square lattices in any dimension D. For $D = 2$, infinite periodic open *honeycomb* lattices show the asymptotic value $\rho_\infty = 48/35$ (Cataldo et al., 2011). Eq. (6) does not apply to non-periodic infinite nanosystems like for the 1-pentagon nanocone, which is a $D = 2$ structure with asymptotic limit $\rho_\infty \approx 1.273$ (Cataldo et al., 2010) or for the tessellation on the sphere made by pentagons and hexagons that, simulated by large icosahedral fullerenes, shows $\rho_\infty \approx 1.009$.

Examples of applications of ρ to define the stability of chemical nanostructures are presented in the next section.

38.4 NANOCHEMICAL APPLICATIONS

Recently introduced invariants ρ and ρ^E have found important applications involving relevant nanostructures such as fullerenes and defective graphenic layers. The topological index gives, in fact, a simple numerical evidence of nanosystems' topological symmetry.

As a first example, we report here the topological studies, based on ρ, aiming to determine the stable cages of the C_{66} fullerene published in Vukicevic et al. (2011). C_{66} fullerene just shows non-IPR isomers, 4478 possible molecules overall. Experimentally this hollow molecule has been produced only as stable metallofullerene endoclusters $Sc_2@C_{66}$, the detected cage having two pairs of fused pentagons. The electronic structures and relative stabilities of metallofullerenes have been investigated predicting $C_{66}(C_2)$ (isomer count #0083) as the most stable endoclusters, although the C_{2v} cage (isomer count #0011) is the actually observed molecule. The theoretical article (Vukicevic et al., 2011) provides the first extended application of ρ to fullerenes and correctly classifies the relative topological stability of the 4478 C_{66} molecules just considering the topological information stored in the chemical graphs made of 12 pentagonal and 23 hexagonal carbon faces differently connected. The joint effect of topological compactness (Wiener index) and topological efficiency correctly sieves as the most stable isomer the C_{66} (C_{2v}) molecule (characterized by topological markers $W = 10,674$ and $\rho = 1.0268$) using a fast (and elegant) computational method for predicting stable molecules for fullerenes of any size. Topological efficiency role ρ as a chemical stability driver for carbon nanosystems has been successfully confirmed by topological simulations concerning C_{50} (Graovac et al., 2014) and C_{84} fullerenes (Khataee et al., 2014). In this last case, invariant ρ^E given in Eq. (4) outperforms every ab-initio methods indicating the stable IPR cages in the middle of 51,592 isomers! Both molecules in fact C_{84}-D_{2d}(#23) and C_{84}-D_2 (#22) present the minimum value $\rho^E = 1.0129$ with distinct structural compactness, expressed by the Wiener index (WIE), that favors W(#23) = 19,718 the formation of the $C_{84} - D_{2d}$(#23) cage, matching in this way the experimental findings.

The relevance of the topological approach also arises from various applications regarding infinite structures like honeycomb planar lattices. See the results in various articles (Cataldo et al., 2010; Putz et al., 2013; Ori and Putz, 2014).

Worth mentioning here are the new results achieved by Ori et al., (2016) on the topological mechanism governing both the formation and healing of structural defects (multi-vacancies) in graphenic layers. The study indicates that the self-healing featured by defective honeycomb networks, corresponds to a universal topological feature arising from the topology of the 3-connected hexagonal network treated as a topological network rather than an effect arising from the fine-tuning of DFT and related arbitrary parametrizations. Defect-generation and annihilation are influenced by the long-range topological properties of the honeycomb mesh. In Figure 38.3, the main steps of the topological mechanism are shown: accumulation of vacancies, for certain vacancy concentration (a,b,c), originates the ladder-behavior of the extreme topological efficiency index ρ^E; in the same intervals with quasi-flat ρ^E values reversible hole-filling is allowed, providing a pure topological explanation for the graphene self-heal effect experimentally reported.

This intriguing and outstanding result evidences the profound link between lattice topology and physical properties, confirming once more the validity of Wiener's pioneer approach to chemistry (Wiener, 1947).

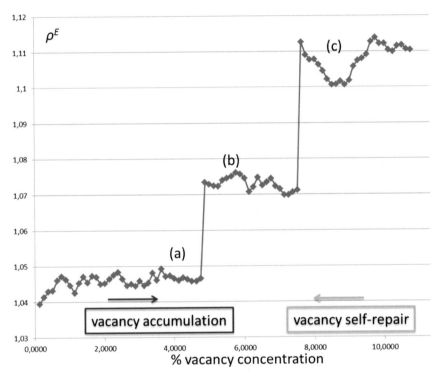

FIGURE 38.3 Response of topological invariants ρ^E in a graphene plane with increasing vacancy concentration. The typical ladder-curve show almost flat regions (a,b,c) where vacancy formation (rightward) and self-heal mechanisms (leftward) arise as inherent topological properties of honeycomb nanosystems.

38.5 MULTI-/TRANS-DISCIPLINARY CONNECTIONS

Innovative applications of topological efficiency invariants are already described in WIE, including the recent applications to neuroscience (Tozzi et al., 2017) the most promising sector in which topological efficiency indices will find useful correlations between system topology and neural functions.

38.6 OPEN ISSUES

Like the Wiener index, topological efficiency invariants represent optimal tools for investigating large nanostructures. Topological methods express the role of long-range interactions and are therefore capable of describing new kinds of phase-transition governed by these interactions. Recent original studies on $D = 1$ graphenic nanoribbons with topological defects (Putz & Ori, 2012) clearly relate long-range interactions to the emergence of 'bondon' quasi-particle in describing the electronic properties low-D nanosystems. These topics indicate the road for future studies based on ρ and ρ^E invariants.

KEYWORDS

- **compression factor**
- **extreme topological efficiency**
- **topological roundness**
- **Wiener index**

REFERENCES AND FURTHER READING

Cataldo, F., Ori, O., & Graovac, A., (2011). Graphene topological modifications. *International Journal of Chemical Modeling, 3*, 45–63.

Cataldo, F., Ori, O., & Iglesias-Groth, S., (2010). Topological lattice descriptors of graphene sheets with fullerene-like nanostructures. *Molecular Simulation, 36*(5), 341–353.

Graovac, A., Ashrafi, A. R., & Ori, O., (2014). Topological efficiency approach to fullerene stability–case study with C_{50}. In: Basak, S. C., Restrepo, G., & Villaveces, J. L., (eds.), *Advances in Mathematical Chemistry and Applications* (Vol. 2, pp. 3–23). Bentham Science Publishers.

Khataee, H., Arefi-Oskoui, S., Khataee, A., & Liew, A. W. C., (2014). Topological Wiener indices and polynomials of C_{84} fullerene. *Nanocage Nanoscience and Nanotechnology Letters, 6*, 532–536.

Ori, O., & D'Mello, M., (1992). A topological study of the structure of the C_{76} fullerene. *Chemical Physics Letters, 197*(1/2), 49–54.

Ori, O., & Putz, M. V., (2014). Isomeric formation of 5|8|5 defects in graphenic systems. *Fullerenes, Nanotubes and Carbon Nanostructures, 22*(10), 887–900.

Ori, O., Cataldo, F., & Graovac, A., (2009). Topological ranking of C_{28} fullerenes reactivity. *Fullerenes, Nanotubes and Carbon Nanostructures, 17*(3), 308–323.

Ori, O., Cataldo, F., Putz, M. V., Kaatz, F., & Bultheel, B., (2016). Cooperative topological accumulation of vacancies in honeycomb lattices. *Fullerenes, Nanotubes and Carbon Nanostructures*, *24*(6), 353–362.

Putz, M. V., & Ori, O., (2012). Bondonic characterization of extended nanosystems: Application to graphene's nanoribbons. *Chemical Physics Letters*, *548*, 95–100.

Putz, M. V., Ori, O., Cataldo, F., & Putz, A. M., (2013). Parabolic reactivity "coloring" molecular topology: Application to carcinogenic PAHs. *Current Organic Chemistry*, *17*(23), 2816–2830.

Tozzi, A., Peters, J. F., & Ori, O., (2017). Fullerenic-topological tools for honeycomb nanomechanics. Toward a fullerenic approach to brain functions. *Fullerenes, Nanotubes and Carbon Nanostructures*, Available online: doi: 10.1080/1536383X.2017.1283618. Available online: URL http://dx.doi.org/10.1080/1536383X.2017.1283618 (accessed on 12 Apr 2017).

Vukicevic, D., Cataldo, F., Ori, O., & Graovac, A., (2011). Topological efficiency of C_{66} fullerene. *Chem. Phys. Lett.*, *501*, 442–445.

Wiener, H., (1947). Structural determination of paraffin boiling points. *J. Am. Chem. Soc.*, *69*, 17–20.

CHAPTER 39

Topological Indices of C$_{10n}$ Fullerene

YASER ALIZADEH[1] and ALI IRANMANESH[2]

[1]*Department of Mathematics, Hakim Sabzevari University, Sabzevar, Iran, E-mail: y.alizadeh@hsu.ac.ir*

[2]*Department of Mathematics, Tarbiat Modares University, Tehran, Iran, E-mail: iranmanesh@modares.ac.ir*

39.1 DEFINITION

A topological index of a molecular graph G is a real number derived from the structure of graph that is invariant under automorphisms of G. A fullerene is a molecule of carbon in the form of a hollow sphere, ellipsoid, tube, and many other shapes. A fullerene graph is a 3-connected, 3-regular planar graph with only pentagonal and hexagonal faces. Structure of C$_{60}$ fullerene is shown in Figure 39.1.

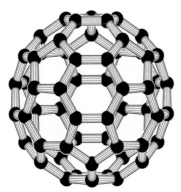

FIGURE 39.1   C$_{60}$ fullerene.

Let G has vertex set $V(G) = \{v_1, v_2 \ldots v_n\}$. Distance between two vertices v_i and v_j, $d(v_i, v_j)$ is defined as the length of the shortest path connecting them.

The diameter of G is the maximum distance between its vertices and denoted *diam(G)*. The eccentricity ecc(u) of u is the maximum distance between u and any other vertex of G. The distance between edges $g = u_1v_1$ and $f = u_2v_2$ is defined as:

$$d_e(g,f) = min\{d(u_1,u_2), d(u_1,v_2), d(v_1,u_2), d(v_1,v_2)\} + 1.$$

Equivalently, this is the distance between the vertices g and f in the line graph of G. The degree of u will be denoted *deg(u)*. For a simple connected graph G, four more applicable topological indices based on distance The Wiener index $W(G)$, the edge Wiener index $W_e(G)$, the eccentric connectivity index $\zeta(G)$, the Szeged index $Sz(G)$ are defined as:

$$W(G) = \sum_{\{u,v\}\subseteq V(G)} d(u,v),$$

$$W_e(G) = \sum_{\{f,g\}\subseteq E(G)} d_e(f,g),$$

$$RW(G) = \frac{1}{2}n(n-1)\mathrm{diam}(G) - W(G),$$

$$\zeta(G) = \sum_{u\in V(G)} ecc(u)\deg(u),$$

$$Sz(G) = \sum_{uv\in E(G)} n_u n_v$$

where n_v (resp.n_u) is the number of vertices that are closer to vertex u (resp. v) than vertex v (resp. u).

39.2 HISTORICAL ORIGIN(S)

The first fullerene molecule to be discovered, and the family's namesake, buckminsterfullerene (C_{60}), was prepared in 1985 by Richard Smalley, Robert Curl, James Heath, Sean O'Brien, and Harold Kroto at Rice University (Kroto et al., 1985).

The name was a homage to Buckminster Fuller, whose geodesic domes it resembles. The structure was also identified some five years earlier by Sumio Iijima, from an electron microscope image, where it formed the core of a "bucky onion."

Fullerenes have since been found to occur in nature. More recently, fullerenes have been detected in outer space. According to astronomer

Letizia Stanghellini, "It's possible that buckyballs from outer space provided seeds for life on Earth."

The discovery of fullerenes greatly expanded the number of known carbon allotropes, which until recently were limited to graphite, diamond, and amorphous carbon such as soot and charcoal. Buckyballs and bucky-tubes have been the subject of intense research, both for their unique chemistry and for their technological applications, especially in materials science, electronics, and nanotechnology. The smallest fullerene is the dodecahedral C$_{20}$. There are no fullerenes with 22 vertices. The number of fullerenes C$_{2n}$ grows with increasing n $=$ 12, 13, 14... roughly in proportion to n^9. For instance, there are 1812 non-isomorphic fullerenes C$_{60}$. Note that only one form of C$_{60}$, the buckminsterfullerene alias truncated icosahedron, has no pair of adjacent pentagons (the smallest such fullerene). Topological indices are used for example in biological activities or physic-chemical properties of alkenes which are correlated with their chemical structure. In a series of papers topological indices of fullerenes were studied. For example see Alizadeh et al., (2012) and Iranmanesh et al., (2009).

39.3   NANO-SCIENTIFIC DEVELOPMENT(S)

Fullerenes are used in certain medical applications nano-medicine. The idea is to use the very small fullerene molecules to easily deliver drugs directly into cells in a highly controlled manner. This is possible because the extremely small diameter of the nanoparticle fullerenes (which act like a cage to hold the drug) allows them to readily pass through cell membranes.

39.4   NANOCHEMICAL APPLICATION(S)

As an example, topological indices such as the Wiener index, the Szeged index, edge Wiener index, PI$_v$ index and eccentric connectivity index of the family of C$_{10n}$ fullerenes are computed. A method for computing topological indices of a family of molecular graphs is described in Alizadeh et al., (2012); Alizadeh & Klavzar (2013). The structure of C$_{10n}$ fullerene is illustrated in Figure 39.2. Topological indices of C$_{10n}$ fullerene for $1 \leq n \leq 9$ is computed in Table 39.1. By interpolation method (Alizadeh et al., 2012; Alizadeh & Klavzar, 2013), the explicit formula is obtained for C$_{10n}$ fullerene. It is proved in that $diam(C_{10n}) = 2n - 1$, $n \geq 5$.

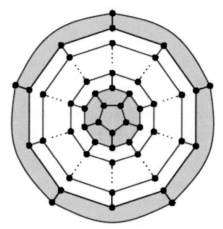

FIGURE 39.2 C_{10n} fullerene. Reprinted from Alizadeha, 2014. Open Access.

TABLE 39.1   Topological Indices of C_{10n} Fullerene $1 \leq n \leq 9$

n	$W(C_{10n})$	$Sz(C_{10n})$	$W_e(C_{10n})$	$PI_v(C_{10n})$	$\zeta(C_{10n})$
3	1435	6655	3420	1090	540
4	3035	16830	7125	2030	840
5	5455	33545	12680	3250	1170
6	8880	58900	20505	4800	1690
7	13505	93535	31030	6650	2130
8	19530	138810	44705	8800	2760
9	27155	196125	61980	11250	3510

Reprinted from Alizadeha (2014). Open Access.

$$W(C_{10n}) = \frac{100}{3}n^3 + \frac{1175}{3}x - 670, \quad n \geq 5$$

$$Sz(C_{10n}) = 250n^3 + 3075n - 13800, \quad n \geq 9$$

$$W_e(C_{10n}) = 75n^3 + 1000n - 1695, \quad n \geq 5$$

$$\zeta(C_{10n}) = 45n^2 - 15, \quad n \geq 8$$

39.5 MULTI-/TRANS-DISCIPLINARY CONNECTION(S)

Many properties of fullerene molecules can be studied using mathematical tools and results. Fullerene graphs were defined as cubic (i.e., 3-regular) planar 3-connected graphs with pentagonal and hexagonal faces. By Euler's formula, the number of pentagonal faces is always twelve.

Grunbaum and Motzkin (1963) showed that fullerene graphs with n vertices exist for all even $n \geq 24$ and for $n = 20$. Although the number of pentagonal faces is negligible compared to the number of hexagonal faces, their layout is crucial for the shape of fullerene graphs. Perfect matching and some structural properties and spectral properties of fullerenes were studied in Doslic (1998), Doslic (2002), and Slanina et al., (2001).

KEYWORDS

- C_{10n} fullerene
- fullerene
- topological index

REFERENCES AND FURTHER READING

Alizadeh, Y., Andova. V., Klavzar, S., Skrekovski, R. (2014) Wiener Dimension: Fundamental Properties and (5,0)-Nanotubical Fullerenes. *MATCH Commun. Math. Comput. Chem. 72*: 279-294

Alizadeh, Y., Iranmanesh, A., & Klavzar, S., (2012). Interpolation method and topological indices: The case of fullerenes C_{12k+4}, *MATCH Commun. Math. Comput. Chem., 68*, 303–310.

Doslic, T., (1998). On lower bounds of number of perfect matchings in fullerene graphs. *J. Math. Chem., 24*, 359–364.

Doslic, T., (2002). On some structural properties of fullerene graphs. *J. Math. Chem., 31*, 187–195.

Grunbaum, B., & Motzkin, T. S., (1963). The number of hexagons and the simplicity of geodesics on certain polyhedral. *Canad. J. Math., 15*, 744–751.

Iranmanesh, A., Alizadeh, Y., & Mirzaie, S., (2009). Computing Wiener polynomial, Wiener index, hyper Wiener index of C_{80} fullerene by GAP program. *Fullerenes, Nanotubes and Carbon Nanostructures, 17*, 560–566.

Kroto, H. W., Heath, J. R., O'Brien, S. C., Curl, R. F., & Smalley, R. E., (1985). C_{60} Buckminsterfullerene. *Nature, 318*, 162–163.

Slanina, Z., Uhlk, F., Lee, S., & Sawa, E., (2001). Geometrical and dynamic approaches to the relative stabilities of fullerene isomers. *MATCH Commun. Math. Comput. Chem., 44*, 335–348.

CHAPTER 40

Topology-Based Molecular Descriptors: The Current Status

RAKESH KUMAR MARWAHA,[1] HARISH DUREJA,[1] and A. K. MADAN[2]

[1]*Faculty of Pharmaceutical Sciences, Maharshi Dayanand University, Rohtak, 124001, India*

[2]*Faculty of Pharmaceutical Sciences, Pt. B.D. Sharma University of Health Sciences, Rohtak, 124001, India, E-mail: madan_ak@yahoo.com*

ABSTRACT

Computational techniques in drug discovery/development are routinely used as a versatile tool to minimize experiments and cost involved at each stage of the process. A large number of different computational aids/methods are currently applied in the rational drug discovery process. These include (quantitative) structure-activity/property/toxicity/pharmacokinetic relationship [(Q)SAR/QSPR/QSTR/QSP$_k$R] models, molecular modeling, homology modeling, similarity searching, pharmacophores, drug discovery databases, machine learning, data mining, network analysis, and data analysis tools, etc. (Q)SAR/QSPR/QSTR/QSP$_k$R approach conserves valuable resources and accelerates the drug development process by transforming searches for lead compounds into a mathematically quantified and computerized form using chemical intuition and experience. Molecular descriptors [MDs] are utilized to extract the structural information of the molecule in numerical form, suitable for model development and serve as a bridge between the molecular structure and physicochemical/biological activities of the molecules. Among various MDs used in (Q)SAR/QSPR/QSTR/QSP$_k$R studies, graph-theoretical invariants or topological indices (TIs) are fundamental in nature and true structural invariants. Due to their conceptual simplicity, these MDs are easily computable using simple mathematical operations. They quantify certain aspects of molecular structure and are generally sensitive to different chemical features of the molecule such as size, shape, branching, symmetry, branching pattern, cyclicity, heterogeneity of atomic neighborhoods, etc. TIs can be classified as topostructural and topochemical depending upon their ability to distinguish among atoms or bond types present in the molecule. Based on their discriminating power, MDs are classified into six generations. The development of new TIs having high discriminating power but devoid of degeneracy continues to be a challenge in computational chemistry.

40.1 INTRODUCTION

The process of drug discovery is a highly complex, time-consuming, very expensive and challenging in spite of recent advances in our comprehension of biological systems and several technical improvements (Ravina, 2011). Out of a large number of new therapeutic candidates discovered every year in laboratories, only a small number of drug candidates successfully pass the clinical trials and the stringent criteria laid down by different drug regulatory agencies (Huang et al., 2010). *In-silico*-based techniques have emerged as promising alternative or complementary tool for the effective screening of a large number of potential drugs and this could lead to a reduction in the cost of drug design and development process significantly (McGee, 2005). The study of (quantitative) structure-activity/property relationship [(Q)SAR/ QSPR] is a vital research area among different *in-silico* techniques and has been widely utilized to predict the physicochemical properties and biological activities of various organic compounds (Ferydoun et al., 2008). The main aim of the (Q)SAR/QSTR/QSPR studies is to find a mathematical relationship between the property of interest and the descriptive parameter(s) called molecular descriptors (MDs). The MDs are derived from the structure of the molecule in such a way that the behavior/desired activity of a series of molecules can be explained and, at the same time, the process can be reversed to obtain new structures that have the desired activity or property (Katritzky et al., 2010).

Different types of MDs, commonly used for QSAR/QSPR include lipophilic descriptors, e.g., Log P, Log D, R_m (retention parameter in thin layer chromatography), Log K (reverse phase liquid chromatography), Hansch substituent constant (π) and Rekker's hydrophobic fragmental constant (f), etc., electronic descriptors, e.g., Hammet constant (σ), resonance parameters, dipole moments, Taft polar constant, Taft inductive constant, charge-transfer constants, molecular spectroscopy based parameters, etc., polarizability descriptors, e.g., Parachor, molar volume and molar refractivity, etc., constitutional descriptors, e.g., molecular weight, average atomic weight, total number of bonds, total number of atoms, number of rings, etc., quantum-chemical descriptors, e.g., quantum molecular energies HOMO and LUMO energies, orbital electron densities, atom-atom polarizabilities, molecular polarizabilities and super-delocalizabilities, etc., steric descriptors, e.g., Taft's steric parameter, Charton steric parameter, molar refractivity, Vander Waal's radii and parameters derived from the STERIMOL program, etc., solubility descriptors, e.g., Hildebrand solubility parameters, mole fraction solubility, molar solubility, etc., geometrical descriptors, e.g., molecular volume, molecular surface area and Shadow indices, etc., electrostatic descriptors, e.g., charged partial surface area indices, etc., topological indices, e.g., Balaban index J, Wiener index W, Hosoya index Z, molecular connectivity indices, eccentric connectivity indices, Kappa shape indices, Kier and Hall connectivity indices, Kier shape indices and detour indices, etc. (Karelson and Lobanov, 1996; Grover et al., 2000).

Topological indices (TIs) are the numbers associated with constitutional formulae by mathematical operations on the graphs representing these formulae, which can be used to characterize and order molecules and predict their properties (Basak et al., 1990). Among thousands of diverse chemical graph theory based MDs, TIs have the advantage of being true structural invariant, which means that their values do not depend on the molecular conformations (Mahmoudi et al., 2006). This conformational independence becomes particularly important in the study of flexible molecules, where the proper conformation of the molecules is not well established (Stanton, 2008).

Depending upon the chemical information encoded in TIs, they can be classified as topostructural and topochemical. Topostructural indices simply encode information based on adjacency or connectedness of atoms within a molecule, whereas topochemical indices encode information related to both molecular topology as well as chemical nature of atoms and bonds in the molecule (Basak and Gute, 1997). Topostructural indices can easily be derived from simple matrices, such as the distance matrix and/or the adjacency matrix, whereas topochemical indices need chemically weighted matrices for their derivation (Dureja et al., 2008).

TIs can also be classified on the basis of their discriminating power into various generations.

First generation TIs are integer numbers derived from integer "local vertex invariants" (LOVIs) assigned for each vertex, such as vertex degrees, topological distances, distance sums, etc. The main limitation of these TIs is their high degeneracy, i.e., the same value for the descriptor for many non-isomorphic graphs. The most important indices of this class are Wiener index, W (Wiener, 1947), Hosoya index, Z (Hosoya, 1971) and the centric indices of Balaban B (Balaban, 1979).

Second generation TIs are the real numbers depending on integer LOVIs such as vertex degrees/distance sums or other integer graph properties. The Randic molecular connectivity index (Randic, 1975), the related Kier-Hall indices (Kier and Hall, 1976), the mean square distance (Balaban, 1983), as well as the Balaban index J (Balaban, 1982) and J_{het} (Balaban, 1986), overall connectivity index by Bonchev (1999), Harary descriptor, also called Harary number, H (Plavsic et al., 1993), etc. are some of the frequently used second-generation TIs.

Third generation TIs are the real numbers based on real-number LOVIs. These TIs have relatively low degeneracy. These indices transform a matrix into a system of linear equations having LOVIs as solutions by including a column vector on the main diagonal and another column vector as the free term. The column vectors may reveal chemical information such as graph-theoretical information, e.g., the vertex degree or the distance sum of the corresponding vertex, or the atomic number of the atom symbolized by the corresponding vertex or simply a constant numerical data, e.g., the number of vertices in the graph, or its square. Information-theoretic indices (Bonchev, 1983; Balaban et al., 1991), the triplet indices (Filip et al., 1987) and hyper-Wiener descriptor or the molecular identification (ID) numbers (Randic, 1984) are the examples of third-generation TIs.

Fourth generation TIs have discriminating power ≥ 100 for structures containing only five vertices [with or without heteroatom(s)]. Discriminating power is the ratio of the highest to lowest index values for all possible structures with the equal number of vertices and is one of the basic and important characteristics of any MD (Dureja and Madan, 2007). Fourth generation TIs are exemplified by augmented eccentric connectivity indices and super-augmented eccentric connectivity indices (Dureja and Madan, 2007).

Fifth generation TIs are the topochemical versions of TIs with the high discriminating power of the order of ≥ 100 for all possible structures containing only five vertices and ≥ 25 in case of pendenticity-based TIs (Dureja and Madan, 2012). These TIs are sensitive to both the presence as well as relative position(s) of heteroatom(s) in molecules. Fifth generation TIs normally exhibit extremely low degeneracy.

Sixth generation TIs take into consideration the nature of bonds in addition to fulfilling various requirements for fifth-generation TIs. These TIs enjoy very high discriminating power and negligible degeneracy (Dureja and Madan, 2012). Development of sixth generation TIs continues to be a challenge for the scientific community.

TIs may also be classified on the basis of distance, adjacency, distance-cum adjacency, centricity, valence electron mobile (VEM) environment and information content.

Distance-based topological indices may further be classified depending on the distance matrix or detour matrix employed to characterize molecular graphs. The *distance matrix* is a real, square, symmetrical matrix of order n, taking the minimum distance traversed in moving from vertex i to vertex j in graph G, whereas the *detour matrix* is real, square, symmetrical matrix of order n, taking the maximum distance traversed in moving from vertex i to vertex j in graph (G).

Adjacency-based TIs employ adjacency matrix, which is based on the consideration of connections between adjacent pairs of atoms in the molecule.

Distance-cum-adjacency-based TIs use *distance matrix/detour matrix* as well as the *adjacency matrix* to characterize molecular graphs. They contain more topological information in a graph, G as compared to other TIs derived from the distance or the adjacency matrix alone.

Centric Graph TIs quantify the degree of compactness of molecules depending on the graph center (Todeschini and Consonni, 2009). The graph center may be a single edge, a single vertex, or a single group of equivalent vertices having the smallest maximal distance to other vertices.

Information theory based TIs have been used in chemical graph theory for describing the chemical structures. The main advantage of these descriptors is their direct use as simple numerical descriptors in a comparison with physical, chemical or biologic parameters of molecules in structure-property and activity relationships. They normally have higher discriminating power for isomers than the respective TIs. The representative examples of different TIs along with the type of matrices used have been described in Table 40.1. Applications of some of the most commonly used TIs in QSAR/QSPR studies have been exemplified in Table 40.2.

Though a large number of TIs have been reported, but only a small number of them have been successfully utilized in the process of drug discovery/design. In order to avoid their hazardous proliferation, Randic (1991) proposed some desirable requirements for TIs, e.g., simplicity, direct structural interpretation, ability to discriminate the isomers, non trivial relationship with other TIs, non-dependence on physicochemical properties of molecules, linearly independence, gradual change in value with gradual change in molecular structure and good correlation with at least one property, etc.

Advantages and limitations of TIs (Gozalbes et al., 2002) are as per the following sections.

TABLE 40.1 Examples of TIs Classified on the Basis of Various Matrices

S. No.	Topological index	Matrix used	References
1.	Zagreb group indices, M_1 and M_2	Adjacency matrix	Gutman and Trinajstic, 1972
2.	Molecular connectivity indices, $^0\chi, \, ^1\chi, \, ^2\chi$	Adjacency matrix	Kier et al., 1975
3.	Hosoya's index, 'Z'	Adjacency matrix	Hosoya, 1971
4.	Extended edge connectivity indices	Adjacency matrix	Estrada et al., 1998
5.	Wiener index	Distance matrix	Wiener, 1947
6.	Mean square distance indices, D, D_1	Distance matrix	Balaban, 1982
7.	Gutman molecular topological descriptor, S_G	Distance matrix	Gutman, 1994
8.	Superpendentic descriptor $\int^p$	Distance matrix	Gupta et al., 1999
9.	Eccentric connectivity index, ECI	Distance and adjacency matrices	Sharma et al., 1997
10	Superaugmented pendentic indices	Distance and adjacency matrices	Dureja and Madan, 2009
11.	Augmented path eccentric connectivity indices	Detour and adjacency matrices	Marwaha et al., 2012

TABLE 40.2 Applications of Some TIs in QSAR/QSPR Studies

S. No.	Topological index	Applications	Application References
1.	Wiener index	Determination of boiling points and other physicochemical properties of paraffins.	Wiener, 1947
2.	Molecular connectivity indices	Biological potencies of nonspecific local anesthetics	Kier et al., 1975
3.	Zagreb group indices M_1 and M_2	Physicochemical properties of alkanes	Bonchev and Trinajstic, 2001
4.	Balaban index 'J'	Prediction of mutagenicity of aromatic and heteroaromatic Amines	Basak et al., 2001

TABLE 40.2   *(Continued)*

S. No.	Topological index	Applications	Application References
5.	Hosoya's index 'Z'	Prediction of boiling points of saturated hydrocarbons	Hosoya, 1971
6.	Reverse Wiener's index '$\varLambda$'	Physico-chemical properties of alkanes	Balaban et al., 2000
7.	Eccentric connectivity index, ζ^c	Predicting Genotoxicity of thiophene derivatives	Mosier et al., 2003
		Predicting HIV-protease inhibitory activity of tetrahydropyrimidin-2-ones	Lather and Madan, 2005
8.	Augmented eccentric connectivity descriptor, $^A\zeta^c$	Predicting anti-HIV activity of 2-pyridinones	Bajaj et al., 2006
9.	Superaugmented pendentic indices, $^{SA}\int^{P-N}$	Predicting anti-HIV activity of dimethylaminopyridin-2-ones	Dureja et al., 2009
10.	Augmented path eccentric connectivity topochemical indices, $^{AP}\zeta^{~C}_{nc}$	Prediction of antitubercular activity of PA-824 analogs	Marwaha and Madan, 2014

40.2   ADVANTAGES

➢ TIs do not need any experimental derived value/measurement for their use. They simply need structural information of the molecules which is easily available in modern databases.

➢ Use of TIs for development of QSAR models does not need any previous knowledge of the receptor structure or mechanism of action of the drugs.

➢ Topological features of almost every chemical structure can be described; therefore TIs can be calculated for compounds with diverse chemical structures, and theoretical models can be built using databases without a common basic chemical structure.

➢ They can be calculated for new and under-development molecules also.

➢ TIs are simple indices/descriptors that can be quickly calculated for a large number of compounds. It is not mandatory to firstly draw/display the chemical structures on graphical workstations.

➢   TIs provide simply 2D information and not 3D, which could be regarded as a limitation. However, it is also an advantage as the conformational and alignment problems associated with 3-D QSAR (as in CoMFA applications) are completely avoided. As a consequence, the results are more reproducible in 2D QSAR.

40.3   LIMITATIONS

➢   Degree of degeneracy and redundancy of some TIs can be very high. TIs are said to be degenerative when they present identical values for two or more different molecular graphs, while redundancy is the duplicated information possessed by several TIs. However, the 4th and higher generation TIs have solved this problem to some extent.

➢   TIs are usually criticized about physicochemical interpretation of their meaning. All TIs doesn't present this difficulty of interpretation, and a good example of this is the E-state descriptors, in which the role of skeletal atoms and substituents is directly linked with electronegativities.

➢   Some researchers have warned about the inherent danger of chance effects because of the usage of large numbers of TIs. Apparently, this is a common statistical problem shared by other descriptors also and not exclusive to TIs. Care should be taken in the indiscriminate use of TIs for sets without a common parent.

40.4   CONCLUSION

Molecular topology and chemical graph theory provide an ideal platform in the form of topology-based molecular descriptors for quantification of chemical structures for the prediction of desired property or biological activity for accelerating drug discovery process in a cost-effective way. The development of non-degenerate sixth generation topology based molecular descriptors with high discriminating power continues to be a challenge for computational chemists. There is a strong need of developing even better TIs that can describe the molecular structure in a more effective way and support the recent advances in drug discovery technologies to accelerate the process of lead discovery and optimization.

KEYWORDS

- **chemical graph theory**
- **degeneracy**
- **discriminating power**
- **graph invariant**
- **matrix**
- **QSAR/QSPR**
- **topochemical indices**
- **topostructural indices**

REFERENCES

Bajaj, S., Sambi, S. S., & Madan, A. K., (2006). Model for prediction of anti-HIV activity of 2-pyridinone derivatives using novel topological descriptor. *QSAR & Comb. Sci., 25*, 813–823.

Balaban, A. T., & Balaban, T. S., (1991). New vertex invariants and topological indices of chemical graphs based on information on distances. *J. Math. Chem., 8*, 383–397.

Balaban, A. T., (1979). Chemical graphs XXXIV. Five new topological indices for the branching of tree-like graphs. *Theoret. Chem. Acta., 53*, 355–375.

Balaban, A. T., (1982). Highly discriminating distance-based topological index. *J. Chem. Phys. Lett., 89*, 399–404.

Balaban, A. T., (1983). Topological indices based upon topological distances in molecular graphs. *Pure Appl. Chem., 55*, 199–206.

Balaban, A. T., (1986). Chemical graphs 48, topological index J for heteroatom-containing molecules taking into account periodicities of element properties. *MATCH Comm. Math. Comput. Chem., 21*, 115–122.

Balaban, A. T., Mills, D., Ivanviuc, O., & Basak, S. C., (2000). Reverse Wiener indices. *Croat. Chem. Acta, 73*, 923–941.

Basak, S. C., Grunwald, G. D., & Niemi, G. J., (1997). Use of graph-theoretic and geometrical molecular descriptors in structure-activity relationships. In: Balaban, A. T., (ed.), *From Chemical Topology to Three Dimensional Molecular Geometry* (pp. 73–116). Plenum Press, New York.

Basak, S. C., Mills, D. R., Balaban, A. T., & Gute, B. D., (2001). Prediction of mutagenicity of aromatic and heteroaromatic amines from structure: A hierarchical QSAR approach. *J. Chem. Inf. Comput., 41*, 671–678.

Basak, S. C., Niemi, G. J., & Veith, G. D., (1990). Optimal characterization of structure for prediction of properties. *J. Math. Chem., 4*, 185–205.

Bonchev, D., & Trinajstić, N., (2001). Overall molecular descriptors. 3. Overall Zagreb indices. *SAR QSAR Environ Res., 12*(1/2), 213–236.

Bonchev, D., (1983). *Information Theoretic Indices for Characterization of Chemical Structures*. Research Study Press, Chichester.

Bonchev, D., (1999). Overall connectivity and molecular complexity: A new tool for QSPR/QSAR. In: Devillers, J., & Balaban, A. T., (eds.), *Topological Indices and Related Descriptors in QSAR and QSPR, Gordon and Breach* (pp. 361–402). The Netherlands.

Dureja, H., & Madan, A. K., (2007). Superaugmented eccentric connectivity indices: New-generation highly discriminating topological descriptors for QSAR/QSPR modeling. *Med. Chem. Res., 16*, 331–341.

Dureja, H., & Madan, A. K., (2009). Predicting anti-HIV activity of dimethyl aminopyridine–2-ones: Computational approach using topochemical descriptors. *Chem. Biol. Drug Des., 73*, 258–270.

Dureja, H., & Madan, A. K., (2012). Pendencity based descriptors for QSAR/QSPR, In: Gutman, I., & Furtula, B., (eds.), *Distance in Molecular Graphs-Applications, Mathematical Chemistry Monographs* (No. 13, pp. 55–80). University of Kragujevac.

Dureja, H., Gupta, S., & Madan, A. K., (2008). Predicting anti-HIV-1 activity of 6-arylbenzonitriles: Computational approach using super-augmented eccentric connectivity topochemical indices. *J. Mol. Graph. Model., 26*, 1020–1029.

Estrada, E., Guevara, N., & Gutman, I., (1998). Extension of edge connectivity index relationship to line graph indices and QSPR applications. *J. Chem. Inf. Comput. Sci., 38*, 428–431.

Ferydoun, A., Ali, R. A., & Najmeh, M., (2008). Study on QSPR method for theoretical calculation of heat of formation for some organic compounds. *Afr. J. Pure Appl. Chem., 2*(1), 006–009.

Filip, P. A., Balaban, T. S., & Balaban, A. T., (1987). A new approach for devising local graph invariants: Derived topological indices with low degeneracy and good correlation ability. *J. Math. Chem., 1*, 61–83.

Gozalbes, R., Doucet, J. P., & Derouin, F., (2002). Application of topological descriptors in QSAR and drug design: History and new trends. *Curr. Drug Targets Infect. Disord., 2*, 93–102.

Grover, M., Singh, B., Bakshi, M., & Singh, S., (2000). Quantitative structure-property relationship in pharmaceutical research–Part 1. *Pharm. Sci. Technol. Today, 1*, 28–35.

Gupta, S., Singh, M., & Madan, A. K., (1999). Superpendentic index: A novel topological descriptor for predicting biological activity. *J. Chem. Inf. Comput. Sci., 39*, 272–277.

Gutman, I., & Trinajstic, N., (1972). Graph theory and molecular orbitals total pi-electron energy of alternant hydrocarbons. *Chem. Phys. Lett., 17*, 535–538.

Gutman, I., (1994). Selected properties of the Schultz molecular topological index. *J. Chem. Inf. Comput. Sci., 34*, 1087–1089.

Hosoya, H., (1971). Topological index, newly proposed quantity characterizing the topological nature of structure of isomers of saturated hydrocarbons. *Bull. Chem. Soc. Jpn., 44*, 2332–2337.

Huang, H. J., Yu, H. W., Chen, C. Y., Hsu, C. H., Chen, H. Y., Lee, K. J., Tsai, F. J., & Chen, C. Y. C., (2010). Current developments of computer-aided drug design. *J. Taiwan Inst. Chem. Eng., 41*, 623–635.

Karelson, M., & Lobanov, V. S., (1996). Quantum-chemical descriptors in QSAR/QSPR studies. *Chem. Rev., 96*, 1027–1043.

Katritzky, A. R., Kuanar, M., Slavov, S., & Hall, C. D., (2010). Quantitative correlation of physical and chemical properties with chemical structure: Utility for prediction. *Chem. Rev., 110*, 5714–5789.

Kier, L. B., & Hall, L. H., (1976). *Molecular Connectivity in Chemistry and Drug Research.* Academic Press, New York, pp. 257.

Kier, L. B., Hall, L. H., Murray, W. J., & Randic, M., (1975). Molecular connectivity I: Relationship to nonspecific local anesthesia. *J. Pharm. Sci., 64*, 1971–1974.

Lather, V., & Madan, A. K., (2005). Topological models for the prediction of HIV-protease inhibitory activity of tetrahydropyrimidin–2-ones. *J. Mol. Graph. Model., 23*, 339–345.

Mahmoudi, N., De Julian-Oritiz, J. V., Ciceron, L., Gálvez, J., Mazier, D., Danis, M., Derouin, F., & García-Domenech, R., (2006). Identification of new anti-malarial drugs by linear discriminant analysis and topological virtual screening. *J. Antimicrob. Chemother., 57*, 489–497.

Marwaha, R. K., & Madan, A. K., (2014). Fourth generation detour matrix-based topological descriptors for QSAR/QSPR–Part–2: Application in development of models for prediction of biological activity. *Int. J. Computational Biology and Drug Design, 7*(1), 1–30.

Marwaha, R. K., Jangra, H., Das, K. C., Bharatam, P. V., & Madan, A. K., (2012). Fourth generation detour matrix-based topological indices for QSAR/QSPR–Part–1: Development and evaluation. *Int. J. Computational Biology and Drug Design, 5*(3/4), 335–360.

McGee, P., (2005). Modeling success with *in silico* tools. *Drug Discov. Today, 8*, 23–28.

Mosier, P. D., Jurs, P. C., Custer, L. L., Durham, S. K., & Pearl, G. M., (2003). Predicting the genotoxicity of thiophene derivatives from molecular structure. *Chem. Res. Toxicol., 16*, 721–732.

Plavsic, D., Nikolic, S., Trinajstic, N., & Nihalic, Z., (1993). On the Harary index for the characterization of chemical graphs. *J. Math. Chem., 12*, 235–250.

Randic, M., (1975). On characterization of molecular branching. *J. Am. Chem. Soc., 97*, 6609–6615.

Randic, M., (1984). On molecular identification numbers. *J. Chem. Inf. Comput. Sci., 24*, 164–175.

Randic, M., (1991). Generalized molecular descriptors. *J. Math. Chem., 7*, 155–168.

Ravina, E., (2011). *Evolution of Drug Discovery: From Traditional Medicines to Modern Drugs.* Wiley-VCH, Weinheim.

Sharma, V., Goswami, R., & Madan, A. K., (1997). Eccentric connectivity index: A novel highly discriminating topological descriptor for structure-property and structure-activity studies. *J. Chem. Inf. Comput. Sci., 37*, 273–282.

Stanton, D. T., (2008). On the importance of topological descriptors in understanding structure-property relationships. *J. Comput. Aided Mol. Des., 22*, 441–460.

Todeschini, R., & Consonni, V., (2009). *Molecular Descriptors for Chemoinformatics.* Wiley VCH, Weinheim.

Trinajstic, N., (1983). *Chemical Graph Theory.* CRC Press, Boca Raton, Florida.

Wiener, H., (1947). Structural determination of paraffin boiling points. *J. Am. Chem. Soc., 69*, 17–20.

CHAPTER 41

Topological Reactivity

MIHAI V. PUTZ,[1,2] OTTORINO ORI,[3] ANA-MARIA PUTZ,[4] and
MARINA A. TUDORAN[1,2]

[1]*Laboratory of Structural and Computational Physical Chemistry for
Nanosciences and QSAR, Biology-Chemistry Department,
West University of Timisoara, Pestalozzi Street No. 44, Timisoara,
RO-300115, Romania, Tel: +40-256-592638, Fax: +40-256-592620,
E-mail: mv_putz@yahoo.com or mihai.putz@e-uvt.ro*

[2]*Laboratory of Renewable Energies-Photovoltaics,
R&D National Institute for Electrochemistry and Condensed Matter,
Dr. A. Paunescu Podeanu Str. No. 144, Timisoara, RO-300569, Romania*

[3]*Actinium Chemical Research, Via Casilina 1626/A, 00133 Rome, Italy*

[4]*Institute of Chemistry Timisoara of the Romanian Academy,
24 Mihai Viteazul Bld., Timisoara 300223, Romania*

41.1 DEFINITION

Topological reactivity method represents an effective mathematical tool in which the topological properties of a chemical system derived from the system adjacency or distance matrix are combined with its reactivity properties obtained with the aid of electronegativity and chemical hardness descriptors.

41.2 HISTORICAL ORIGIN(S)

In the science field, especially in physics and physical chemistry, there are phenomena which can not be described and interpreted only by using basic quantitative theories, without external parameters. In order to correct this problem, the classical molecular mechanics methods were developed, where the molecules were considered as a tiny group of moving balls attached by massless springs to each other. Years later, on the second half of the 19th century, the chemical graph theory begins its developing as both absolute and semiempirical theory, so that chemical problems (such as organic isomers enumeration) were successfully solved (Garcia-Domenech et al., 2008).

When combined with mathematics, the graph theory develops the so-called topological indices, defined as molecular descriptors which can be used finding and analyzing the physicochemical properties of molecules, including those with heteroatoms, and biological activity of drugs, in searching for pharmacophores, in isomers enumeration, drug design, in characterization of molecular shape and chirality, characterization of DNA primary sequences, t-RNA and folded proteins, in numerical characterization of proteomic maps, etc. (Randic, 2003).

The graph theoretical indices are very used as molecular descriptors in QSAR and QSPR studies. In their works, Galvez and his collaborators combine topological indices with quantum chemical parameters in an attempt to obtain a discriminate function to study the antibacterial activity (Mishra & Galvez, 2001). Liu and his co-workers (Liu et al., 2001) developed the MEDV-13 (molecular electronegativity distance vector) based on thirteen atomic types with applicability in QSAR studies in predicting the molecular activity. The information depicted by MEDV-13 were mostly referring to an element atom type, chemical bond type and valence electron state from the 2D molecular topology, being predictable, reproducible, and easy to use for QSAR studies (Randic, 2003). In their work, Estrada and Molina (2001) developed the topographic (3D) molecular connectivity indices starting from the molecular graphs improved with quantum chemical parameters in QSAR and QSPR studies on 2-furyl ethylene derivatives activity classification. When compared to quantum chemical descriptors and 2D connectivity indices, the 3D connectivity indices produced a significant improvement in the predictive capacity in modeling the partition coefficient (LogP) and produced the best discriminate model in the antibacterial activity classification for the working set (Randic, 2003).

Over the years, even if the graph theory importance highly recognized, there are still several areas in which is not yet accepted as a suitable theoretical tool, such as physical chemistry (Schmalz et al., 1992) or polymer chemistry (Corey, 1971, 1989; Zefirov & Gordeeva, 1987; Randic, 2003). However, the graph theory is successfully applied in hydrocarbons characterization and, based on the fact that if a method cannot be used to characterize some properties of hydrocarbons is less possible to be applied to other molecules, the graph theory is still an interesting topic in the scientific world, being expected that in the future, that more new topological indices with new areas of applicability to be discovered (Randic, 2003).

41.3   NANO-SCIENTIFIC DEVELOPMENT(S)

Recent studies in graph theory field proved that leads to the idea that the chemical properties of hexagonal ring systems can be influenced by the topology through the topological indices. In their work, Putz and his co-workers (2013a, 2013b) proposed a new topo-reactive method which combined the chemical information of atom-in-molecule from the electronegativity and chemical hardness with the topological information of compactness given by the Wiener index.

For a given graph $G(N)$ having N atoms, the Wiener index $W(N)$ is used to measure the average topological compactness (Wiener, 1947) and is defined as the semi-sum of the minimum chemical distance (d_{ij}) with the formula (Putz et al., 2013a):

$$W(N) = \frac{1}{2}\sum_{ij} d_{ij} \quad \text{with} \quad \begin{cases} d_{il} = 0 \\ i, j = 1, 2, \dots N-1, N \end{cases} \tag{1}$$

where d_{ij} represents the distance matrix $N \times N$ of the graph $\hat{D} = \left[d_{ij} \right]$.

Another parameter used in topological studies is the eccentricity (ε_i) and represents the largest distance between a given vertex vi and any other vertex of the graph G with the property that $M = \max\{\varepsilon_i\}$, i.e., the maximum eccentricity is equal to the graph diameter M. In this context, one can define the eccentric connectivity index $\xi(N)$ with the formula (Sharma et al., 1997):

$$\xi(N) = \frac{1}{2}\sum_i b_{il}\varepsilon_i \tag{2}$$

where b_{il} represents the atom v_i bonds number (Putz et al., 2013a).

The Timisoara eccentricity index (TM-EC) was proposed by Putz and Ori (2014) as an alternative to the original eccentricity and had the formula (Putz et al., 2015):

$$TM - EC = \frac{1}{2}\sum_{i=1}^{n} \tau_i \varepsilon_i \delta_i \tag{3}$$

where τ_i represents the number of atoms which are at the ε_i distance from atom *i*.

Other parameters used in planar hexagonal systems like PAHs (polycyclic aromatic hydrocarbons) are the topological efficiency index ρ (4) which was successfully applied in graphenic lattices investigations (Cataldo et al., 2010) and the extreme topological efficiency index ρ^E (5) which was used in topological stability classification of schwarzitic infinite lattices (Schwerdt-feger et al., 2013; De Corato et al., 2012):

$$\rho = W / N\underline{w}, \ \rho \geq 1 \tag{4}$$

$$\rho^E = \underline{\underline{w}} / \underline{w}, \ \rho^E \geq 1 \tag{5}$$

where $\underline{w} = \min\{w_i\}$ and $\underline{\underline{w}} = \max\{w_i\}$ and w_i is the atom i contribution to the overall Wiener index (Putz et al., 2015).

The general topological descriptors are based on the adjacency matrix (entries in binary form) and the distance matrix (entries represents the sums of binary elements). The newly proposed method of colored molecular topology by Putz and his co-workers (Putz et al., 2013b) improve the binary topology for the configurational differences which depends on the topological environment and affects locally the electronegativity and the chemical hardness. The chemical reactivity can be modeled using the density functional theory through the compact finite difference (CFD) of the electronegativity and chemical hardness after the generic formula (Putz et al., 2004, 2013a; Putz, 2010, 2011a, 2011b):

$$\chi_{CFD} = -\left[a_1\left(1-\alpha_1\right)+\frac{1}{2}b_1+\frac{1}{3}c_1\right]\frac{\varepsilon_{HOMO(1)}+\varepsilon_{LUMO(1)}}{2}$$
$$-\left[b_1+\frac{2}{3}c_1-2a_1\left(\alpha_1+\beta_1\right)\right]\frac{\varepsilon_{HOMO(2)}+\varepsilon_{LUMO(2)}}{4} \tag{6}$$
$$-\left(c_1-3\alpha_1\beta_1\right)\frac{\varepsilon_{HOMO(3)}+\varepsilon_{LUMO(3)}}{6}$$

$$\eta_{CFD} = \left[a_2\left(1-\alpha_2+2\beta_2\right)+\frac{1}{4}b_2+\frac{1}{9}c_2\right]\frac{\varepsilon_{HOMO(1)}-\varepsilon_{LUMO(1)}}{2}$$
$$+\left[\frac{1}{2}b_2+\frac{2}{9}c_2-2a_2\left(\beta_2-\alpha_2\right)\right]\frac{\varepsilon_{HOMO(2)}-\varepsilon_{LUMO(2)}}{4} \tag{7}$$
$$+\left[\frac{1}{3}c_2-3\alpha_2\beta_2\right]\frac{\varepsilon_{HOMO(3)}-\varepsilon_{LUMO(3)}}{6}$$

There are nine topological classes with increase complexity which can be obtained for electronegativity and chemical hardness: the compact finite second (2C)-, fourth (4C), and sixth (6C)-order central differences, the standard Pade (SP), the sixth (6T)- and eight (8T)-order tridiagonal schemes, and the eighth (8P)- and tenth (10P)-order pentadiagonal schemes, up to spectral-like resolution (SLR) schemes (Putz et al., 2013b).

In the next step, the newly topo-reactivity index is obtained by coloring its topological matrix with the values of electronegativity nodes by respecting the Timisoara-Parma rule (Putz et al., 2013a), i.e., the reactivity indices χ and η values are assigned starting from the "central" most populated node with bonds and frontier electrons and continued by decreasing the CFD descriptors values until all molecular colored (Putz et al., 2013b). In order to establish a hierarchy on the CFD classes, one need to asses the best correlation model between the π-reactive energy (8) and π-parabolic energy (9):

$$E_\pi\left(\text{molecule}\right) \cong E_{BIND}\left(\text{molecule}\right) + E_{HEAT}\left(\text{molecule}\right) - E_{TOTAL}\left(\text{molecule}\right) \quad (8)$$

$$E_\pi\left(\chi,\eta\right) \cong -\chi N_\pi + \eta N_\pi^2 \quad (9)$$

resulting in the following CDF hierarchy (Putz et al., 2013a):

$$SLR > 10P > C2 > 6T > 8P > C4 > SP > 8T > C6 \quad (10)$$

Using the CFD hierarchy the topological invariants based on the distance matrix were also improved, by coloring the chemical graph G of the PAH molecule being generated a new class of invariants in order to determine the π-parabolic-energy, the electronegativity, and the chemical hardness. The advantage of this new method is given by the fact that two new colors are assigned to each vertex v_i and depict a portion of the overall molecular χ_i and η_i. This new method can also lead to obtaining the Wiener reactive forms of both electronegativity and chemical hardness as following:

$$E_\pi^W\left(\chi,\eta\right) \cong -W\left(\chi\right)N_\pi + W\left(\eta\right)N_\pi^2 \quad (11)$$

After several mathematical calculations, one will obtain the new reactive topological descriptors:

$$W\left(\chi\right) = \det{}^{1/N_\pi}\left\|\hat{W}_\chi\right\| \quad (12)$$

$$W\left(\eta\right) = \det{}^{1/N_\pi}\left\|\hat{W}_\eta\right\| \quad (13)$$

For atoms considered in the path $i \rightarrow j$ with a geometric average of $1 + d_{ij}$ colors one will obtain the final formulae of the Wiener-reactive matrix elements for electronegativity (14) and chemical hardness (15):

$$w_{ij}\left(\chi\right) = \left[\hat{W}_\chi\right]_{ij} = \left(\prod_{i \xrightarrow{a} j} \chi_\alpha\right)^{1/\left(1+d_{ij}\right)} \quad (14)$$

$$w_{ij}(\eta) = \left[\hat{W}_\chi\right]_{ij} = \left(\prod_{i \xrightarrow{\alpha} j} \eta_\alpha\right)^{1/(1+d_{ij})} \tag{15}$$

This way is emphasized, once again, the important role of the molecular topological information to predict the molecule chemical-physical properties (Putz et al., 2013b).

41.4 NANOCHEMICAL APPLICATION(S)

Several PAHs structures were analyzed by Putz and his co-workers (2013b) using the electro-topological coloring algorithm (Figure 41.1). In the beginning, the first three HOMOs and LUMOs values were computed for each molecule of the working set, followed by the evaluation of the electronegativity and chemical hardness compact finite difference (Putz et al., 2013b). Next, the CFD hierarchy was applied to the PAHs molecules with respecting the resulted parabolic energies with pi-energies best fitting. The topological descriptors were colored with reactivity contents and the eccentric connectivity (ξ), topological (ρ) and extreme topological (ρ^E) efficiency and the parabolic-reactive counterpart of the Wiener index were computed for a given coloring scheme. Also, the molecular weight (MW), retention indices (RI) and structural-experimental octano-water partition coefficients values ($\log K_{OW}$) were computed (Putz et al., 2013b).

As a result, the authors obtained a good correlation between the reactive colored Wiener index (WEp) and the classical Wiener index (W). Then the experimental properties are considered (i.e., MW, RI, and $\log K_{OW}$), WEp correlates above W performances. However, it appears that WEp correlates below W when the structural properties are considered (i.e., $\xi(N)$, ρ, and ρ^E) in studies (Putz et al., 2013b).

Even if the topologic-reactivity coloring scheme is able to provide more structure-stability information about physical chemical working indices, it can be further improved by creating connections with experimental data such as carcinogenicity, ecotoxicity or chemical activity with applications in modeling the chemical ligand systems (Putz et al., 2013b).

In the same context, using the size-independent topological efficiency index on a set of PAHs molecules, Putz and his collaborators (2015) determined that the molecular reactivity increase can be associated with the molecular conformation. Results obtained for Wiener index, eccentric connectivity index and Timisoara eccentricity index on the working set, determine that naphto[o]5-helicene is the most reactive one, while the less reactive one appears to be pyrene (Putz et al., 2015).

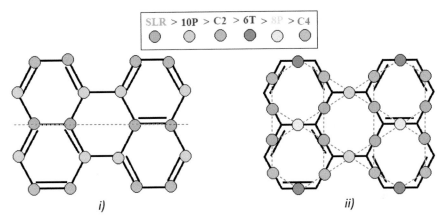

FIGURE 41.1 Perylene molecule: (i) carbon atoms are colored by SRL parameters belonging to the molecules central axis; (ii) colored with Timisoara-Parma rule respecting the horizontal-middle/most "dense" symmetrical axis. Redrawn and adapted after (i) Putz et al. (2013a); and (ii) Putz et al. (2013b).

41.5 MULTI-/TRANS-DISCIPLINARY CONNECTION(S)

Developing a new method of quantitative estimation of lipophilicity parameter has become a necessity due to its importance in studies of medical chemistry and prediction of chemical toxic effects (Agrawal et al., 2002, 2005). There are two types of topological indices which were used in studying the organic compounds in order to relate their structure with the biological/chemical properties (Agrawal et al., 2001, 2002, 2003a, 2003b, 2003c, 2003d, 2003e, Agrawal & Khadikar, 2003; Balaban et al., 2004): fundamental indices, such as Wiener (W), Randic ($^m\chi$), Szeged (Sz), Kier and Hall (χ^{mv}) (Wiener, 1947; Khadikar et al., 1995; Das et al., 1997; Randic, 1985; Kier & Hall, 1986, 1999; Kier, 1980), and less fundamental indices, designed to express the molecular shape, polarity and reactivity with applications in modeling physiological activity and physicochemical properties of drugs (Agrawal et al., 2005).

In their work, Agrawal and his co-workers (2005) modeled the lipophilicity (log P) for a set of 233 heterogeneous compounds using a larger set of topological indices based on the distance matrix. They start from the topological conditioning of reactivity, i.e. the principle of molecular structure, which states that the molecules can be represented as isolated objects with a permanent and rigid location of nuclei/atoms, which are interconnected through specific and highly localized electronic forces, also known as chemical bond (Agrawal et al., 2005).

In modeling the lipophilicity (Agrawal et al., 2005), a regression analysis was effectuated, using the maximum R^2-method, and the three parametric regression containing Ip2, Ip3, and $^1\chi$ returned the best results:

$$\log P = 0.4467 + 0.3672(\pm0.0594)\,^1\chi$$
$$+1.4262(\pm0.2380)\,Ip_2 - 1.3032(\pm0.2354)\,Ip_3 \tag{16}$$

with $n = 223$ is the number of compounds, $R = 0.7740$ is the multiple correlation coefficient, $R^2_A = 0.5935$ is the adjusted R square, Se = 1.2187 is the standard error of estimation, $Q = 0.6315$ is the quality factor and $F = 109.053$ is the F statistics.

In order to obtain better results (Agrawal et al., 2005), several parameters were added in this order: Sz index, Ip1, χ_{eq}, J, W, the regression containing now eight parameters, resulting the following model:

$$\log P = 4.8429 + 0.3172(\pm0.5292)\,^1\chi - 0.0027(\pm0.0019)W$$
$$+0.3989(\pm0.1840)J + 0.0020(\pm6.3368\times10^{-4})Sz$$
$$+2.2060(\pm0.5292)\chi_{eq} + 1.0781(\pm0.1625)Ip_1 \tag{17}$$
$$+1.1993(\pm0.2255)Ip_2 - 1.3446(\pm0.1815)Ip_3$$

with $n = 210$, $R^2_A = 0.7496$, Se = 0.9202, $F = 77.171$, $R = 0.8686$, and $Q = 0.9439$.

Results obtained by Agrawal et al., (2005) for a specific set of compounds, proved that an alternative method in determining the logP value for a large set of heterogeneous molecules could be obtained by combining indicator parameters with topological indices based on the distance in the molecular graph along with the equalized electronegativity. In their work, the authors propose a model with eight parameters, the future aim being to reduce the number of those indices at only one (Agrawal et al., 2005).

41.6   OPEN ISSUES

Mutagenic agents (Popelier et al., 2004) are known for their property to induce mutations or genetic changes, due to their dangerous effect being more and more in the scientist attention over the years being developed several analysis methods which aim to be inexpensive and reproducible. In their work, Popelier and his co-workers (2004) propose a new method the quantum topological molecular similarity (QTMS) in order to test two sets of mutagenic compounds, i.e. halogenated hydroxyquinones (Tuppurainen & Lotjonen, 1993) and dimethyl heteroaromatic triazenes (Shusterman et al., 1989; Popelier et al., 2004). QTMS represents a method used to predict the compounds activity and is based on comparing and differentiating molecules using quantum chemical topology (QCT) descriptors. Being developed in the same context as the QSAR/QSPR methods (Popelier, 1999), QTMS delivered novel QSARs in ecological context, such as toxicity of polychlorinated dibenzo-p-dioxins (PCDDs), ^{13}C NMR chemical shifts in m- and p-substituted benzonitriles along with biodegradability and toxicity of phenols or hydrolysis rate constants prediction for polar esters (O'Brien & Popelier, 2000; Popelier et al., 2002; Chaudry & Popelier, 2003), and in medicinal context, such as the steroid binding activity of CBG, the anti-tumor activity of nitrofuran derivatives and (E)-1-phenyl-but-3-en-ones (O'Brien & Popelier, 2002; Smith & Popelier, 2004). Applied to hydroxyfuranones, QMTS is used to resolve the mechanistic ambivalence between a one-electron transfer process and the nucleophilic addition to the unsaturated ring, while for triazenes case, QMTS can be used in resolving the mechanistic issue in choosing in favor of one between two proposed alternative pathways (Popelier et al., 2004). The advantage of using QTMS is given by the fact that it selects the bond which is involved directly in the (re)activity, being able to generate bond descriptors starting from the geometry-optimized *ab initio* wave functions, and can be successfully used to resolve mechanistic ambiguities due to its ability to highlight the mutagens active center. However, QSTM is not able to generate steric descriptors or to predict log P values, meaning that there are still several improvements which need to be made on this method (Popelier et al., 2004).

ACKNOWLEDGMENT

This contribution is part of the Nucleus-Programme under the project "Deca-Nano-Graphenic Semiconductor: From Photoactive Structure to The Integrated Quantum Information Transport," PN-18-36-02-01/2018, funded by the Romanian National Authority for Scientific Research and Innovation (ANCSI).

KEYWORDS

- **chemical hardness**
- **coloring molecular topology**
- **electronegativity**
- **Timisoara-Excentricity index**
- **Timisoara-Parma rule**
- **Wiener index**

REFERENCES AND FURTHER READING

Agrawal, V. K., & Khadikar, P. V., (2003). Modeling of carbonic anhydrase inhibitory activity of sulfonamides using molecular negentropy. *Bioorganic & Medicinal Chemistry Letters, 13*(3), 447–453.

Agrawal, V. K., Bano, S., & Khadikar, P. V., (2003a). Topological approach to quantifying molecular lipophilicity of heterogeneous set of organic compounds. *Bioorganic & Medicinal Chemistry, 11*(18), 4039–4047.

Agrawal, V. K., Chaturvedi, S., Khadikar, P. V., & Abraham, M. H., (2003b). QSAR Study on tadpole narcosis. *Bioorganic & Medicinal Chemistry, 11*(20), 4523–4533.

Agrawal, V. K., Gupta, M., Singh, J., & Khadikar, P. V., (2005). A novel method of estimation of lipophilicity using distance-based topological indices: Dominating role of equalized electronegativity. *Bioorganic & Medicinal Chemistry, 13*, 2109–2120.

Agrawal, V. K., Karmarkar, S., Khadikar, P. V., Shrivastva, S., & Lukovits, I., (2003c). Use of distance-based topological indices in modeling antihypertensive activity: Case of 2-aryl-imino-imidazolidines. *Indian Journal of Chemistry-Section A., 42A*, 1426–1435.

Agrawal, V. K., Mishra, K., & Khadikar, P. V., (2003e). Multivariate analysis for modeling some antibacterial agents. *Oxidation Communications, 26*, 14–21.

Agrawal, V. K., Sharma, R., & Khadikar, P. V., (2002). Quantitative structure–Activity relationship studies on 5-phenyl–3-ureido–1,5-benzodiazepine as cholecystokinin-A receptor antagonists. *Bioorganic & Medicinal Chemistry, 10*(11), 3571–3581.

Agrawal, V. K., Singh, J., & Khadikar, P. V., (2002). On the topological evidences for modeling lipophilicity. *Bioorganic & Medicinal Chemistry, 10*(12), 3981–3996.

Agrawal, V. K., Sohgaura, R., & Khadikar, P. V., (2001). QSAR study on inhibition of brain 3-hydroxy-anthranilic acid dioxygenase (3-HAO): A molecular connectivity approach. *Bioorganic & Medicinal Chemistry, 9*(12), 3295–3299.

Balaban, A. T., Basak, S. C., Beteringhe, A., Mills, D., & Supuran, C. T., (2004). QSAR study using topological indices for inhibition of carbonic anhydrase II by sulfanilamides and Schiff bases. *Molecular Diversity, 8*(4), 401–412.

Cataldo, F., Ori, O., & Iglesias-Groth, S., (2010). Topological lattice descriptors of graphene sheets with fullerene-like nanostructures. *Molecular Simulation, 36*, 341–353.

Chaudry, U. A., & Popelier, P. L. A., (2003). Ester hydrolysis rate constant prediction from quantum topological molecular similarity descriptors. *Journal of Physical Chemistry A., 107*(22), 4578–4582.

Corey, E. J., (1971). Centenary lecture, Computer-assisted analysis of complex synthetic problems. *Quarterly Reviews, Chemical Society, 25,* 455–482.

Corey, E., (1989). *The Logic of Chemical Synthesis.* Wiley, New York.

Das, A., Dömötör, G., Gutman, I., Joshi, S., Karmarkar, S., Khaddar, D., et al., (1997). A comparative study of the Wiener, Schultz and Szeged indices of cycloalkanes. *Journal of the Serbian Chemical Society, 62*(3), 235–239.

De Corato, M., Benedek, G., Ori, O., & Putz, M. V., (2012). Topological study of schwarzitic junctions in 1D lattices. *International Journal of Chemical Modeling, 4,* 105–113.

Estrada, E., & Molina, E., (2001). 3D connectivity indices in QSPR/QSAR Studies. *Journal of Chemical Information and Computer Sciences, 41*(3), 791–797.

Garcia-Domenech, R., Galvez, J., De Julian-Ortiz, J. V., & Pogliani, L., (2008). Some new trends in chemical graph theory. *Chemical Reviews, 108,* 1127–1169.

Khadikar, P. V., Deshpandey, N. V., Kale, P. P., Dubrynin, A., Gutman, I., & Domotor, G., (1995). The Szeged index and an analogy with the Wiener Index. *Journal of Chemical Information and Computer Sciences, 353*(3), 547–550.

Khadikar, P. V., Srivastava, A., Agrawal, V. K., & Shrivastava, S., (2003d). Topological designing of 4-piperazinyl quinazoline as antagonists of PDGFR tyrosine kinase family. *Bioorganic & Medicinal Chemistry Letters, 13*(18), 3009–3018.

Kier, L. B., & Hall, L. H., (1986). *Molecular Connectivity in Structure-Activity Relationship.* Wiley, New York.

Kier, L. B., & Hall, L. H., (1999). *Molecular Structure Description.* Academic, New York.

Kier, L. B., (1980). Use of molecular negentropy to encode structure governing biological activity. *Journal of Pharmaceutical Sciences, 69*(7), 807–810.

Liu, S. S., Yin, C. S., Li, Z. L., & Cai, S. X., (2001). QSAR study of steroid benchmark and dipeptides based on MEDV–13. *Journal of Chemical Information and Computer Sciences, 41*(2), 321–329.

Mishra, R. K., & Galvez, J., (2001). Getting discriminant functions of antibacterial activity from physicochemical and topological parameters. *Journal of Chemical Information and Computer Sciences, 41*(2), 387–393.

O'Brien, S. E., & Popelier, P. L. A., (2000). Quantum molecular similarity: Use of atoms in molecules derived quantities as QSAR variables. In: *European Congress on Computational Methods in Applied Sciences and Engineering (ECCOMAS),* 11-14 September, Barcelona, Spain.

O'Brien, S. E., & Popelier, P. L. A., (2002). Quantum topological molecular similarity. Part 4. A QSAR study of cell growth inhibitory properties of substituted (E)–1-phenylbut–1-en-3-ones. *Journal of the Chemical Society, Perkin Transactions, 23,* 478–483.

Popelier, P. L. A., (1999). Quantum molecular similarity. 1. BCP space. *Journal of Physical Chemistry A., 103*(15), 2883–2890.

Popelier, P. L. A., Chaudry, U., & Smith, P. J., (2002). Quantum topological molecular similarity. Part 5. Further development with an application to the toxicity of polychlorinated dibenzo-p-dioxins (PCDDs). *Journal of the Chemical Society, Perkin Transactions, 27,* 1231–1237.

Popelier, P. L. A., Smith, P. J., & Chaudry, U. A., (2004). Quantitative structure-activity relationships of mutagenic activity from quantum topological descriptors: Triazenes and halogenated hydroxyfuranones (mutagen-X) derivatives. *Journal of Computer-Aided Molecular Design, 18*, 709–718.

Putz, M. V., & Ori, O., (2014). *Timioara Eccentricity Index (TM-EC), Personal Communication*. West University of Timisoara: Timisoara.

Putz, M. V., (2010). On absolute aromaticity within electronegativity and chemical hardness reactivity pictures. *MATCH Communications in Mathematical and in Computer Chemistry, 64*, 391–418.

Putz, M. V., (2011a). Electronegativity and chemical hardness: Different patterns in quantum chemistry. *Current Physical Chemistry, 1*, 111–139.

Putz, M. V., (2011b). In: Putz, M. V., (ed.), *Quantum Frontiers of Atoms and Molecules* (pp. 251–270). NOVA Science Publishers, Inc., New York.

Putz, M. V., Ori, O., Cataldo, F., & Putz, A. M., (2013a). Parabolic reactivity "coloring" molecular topology: Application to carcinogenic PAHs. *Current Organic Chemistry, 17*, 2816–2830.

Putz, M. V., Ori, O., De Corato, M., Putz, A. M., Benedek, G., Cataldo, F., & Graovac, A., (2013b). Introducing "colored" molecular topology by reactivity indices of electronegativity and chemical hardness. In: Cataldo, F., Iranmanesh, A., & Ori, O., (eds.), *Topological Modeling of Nanostructures and Extended Systems, Carbon Materials: Chemistry and Physics* (Vol. 7, pp. 265–286). Springer Netherlands, Chapter 9.

Putz, M. V., Russo, N., & Sicilia, E., (2004). On the application of the HSAB principle through the use of improved computational schemes for chemical hardness evaluation. *Journal of Computational Chemistry, 25*, 994–1003.

Putz, M. V., Tudoran, M. A., & Ori, O., (2015). Topological organic chemistry: From distance matrix to Timisoara eccentricity. *Current Organic Chemistry, 19*, 249–273.

Randic, M., (1985). Graph theoretical approach to structure–activity studies: Search for optimal antitumor compounds. In: Alan, R., (ed.), *Molecular Basis of Cancer Part A, Macromolecular Structure* (pp. 309–318). Carcinogens and Oncogens, NC: New York.

Randic, M., (2003). Chemical graph theory–facts and fiction. *Indian Journal of Chemistry, 42A*, 1207–1218.

Schmalz, T. G., Klein, D. J., & Sandleback, B. L., (1992). Chemical graph-theoretical cluster expansion and diamagnetic susceptibility. *Journal of Chemical Information and Modeling, 32*(1), 54–57.

Schwerdtfeger, P., Wirz, L., & Avery, J., (2013). *Program FULLERENE - A Fortran/C++ Program for Creating Fullerene Structures and for Performing Topological Analyses (Version 4.4)*. Massey University Albany, Auckland, New Zealand.

Sharma, V., Goswami, R., & Madan, A. K., (1997). Eccentric connectivity index: A novel highly discriminating topological descriptor for structure-property and structure-activity studies. *Journal of Chemical Information and Computer Sciences, 37*, 273–282.

Shusterman, A. J., Debnath, A. K., Hansch, C., Gregory, W. H., Frank, R. F., Greene, A. C., & Watkins, S. F., (1989). Mutagenicity of dimethyl heteroaromatic triazenes in the Ames test: The role of hydrophobicity and electronic effects. *Molecular Pharmacology, 36*(6), 939–944.

Smith, P. J., & Popelier, P. L. A., (2004). Quantitative structure–activity relationships from optimized Ab initio bond lengths: Steroid binding affinity and antibacterial activity of nitrofuran derivatives. *Journal of Computer-Aided Molecular Design, 18*(2), 135–143.

Tuppurainen, K., & Lotjonen, S., (1993). On the mutagenicity of MX compounds. *Mutation Research, 287*(2), 235–241.

Wiener, H., (1947). Structural determination of paraffin boiling points. *Journal of the American Chemical Society, 69*(1), 17–20.

Zefirov, N. S., & Gordeeva, E. V., (1987). Computer-assisted Synthesis. *Uspekhi Khimii, 56*(10), 1753–1773.

CHAPTER 42

Unbranched Chain Reactions in Solids and in Viscous Media: Nontraditional Methods and Analysis

A. M. KAPLAN[1] and N. I. CHEKUNAEV[2]

[1]*N.N. Semenov's Institute of Chemical Physics, Russian Academy of Sciences, Russian Federation, E-mail: amkaplan@mail.ru*

[2]*Kosygin Street 4, Moscow 119991, Russian Federation*

42.1 DEFINITION

On the example of the radiation-induced post-polymerization reaction of *crystalline acrylonitrile* (AN), it was first discovered the effect of continuation of the chain reaction after its termination at a specific temperature T_1. This effect was not consistent with traditional views. The continuation of such reaction was observed at the temperature $T_1 + 15K$. This effect was called the *phenomenon of "congealing" and "reanimation" of polymer chains in the solid state* (phenomenon). Later, this phenomenon was observed repeatedly in the study of solid-phase polymerization of other monomers and *unbranched chain reactions in solids* (UChSR).

An original kinetic model was proposed, which could explain the marked phenomenon and other nontrivial kinetic features of UChSR.

In 1975, an unconventional method of implementing an effective radiation-induced radical post-polymerization was proposed – the method consisted in slowly heating of an γ-irradiated at 77K monomer without any additives from the temperature $T = T_g - (10 \div 15)K$ to $T = 300K$. T_g is the glass transition temperature of the studied monomer. It was shown that the proposed method provided the conditions for the continued growth of the polymer chains in the absence of their termination. Thus, on the example of post-polymerization of *butyl methacrylate* (BMA) was first discovered *living radical polymerization* (LRP) at temperatures $T > T_g$ to produce polymers with improved properties.

42.2　HISTORICAL ORIGIN AND SCIENTIFIC DEVELOPMENT

The introduction in the chemical science of the authors' concept has been dictated by the necessity of an explanation of polymer chains "congealing" and "reanimation" during the AN *post-polymerization* (PP) (Kaplan Anatoly, 1996).

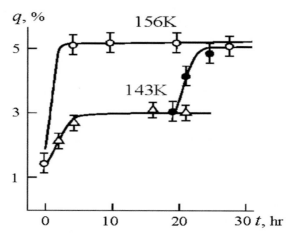

FIGURE 42.1　Kinetics of post-radiation polymerization of AN at $T = 143K$ and $T = 156K$. The points (●) mark the polymer yield (q) in AN specimens placed into thermostat with $T = 156K$ after holding the specimens during 20 h at $T = 143K$.

On a proposal by Kaplan Anatoly an original experiment was carried out: γ-irradiated at 77K AN sample was initially kept at $T_1 = 143K$ for 20 hours, which is enough time to reach the limit of the polymer yield ($q = 3.0 \pm 0.4\%$). Then, after a fast transfer of the AN sample from the thermostat with $T_1 = 143K$ into thermostat with $T_2 = 156K$ researchers continued the process observation. Polymerization at $T = 156K$ was resumed (see dark points (●) in Figure 42.1). In this process, there was an increase of the polymer yield up to the new limit ($q = 5.0 \pm 0.4\%$). According to traditional views reaching the limit yield of AN PP ($q = 3.0 \pm 0.4\%$) at 143K should be considered as a result of the doom of polymerization active centers. But without the active centers, the chain polymerization is not possible (see Gimblett, 1970).

　　Several approaches to the qualitative explanation of the mentioned phenomenon are discussed in Chapter 23.

The most clear explanation of the phenomenon and others *non-trivial kinetic particularities* (NTKP) of unbranched chain reactions in solids and viscous media on the quantitative level was given in 2008. This became possible taking into account the formation of *friable zones* (FZs) near nano-defects with an excess of free volume in solids. It was shown that only in FZs the steric hindrances to the chemical interaction of CAC with neighboring particles reagents are removed.

Proper accounting of the concentration evolution of mentioned FZs in the studied systems allowed to explain the nontrivial kinetic features of UChSR on the example of the *solid phase polymerization* (SP) of different monomers (Kaplan & Chekunaev, 2012).

42.3 NANOCHEMICAL APPLICATION

The inclusion of this chapter was justified with the presentation of the active role of nanoparticles [mobile defects (MD) and living active centers (LAC)] in the UChSR implementation.

The definition of the LAC and main LAC actions can be found in Chapter 23 of this book.

42.3.1 *KINETIC SCHEME OF SOLID-STATE POLYMERIZATION IN MICRO-HETEROGENEOUS SOLID MONOMER*

The authors proposed a model describing the kinetics of SP in the form of the original system of equations. The detailed analysis of the application of this system of equations to represent the kinetic features of SP can be found in the works of Kaplan & Chekunaev (2009, 2012).

Below is given a brief kinetic analysis of such a system in the study of SP at a constant temperature.

Equation (1) of this system describes the evolution of the PP rate.

$$-\frac{dC_M(t)}{dt} = k_P^{A*} C_{A*}(t) C_M(t) \tag{1}$$

Here: $k_P^{A*} = k_0^{A*} \cdot \exp(-E_p/RT)$ is a *growth rate constant* (GRC) of polymer chains on *living active centers*, A*; $C_A(t)$, $C_{A*}(t)$, and $C_M(t)$ – current concentrations of CACs, LACs, and monomer. GRC of CAC is equal $k_P^A = 0$.

LAC(A*) is CAC localized in the "friable" zone of a structural defect that has excess free volume.

As a rule, the concentrations ratio is $C_{A*}(t)/C_A(0) < 3 \times 10^{-3}$. From experimental data (Chachaty & Forchioni, 1972), Kaplan (1996) and Kaplan & Chekunaev (2009) follows that the effect of "congealing" of polymer chains in the course of SP occurs at almost constant concentration of CAC: $(1-C_A(t)/C_A(0) < 0.1)$. On this basis, to simplify computation, in the equation (5) the concentrations of CAC were assumed equal $C_A(0)$.

More information about the difference between traditional and living active centers can be found in Chapter 23.

To analyze the kinetics of SP in micro-heterogeneous solid monomers, the following equations were used:

$$f(E_i, t = 0) = B\exp[(-E_i + E_{min})/\Delta)] \tag{2}$$

$$dC_D^i(E_i, t = 0) = C_D(0)f(E_i, t = 0)dE \tag{3}$$

$$\frac{dC_D^i(E_i, t)}{dt} = -\frac{k_i}{1+\beta_1}C_D^i(E_i, t)\sum_l C_D^l(E_i, t) \tag{4}$$

$$\frac{dC_{A*}^i(E_i, t)}{dt} = \frac{k_i^+}{1+\beta_2}C_A(0)C_D^i(E_i, t) - k_t C_{A*}^i(E_i, t)C_M(t) \tag{5}$$

$$C_{A*}(t) = \sum_i C_{A*}^i(E_i, t) \tag{6}$$

Following Emanuel & Buchachenko (1987), the authors took into accounting in Eqs. (2)–(6) the presence in the studied system of defects (Vac) of different mobility. Vacancies (Vac) were chosen for the analysis because their concentration by several orders exceeds the concentration of other defects in solids.

The initial distribution function of vacancies depending on their diffusion activation energy (E_i) was introduced in Eq. (2) from the set of experimental data specified in Alpha & Alexander, (1972) and Nikolay & Anatoly, (1987). Equations (3) and (4) allow us to calculate the kinetics of a doom of certain Vac groups with different values of *diffusion activation energy* (DAE) $E_i = E_{min} + (E_{max}-E_{min}) \cdot i/n$. In the calculations, the authors usually had taken the value of $n = 15$ and i from 0 to n.

The rate constant of the Vac doom with DAE E_i with good accuracy equals $k_i \approx 4\pi\lambda D_0\exp(-E_i/RT)$ (Kaplan & Chekunaev, 2009). Here λ is the intermolecular distance, parameter D_0 (pre-exponential factor of the diffusion coefficient) was taken constantly for all vacancies in the system. The difference in the mobility of different Vac groups the authors attributed in accordance with Kaplan (1996) with various values of DAE (E_i) of such groups.

The first term on the right-hand side of Eq. (5) describes the evolution of that part of the LAC (with concentration $C^i_{A*}(E_i,t)$), which is generated at the approach of the vacancies portion with concentration $C^i_D(E_i,t)$ (DAE of which is equal to E_i) with CAC having concentration $C_A(0)$. Justification for the use in Eq. (5) $C_A(0)$ instead of $C_A(t)$ is given above. The second term in the right side of this equation describes the doom of the mentioned portion of the LAC mainly by the mechanism of "chemical diffusion," described in details in the authors' articles (Anatoly & Nicholas, 2009) and Chapter 23 of this book. Equation (6) allows you to calculate the concentration of all LAC, generated by all mobile vacancies at a constant temperature.

42.3.2. "CONGEALING" AND "REANIMATION" OF KINETIC CHAINS IN SOLID-STATE UNBRANCHED CHAIN REACTIONS

For the calculation of SP kinetics with the use of Eqs. (1)–(6), the following parameters of the process are necessary: the initial concentrations of LAC ($C_{A*}(0)$) and MD ($C_D(0)$), GRC of LAC ($k_P^{A*} = k_0^{A*} \cdot \exp(-E_p/RT)$), DAEs of Vac (from E_{min} to E_{max}), the defects scattering coefficients β_1 and β_2, the size of FZ in the matrix near Vac ($r = m\lambda$, $m = 2.5 \div 3$, $\lambda = 0.45 \div 0.5$ nm). In this chapter, the possibilities to describe NTKPs of SP are demonstrated by solid-phase post-polymerization of AN. The numerical values of the above-mentioned parameters for a specified object can be found in Anatoly & Nicholas (2009). Figure 42.2 shows the results of computation using the Eqs. (1)–(6) of the post-polymerization kinetics of γ-irradiated at 77K AN samples at different methods of their heating. More importantly, according to the author's model "reanimation" of AN polymer chains at $T = 156$K is a result of the significant increase of the mobility vacancies at $T = 156$K, which were immobile at $T = 143$RK and could not participate in converting CAC in LAC at 143K (see Eq. (5)).

Comparison of experimental data of the post-polymerization kinetics of AN in Figure 42.1 with results of the calculation in Figure 42.2 shows the physically grounded and clear explanation via the authors' model of the "congealing" and "reanimation" of polymer chains phenomenon.

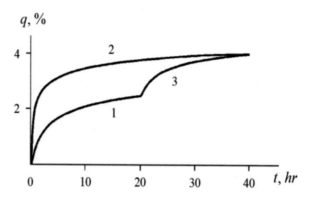

FIGURE 42.2 The calculated curves (1, 2, 3) of post-polymerization kinetics in γ-irradiated at 77K samples of AN. 1) $T = 143$K; 2) $T = 156$K; 3) at $T = 156$K after 20 h of AN sample storage at $T = 143$K.

This phenomenon was also observed in the study of the chain radiation-induced low-temperature reaction of ethylene hydrobromination (Dmitry & Igor, 1979). Figure 42.3 shows the variation of the rate of γ-induced hydrobromination of ethylene vs. time at 50K.

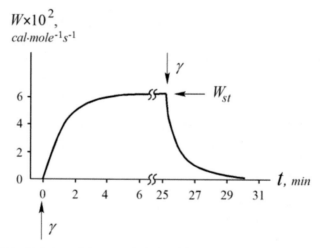

FIGURE 42.3 Variation of the rate of γ-induced hydrobromination of ethylene vs. time at 50K. Arrows indicate the switching of Co^{60} γ-radiation on and off. Dose rate $I = 25$ Rad/sec. Reprinted with permission from Kiryukhin et al., 1979.

Two effects of this reaction contradict to traditional views. First, there is a linear dependence of the rate of this reaction on the intensity (I) γ-irradiation of the studied system. Second, after the cessation of exposure at 50K the rate of the process falls to zero. However, at higher temperatures, there is a revival of the chain process out of the zone of radiation exposure. Both of these effects were well explained by the use of the authors' model. When considering the process which happens during radiation exposure on the analyzed system Eq. (5) is modified in Eq. (7).

$$\frac{dC^i_{A*}(E_i,t)}{dt} = gI \frac{C^i_D(0)}{C_M(t)} - k_t C^i_{A*}(E_i,t)C_M(t) + k^+_i gIt C^i_D(E_i,t) \tag{7}$$

In the right side of Eq. (7), there is an additional first term. It describes the rate of formation of loose zones, which is equal to the rate of LAC (A*) production. Here $C^i_D(0)$ is the initial concentration of defects with excessive volume. At relatively small exposure times, indicated in Figure 42.3, the dose of radiation exposure $D_{rad} = 6 \times 10^{-3}$ Mrad so small that at such dose concentration of CAC increases linearly. And at such low doses, the last term in Eq. (7) is substantially smaller than the first one. This follows from the data on the DAE reagents in solids $E_i > 60$ kJ/mole and with the typical ratio $C^i_D(0)/C_M > 10^{-10}$ in unannealed solids (see Nikolay & Anatoly, 1987).

When the noted above condition is a satisfied value (A*) and the reaction rate (V ~ A*) will first grow from zero to achieve of steady-state values of these characteristics (A*$_{st}$ and V*$_{st}$). When you turn off the irradiation, the second term in the right side of Eq. (7) leads to the doom of LAC (A*) and to decline of reaction rate V (within experimental error) to zero. After that, the system will retain the concentration of CAC $A = gIt_0$, where $t_0 = 25$ min (see Figure 42.3 and Eq. (7)). When the temperature of the test sample increases from 50K to 90K the rate constant for the interaction of the CAC with vacancies $k^+_i = 4\pi(m+0.5)\lambda D_0 \exp(E_i/RT)$ will increase by many orders of magnitude (because the value of E > 60 kJ/mole). This will lead to increase the rate of growth of A* concentration (see the last term on the right side of Eq. (7)) and it will provide "reanimation" reaction.

42.3.3 NONTRADITIONAL METHOD OF LIVING RADICAL POLYMERIZATION

Living radical polymerization (LRP) for the first time was performed in 1972 on the example of post-polymerization of *n-butyl methacrylate* (BMA) in admixture with $ZnCl_2$. This effect was discovered by Kaplan Anatoly together with the researchers group of the Chemistry Department of Lomonosov Moscow State University (MGU) (see Zaremski & Golubev, 2001).

The rate of BMA polymerization was measured directly during this process at low temperatures (90–250K) via developed by Kaplan Anatoly calorimeter. The conversion of BMA in polyBMA was near 100% on heating of γ-irradiated at 77 mixture of BMA with $ZnCl_2$ at temperatures above the glass transition temperature of the mixture, when the heating rate of investigated system was $v = 1°C/min$. ESR investigations, conducted under the guidance of MGU Professor Golubev Vladimir have shown that most of the polymer is formed at a constant concentration of active centers (radicals). Thus, the use of the combination of low-temperature calorimetry with ESR method has allowed for the first time to observe the process of "living radical polymerization." Get the polymer by similar manner using pure BIA failed.

However, as indicated in Michael & Vladimir (2001), Kaplan Anatoly was first who has carried out BMA post-polymerization with conversion up to 80–100% at a constant concentration of radicals in pure (without impurities) monomer. To do this, it was necessary to reduce the rate of heating of γ-irradiated BMA to the value $v = 1°C$ per hr. Thus, the possibility of obtaining polymers by mechanism "living radical polymerization" in pure monomer was demonstrated. From this method, it is obtained that the polymer product has less dispersion of polymer chains ($M_w/M_n \approx 1.4$–1.6) compared with the products of conventional radical polymerization. This leads to better performance of the polymers obtained by LRP (Zaremski & Golubev, 2001).

KEYWORDS

- **classic and living active centers**
- **congealing**
- **mobile and immobile nanodefects**
- **reanimation**

REFERENCES

Abkin, A. D., Scheinker, A. P., & Gerasimov, G. N., (1973). Chapter 1. Radiation polymerization. In: Kargin, V. A., (ed.), *Radiation Chemistry of Polymers* (pp. 7–107). Publishing house "Science." Moscow.

Chachaty, C., & Forchioni, A., (1972). NMR and ESR studies of the solid state polymerization of vinyl monomers. *J. Polymer Sci., Part A–1, 10*, 1905.

Emanuel, N. M., & Buchachenko, A. L., (1987). *Chemical Physics of Polymer Degradation and Stabilization (New Concepts in Polymer Science)* (p. 336). Publisher: VSP International Science Publishers.

Gimblett, F. G. R., (1970). *Introduction to the Kinetics of Chemical Unbranched Chain Reactions* (p. 199). Published: London, New York, McGraw-Hill, Description.

Kaplan, A. M., (1996). Thesis for the degree of doctor of chemical sciences. "Peculiarities of radiation and mechanical-stimulated processes in nonequilibrium condensed systems." *Inst. Chem. Phys. RAS, Moscow, 14*(3).

Kaplan, A. M., & Chekunaev, N. I., (2012). About the role played by mobile nanovoids in the stimulation of solid-state processes. *Russian Journal of Physical Chemistry B, 6*(3), 407–415. © Pleiades Publishing, Ltd.

Kaplan, A. M., & Chekunaev, N. I., (2009). Model of living active centers in kinetics of the solid-phase polymerization. *Russian Chemical Bulletin, International Edition, 58*(8), 1616–1622.

Kiryukhin, D. P., Barkalov, I. M., & Goldanskii, V. I., (1979). Low-temperature limit of the rate of propagation of chains in the hydrobromination of ethylene. *J. de Chimie Physique, 76*(11/12), 1013–1015.

Zaremski, M., & Golubev, V., (2001). Reversible inhibition in radical polymerization. *Polymer Science–Series C., 43*(1), 81–116.

CHAPTER 43

Vapor Phase Epitaxy

GIOVANNI ATTOLINI, MATTEO BOSI, and MARCO NEGRI

*IMEM-CNR Institute, Parco Area delle Scienze, 37A 43124 PARMA, Italy,
E-mail: giovanni.attolini@imem.cnr.it*

43.1 DEFINITION

Vapor phase epitaxy is a process where a layer deposition occurs on a substrate, reproducing the same crystalline structure. In this technique, the nutrient phase is gaseous.

Generally, the deposition needs that in the epitaxial reactor/growth chamber both homogeneous chemical reactions (with compounds in the same phase) and heterogeneous reactions (with compounds in different phases: solid and gaseous) occur.

43.2 HISTORICAL ORIGIN(S)

The vapor phase epitaxy has been mainly developed to prepare the main semiconductor compounds of the III–V, II–VI, IV–VI, IV, and group-III nitrides.

However, many other materials can be obtained using this technique, and several processes have been studied to deposit semiconductor compounds, for III-V alloys solid solutions the first techniques where chloride–VPE and Hydride–VPE.

The terminology is due to the precursor of the element of the V group introduced into the rector as chloride ($AsCl_3$, PCl_3) or as hydride (AsH_3, PH_3).

The term epitaxy derives from Greek roots with the meaning "to arrange upon." In semiconductor growth, epitaxy means to grow a layer which reproduces the structure of the substrate and generally refers to the growth of single crystalline film over a single crystalline substrate, both having similar lattice parameters.

Vapor phase epitaxy uses a gaseous phase composed of elements of the materials to be obtained.

At a fixed temperature the vapor phase precursors decompose and react at the substrate surface, giving the desired layer.

Generally, the growth rate depends on the reagent concentrations in the gas phase and on temperature. In fact, when the temperature increases the amount of material that can react increases.

43.3   NANO-SCIENTIFIC DEVELOPMENT(S)

Vapor phase epitaxy was initially developed using chloride and hydride compounds, and later improved by the introduction of new precursors and methodologies such as metal-organic compounds and atomic layer epitaxy (ALE)

43.3.1   *CHLORIDE VAPOR PHASE EPITAXY (CL-VPE)*

This process appeared in literature for the first time in 1965 [Effer], for the preparation of gallium arsenide (GaAs). The element of the V-group of the periodic table is in the form of chloride ($AsCl_3$).

The $AsCl_3$ diluted in hydrogen is flowed over the Ga or GaAs source: at high temperature, the arsenic chloride decomposes completely with the reaction $AsCl_3 + 3/2H_2 = 1/4As_4 + 3HCl$.

HCl, as a reaction by-product, reacts with the Ga or GaAs sources to form $GaCl_3$ according to the reaction $GaAs + 3HCl = GaCl_3 + 1/4As_4 + 3/2H_2$. In the end, $GaCl_3$ reacts with arsenic to give GaAs layer.

Phosphorous trichloride (PCl_3) is usually used for the growth of GaP.

The main drawback of this method is the impossibility to change the gas phase composition due to the required saturation/desaturation of the metallic source when PCl_3 or $AsCl_3$ fluxes are changed; the advantage of trichloride VPE is the growth of high purity epilayers (Figure 43.1).

43.3.2   *Hydride Vapor Phase Epitaxy (H-VPE)*

H-VPE was introduced for the first time in 1965 [Tietjen] to prepare structures for the light emitting diode (LED) and laser based on GaAsP. Since then it has been used to prepare a wide range of III–V alloys.

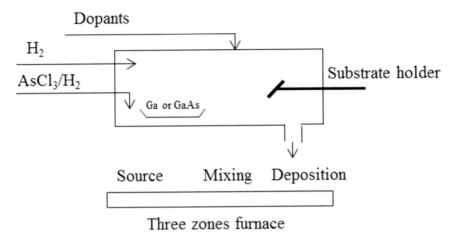

FIGURE 43.1   Chloride vapor phase epitaxy for gallium arsenide.

In this technique, the elements of the III-group are pure metals, while the elements of V-group are in the form of hydrides (AsH_3, PH_3).

The metals are heated at high temperature, ranging from 800-850°C, and a flow of HCl is used to form the III-chlorides. They are transported in the mixing zone thanks to a hydrogen flow (carrier) where the hydrides (AsH_3 and/or PH_3) are introduced.

After mixing all the gaseous species flow to the deposition zone where they react to deposit the desired layer on the substrate.

With this technique, it is possible to obtain a high growth rate, but the disadvantage is the difficulty to control the saturation of metals with HCl flow, because it is necessary a long resident time over their surfaces. For this reason, it is hard to change the composition of the vapor and solid phases, and this makes the preparation of multilayer structures very difficult (Figure 43.2).

43.3.3   METALORGANIC VAPOR PHASE EPITAXY (MOVPE)

MOVPE is based on the principle that metalorganics (MO) and hydrides precursors are thermally cracked on a hot substrate where the deposition takes place.

These precursors are transported in the growth chamber by a carrier gas (usually hydrogen) and reach the substrate.

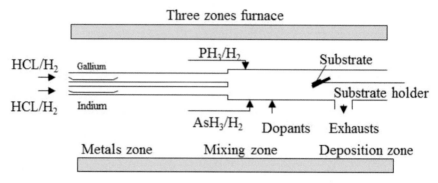

FIGURE 43.2 Hydride vapor phase epitaxy for III-V alloys.

This technique has been used for the first time by Manasevit in 1968 to grow gallium arsenide on sapphire and can be used to grow a large number of compounds (II–VI, III–V and nitrides).

For III-V compounds deposition metalorganics of III-group such as trimethylgallium (($CH_3)_3$Ga. TMGa), trimethylindium (($CH_3)_3$In, TMI), trimethylaluminum (($CH_3)_3$Al, TMA) and hydride of V-group (AsH$_3$, PH$_3$, NH$_3$) are commonly used as precursors. Nowadays, alternative metal organic precursors of arsenic and phosphor are also available, with lower toxicity. Alkyl substituted of arsine and phosphine (tertiary butyl arsine, TBAs; trimethyl arsenic, TMAs; tertiary butyl phosphine, TBP) have low vapor pressure and can be stored and easily handled in comparison to toxic gases.

Dopants are also available in the metalorganic form: diethylzinc (($C_2H_5)_2$Zn), diethylmagnesium (($C_2H_5)_2$Mg), diethyltellurium (($C_2H_5)_2$Te) or gases: hydrogen selenide (H$_2$Se) and silane (SiH$_4$).

In the MOVPE technique, the growth is not in equilibrium, and the growth rate is a function of the arrival time of the precursors on the growth surface rather than the surface reaction between the gas and solid phases, depending upon temperature (Figure 43.3).

43.3.4 ATOMIC LAYER EPITAXY (ALE)

Atomic layer epitaxy was developed by Suntola. In this technique, the precursors are introduced alternately into the growth chamber, followed by a cleaning/purge step with a carrier gas. During the introduction of each precursor, a single layer is deposited (chemisorbed) on the substrate surface.

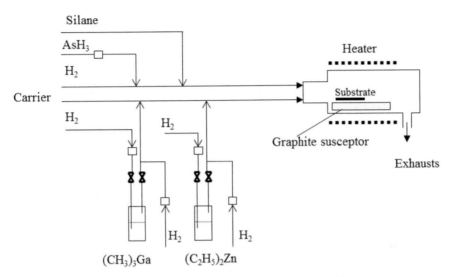

FIGURE 43.3 Metal-organic vapor phase epitaxy for gallium arsenide.

This technique is characterized by single layer deposition of the desired compound after each deposition cycle. With ALE is possible to growth layers with good morphology and crystal structure, but the growth rate is very low.

43.4 NANOCHEMICAL APPLICATION(S)

VPE is widely used to prepare nanomaterials (1D, one-dimensional) such as nanowires, nanoparticles, quantum dots, etc.

Many efforts were focused on using the MOVPE technique. This method is based on the use of chemical elements to form an alloy with the substrate, with the alloy acting as a catalyst to grow the nanostructures.

The nanomaterials offer a large surface area with respect to the mass materials produced in a larger form. They are more chemically reactive, with better electrical and mechanical properties; when their size is below 50 nm quantum effects are observed, giving characteristics (optical, electrical, thermal, etc.) different from the bulk state.

The above characteristics permit to realize novel nanodevices such as field effect transistors, sensors with high performance in many application fields such as medicine and life-science, electronic, optoelectronic, food and agriculture, drug delivery, energy, etc.

43.5   MULTI-/TRANS-DISCIPLINARY CONNECTION(S)

The vapor phase epitaxial growth technique is very complex when it is viewed at the atomic scale.

The complexity is due to the presence in the vapor phase of multicomponent, multi-phases, and to the dynamicity and inhomogeneity of the system.

Many efforts have been devoted to understand what happened in the growth chamber during the process, systematic studies to understand this system from a fundamental point of view involve thermodynamic, fluid dynamic and kinetics of the chemical reactions occurring in the gas phase (homogeneous) and at the surface (heterogeneous).

43.6   OPEN ISSUES

It is not easy to apply the VPE for large productions; however, it offers the possibility of obtaining materials with high purity with good properties.

Another disadvantage is that in general the equipment is complex and for certain compounds, some toxic precursors are required.

Due to their sizes, the main risks of nanomaterials are:

- There is much discussion about the risks for workers and consumers because exposure to nanomaterials can cause adverse respiratory and cardiovascular.
- Environmental risks to release nanomaterials into the environment.
- For the above reasons, it is necessary that the risk evaluation take adequate safety measures.

Nanomaterials are embedded in the final product which are not directly in contact with workers, consumers, and the environment, and so they are safer. But the problem may remain when products are destroyed and discharged.

KEYWORDS

- **epitaxy**
- **semiconductors**
- **vapor phase**

REFERENCES AND FURTHER READING

Effer, D., (1965). Epitaxial growth of doped and pure gas in an open flow system. *J. Electrochem. Soc. 112*, 1020–1025. doi: 10.1149/1.2423334.

Herman, M. A., Richter, W., & Sitter, H., (2004). *Epitaxy: Physical Principles and Technical Implementation.* Springer series in materials science ISBN 978-3-662-07064-2. Springer-Verlag Berlin Heidelberg.

Manasevit, H. M., & Simpson, W. I., (1969). The use of metal-organics in the preparation of semiconductor materials: 1. Epitaxial gallium- v compounds. *J. Electrochem. Soc., 116,* 1725–1732. doi: 10.1149/1.2411685.

Manasevit, H. M., (1968). Single crystal gallium arsenide on insulating gallium arsenide. *Appl. Phys. Lett., 12,* 156–159. doi: 10.1063/1.1651934.

Scott, T. R., King, G., & Wilson, J. M., (1954). U.K. Patent 778.383.8.

Stringfellow, G. B., (1990). *Organometallic Vapor Phase Epitaxy–Theory and Practice.* Academic Press, New York. ISBN–13, 978–0126738421.

Suntola, T., & Auston, J., (1974). Finnish Patent n° 52359, (1974) U.S. Patent n° 4058430.

Tietjen, J. J., & Amick, J. A., (1966). Preparation and properties of vapor-deposited epitaxial $GaAs_{1-x}P_x$ using arsine and phosphine. *J. Electrochem. Soc., 113*, 724–728. doi: 10.1149/1.2424100.

CHAPTER 44

Wiener Index

OTTORINO ORI[1,2,3]

[1]*Actinium Chemical Research, 00133 Rome, Italy*

[2]*Laboratory of Computational and Structural Physical Chemistry for Nanosciences and QSAR, West University of Timişoara, 300115, Timişoara, Romania*

[3]*Laboratory of Renewable Energies-Photovoltaics, National Institute for R&D in Electrochemistry and Condensed Matter INCEMC-Timisoara, 300569 Timisoara, Romania, E-mail: Ottorino.Ori@gmail.com*

44.1 DEFINITION

Wiener index belongs to the set of graph invariants, and it is remembered as the first molecular-graph descriptor applied in 1947 to investigate the correlations between physicochemical properties and the topological structure of the organic compounds. The Wiener index (commonly labeled with W) corresponds to the semisum of the distances between all pairs of vertices in the graph and it provides an average measure of the topological compactness of the molecular structure.

44.2 HISTORICAL ORIGINS

In 1947, Harry Wiener (by publishing his seminal work on "Structural Determination of Paraffin Boiling Points") introduced a molecular descriptor, known nowadays as the Wiener index W, just considering the molecular skeleton made by the sole carbon-carbon bonds of the molecules. In doing this, he established the basal link between molecular graph and physicochemical characteristics of organic compounds. However, this idea did not encounter the immediate favor of the scientific community until the early 70s, when

Hosoya gave the general definition of W. Nevertheless, Wiener retains the merit for having unveiled the profound connection between topology and chemistry. This fruitful concept, not yet fully exploited, originated a new branch of Theoretical Chemistry named Mathematical Chemistry gathering contributions from several teams of scientists worldwide, being worth mentioning the invigorating results coming from East European and Iranian theoreticians.

44.3   NANO-SCIENTIFIC DEVELOPMENTS

The visionary assumption originally made by Wiener (1947) consisted in describing an organic molecule just like a graph, a simple mathematical object made by two sole elements, vertices (the carbon atoms) and edges (the molecular bonds). In this way, molecules are merely represented as networks of bonds. In Figure 44.1 the diagram of naphthalene is given as an example of chemical graph G formed by the sole carbon skeleton. The length, expressed by the number of carbon-carbon bonds connecting atom i and j along the shortest path, gives the chemical distance d_{ij} separating the two atoms. Integers d_{ij} are the basic graph-invariant that enter in the construction of the Wiener index W. In this context, a graph-invariant is a quantity which is not affected by an arbitrary relabeling of the atoms of G. One has $d_{ii} = 0$ and, by symmetry, $d_{ij} = d_{ji}$ (see some examples of d_{ij} in Figure 44.1). For a simple graph G with n vertices W is defined as

$$W(G,n) = \sum_{i<j}^{n} d_{ij} \tag{1}$$

W was the first distance-based topological descriptor appeared on the theoretical chemistry scene. The General definition, given in Eq. (1) (Hosoya, 1971), is applicable to every kind of molecule providing a measure of the topological compactness of such a chemical the structure. W has been successfully applied to investigate several chemical systems (Diudea & Gutman, 1998) with some recent works reported in literature concerning the stability of fullerenes and other carbon nanosystems (Iranmanesh et al., 2012) or the structure-reactivity properties of polycyclic aromatic hydrocarbons (PAHs) (Putz et al., 2013). The $C_{60}(I_h)$ fullerene shows the minimum Wiener index $W = 8340$ among all other C_{60} isomers (Ori & D'Mello, 1992) having stable fullerene cages tendency to minimize W (fullerene stability is treated in the dictionary at the voice (Topological Efficiency Index (TEI)).

Pioneer studies of Bonchev & Mekenyan (1980) indicated an extraordinary property of W making this quantity easy to compute also for infinite systems: when it is computed for a $D = 1$ infinite crystalline structure like a polymer with a carbon-skeleton made by n atoms, index W follows a cubic curve in the large n limit.

$$W(G,n) \cong a_3 n^3 + a_2 n^2 + a_1 n + a_0 \tag{2}$$

Thirty years later, this rule has been generalized (Cataldo et al., 2010; Ori et al., 2010) to include D-dimensional crystal structures for $D \geq 1$, obtaining for W the characteristic asymptotic form consisting of a polynomial of powers of $n^{1/D}$ of grade $2D + 1$

$$\begin{aligned} W(G,n) &\cong a_{2D+1} n^{(2D+1)/D} + a_{2D} n^D + a_{2D-2} n^{(2D-2)/D} + \\ &a_{2D-2} n^{(2D-2)/D} + \ldots + a_1 n^{1/D} + a_0 \end{aligned} \tag{3}$$

Eq. (3) establishes the crucial, asymptotic dependence between W and the number of atoms n forming the nanostructure $W \approx n^s$ with

$$s = 2 + D^{-1} \tag{4}$$

In Eq. (4) dimensionality D is also named Wiener dimensionality d_W (Ori et al., 2010) and it may be extracted from W in several intriguing computational ways as the following

$$d_W^{-1} == Lim_{n \to \infty} \ln_n \frac{W(G,n)}{n^2} \tag{5}$$

A comprehensive study of the Wiener dimensionality d_W and Wiener compression factor f_W, see Eq. (6) below, is still required in order to reach a full understanding of their properties from both sides, theoretical and practical.

In case of $D = 2$, Eq. (4) produces the peculiar dependence $W \approx n^s \approx n^{5/2}$, the topological fingerprint of bidimensional infinite structures. As an example, the asymptotic polynomials for square and honeycomb (graphene) lattices are $W_S = (n^{5/2} - n^{3/2})/3$ and $W_H = (6n^{5/2} - 5n^{3/2} - n^{1/2})/15$ respectively (Cataldo et al., 2011). When periodic closed boundary conditions are imposed on the lattices assumes more compact values indicated by W^C. The compression factor for the Wiener index f_W is an asymptotic invariant defined as

$$f_W = Lim_{n \to \infty} \frac{W^C(G,n)}{W(G,n)} \tag{6}$$

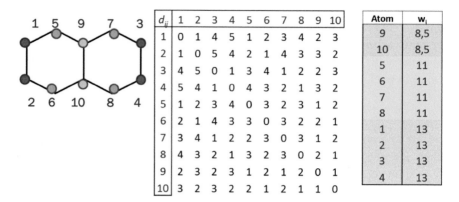

d_{ij}	1	2	3	4	5	6	7	8	9	10
1	0	1	4	5	1	2	3	4	2	3
2	1	0	5	4	2	1	4	3	3	2
3	4	5	0	1	3	4	1	2	2	3
4	5	4	1	0	4	3	2	1	3	2
5	1	2	3	4	0	3	2	3	1	2
6	2	1	4	3	3	0	3	2	2	1
7	3	4	1	2	2	3	0	3	1	2
8	4	3	2	1	3	2	3	0	2	1
9	2	3	2	3	1	2	1	2	0	1
10	3	2	3	2	2	1	2	1	1	0

Atom	w_i
9	8,5
10	8,5
5	11
6	11
7	11
8	11
1	13
2	13
3	13
4	13

FIGURE 44.1   The chemical graph of naphthalene is represented, together with its matrix of the chemical distances d_{ij} and the list of w_i; (see Eq. (8) for each atom: minimal (maximal) nodes are represented in green (red) corresponding to the less (more) reactive atoms in the molecule.

that corresponds to the gain in topological compactness in passing from the open to the closed lattice. It has many interesting properties the basic being the following intriguing inequality

$$f_W \leq \frac{3}{4} \tag{7}$$

The universal behavior expressed by Eq. (7) is conjectured to be valid for any pair G and G^C of open/closed graphs, irrespectively from graph complexity or dimensionality. In particular, it has been shown that the $f_W = 0.75$ upper limit strictly holds for any kind of $D = 1$ lattices and any D-dimensional cubic lattices. Whatever is its fine topological structure, a D-dimensional torus is an at least 25% more compact than the respective open structure. No other general properties regarding f_W have been derived so far to confirm or dismantle this conjecture. Compression factors for the previous $D = 2$ cases are quickly derived by definition Eq. (6) by considering the closed forms $W^C_S = (n^{5/2} - n^{3/2})/4$ and $W^C_H = (7n^{5/2} - 4n^{3/2})/24$. For the square grid, the $f_W = 3/4$ values are confirmed instead for graphene one obtains the value $f_W = 35/48 = 0.73$ that evidences, compared to the square grid, the greater efficiency of honeycomb lattices in tiling a torus in a more compact way. Current calculations confirm that the asymptotic dependence $W \approx n^{2+1/D}$ given in Eqs. (3) and (4) also holds for infinite nanosystems without a real crystal structures (i.e., systems without a unit cell) like the one-pentagon nanocone or fractals such the Sierpinski triangle having fractal dimension $D = \ln(3)/\ln(2)$ (Ori et al., 2010). Finally, it is worth noticing that in Eq. (4) the term depending on the dimensionality plays a $1/D$ role posing interesting questions about the effective relevance of high dimensions in modeling physical systems.

44.4   NANOCHEMICAL APPLICATIONS

Outstanding applications of W in the nanosystems field concern defective graphenic layers. In this sector, in fact, the relevance of the topological approach arises with glaring evidence producing important results that compete with ab-initio methods. Worth to mention is the results achieved by (Ori et al., 2011) on the topological mechanism governing both the formation and propagation of Stone-Wales defects that are in fact preferably generated and propagated along the "diagonal" of nanostructures such carbon nanotubes and graphene nanoribbons.

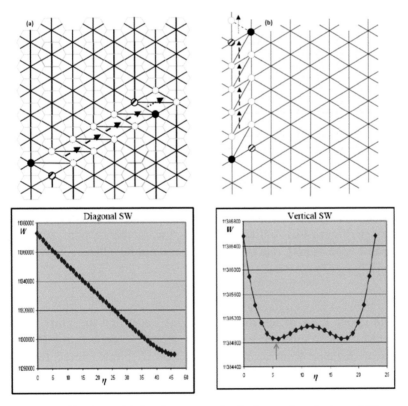

FIGURE 44.2   (a) Diagonal defect free propagation in the graphene lattice and W response (bottom). (b) Vertical rearrangements are instead opposed by W barrier (bottom) with propagation length evidenced by the arrowed W minimum.

The use of W to analyze the response of the lattice to the presence of defects greatly simplify the computational tasks and leads to the clear understanding of the topological roots of the peculiar anisotropy shown by various properties (structural, electronic and magnetic, etc.) of honeycomb nano-structures. Figure 44.2 gives the curves of W for diagonal and vertical propagation of defects for a graphene made by $n = 1250$ hexagons. Closed periodic conditions are imposed on the graph. The diagonal defects flow in the lattice without encountering W barriers whereas the vertical rear-rangements are stopped by W increase evidencing furthermore a maximum propagation distance in correspondence of the W minimum. In this case, the topological simulations are not only able of distinguishing the correct anisotropy behaviors in systems with thousands of carbon atoms, but also define precise length-scale (like the propagation length in Figure 44.2) that represent a useful aid for experimental studies on nanosystems. The propagation mechanism of the Stone-Wales dipole, boosted by the collec-tive minimization of the Wiener index of the whole lattice, has been named Stone-Wales wave and has been used in studying the isomeric generation of pentagon-octagon-pentagon defects in graphene determined only by specific sequences of topological modifications of the lattice (Ori & Puzt 2014).

Other important applications of the Wiener index derives by the compar-ison of the contribution to W coming from the single atom i

$$w_i = \frac{1}{2} \sum_{j=1}^{n} d_{ij} \qquad (8)$$

Therefore, $W = \sum w_i$. Invariants w_i are able to sort the topological-reac-tivity of the nodes of the molecular carbon skeleton being the nodes with minimal (maximal) w_i the ones with bigger (smaller) stability. See a practical example in Figure 44.1. Indicating with b_{ik} the number of atoms placed at the distance k from atom i, one may usefully rewrite Eq. (8) as

$$w_i = \frac{1}{2} \sum_{k=1}^{M} k b_{ik} \qquad (9)$$

The upper limit M appearing in Eq. (9) is called the graph diameter $M = \max\{\varepsilon_i\}$ with atom eccentricity $\varepsilon_i = \max\{d_{ij}\}$ representing the longest chemical distance from atom i. Quantities $\{b_{ik}\}$ are named the Wiener weights (WW). For each atom i the WW follow the simple normalization rule

$$n - 1 = \sum_{k=1}^{M} b_{ik} \qquad (10)$$

The number of edges B of the graph G, i.e., the number of bonds of the molecule is computed from the WW

$$B = \frac{1}{2}\sum_{i=1}^{n} b_{i1} \tag{11}$$

The basal idea underlying the assumption made by Wiener may be expressed in the following way: quantities WW fully represent the surroundings of the i-th molecular site, carrying the full topological information about long-range connectivity of the vertex i with the atoms placed in all of its coordination shells. From the structural point of view, two symmetry-equivalent atoms i and j sharing the same WW set are completely equivalent in respect to their physicochemical properties, for example, contributing to the same ^{13}C NMR resonance line. Sorting the atoms by their WW sets gives at the end a fast (although approximated) definition of the molecular symmetry. This beautiful and profound link between molecular topology and geometry represents a mathematical jewel due to Wiener's pioneer approach to chemistry.

44.5 MULTI-/TRANS-DISCIPLINARY CONNECTIONS

By their nature, topological invariants may be applied every time the system under investigation may be represented by the mean of a graph. Then beside chemistry and material science, pure mathematics and computer science are the fields in which these instruments find many applications. Researches also point out the important role of topological descriptors in explaining the large-scale structure of the cosmic web of galaxies. An interesting application in the neuroscience realm has been just reported (Tozzi et al., 2017) showing how to model the signal propagation in the neutral network, another promising sector in which the use of W will certainly surprise the scientists with a cascade of correlations between system topology and functionality.

44.5 OPEN ISSUES

As reported, the Wiener index may be used to investigate nanosystems with structural and topological defects. Recent original studies on $D = 1$ graphenic nanoribbons with topological defects (Putz & Ori, 2012) clearly evidenced the role of the 'bondon' in allowing phase-transition effects governed by long-range interactions. The presence of this bosonic quasi-particle is very well approximated by the inherent long-range topological nature of W. Future researches are planned for extending these kinds of fundamental studies to $D = 2$ defective nanosystems, such as large fullerenes and graphenic layers.

KEYWORDS

- **asymptotic forms**
- **compression factor**
- **minimal vertices**
- **Wiener-dimensionality**
- **Wiener-weight**

REFERENCES AND FURTHER READING

Bonchev, D., & Mekenyan, O., (1980). A topological approach to the calculation of the p-electron energy and energy gap of infinite conjugated polymers. *Zeitschrift für Naturforschung, 35a,* 739–747.

Cataldo, F., Ori, O., & Graovac, A., (2011). Graphene topological modifications. *International Journal of Chemical Modeling, 3,* 45–63.

Cataldo, F., Ori, O., & Iglesias-Groth, S., (2010). Topological lattice descriptors of graphene sheets with fullerene-like nanostructures. *Molecular Simulation, 36*(5), 341–353.

Diudea, M. V., & Gutman, I., (1998). Wiener-type topological indices. *Croatica Chemica Acta, 71*(1), 21–51.

Hosoya, H., (1971). Topological index. A newly proposed quantity characterizing the topological nature of structural isomers of saturated hydrocarbons. *Bull. Chem. Soc. Japan, 44,* 2332–2339.

Iranmanesh, A., Ashrafi, A. R., Graovac, A., Cataldo, F., & Ori, O., (2012). Wiener index role in topological modeling of hexagonal systems–from fullerenes to graphene. In: Gutman, I., & Furtula, B., (eds.), *Distance in Molecular Graphs–Applications* (pp. 135–155). Univ. Kragujevac: Kragujevac, Serbia.

Ori, O., & D'Mello, M., (1992). A topological study of the structure of the C_{76} fullerene. *Chemical Physics Letters, 197*(1/2), 49–54.

Ori, O., & Putz, M. V., (2014). Isomeric formation of 5|8|5 defects in graphenic systems. *Fullerenes, Nanotubes and Carbon Nanostructures, 22*(10), 887–900.

Ori, O., Cataldo, F., & Putz, M. V., (2011). Topological anisotropy of Stone-Wales waves in graphenic fragments. *Int. J. Mol. Sci., 12*(11), 7934–7949.

Ori, O., Cataldo, F., Vukicevic, D., & Graovac, A., (2010). Wiener way to dimensionality. *Iranian Journal of Mathematical Chemistry, 1*(2), 5–15.

Putz, M. V., & Ori, O., (2012). Bondonic characterization of extended nanosystems: Application to graphene's nanoribbons. *Chemical Physics Letters, 548,* 95–100.

Putz, M. V., Ori, O., Cataldo, F., & Putz, A. M., (2013). Parabolic reactivity "coloring" molecular topology: Application to carcinogenic PAHs. *Current Organic Chemistry, 17*(23), 2816–2830.

Rouvray, D. H., (2002). Harry in the limelight: The life and times of harry wiener. In: Rouvray, D. H., & King, R. B., (eds.), *Topology in Chemistry: Discrete Mathematics of Molecules* (pp. 1–15). Horwood: Chichester, U.K.

Tozzi, A., Peters, J. F., & Ori, O., (2017). Fullerenic-topological tools for honeycomb nanomechanics. Toward a fullerenic approach to brain functions. *Fullerenes, Nanotubes and Carbon Nanostructures*, Available online: doi: 10.1080/1536383X.2017.1283618. Available online: URL http://dx.doi.org/10.1080/1536383X.2017.1283618 (accessed on 12 Apr 2017).

Wiener, H., (1947). Structural determination of paraffin boiling points. *J. Am. Chem. Soc., 69*, 17–20.

CHAPTER 45

Zagreb Indices

ASMA HAMZEH and ALI IRANMANESH

Department of Mathematics, Faculty of Mathematical Sciences,
Tarbiat Modares University, P.O. Box: 14115-137, Tehran, Iran,
Tel: +982182883447; Fax: +982182883493;
E-mail: iranmanesh@modares.ac.ir

45.1 DEFINITION

Graph theory has provided chemists with a variety of very useful tools, and one of such tools is the topological indices. In the field of chemical graph theory and mathematical chemistry, a topological index also known as a connectivity index is a type of a molecular descriptor that is calculated based on the molecular graph of a chemical compound. Topological indices are numerical parameters that can be used to characterize some properties of graphs. Among the oldest and most famous topological index, the first and the second are Zagreb indices.

Let $G = (V; E)$ be a simple graph with vertex set $V(G)$ and edge set $E(G)$. We let $d_G(u)$ denote the degree of vertex u of G. The first Zagreb index $M_1(G)$ and the second Zagreb index $M_2(G)$ are defined as follows:

$$M_1(G) = \sum_{uv \in E(G)} (d_G(u) + d_G(v)),$$

$$M_2(G) = \sum_{uv \in E(G)} (d_G(u) d_G(v)).$$

The first Zagreb index can be also expressed as:

$$M_1(G) = \sum_{u \in V(G)} d_G(u)^2,$$

45.2   HISTORICAL ORIGIN(S)

Analyzing the structure-dependency of total π-electron energy (Gutman et al., 1972), an approximate formula was obtained in which terms of the form occurred (Zagreb indices). It was immediately recognized that these term increase with the increasing extent of branching of the carbon-atom skeleton, (Gutman et al., 1972; Gutman et al., 1975), i.e., that these provide quantitative measures of molecular branching. Ten years later, in a review article (Balaban et al., 1983) included M_1 and M_2 among topological indices and named them "Zagreb group indices." With regard to this, some explanation is needed. First, in the early 1980s, only a handful of topological indices were known, and the authors of the review (Balaban et al., 1983) needed as many of them as possible. Second, in that time both authors of the paper (Gutman et al., 1972) were members of the Theoretical Chemistry Group of the "Ruđer Bošković" Institute in Zagreb. Balaban et al., probably wanted to avoid calling M_1 and M_2 by the name(s) of the discoverers, which otherwise is the usual practice (recall the Wiener, Hosoya, Balaban, Merrifield-Simmons, Narumi-Katayama indices, in addition to the Randić index). The name "Zagreb group index" was soon abbreviated to "Zagreb index," and nowadays M_1 is referred to as the "first Zagreb index" whereas M_2 as the "second Zagreb index."

45.3   NANO-SCIENTIFIC DEVELOPMENT(S)

Zagreb indices possess many interesting properties. Here we will describe a few of them. There is a relationship between M_1 and the number of vertices having degree 1 (corresponding to the number of primary carbon atoms, P), degree 2 (corresponding to the number of secondary carbon atoms, S), degree 3 (corresponding to the number of tertiary carbon atoms, T) or degree 4 (corresponding to the number of quaternary carbon atoms, Q) for trees representing alkanes:

$$M_1 = P + 4S + 9T + 16Q \tag{1}$$

Eq. (1) transforms into:

$$M_1 = 4V + 2T + 6Q - 6,$$

by means of the following equations:

$$E = V - 1$$
$$V = P + S + T + Q,$$
$$E = (P + 2S + 3T + 4Q)/2,$$

where V stands for the number of vertices in a tree (Nikolić et al., 2003).

M_1 is also equal to the number of walks of length $2(mwc_2)$ in a graph (Nikolić et al., 2003):

$$M_1 = mwc_2,$$

and is related to the number of self-returning walks of length 4 (srw_4) (Nikolić et al., 2003):

$$M_1 = (srw_4 + 2 E)/2.$$

The M_1 and M_2 indices can be given in a closed form for homologous structures. Here we give analytical formulas for several classes of regular structures:

i) *n*-alkanes (*n* stands for normal, that is, unbranched alkanes):

$$M_1 = 4 (V - 2) + 2$$
$$M_2 = 4 (V - 2) \text{ for } V > 2.$$

ii) *V*-cycloalkanes (*V* stands for the size of cycloalkanes in terms of the number of vertices):

$$M_1 = M_2 = 4 V.$$

iii) polyacenes

$$M_1 = 26 R - 2$$
$$M_2 = 33 R - 9.$$

iv) polyphenanthrenes (zig-zag benzenoids)

$$M_1 \text{ (polyphenanthrene)} = M_1 \text{ (polyacene)}$$
$$M_2 = 34 R - 11$$

where R is the number of hexagons in a polyacene or polyphenanthrene.

The equality $M_1 = M_2$ for V-cycloalkanes is a consequence of the fact that the number of carbon atoms in these molecules is equal to the number of carbon-carbon bonds, that is, $V = E$. The equality M_1 (polyphenanthrene) = M_1 (polyacene) is a consequence of the definition of the M_1 index. Index M_1 depends only on the valencies of atoms (degrees) in a molecule (graph), and these are the same in isomeric benzenoids (Nikolić et al., 2003).

In Nikolić et al., (2003) authors tested the use of the following Zagreb indices: M_1, M_2, in modeling boiling points of 38 C_3–C_8 alkanes. These compounds and their boiling points were selected because several QSPR studies that can be used for comparison already exist in the literature. In Table 45.1, besides the Zagreb indices, the total walk count (*twc*) and their experimental boiling points (bp) are computed. The latter this index was included in our modeling for the following reasons: the *twc* index has been shown to be very useful in producing multilinear structure-boiling point correlations. Their structure-boiling point modeling is based on the CROMRsel procedure. This is a multivariate procedure that picks up the best possible model among the set of models obtained for a given number of parameters; the criterion of the model goodness being the standard error of estimate (for more details see Nikolić et al., 2003).

TABLE 45.1 C_3-C_8 Alkanes, Their Zagreb Indices, Total Walk Counts (twc) and Their Experimental Boiling Points (bp)

	Alkane	M_1	M_2	twc	bp
1	Propane	6	4	5	−42.1
2	Butane	10	8	16	−0.5
3	2-Methylpropane	12	9	18	−11.7
4	Pentane	14	12	44	36.0
5	2-Methylbutane	16	14	53	27.8
6	2,2-Dimethylpropane	20	16	70	9.5
7	Hexane	18	16	111	68.7
8	2-Methylpentane	20	18	134	60.3
9	3-Methylpentane	20	19	142	63.3
10	2,2-Dimethylbutane	24	22	185	49.7
11	2,3-Dimethylbutane	22	21	165	58.0
12	Heptane	22	20	268	98.5
13	2-Methylhexane	24	22	329	90.0
14	3-Methylhexane	24	23	354	92.0
15	3-Ethylpentane	24	24	378	93.5
16	2,2-Dimethylpentane	28	26	489	79.2
17	2,3-Dimethylpentane	26	26	436	89.8
18	2,4-Dimethylpentane	26	24	399	80.5
19	3,3-Dimethylpentane	28	28	526	86.1
20	2,2,3-Trimethylbutane	30	30	588	80.9
21	Octane	26	24	627	125.7
22	2-Methylheptane	28	26	764	117.6
23	3-Methylheptane	28	27	838	118.9
24	4-Methylheptane	28	27	856	117.7
25	3-Ethylhexane	28	28	928	118.5
26	2,2-Dimethylhexane	32	30	1142	106.8
27	2,3-Dimethylhexane	30	30	1068	115.6

TABLE 45.1 *(Continued)*

Alkane		M_1	M_2	twc	bp
28	2,4-Dimethylhexane	30	29	997	109.4
29	2,5-Dimethylhexane	30	28	911	109.1
30	3,3-Dimethylhexane	32	32	1301	112.0
31	3,4-Dimethylhexane	30	31	1136	117.7
32	3-Ethyl-2-methylpentane	30	31	1152	115.6
33	3-Ethyl-3-methylpentane	32	34	1441	118.2
34	2,2,3-Trimethylpentane	34	35	1536	109.8
35	2,2,4-Trimethylpentane	34	32	1317	99.2
36	2,3,3-Trimethylpentane	34	36	1609	114.8
37	2,3,4-Trimethylpentane	32	33	1296	113.5
38	2,2,3,3-Tetramethylbutane	38	40	2047	106.5

Adapted from Nikolić, et al. (2003). Open Access.

45.4 NANOCHEMICAL APPLICATION(S)

Mathematical chemistry is a branch of theoretical chemistry for discussion and prediction of the molecular structure using mathematical methods without necessarily referring to quantum mechanics. Chemical graph theory is a branch of mathematical chemistry which applies graph theory to mathematical modeling of chemical phenomena. This theory had an important effect on the development of the chemical sciences.

A topological index is a numeric quantity from the structural graph of a molecule. Usage of topological indices in chemistry began in 1947 when chemist Harold Wiener developed the most widely known topological descriptor, the Wiener index, and used it to determine physical properties of types of alkanes known as paraffin.

Diudea and his co-authors was the first scientist considered topological indices of nanostructures into account. In some research paper, he and his team computed the Wiener index of armchair and nanotubes. In this section, we bring some results of Zagreb indices, for chemical graphs and nanostructures.

We denote by $\alpha_G(u)$, the sum of degrees of all neighbors of the vertex u in G. Two vertices v and w of a hexagon H are said to be in ortho-position if there are adjacent in H. If two vertices v and w are at distance two, then they are said to be in meta-position, and if two vertices v and w are at distance three, then they are said to be in para-positions. Examples of vertices in the above three types of positions are illustrated in Figure 45.1.

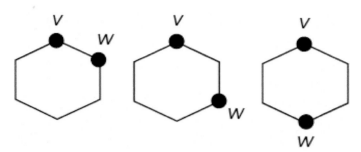

FIGURE 45.1   Ortho-, meta- and para-positions of vertices in a hexagon.

An internal hexagon H in a polyphenyl chain is said to be an ortho-hexagon, meta-hexagon, and para-hexagon, respectively, if two vertices of H incident with two edges which connect other two hexagons are in ortho-, meta-, and para-position. A polyphenyl chain of h hexagons is ortho-PPC_h and is denoted by O_h, if all its internal hexagons are ortho-hexagons. In a fully analogous manner, we define meta-PPC_h (denoted by M_h) and para-PPC_h (denoted by L_h) (see Figure 45.2).

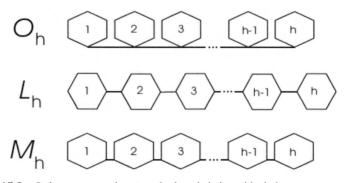

FIGURE 45.2   Ortho-, para-, and meta-polyphenyl chains with six hexagons.

In polyphenyl chains O_h, M_h, and L_h since all vertices of C_6 are of degree two, it is $M_1(C_6) = M_2(C_6) = 24$, $v = \omega = 2$, and $\alpha_{C_6}(v) = \alpha_{C_6}(w) = 4$. So $M_1(O_h) = M_1(M_h) = M_1(L_h) = 34h - 10$. Note that v and w are adjacent in O_h, but are not adjacent in M_h and L_h. Thus $M_2(O_h) = 42h-19$ and $M_2(M_h) = M_2(L_h) = 41h - 17$ (Azari et al., 2013).

In Azari et al., (2013) authors considered the molecular graph of the nanostar dendrimer D_n shown in Figure 45.3 and given formulas for Zagreb indices of this nanostar dendrimer.

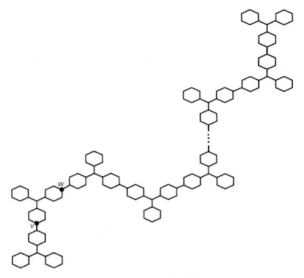

FIGURE 45.3 The molecular graph of nanostar dendrimer D_n.

The graph G is the graph depicted in Figure 45.4, v and w are the vertices shown in Figure 45.3, and n is the number of repetition of the fragment G.

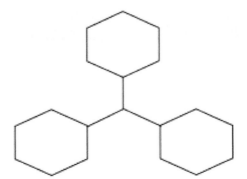

FIGURE 45.4 The graph of nanostar dendrimer D_n for $n = 1$.

So $M_1(G) = 96$ and $M_2(G) = 111$. Thus $M_1(D_n) = 106n - 10$ and $M_2(D_n) = 128n - 17$.

Also, in Azari et al., (2013), authors considered another class of dendrimers and given formulas for Zagreb indices of this graph. This molecular structure can be seen in some of the dendrimer graphs such as tertiary phosphine dendrimers. Let D_0 be the graph of Figure 45.5.

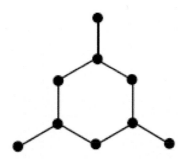

FIGURE 45.5   The dendrimer graph D_0.

For positive integers d and k, suppose $D_{d,k}$ be a series of dendrimer graphs obtained by attaching d pendent vertices to each pendant vertex of $D_{d,k-1}$ and set $D_{d,0} = D_0$. Some examples of this class of dendrimer graphs are shown in Figure 45.6.

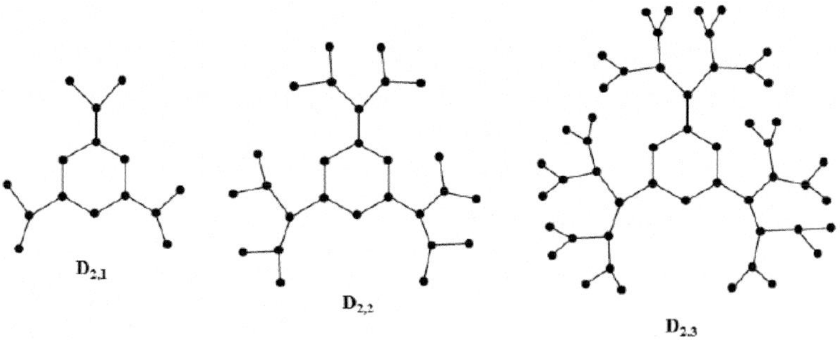

FIGURE 45.6   Dendrimer graphs $D_{d,k}$ for d = 2, k = 1,2,3.

The first and second Zagreb indices of the dendrimer graph $D_{d,k}$, $k \geq 1$ are given by:

$$M_1(D_{d,k}) = 42 + 3d(d+3)\sum_{s=0}^{k-1} d^s,$$

$$M_2(D_{d,k}) = 3d^2 + 12d + 45 + 6d^2(d+1)\sum_{s=0}^{k-2} d^s.$$

45.5   MULTI-/TRANS-DISCIPLINARY CONNECTION(S)

Recently, the Zagreb indices and their variants have been used to study molecular complexity (Bertz et al., 1988) chirality, ZE-isomerism and heterosystems whilst the overall Zagreb indices exhibited a potential applicability for deriving multilinear regression models. Zagreb indices are also used by various researchers in their QSPR and QSAR studies (Devillers et al., 1999). Mathematical properties of the Zagreb indices have also been studied (Estrada et al., 1997). Zagreb indices are referred to in most books reporting topological indices and their uses in QSPR and QSAR (Devillers et al., 1999). They are also included in a number of programs used for the routine computation of topological indices, such as POLLY, OASIS, DRAGON, Cerius, TAM, DISSIM, etc.

KEYWORDS

- **dendrimer**
- **first Zagreb index**
- **second Zagreb index**

REFERENCES AND FURTHER READING

Azari, M., & Iranmanesh, A., (2013). Chemical graphs constructed from rooted product and their Zagreb indices. *MATCH Commun. Math. Comput. Chem., 70*, 901–919.

Azari, M., Iranmanesh, A., & Gutman, I., (2013). Zagreb indices of bridge and chain graphs. *MATCH Commun. Math. Comput. Chem., 70*, 921–938.

Balaban, T., Motoc, I., Bonchev, D., & Mekenyan, O., (1983). Topological indices for structure-activity correlations. *Topics Curr. Chem., 114*, 21–55.

Bertz, S. H., & Wright, W. F., (1988). The graph theory approach to synthetic analysis: Definition and synthetic complexity. *Graph Theory Notes of New York, 35*, 32–48.

Devillers, J., & Balaban, A. T., (1999). *Topological Indices and Related Descriptors in QSAR and QSPR* (3rd edn.). Publisher: Gordon & Breach, Amsterdam.

Estrada, E., Rodriguez, L., & Gutierrez, A., (1997). Matrix algebraic manipulation of molecular graphs. 1. Distance and vertex-adjacency matrices. *MATCH Commun. Math. Comput. Chem., 35*, 145–156.

Gutman, I., & Trinajstić, N., (1972). Graph theory and molecular orbitals. Total π-electron energy of alternant hydrocarbons. *Chem. Phys. Lett., 17*, 535–538.

Gutman, I., Ruščić, B., Trinajstić, N., & Wilcox, C. F., (1975). Graph theory and molecular orbitals, XII. Acyclic polyenes. *J. Chem. Phys., 62*, 3399–3405.

Nikolić, S., Kovačević, G., Miličević, A., & Trinajstić, N., (2003). The Zagreb indices 30 years after. *Croat. Chem. Acta, 76*, 113–124.

CHAPTER 46

Zeolite

ADRIANA URDĂ and IOAN-CEZAR MARCU

Laboratory of Chemical Technology and Catalysis,
Department of Organic Chemistry, Biochemistry and Catalysis,
Faculty of Chemistry, University of Bucharest, 4-12, Blv. Regina Elisabeta,
030018 Bucharest, Romania, Tel: +40213051464, Fax: +40213159249,
E-mail: ioancezar.marcu@chimie.unibuc.ro,
ioancezar_marcu@yahoo.com

46.1 DEFINITION

Zeolites are crystalline aluminosilicates with a three-dimensional framework containing pores with uniform sizes, in the range of molecular dimensions. Zeolites are included in the group of molecular sieves, because they preferentially adsorb inside the pore structure molecules with a smaller diameter than their pore size, while excluding larger molecules. Their framework consists of crystalline silica (SiO_2) in which some Si^{4+} are replaced by Al^{3+} ions, the resulting negative charges being compensated by loosely held cations in order to preserve the electroneutrality of the solid. The cations can be replaced by ionic exchange, leading to solids with tunable properties that can have many applications, from petrochemical industry to laundry detergents, and from adsorbents to agriculture. There are more than 220 known zeolite framework types, approved by the Structure Commission of the International Zeolite Association (IZA).

46.2 HISTORICAL ORIGIN(S)

Axel F. Cronstedt, a Swedish mineralogist, discovered in 1756 the mineral stilbite, a hydrated aluminosilicate of calcium and sodium (Flanigen, 2001). Because the solid formed a frothy mass when heated, Cronstedt named it (and the whole class of related aluminosilicates) *zeolite*, from the Greek words *zeo* (to boil) and *lithos* (stone). The reversible dehydration and ion exchange of zeolite crystals were observed by Damour in 1840 (Turkevich, 1968) and Eichhorn in 1858, respectively. In 1862, the first hydrothermal synthesis of a zeolite (levynite) was reported, and in 1896 the spongy framework of dehydrated zeolites was postulated by Friedel (Flanigen, 2001). The molecular sieve effect was reported in 1925 by Weigel and Steinhoff, while the structures of zeolites were determined for the first time in 1930 by Taylor and Pauling. R. M. Barrer studied adsorption by zeolites and their synthesis during 1930s and 1940s, and presented the first classification of known zeolites in 1945. R. M. Milton and D. W. Breck, searching for new materials for the separation and purification of air, studied zeolite synthesis and discovered zeolites A, X, and Y, used for the first time for drying of refrigerant and natural gases. Later, in the 1950s to 1970s, synthetic zeolites were introduced as catalysts for the industrial hydrocarbon processing (such as isomerization of alkanes, cracking), and ion exchange separations. Nowadays, zeolites are used as materials for adsorption, purification, catalysis, ion exchange and as functional powders (e.g., in odor removal or plastic additives).

46.3   NANO-SCIENTIFIC DEVELOPMENT(S)

Zeolites are represented by the general formula: $M_{m/z}[mAlO_2 \cdot nSiO_2] \cdot qH_2O$, where z is the valence of cation M, m and n are the total numbers of tetrahedra per unit cell, while q is the number of water molecules per unit cell (Turkevich, 1968; Meier, 1986; Flanigen, 2001). Many zeolites occur as natural minerals, but some of the synthetic ones, in their acidic form, are probably the most important heterogeneous catalysts used by the chemical industry (see Chapter 5: Catalytic Material) (Cundy et al., 2003).

Today, the zeolite chemistry topics also deal with other zeolite-type (zeotype) structures besides aluminosilicates, but this contribution discusses only those solids containing Si and Al as T atoms in the structure.

The structure of zeolites has a strong influence on their properties: the sorption characteristics depend on the pore sizes and void volumes inside the structure, ion exchange is influenced by the number, type and accessibility of cation sites, while catalytic properties depend on the framework type (pore openings, type and size of channels and void volumes available for the reaction), cationic and/or redox sites, etc. (McCusker & Baerlocher, 2001).

All zeolite types have an open 3-dimensional structure formed by TO_4 tetrahedra linked to each other by sharing the oxygen ions in the corners, as shown in Figure 46.1 (Meier, 1986; Flanigen, 2001), where T is a Si or Al cation (or other cations in zeotypes).

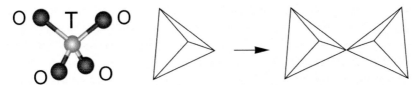

FIGURE 46.1   Corner-sharing TO$_4$ tetrahedra in zeolites (adapted from http://www.bza.org/zeolites/).

Aluminum is incorporated in the framework, isomorphously replacing silicon in tetrahedral coordination with oxygen. In order to form four bonds, an additional electron is necessary. Therefore, each [AlO$_4$] tetrahedron has a negative charge, compensated by an extra-framework cation (such as Na$^+$, K$^+$, Mg^{2+}, Ca^{2+}, NH$_4^+$, or H$^+$) that is placed in the void spaces of the structure, together with water molecules (Flanigen, 2001). The cations are mobile and easily replaced by ion exchange, while the water molecules can be reversibly removed by heating.

The TO$_4$ tetrahedra are the basic building units (BBU), from which secondary building units (SBU) can be imagined to form (van Koningsveld, 1991), and by joining these SBU the zeolite framework can be constructed (Figure 46.2).

In these representations, T atoms are found at intersections and terminations of lines (black dots), while O atoms, not shown, are located midway between T atoms (van Koningsveld, 1991). Apart from SBU, which are oligomer units, zeolite frameworks can be constructed from infinite units such as chains and layers (van Koningsveld, 2001).

In zeolite frameworks, aluminum distribution is not entirely random, and it is considered that Al–O–Al links are forbidden because they lead to framework instability (Turkevich, 1968). This is known as the Lowenstein rule and explains why the Si/Al ratio is higher than unity, although experimental evidence for direct Al–O–Al links has been found (Pavón et al., 2014).

The *framework type* describes the way tetrahedral T atoms are connected (the topology) and defines the size of pore openings, their shape, the dimensionality of the channel system, the void volumes (channels, their interconnections and cages) and the available cation sites (McCusker & Baerlocher, 2001), without considering the composition, symmetry and unit cell dimensions. Therefore, many materials can be classified under the same designation: for example, the MFI framework type contains more than 25 related materials, including ZSM-5 zeolite and silicalite. The three letter code (e.g., MFI) is assigned by the Structure Commission of the International Zeolite Association to framework types that are confirmed according to IUPAC rules, and is derived from the name of the zeolite [e.g., MFI from Zeolite Socony Mobil-5 (ZSM-5)] (McCusker & Baerlocher, 2001; Database of Zeolite Structures). The framework type can be determined by X-ray powder diffraction, even if diffraction patterns are sensitive to changes in framework conformation, symmetry, etc. (Meier, 1986).

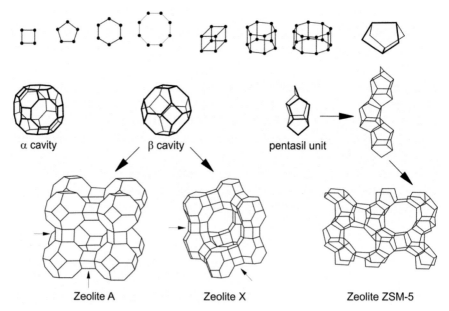

α cavity   β cavity   pentasil unit

Zeolite A   Zeolite X   Zeolite ZSM-5

FIGURE 46.2 Examples of building blocks and the formation of zeolite frameworks (adapted from McCusker & Baerlocher, 2001).

The *framework structure*, as opposed to the framework *type*, takes into consideration the composition (framework and extra-framework cations), organic species, structural defects and sorbed molecules (McCusker & Baerlocher, 2001). The framework composition (the Si/Al ratio) is extremely important for zeolites, determining many of their properties. Since the framework is anionic due to the presence of aluminum ions, cations such as Na^+ or H^+ are necessary in order to preserve neutrality, turning the zeolites into interesting materials for catalysis. As a consequence, many chemical elements can be incorporated in the zeolitic structures. Extra-framework species (e.g., the cations that balance the negative charge of the framework, removable water and/or organic molecules) usually exist in the channels and cages of the zeolites. They are introduced during the synthesis of the solid or by a post-synthesis treatment, and their presence strongly influences the sorption properties of the zeolites.

Pore sizes are determined by the crystallographic structure, namely by the size of the *n*-ring that defines the pore mouth, where *n* is the number of T atoms in the ring. An 8-ring (with a free diameter of ca. 4.1 Å, calculated using the oxygen radius of 1.35 Å) is considered to be rather small, while a 10-ring (ca. 5.5 Å) is a medium one, and a 12-ring (ca. 7.4 Å) is large (McCusker & Baerlocher, 2001). The pore system can be mono-, bi-, or three-dimensional (Maxwell & Stork, 2001). The pore openings can be changed (pore size engineering) by ion exchange, post-impregnation or chemical reaction with specific compounds (Kouwenhoven & de Kroes, 2001). The uniform size of pores in zeolites is in contrast with the broader distribution and larger mean pore diameters observed in other porous solids (Flanigen, 2001).

Based on their Si/Al ratio, zeolites are divided into four categories: "low," "intermediate," "high" silica zeolites, and "silica" molecular sieves (Flanigen, 2001). The first category includes zeolites with Si/Al ratios from 1 to ca. 1.5 (such as types A and X). The intermediate ratios (from ca. 2 to 5) are found, for example, in natural zeolites clinoptilolite and mordenite, and in synthetic zeolites such as Y and omega. "High" silica zeolites (with ratios from ca. 10 to 100) are represented by ZSM-5, beta zeolite, or highly siliceous Y zeolite. Si/Al ratios higher than 100 are found in silicalite, a ZSM-5 type of zeolite with very low concentrations of Al in the structure.

The increase in Si/Al ratio is accompanied by a change in properties such as thermal stability (that increases from below 700°C up to 1300°C), surface selectivity (that changes from hydrophilic at low ratios to hydrophobic at high values), acidity (that increases in strength simultaneously with a decrease in number of acid centers), cation concentration and ion exchange capacity (both strongly decreasing), and structure (changing from predominantly 4-, 6- and 8-member rings to 5-member) (Flanigen, 2001). As a consequence, the zeolites with low and intermediate Si/Al ratio are able to remove water from organics and perform separations and catalysis on dry reactants, while the structures with high ratios can recover organics from water and perform separations and catalysis in the presence of water (Flanigen, 2001).

One of the main properties of zeolites (and other molecular sieves) is the shape selectivity, arising from the confined environment in which the molecules are converted (Martens & Jacobs, 2001). This principle was first demonstrated in 1962 by Weisz and co-workers, and in 1972, Csicsery defined three types of this effect (Whyte & Dalla Betta, 1982): (1) the reactant shape selectivity, observed when only some of the molecules from the feed are able to enter the pores of the zeolite, while the rest are too large to diffuse inside; (2) the product shape selectivity that occurs when some of the products formed inside the pores have slower rates of diffusion compared to other products due to their size, so they undergo further transformation on catalytic sites inside the pore system (including coking); (3) transition state shape selectivity, when the transition state complex for a certain reaction is too bulky to be formed inside the pore system, therefore the product of that reaction will not be formed. However, the reactant, transition state, and product shape selectivity are sometimes difficult to prove separately, and still remain intertwined (Bhan & Iglesia, 2008). Two other shape selectivity effects were reported in the early 1980s: the molecular traffic control (Derouane & Gabelica, 1980), and shape selective constraints that affect local site kinetics prior to reaction (Whyte & Dalla Betta, 1982).

Another important property of zeolites is their acidity: when the compensating cation is the proton, H^+, zeolites act as solid acid catalysts. The catalytic acid sites in zeolites are complex, involving hydroxyl groups bridging Si and Al atoms, Si–O(H)–Al, with strong Brønsted acid properties, and also oxo bridges that have Lewis base properties (Martens & Jacobs, 2001) (Figure 46.3).

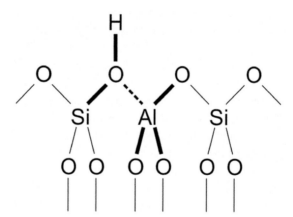

FIGURE 46.3   Active sites in zeolites (adapted from Martens & Jacobs, 2001).

The hydroxyl group bridging Si and Al was first described as a SiOH (silanol) group activated by the presence of a Al^{3+} neighbor acting as a Lewis acid center, but later it was observed that the three bonds, O–Al, O–Si and

O–H are all strong covalent bonds, and the oxygen atom has a threefold coordination similar with that in H_3O^+ cation, with angles having ca. 120° (Boronat & Corma, 2015). The threefold coordination of the oxygen atom, together with the geometry restrictions imposed by the crystalline structure, lead to the strong acidity of the bridged hydroxyl groups.

It was considered that these sites catalyze hydrocarbon conversions through carbocations (as reaction intermediates or transition states): a Brønsted acid site transfers a proton, either to an unsaturated molecule (alkene) to form an alkylcarbenium ion (tri-coordinated carbocation), or to an alkane, leading to an alkylcarbonium ion (penta-coordinated carbocation, the five bonds of the charged C-atom containing only eight electrons); the Lewis acid sites generate carbocations by extracting hydride ions H^- from hydrocarbon molecules. After carbocations are formed, they are further converted by rearrangements, scission, alkylation, and hydrogen transfer reactions (Martens & Jacobs, 2001). These reactions also lead to some heavy, unsaturated cyclic and acyclic molecules that deactivate the zeolites (coking). However, experimental techniques showed that simple carbenium ions are not stable in zeolite channels, and surface alkoxide species formed by the interaction of alkyl groups with oxygens in the framework are more likely the reaction intermediates (Boronat & Corma, 2015). The carbonium ions were proved by computational methods to exist as intermediates only if their positive charge is delocalized and sterically inaccessible to framework oxygens; otherwise, the proton is transferred to oxygen from the framework, and the intermediates are neutral molecules adsorbed on Brønsted sites (Boronat & Corma, 2015).

The acid strength of these sites depends on the Si/Al ratio, but at high Si/Al values (higher than 10), the isolated OH groups have practically constant acid strength (Bhan & Iglesia, 2008). The strength of the OH bond can be measured by different methods, such as spectroscopic techniques (IR and NMR), adsorption calorimetry (for the interaction with bases), TPD of a base (usually ammonia, pyridine, etc.), but only some of the methods can discriminate between Brønsted and Lewis acidity (Boronat & Corma, 2015).

Among the many zeolite types, only a few are used in large amounts in industry: A, Y, and ZSM-5 zeolites are the most important of them.

Zeolite A (LTA structure, $1 \leq Si/Al \leq 3$) has small 8-ring channels (4.1 Å) connecting large spherical cavities (11.4 Å). It is mainly used in laundry detergent formulations, for water softening.

Y zeolite is a faujasite-type zeolite (FAU, Si/Al ratio ≥ 2.5), with large spherical cavities (called supercages, 11.8 Å) connected by large 12-ring channels (7.4 Å). Modified with rare-earth ions and thermally stabilized by dealumination, it is used in large amounts as a catalyst for the catalytic cracking process.

ZSM-5 zeolite (MFI structure) is a member of the "high" silica zeolites group. It is interesting because of its 10-ring channel structure (5.2 x 5.7 Å straight channels, connected by 5.3 x 5.6 Å sinusoidal channels), with intermediate size between wide-pore zeolites (e.g., faujasite) and narrow-pore zeolites (like A or erionite). This feature prevents molecules with large dimensions (such as aromatics higher than C_{10}, which are important coke precursors) to be formed inside the channels. Therefore, its deactivation by coke deposition is less intense than for other zeolite types and, together with its strong acidity, explains the use of this particular zeolite structure as a catalyst in many hydrocarbon conversion processes.

Due to their use in so many applications, systematic research was devoted to the synthesis of zeolite structures with tuned properties for specific processes.

The synthesis of zeolites requires several reagents: a SiO_2 source for the primary building units of the framework, an AlO_2^- source for the framework charge, a mineralizer (usually HO^-, acting also as a guest species), a counterion for the negative charge (usually an alkali/ammonium cation or an organic cation/template) and a solvent (water, acting also as guest molecule) (Jansen, 2001). Many Si-sources are mentioned in the literature, such as tetramethyl or tetraethyl ortosilicate (TMeOS or TEOS), Na_2SiO_3, water glass, colloidal or fumed silica, while Al-sources include $NaAlO_2$, $AlOOH$ (pseudo-boehmite), $Al(OH)_3$ (gibbsite) or $Al(NO_3)_3$. Since Na ion is often present in the Si- and Al-sources, this is the usual counterion in the zeolite framework after synthesis. Numerous organic templates (both charged and neutral molecules) are used, most often alkylammonium salts, amines or alcohols. The reactant ratios, such as H_2O/SiO_2, HO^-/SiO_2, SiO_2/Al_2O_3 and, H_2O/template, characterize the concentration and solubility of the precursors, and influence the expected zeolite type that is formed as a product.

Several factors affect the outcome of the zeolite synthesis: type of reactants, the way the reactant mixture is made (addition sequence, stirring and gel aging), the pH and the temperature (heating rate and final synthesis temperature) (Jansen, 2001). The rate of Si- and Al-source dissolution influences the rate of nucleation and crystallization; therefore the product formation is affected, and the impurities present in the precursors can change the crystal form and the chemical and/or catalytic properties of the product. Also, usually, higher pH (for syntheses in the presence of HO^-) and temperature values lead to higher crystal growth rate.

The first synthesis method, used by R. M. Milton in the 1940s, was the hydrothermal crystallization of alkali aluminosilicate gels at low pressures and temperatures (ca. 100°C) and high pH (10-14), using cation templating (Flanigen, 2001). The alkali cation is used for "templating" and stabilizing the sub-units in the zeolite structure during synthesis. The use of quaternary ammonium cations in the 1960's made possible the discovery of intermediate and high silica zeolites, with the crystallization step occurring at higher temperatures (up to ca. 200°C). Later, F$^-$ ion used as mineralizer and template allowed the synthesis to be carried out in acidic pH, with the zeolite product having larger and more defect-free crystals, and even new structures.

Several stages occur during the zeolite synthesis: gel formation upon mixing of reactant solutions (at low temperature), followed by the rearrangement and dissolution of the gel, dissociation of silicate (with increasing temperature), nucleation and crystallization of the zeolite product (at high temperature) and, finally, the isolation (by decantation/ centrifugation/ filtration) of the product after cooling (Jansen, 2001). The process was studied by IR-, Raman, and ^{29}Si-NMR spectroscopies, showing the formation of monomers by gel dissolution, followed by condensation reactions leading to oligomers and then higher order rings, which regroup around alkali cations and organic templates (Cundy et al., 2003). After an induction period, either primary (heterogeneous or homogeneous) or secondary nucleation (with seed crystals) takes place, followed by precipitation (crystallization) in the supersaturated solution generated during previous stages. If a metastable zeolite phase is formed, it may transform into a more stable phase according to Ostwald rule of successive transformations. The template molecules have to be removed from the structure, by calcination, before using the zeolite (Kouwenhoven & de Kroes, 2001).

Some zeolite types can form intergrown forms, with an infinite number of intermediate structures (van Koningsveld, 2001). For example, zeolites ZSM-5 and ZSM-11 (framework types MFI and MEL, respectively) form intermediate, disordered structures that contain parallel pentasil layers formed by left- and right-hand chains (related by a mirror plane) along the z-axis.

When the desired properties are not achieved by direct synthesis, several secondary or post-synthesis treatments can be used in order to improve the catalytic activity, thermal or hydrothermal stability and modified textural properties (such as porosity) (Szostak, 2001).

One of the frequently used post-synthesis treatments is the ion exchange of the zeolite with dilute mineral acids, from the initial counterion to proton, leading to a solid acid with interesting catalytic properties. Aluminum-rich zeolites cannot be directly ion exchanged with protons, due to the loss in crystallinity as the pH decreases (Szostak, 2001; Townsend & Coker, 2001). Therefore, after being synthesized in the Na^+ form, these zeolite solids can be exchanged with NH_4^+ ions and subjected to thermal decomposition, leading to the protonic (H^+, Brønsted acid) form (Turkevich & Ono, 1969; Szostak, 2001; Townsend & Coker, 2001). For each zeolite structure, the maximum acidity per site is reached at a certain Si/Al ratio, when each Al atom is isolated and has no next nearest T site occupied by Al atoms (Kouwenhoven & de Kroes, 2001).

Other treatments aim at the improvement in the thermal and hydrothermal stability by increasing the Si/Al ratio, resulting also in increased catalytic activity. Few zeolites can be prepared with the desired Si/Al ratios only by direct synthesis; therefore secondary methods are used, such as hydrothermal, chemical treatment (with chelating agents, $SiCl_4$, etc.), or a combination of them (Szostak, 2001). However, these methods may result in inhomogeneities within the zeolite crystals and textural changes, such as the appearance of mesopores. The aluminum that is removed from the framework can be washed out of the structure with dilute acids. Thermal and hydrothermal stabilities are also achieved by ion exchange with multivalent cations, e.g., lanthanum (Szostak, 2001).

The reverse process is also possible–the insertion of aluminum in the framework, either from an external source (such as aluminum halides, $NaAlO_2$, etc.), from the binder or non-framework aluminum, resulting in increased number of Brønsted acid sites. High temperatures and/or alkaline conditions facilitate the insertion (Szostak, 2001).

For bifunctional catalysts, a second function (besides the acidic one represented by the zeolite) is introduced, i.e., hydrogenation/dehydrogenation represented by a transition metal (Kouwenhoven & de Kroes, 2001). The metal can be introduced either by ion exchange or by impregnation, taking care for a good dispersion and an even distribution on the zeolite support.

46.4 NANOCHEMICAL APPLICATION(S)

Using the pore-controlled reactivity in zeolite, catalyzed reactions have evolved rapidly in the past 50 years, from laboratory research to large-scale industrial applications, although only a small part of the known zeolites are used in practice (Bhan & Iglesia, 2008).

The well-defined, open crystalline structure of zeolites, with a regular pore distribution, has properties that are appreciated in industrial catalysis since the early 1960s. As solid catalysts, they contain a distribution of acid sites with intrinsic strength associated to different local structures, chemical environment or location inside the channels (Boronat & Corma, 2015). However, their strong acidity is not the most important of their properties: the microporous channels that induce selectivity effects related to the diffusion of reactants and products, the electrostatic interactions of the charged species with the zeolite framework and site specificity for particular reactions play a more important role in their catalytic performance.

The majority of the commercial applications use A, X or other zeolite structures with Si/Al ratios close to unity for ion exchange purposes (Townsend & Coker, 2001), while Y, mordenite or ZSM-5 zeolites are used as catalysts for hydrocarbon processing, due to their reduced tendency to form coke compared to other catalysts (Heinemann, 1981; Maxwell & Stork, 2001).

In ion exchange applications, zeolites are used for water softening, radionuclide separations and wastewater treatments (Townsend & Coker, 2001). Large amounts of low Si/Al ratio zeolites (mainly A type) are used in detergent formulations for water softening, i.e., replacing their exchangeable cations (e.g., Na^+) for Ca^{2+} and Mg^{2+} ions from water and reducing the environmental impact that previously used sodium triphosphate had. For the $^{90}Sr^{2+}$ and $^{137}Cs^+$ radionuclide separations from low- and medium-level nuclear waste, natural zeolites are of great interest due to their abundance and low cost, but synthetic ones are also highly efficient. They have to remove the nuclides in the presence of significant amounts of competing ions, such as Ca^{2+}, Mg^{2+}, Na^+, etc.; therefore highly selective exchangers are needed. In the wastewater treatment, zeolites are used for the removal of ammonia and ammonium salts, but also heavy metal cations or other transition metal ions, in certain conditions (Townsend & Coker, 2001).

Zeolites are also used as selective adsorbents, e.g., for drying and purification of reactants. Zeolites with low Si/Al (e.g., zeolite A) are hydrophilic due to their high cation content, with a high saturation capacity, and are used for drying of liquids and gases (Hölderich & van Bekkum, 2001). Another important application is the separation of products using zeolites: for example, p-xylene is separated from C_8 aromatics (o-, m-, p-xylene, and ethylbenzene) for further conversion to terephthalic acid and polyethylene terephthalate, while the rest of the mixture isomerizes again to the equilibrium composition (see also below, xylenes isomerization). The separation is possible due to different diffusion rates of xylene isomers in the zeolite channels (product shape selectivity). It is to be noted that an important potential application of zeolites as adsorbents is the flue gas clean up (Liu et al., 2010).

Industrial conversion of hydrocarbons over acidic zeolites includes catalytic cracking, hydrocracking, isomerization, alkylation, and oligomerization, among other processes (Martens & Jacobs, 2001). The process using the largest amount of zeolites is catalytic cracking (Heinemann, 1981; Whyte & Dalla Betta, 1982). It employs Y zeolite partially exchanged with rare-earth ions and thermally treated to improve its stability (REY and USY) incorporated in a silica-alumina matrix (Heinemann, 1981; Maxwell and Stork, 2001). A low amount of platinum (ppm levels) may be used to improve catalyst regeneration by coke combustion. The use of USY leads to much greater selectivity to gasoline compared to previous amorphous silica-alumina catalysts (Heinemann, 1981).

Hydrocracking also benefits from the use of bifunctional catalysts containing zeolites (USY, erionite or mordenite) for the production of a large range of distillate fuels, starting from high-boiling hydrocarbons, in the presence of hydrogen (Heinemann, 1981; Whyte & Dalla Betta, 1982).

Gasoline octane numbers can be increased by several methods. One process converts C_5–C_6 alkanes by isomerization on zeolites such as mordenite, modified with noble metals (Turkevich & Ono, 1969; Heinemann, 1981). Another process is the alkene/alkane (e.g., isobutene/isobutane) alkylation to obtain the so-called alkylate. This is a mixture of methyl-branched alkanes with high octane number, free of aromatics, alkenes, and sulfur, which is used for reformulated gasoline. The process uses zeolites, avoiding the disadvantages of liquid acid catalysts (HF, H_2SO_4) used in other industrial processes (Maxwell & Stork, 2001; Feller & Lercher, 2004). Isobutane and C_3–C_5 alkenes are used as raw materials, with trimethylpentanes having high octane numbers as the main desired components of the alkylate mixture.

In shape selective processes, linear or slightly branched alkanes from the reformate gasolines are selectively hydrocracked using zeolites such as ZSM-5, resulting in high-octane gasolines and liquefied petroleum gases (LPG) (Whyte & Dalla Betta, 1982). Alkylbenzenes are also obtained on the same zeolites, from the alkylation of aromatics with the cracked fragments (Kaeding et al., 1984). Examples of processes using the shape selectivity of zeolites are catalytic iso-dewaxing, LPG conversion to aromatics (CYCLAR process), methanol-to-olefins, deep catalytic cracking (DCC) (Maxwell & Stork, 2001). In the iso-dewaxing process, zeolites are used for the isomerization of n-alkanes into branched alkanes, with higher octane number (Maxwell and Stork, 2001). The CYCLAR process uses ZSM-5 zeolite modified with gallium to convert C_3–C_5 alkanes into high octane gasoline and BTX (benzene, toluene, and xylene) aromatics (Maxwell & Stork, 2001).

Many industrial processes use zeolite catalysts to increase the value of different petroleum cuts. One example is the alkene oligomerization process that transforms alkenes from the catalytic cracking into gasoline or middle distillates, using catalysts such as ZSM-5 (Maxwell and Stork, 2001).

ZSM-5 zeolite is also used for xylenes isomerization, a process designed to obtain p-xylene (for terephthalic acid) from ethylbenzene, o- and m-xylenes, with improved efficiency, compared to the previously used $Pt-Al_2O_3$ catalyst, due to the shape selectivity effect (Heinemann, 1981; Whyte & Dalla Betta, 1982; Kaeding et al., 1984). Combined with aromatics (toluene, xylenes, and ethylbenzene) disproportionation and alkylation (Heinemann, 1981; Kaeding et al., 1984), the desired aromatic hydrocarbons can be obtained using a small number of zeolite structures.

The methanol-to-hydrocarbons (MTH) process is a key step in producing fuel and alkenes from alternative carbon sources like biomass, coal, natural gas and CO_2 (Olsbye et al., 2012). It can be turned into the production of gasoline-rich (MTG) or alkene-rich (MTO) product mixtures by the proper choice of catalyst and reaction conditions. In both cases, the selectivities to the desired products reach 80%. In the methanol-to-gasoline (MTG) or the methanol to gasoline and distillate conversions (Heinemann, 1981; Mokrani & Scurrell, 2009), branched chain alkanes in the gasoline range, with high octane numbers, are obtained in the presence of hydrogen, taking advantage of the shape selectivity effects on zeolites such as ZSM-5 (Whyte & Dalla Betta, 1982). The methanol-to-olefins (MTO) process is used for the synthesis of C_2–C_4 alkenes from methanol or even from syngas, via methanol and/or dimethyl ether (Mokrani & Scurrell, 2009; Tian et al., 2015). The process uses zeolites like ZSM-5 or zeolite-like molecular sieves as catalysts.

The same ZSM-5 zeolite, as such or modified with zinc, gallium or silver, is used for the conversion of light alkanes or alkenes (C_3–C_4) to aromatic hydrocarbons, a process known as aromatization (Ono, 1992; Mokrani & Scurrell, 2009).

This brief review on the industrial use of zeolites is by no means exhaustive, as new structures and processes are developed for new applications, and improvement of the older ones is needed for better economic results.

46.5 MULTI-/TRANS-DISCIPLINARY CONNECTION(S)

The zeolites study started more than a century ago as a curiosity for geologists and became nowadays a complex research topic involving the cooperation of several scientific disciplines such as materials chemistry, physical chemistry, physics, chemical engineering, etc. in order to control their synthesis and structure, to design their properties and to find new applications or to improve the existing ones.

46.6 OPEN ISSUES

The search for other zeolite structures to be used in hydrocarbon processing for the production of aromatics with a cut-off at C_9, to avoid deactivation by coking, or for completely avoiding the formation of aromatics will continue. Irreversible deactivation during hydrocarbon processing reaction-regeneration cycles arises from materials degradation, either by loss of active sites, dealumination with the formation of extra-framework aluminum species or by phase change leading to other zeolite structures, amorphous and/or dense phases. Therefore, research in this area of materials chemistry is continuously expanding, as long-term zeolite stability is an important issue.

The rational design of improved zeolite catalysts is still an important objective, as more technologies and applications are developed in the hydrocarbon processing, environmental catalysis and other areas of interest. This design should be based on the knowledge of reaction mechanisms; therefore, research in this topic is of great interest.

KEYWORDS

- **adsorbents**
- **aluminosilicates**
- **ion exchangers**
- **shape selectivity**
- **solid acid catalysts**

REFERENCES AND FURTHER READING

Bhan, A., & Iglesia, E., (2008). A link between reactivity and local structure in acid catalysis on zeolites. *Accounts of Chemical Research, 41*, 559–567.
Boronat, M., & Corma, A., (2015). Factors controlling the acidity of zeolites. *Catalysis Letters, 145*, 162–172.

British Zeolite Association, http://www.bza.org/zeolites/ (accessed on 23 October 2015).

Cundy, C. S., & Cox, P. A., (2003). The hydrothermal synthesis of zeolites: History and development from the earliest days to the present time. *Chemical Reviews, 103*, 663–701.

Database of Zeolite Structures. http://www.iza-structure.org/databases/ (accessed on 23 October 2015).

Derouane, E. G., & Gabelica, Z., (1980). A novel effect of shape selectivity: Molecular traffic control in zeolite ZSM–5. *Journal of Catalysis, 65*, 486–489.

Feller, A., & Lercher, J. A., (2004). Chemistry and technology of isobutane/alkene alkylation catalyzed by liquid and solid acids. *Advances in Catalysis, 48*, 229–295.

Flanigen, E. M., (2001). Zeolites and molecular sieves: An historical perspective. In: Van Bekkum, H., Flanigen, E. M., Jacobs, P. A., & Jansen, J. C., (eds.), *Introduction to Zeolite Science and Practice* (2nd edn., Vol. 137, pp. 11–35). Elsevier Science, B. V., Amsterdam, The Netherlands, studies in surface science and catalysis.

Heinemann, H., (1981). Technological applications of zeolites in catalysis. *Catalysis Reviews, 23*, 315–328.

Hölderich, W. F., & Van Bekkum, H., (2001). Zeolites and related materials in organic syntheses. Brønsted and Lewis Catalysis. In: Van Bekkum, H., Flanigen, E. M., Jacobs, P. A., & Jansen, J. C., (eds.), *Introduction to Zeolite Science and Practice* (2nd edn., Vol. 137, pp. 821–910.). Elsevier Science, B. V., Amsterdam, The Netherlands, Studies in Surface Science and Catalysis.

Jansen, J. C., (2001). Synthesis of zeolites. In: Van Bekkum, H., Flanigen, E. M., Jacobs, P. A., & Jansen, J. C., (eds.), *Introduction to Zeolite Science and Practice* (2nd edn., Vol. 137, pp. 175–227.). Elsevier Science, B. V., Amsterdam, The Netherlands, Studies in Surface Science and Catalysis.

Kaeding, W. W., Barile, G. C., & Wu, M. M., (1984). Mobil zeolite catalysts for monomers. *Catalysis Reviews, 26*, 597–612.

Kouwenhoven, H. W., & De Kroes, B., (2001). Preparation of zeolite catalysts. In: Van Bekkum, H., Flanigen, E. M., Jacobs, P. A., & Jansen, J. C., (eds.), *Introduction to Zeolite Science and Practice* (2nd edn., Vol. 137, pp. 673–706). Elsevier Science, B. V., Amsterdam, The Netherlands, studies in surface science and catalysis.

Liu, Y., Bisson, T. M., Yang, H., & Xu, Z., (2010). Recent developments in novel sorbents for flue gas clean up. *Fuel Processing Technology, 91*, 1175–1197.

Martens, J. A., & Jacobs, P. A., (2001). Introduction to acid catalysis with zeolites in hydrocarbon reactions. In: Van Bekkum, H., Flanigen, E. M., Jacobs, P. A., & Jansen, J. C., (eds.), *Introduction to Zeolite Science and Practice* (2nd edn., Vol. 137, pp. 633–671.). Elsevier Science, B. V., Amsterdam, The Netherlands, Studies in Surface Science and Catalysis.

Maxwell, I. E., & Stork, W. J. H., (2001). Hydrocarbon processing with zeolites. In: Van Bekkum, H., Flanigen, E. M., Jacobs, P. A., & Jansen, J. C., (eds.), *Introduction to Zeolite Science and Practice* (2nd edn., Vol. 137, pp. 747–819.). Elsevier Science, B. V., Amsterdam, The Netherlands, Studies in surface science and catalysis.

McCusker, L. M., & Baerlocher, C., (2001). Zeolite structures. In: Van Bekkum, H., Flanigen, E. M., Jacobs, P. A., & Jansen, J. C., (eds.), *Introduction to Zeolite Science and Practice* (2nd edn., Vol. 137, pp. 37–67). Elsevier Science, B. V., Amsterdam, The Netherlands, Studies in Surface Science and Catalysis.

Meier, W. M., (1986). Zeolite and zeolite-like materials. *Pure & Applied Chemistry, 58*, 1323–1328.

Mokrani, T., & Scurrell, M., (2009). Gas conversion to liquid fuels and chemicals: The methanol route–catalysis and process development. *Catalysis Reviews, 51,* 1–145.

Olsbye, U., Svelle, S., Bjørgen, M., Beato, P., Janssens, T. V. W., Joensen, F., Bordiga, S., & Lillerud, K. P., (2012). Conversion of methanol to hydrocarbons: How zeolite cavity and pore size controls product selectivity. *Angewandte Chemie International Edition, 51,* 5810–5831.

Ono, Y., (1992). Transformation of lower alkanes into aromatic hydrocarbons over ZSM–5 zeolites. *Catalysis Reviews, 34,* 179–226.

Pavón, E., Osuna, F. J., Alba, M. D., & Delevoye, L., (2014). Direct evidence of Lowenstein's rule violation in swelling high-charge micas. *Chemical Communications, 50,* 6984–6986.

Szostak, R., (2001). Secondary synthesis methods. In: Van Bekkum, H., Flanigen, E. M., & Jansen, J. C., (eds.), *Introduction to Zeolite Science and Practice* (Vol. 137, pp. 261–297). Elsevier Science, B. V., Amsterdam, The Netherlands, Studies in surface science and catalysis.

Tian, P., Wei, Y., Ye, M., & Liu, Z., (2015). Methanol to olefins (MTO): From fundamentals to commercialization. *ACS Catalysis, 5,* 1922–1938.

Townsend, R. P., & Coker, E. N., (2001). Ion exchange in zeolites. In: Van Bekkum, H., Flanigen, E. M., & Jansen, J. C., (eds.), *Introduction to Zeolite Science and Practice* (Vol. 137, pp. 467–524). Elsevier Science, B. V., Amsterdam, The Netherlands, Studies in surface science and catalysis.

Turkevich, J., & Ono, Y., (1969). Catalytic research on zeolites. *Advances in Catalysis, 20,* 135–152.

Turkevich, J., (1968). Zeolites as catalysts. I. *Catalysis Reviews, 1,* 1–35.

Van Koningsveld, H., (1991). Structural subunits in silicate and phosphate structures. In: Van Bekkum, H., Flanigen, E. M., & Jansen, J. C., (eds.), *Introduction to Zeolite Science and Practice* (Vol. 58, pp. 35–76). Elsevier Science, B. V., Amsterdam, The Netherlands, Studies in Surface Science and Catalysis.

Van Koningsveld, H., (2001). How to build zeolites. In: Van Bekkum, H., Flanigen, E. M., Jacobs, P. A., & Jansen, J. C., (eds.), *Introduction to Zeolite Science and Practice* (2nd edn., Vol. 137, pp. 69–173), Elsevier Science, B. V., Amsterdam, The Netherlands, Studies in surface science and catalysis.

Whyte, Jr., T. E., & Dalla, B. R. A., (1982). Zeolite advances in the chemical and fuel industries: A technical perspective. *Catalysis Reviews, 24,* 567–598.

Index

NEW FRONTIERS IN NANOCHEMISTRY

Concepts, Theories, and Trends

Volume 2
Topological Nanochemistry